Homological and Homotopical Aspects of Torsion Theories

MEMOIRS
of the
American Mathematical Society

Number 883

Homological and Homotopical Aspects of Torsion Theories

Apostolos Beligiannis
Idun Reiten

July 2007 • Volume 188 • Number 883 (end of volume) • ISSN 0065-9266

American Mathematical Society
Providence, Rhode Island

2000 *Mathematics Subject Classification.*
Primary 18E40, 18E30, 18E35, 18G55, 18G60; Secondary 16G10, 18E30, 18E10, 20C05, 20J05, 55U35.

Library of Congress Cataloging-in-Publication Data

Beligiannis, Apostolos, 1969-
Homological and homotopical aspects of torsion theories /Apostolos Beligiannis, Idun Reiten.
p. cm. — (Memoirs of the American Mathematical Society, ISSN 0065-9266 ; no. 883)
"July 2007, volume 188, number 883 (end of volume)."
Includes bibliographical references.
ISBN 978-0-8218-3996-6 (alk. paper)
1. Torsion theory (Algebra). 2. Algebra, Homological. 3. Homotopy theory. I. Reiten, Idun, 1942– II. Title.

QA251.3.B46 2007
516.3′6—dc22 2007060757

Memoirs of the American Mathematical Society

This journal is devoted entirely to research in pure and applied mathematics.

Subscription information. The 2007 subscription begins with volume 185 and consists of six mailings, each containing one or more numbers. Subscription prices for 2007 are US\$649 list, US\$519 institutional member. A late charge of 10% of the subscription price will be imposed on orders received from nonmembers after January 1 of the subscription year. Subscribers outside the United States and India must pay a postage surcharge of US\$38; subscribers in India must pay a postage surcharge of US\$43. Expedited delivery to destinations in North America US\$53; elsewhere US\$130. Each number may be ordered separately; *please specify number* when ordering an individual number. For prices and titles of recently released numbers, see the New Publications sections of the *Notices of the American Mathematical Society.*

Back number information. For back issues see the *AMS Catalog of Publications.*

Subscriptions and orders should be addressed to the American Mathematical Society, P. O. Box 845904, Boston, MA 02284-5904, USA. *All orders must be accompanied by payment.* Other correspondence should be addressed to 201 Charles Street, Providence, RI 02904-2294, USA.

Memoirs of the American Mathematical Society is published bimonthly (each volume consisting usually of more than one number) by the American Mathematical Society at 201 Charles Street, Providence, RI 02904-2294, USA. Periodicals postage paid at Providence, RI. Postmaster: Send address changes to Memoirs, American Mathematical Society, 201 Charles Street, Providence, RI 02904-2294, USA.

This publication is indexed in *Science Citation Index®*, *SciSearch®*, *Research Alert®*, *CompuMath Citation Index®*, *Current Contents®/Physical, Chemical & Earth Sciences.*
Printed in the United States of America.

∞ The paper used in this book is acid-free and falls within the guidelines established to ensure permanence and durability.
Visit the AMS home page at `http://www.ams.org/`

10 9 8 7 6 5 4 3 2 1 12 11 10 09 08 07

Contents

Introduction 1

Chapter I. **Torsion Pairs in Abelian and Triangulated Categories** 8
1. Torsion Pairs in Abelian Categories 8
2. Torsion Pairs in Triangulated Categories 10
3. Tilting Torsion Pairs 19

Chapter II. **Torsion Pairs in Pretriangulated Categories** 22
1. Pretriangulated Categories 22
2. Adjoints and Orthogonal Subcategories 28
3. Torsion Pairs 32
4. Torsion Pairs and Localization Sequences 35
5. Lifting Torsion Pairs 37

Chapter III. **Compactly Generated Torsion Pairs in Triangulated Categories** 43
1. Torsion Pairs of Finite Type 43
2. Compactly Generated Torsion Pairs 44
3. The Heart of a Compactly Generated Torsion Pair 50
4. Torsion Pairs Induced by Tilting Objects 55

Chapter IV. **Hereditary Torsion Pairs in Triangulated Categories** 60
1. Hereditary Torsion Pairs 60
2. Hereditary Torsion Pairs and Tilting 66
3. Connections with the Homological Conjectures 70
4. Concluding Remarks and Comments 77

Chapter V. **Torsion Pairs in Stable Categories** 79
1. A Description of Torsion Pairs 79
2. Comparison of Subcategories 84
3. Torsion and Cotorsion pairs 88
4. Torsion Classes and Cohen-Macaulay Objects 93
5. Tilting Modules 99

Chapter VI. **Triangulated Torsion(-Free) Classes in Stable Categories** 102
1. Triangulated Subcategories 102
2. Triangulated Torsion(-Free) Classes 104
3. Cotorsion Triples 108
4. Applications to Gorenstein Artin Algebras 112

Chapter VII. **Gorenstein Categories and (Co)Torsion Pairs** 117
1. Dimensions and Cotorsion Pairs 117
2. Gorenstein Categories, Cotorsion Pairs and Minimal Approximations 121
3. The Gorenstein Extension of a Cohen-Macaulay Category 125
4. Cohen-Macaulay Categories and (Co)Torsion Pairs 128

Chapter VIII. **Torsion Pairs and Closed Model Structures** 132
1. Preliminaries on Closed Model Categories 132
2. Closed Model Structures and Approximation Sequences 134
3. Cotorsion Pairs Arising from Closed Model Structures 138
4. Closed Model Structures Arising from Cotorsion Pairs 143
5. A Classification of (Co)Torsion Pairs 154

Chapter IX. **(Co)Torsion Pairs and Generalized Tate-Vogel Cohomology** 163
1. Hereditary Torsion Pairs and Homological Functors 163
2. Torsion Pairs and Generalized Tate-Vogel (Co-)Homology 167
3. Relative Homology and Generalized Tate-Vogel (Co)Homology 174
4. Cotorsion Triples and Complete Cohomology Theories 181

Chapter X. **Nakayama Categories and Cohen-Macaulay Cohomology** 186
1. Nakayama Categories and Cohen-Macaulay Objects 186
2. (Co)Torsion Pairs Induced by (Co)Cohen-Macaulay Objects 190
3. Cohen-Macaulay Cohomology 194

Bibliography 200

Index 204

ABSTRACT. In this paper we investigate homological and homotopical aspects of a concept of torsion which is general enough to cover torsion and cotorsion pairs in abelian categories, t-structures and recollements in triangulated categories, and torsion pairs in stable categories. The proper conceptual framework for this study is the general setting of pretriangulated categories, an omnipresent class of additive categories which includes abelian, triangulated, stable, and more generally (homotopy categories of) closed model categories in the sense of Quillen, as special cases.

The main focus of our study is on the investigation of the strong connections and the interplay between (co)torsion pairs and tilting theory in abelian, triangulated and stable categories on one hand, and universal cohomology theories induced by torsion pairs on the other hand. These new universal cohomology theories provide a natural generalization of the Tate-Vogel (co)homology theory. We also study the connections betweeen torsion theories and closed model structures, which allow us to classify all cotorsion pairs in an abelian category and all torsion pairs in a stable category, in homotopical terms. For instance we obtain a classification of (co)tilting modules along these lines. Finally we give torsion theoretic applications to the structure of Gorenstein and Cohen-Macaulay categories, which provide a natural generalization of Gorenstein and Cohen-Macaulay rings.

The main part of this work was done during summer 1998 at the University of Bielefeld where both authors were visitors and during winter 1999 and spring 2001 at the Norwegian University of Science and Technology, where the first named author was visiting the second. The present version was completed in the fall 2002. The first named author thanks his coauthor for the warm hospitality, and gratefully acknowledges support and hospitality from the Norwegian University of Science and Technology and the TMR-network "Algebraic Lie Representations". [1]

2000 *Mathematics Subject Classification.*
Primary: 18E40, 18E30, 18E35, 18G55, 18G60;
Secondary: 16G10, 18E30, 18E10, 20C05, 20J05, 55U35.

Key words and phrases. Torsion pairs, Cotorsion Pairs, Abelian categories, t-structures, Triangulated Categories, Compact objects, Tilting theory, Derived Categories, Contravariantly finite subcategories, Approximations, Stable categories, Reflective subcategories, Resolving subcategories, Closed Model Structures, Cohen-Macaulay Modules, Tate-Vogel Cohomology, Gorenstein Rings.

[1]Received by the editors: June 25 2001; in revised form: March 8 2003.

Introduction

The concept of torsion is fundamental in algebra, geometry and topology. The main reason is that torsion-theoretic methods allow us to isolate and therefore to study better, important phenomena having a local structure. The proper framework for the study of torsion is the context of torsion theories in a homological or homotopical category. In essence torsion theories provide a successful formalization of the localization process. The notion of torsion theory in an abelian category was introduced formally by Dickson [**41**], although the concept was implicit in the work of Gabriel and others from the late fifties, see the books of Stenström [**100**] and Golan [**56**] for a comprehensive treatment. Since then the use of torsion theories became an indispensable tool for the study of localization in various contexts. As important examples of localization we mention the localization of topological spaces or spectra, the localization theory of rings and abelian categories, the local study of an algebraic variety, the construction of perverse sheaves in the analysis of possibly singular spaces, and the theory of tilting in representation theory. The omnipresence of torsion suggests a strong motivation for the development of a general theory of torsion and localization which unifies the above rather unrelated concrete examples.

In this paper we investigate homological and homotopical aspects of a concept of torsion which is general enough to cover the situations mentioned above. More importantly we provide new connections between different aspects of torsion in various settings, and we present new classes of examples and give a variety of applications. We study their interplay in the general working context of *pretriangulated categories.* This class of categories gains its importance from the fact that it includes the following classes of homological or homotopical categories as special cases:

- Abelian categories.
- Triangulated categories.
- Stable categories.
- Closed model categories in the sense of Quillen and their homotopy categories.

It is well-known that the proper framework for the study of homological algebra and for large parts of representation theory is the context of abelian categories. In recent years triangulated categories, and in particular derived and stable categories, entered into the picture of homological representation theory in a very essential way, offering new invariants and classification limits, through the work of Happel [**57**], Rickard [**91**], Keller [**69**], Neeman [**86**], Happel-Reiten-Smalø [**60**], Krause [**77**] and others. There is a formal analogy between abelian categories and triangulated categories. In the first case we have exact sequences and in the second case we have triangles, which can be regarded as a reasonable substitute for the exact sequences. Pretriangulated categories, which can be regarded as a common generalization of abelian and triangulated categories, incorporate at the same time also the stable categories, which are very useful for the study of the behavior of several stable phenomena occurring in homological representation theory.

Recall that a pretriangulated category is an additive category $\mathcal{C}$ equipped with a pair (Σ, Ω) of adjoint endofunctors, where Ω is the loop functor and Σ is the suspension functor, and in addition with a class of left triangles Δ and a class ∇ of right triangles which are compatible with each other and with Σ and Ω. Important examples are abelian categories (in which case $\Sigma = 0 = \Omega$) and triangulated categories (in which case Σ or Ω is an equivalence). A central source of examples of pretriangulated categories is an abelian category $\mathcal{C}$ with a functorially finite subcategory ω. Then the stable category $\mathcal{C}/\omega$ becomes in a natural way a pretriangulated category; the functors Ω, Σ and the class of left and right triangles are defined via left and right ω-approximations of objects of $\mathcal{C}$, in the sense of Auslander-Smalø [**13**] and Enochs [**46**]. As a special case we can choose $\mathcal{C} = \text{mod}(\Lambda)$, the category of finitely presented modules over an Artin algebra, and let ω be the additive subcategory generated by a (tilting or cotilting) module.

Another important source of examples of pretriangulated categories is coming from homotopical algebra. It is well known that the proper framework for doing homotopy theory is the context of closed model categories in the sense of Quillen [**88**]. The homotopy category of a closed model category has a rich structure and in particular is in a natural way a pretriangulated category, see the book of Hovey [**64**] for a comprehensive treatment. Actually the stable category $\mathcal{C}/\omega$ mentioned above can be interpreted as a homotopy category of a suitable closed model structure, see [**24**].

Torsion theories play an important role in the investigation of an abelian category. There is a natural analogous definition for triangulated categories which is closely related to the notion of a t-structure. These concepts of torsion generalize naturally to the setting of pretriangulated categories. Our main interest lies in the investigation of these generalized torsion theories on pretriangulated, and especially on triangulated or stable categories. In this way we are provided with a convenient conceptual umbrella for the study of various aspects of torsion and their interplay. Our results indicate that torsion theories in this general setting can be regarded as generalized tilting theory.

We would like to stress that there is an interesting interplay between the different settings where we have torsion theories. An abelian category $\mathcal{C}$ is naturally embedded in interesting triangulated categories like the bounded derived category $\mathbf{D}^b(\mathcal{C})$. A torsion theory in $\mathcal{C}$ induces in an natural way a torsion theory (t-structure) in $\mathbf{D}^b(\mathcal{C})$, a fact which was important in the investigation of tilting in abelian categories [**60**]. Another important connection is the fact that a torsion theory (t-structure) in a triangulated category gives rise to an abelian category, the socalled heart [**18**]. Applying both constructions, an abelian category with a given torsion theory gives rise to a new abelian category, actually also with a distinguished torsion theory. Further, given any pretriangulated category, there are naturally associated with it two triangulated categories, and we show that through this construction there is a close relationship between (hereditary) torsion theories, a fact which is used for constructing new (co)homology theories in the last chapter.

Tilting theory, a central topic in the representation theory of Artin algebras, is intimately related to torsion theories in several different ways. When T is a tilting module with $\operatorname{pd} T \leq 1$, that is $\operatorname{Ext}^1(T,T) = 0$ and there is an exact sequence $0 \to \Lambda \to T_0 \to T_1 \to 0$ with T_0 and T_1 summands of finite direct sums of copies of T, there is an associated torsion theory $(\mathcal{T}, \mathcal{F})$ where $T = \operatorname{Fac} T$ (the factors of finite direct sums of copies of T). This torsion theory plays an important role in tilting theory, and is closely related to a torsion theory for the endomorphism algebra $\Gamma = \operatorname{End}(T)^{\mathrm{op}}$. The tilting in abelian categories referred to above gave a way of constructing $\operatorname{mod}(\Gamma)$ from $\operatorname{mod}(\Lambda)$ via the torsion theory $(\mathcal{T}, \mathcal{F})$ and the one induced in $\mathbf{D}^b(\operatorname{mod}(\Lambda))$. When T is more generally a tilting module with $\operatorname{pd} T < \infty$, there is no natural associated torsion theory in $\operatorname{mod}(\Lambda)$ which plays a similar role. But it is interesting that in this case we can show that T generates a torsion theory in $\mathbf{D}^b(\operatorname{mod}(\Lambda))$, whose heart is equivalent to $\operatorname{mod}(\Gamma)$. There is a still more general concept of tilting module, called Wakamatsu tilting module. We conjecture that any Wakamatsu tilting module of finite projective dimension is in fact a tilting module. We give interesting reformulations of this conjecture. One of them is that if $(\mathcal{X}_T, \mathcal{Y}_T)$ is a hereditary torsion theory in $\mathbf{D}(\operatorname{Mod} \Lambda)$ generated by a Wakamatsu tilting module T of finite projective dimension, then $\mathcal{X}_T$ is closed under products.

It is well-known that tilting theory can be regarded as an important generalization of classical Morita theory, which describes when two module categories are equivalent. During the last fifteen years investigations of many authors extended Morita theory for module categories to derived categories, thus offering new invariants and levels of classification. These investigations culminated in a Morita Theory for derived categories of rings and DG-algebras, which describes explicitly when derived categories of rings or DG-algebras are equivalent as triangulated categories. This important generalization can be regarded as a higher analogue of tilting theory and plays a fundamental role in representation theory, providing interesting connections with algebraic geometry and topology. In this paper we interpret Morita theory for derived categories in torsion theoretic terms, and we give simple torsion theoretic proofs of (slight generalizations of) central results of Happel [**57**], Rickard [**91**] and Keller [**69**], concerning the construction of derived equivalences. In addition this interpretation via torsion theories gives interesting reformulations of several important open problems in various contexts, providing new ways for their investigation.

On the other hand there are the notions of contravariantly, covariantly and functorially finite subcategories of an additive category as introduced by Auslander and Smalø in [**13**] and independently by Enochs in [**46**]. Special cases of these are subcategories for which there exist a right or left adjoint of the inclusion. Contravariantly or covariantly finite subcategories play a fundamental role in the representation theory of Artin algebras. They provide a convenient setting for the study of several important finiteness conditions in various settings and there is again a strong connection to tilting theory. For Artin algebras these categories occur in pairs $(\mathcal{X}, \mathcal{Y})$, under some natural additional assumptions, namely $\mathcal{X}$ is contravariantly finite, closed under extensions and contains the projectives and $\mathcal{Y}$ is covariantly finite, closed under extensions and contains the injectives. Then $\mathcal{X}$ and $\mathcal{Y}$ are in one-one correspondence, via the vanishing of Ext^1 rather than of Hom. This correspondence was observed by Auslander and Reiten in [**9**] in the setting of Artin algebras and by Salce in the setting of abelian groups in [**97**]. By definition the subcategories $\mathcal{X}$ and $\mathcal{Y}$ involved in the above correspondence form a cotorsion pair $(\mathcal{X},\mathcal{Y})$. Cotorsion pairs have been studied recently by many people leading to the recent proof of the Flat Cover Conjecture by Enochs, Bashir and Bican [**17**]. A central concept which is omnipresent in the paper and that it is closely related to cotorsion pairs is that of a (relative) (Co)Cohen-Macaulay object. Their importance follows from the fact that for any cotorsion pair $(\mathcal{X},\mathcal{Y})$ in an abelian category, the cotorsion class $\mathcal{X}$, resp. the cotorsion-free class $\mathcal{Y}$, consists of relative Cohen-Macaulay, resp. CoCohen-Macaulay, objects with respect to $\mathcal{X}\cap\mathcal{Y}$. We would like to stress that prominent examples of cotorsion theories are coming from Gorenstein rings, and, more generally, from tilting or cotilting modules, where Cohen-Macaulay modules play an important role. Generalizing, we define and investigate in this paper Gorenstein and Cohen-Macaulay abelian categories and we prove structure results for them in the context of (co)torsion theories.

For abelian and triangulated categories (in the latter case first formulated in the language of t-structures), there is a close connection between $\mathcal{X}$ being a torsion class of a torsion theory and $\mathcal{X}$ admitting a right adjoint (and between $\mathcal{Y}$ being a torsion-free class of a torsion theory and $\mathcal{Y}$ admitting a left adjoint). There is a similar connection in the setting of pretriangulated categories, although the extension of the results does not work so smoothly in general. The main reason is that the compatibility of the left and right triangles is not so well behaved as in the abelian or triangulated case. However we prove the corresponding result under some additional assumptions. Of particular interest for us is the case of a stable category $\mathcal{C}/\omega$. Here we obtain at the same time the correspondence between contravariantly finite and covariantly finite subcategories via Ext^1 discussed above and the torsion theory correspondence between the associated subcategories of $\mathcal{C}/\omega$, where $\omega=\mathcal{X}\cap\mathcal{Y}$.

As already mentioned such a stable category $\mathcal{C}/\omega$ is the homotopy category of a closed model structure in $\mathcal{C}$ in the sense of Quillen, which is defined via left/right approximations of objects of $\mathcal{C}$ by objects from ω. In this paper we show that there is a strong connection between the pairs of subcategories realizing the correspondence described above, and closed model structures in the sense of Quillen in $\mathcal{C}$. Recently a similar connection from a different viewpoint was observed independently by Hovey [**65**]. More precisely our results give a classification of all cotorsion pairs $(\mathcal{X},\mathcal{Y})$ in an abelian category $\mathcal{C}$ with $\mathcal{X}\cap\mathcal{Y}$ contravariantly or covariantly finite, in terms of suitable closed model structures. This leads to a classification of all torsion theories in a stable category of an abelian category. As a consequence we obtain an interesting connection between (co)tilting modules and closed model structures.

In general it is of central importance for the structure of the pretriangulated category in question, that a reasonable subcategory is a torsion or torsion-free class. For instance when we have a torsion theory in a stable pretriangulated category of an abelian category, we define new universal (co)homology theories on the abelian category which are complete with respect to the torsion or torsion free class in an appropriate sense. These new (co)homology theories are generalizations of the Tate-Vogel (co)homology theories to the non-Gorenstein case. This procedure provides a useful interplay between torsion-theoretic properties of the stable category and homological properties of the abelian category. In this setting there is a strong connection between the relative homological algebra, in the formulation proposed by Butler-Horrocks and later by Auslander-Solberg, induced by the torsion theory and the behavior of the (co)homology theories. This connection has important consequences for the homological structure of the abelian category. For instance these investigations allow us to generalize recent results of Avramov-Martsinkovsky, see [**16**], from finitely generated modules of finite Gorenstein dimension, to arbitrary modules, provided that a suitable torsion theory exists in the stable module category. We show that such torsion theories exist in many cases, and in particular in the stable module category of, not necessarily finitely generated, modules over an Artin algebra. In this case we construct big Cohen-Macaulay modules and modules of virtually finite projective dimension, and we show that these classes of modules define a new relative cohomology theory for Artin algebras, called Cohen-Macaulay cohomology, which has intimate connections with many homological conjectures.

We have provided background material, together with motivation, through recalling previous developments. Some of the main results of the paper have been mentioned along the way. But to make it easier to focus on the essential points, we collect a list of the main features below.

- Unify various concepts of torsion in different settings, and provide a general framework for studying torsion theories.
- Develop a general theory of torsion in pretriangualated categories, including interplay between the various settings.
- Provide methods for constructing torsion theories.
- Give applications to tilting theory.
- Give simpler (torsion theoretic) proofs of (slightly more general) results in the Morita theory of derived categories.
- Give interesting reformulations of homological conjectures for Artin algebras in terms of properties of torsion theories in the unbounded derived category of all modules.
- Establish relationsip between cotorsion theories $(\mathcal{X}, \mathcal{Y})$ in an abelian category $\mathcal{C}$ and torsion theories in the stable category $\mathcal{C}/\mathcal{X} \cap \mathcal{Y}$.
- Give connections with closed model structures and classification of torsion and cotorsion theories in terms of these.
- Give structure results for (generalizations of) Gorenstein rings and, suitably defined, Gorenstein and Cohen-Macaulay abelian categories, in the context of (co)torsion theories.
- Give methods for constructing Gorenstein Categories out of certain Cohen-Macaulay categories.
- Construct interesting (co)torsion theories in abelian categories equipped with suitable Nakayama functors, where the Cohen-Macaulay objects constitute the (co)torsion class and the CoCohen-Macaulay objects constitute the (co)torsion-free class.

- Construct new (co)homology theories generalizing Tate-Vogel (co)homology using torsion theories.

The article is organized as follows.

In Chapter I, which is of preliminary nature, we recall some definitions and results, most of them well-known, concerning (hereditary/cohereditary) torsion theories in an abelian or triangulated category, which serve as motivation for the results of the rest of the paper. Here we observe that in the triangulated case the concept of a torsion theory essentially coincides with the concept of a t-structure in the sense of Beilinson-Bernstein-Deligne [**18**]. We discuss briefly the connection between torsion theories in abelian and derived categories investigated in [**60**] and indicate the relationship with tilting theory.

In Chapter II we introduce the fairly general concept of torsion theory $(\mathcal{X}, \mathcal{Y})$ in a pretriangulated category. But first we recall basic results on left/right triangulated and pretriangulated categories, and provide a rather large source of examples. We give basic properties of torsion theories, and show that the notion specializes to the concept of torsion theory in abelian and triangulated categories discussed in the previous chapter. We also show that we can lift a hereditary or cohereditary torsion theory in a pretriangulated category to such a pair in the left or right stabilization.

In Chapter III we give a method for constructing torsion theories in triangulated categories containing all small coproducts. This is accomplished by starting with a set of compact objects. We investigate properties of the heart of this torsion theory, and give conditions for the heart to be a module category. Then we use these results to give applications to tilting theory, and to proving (a slight generalization of) a result of Rickard on Morita theory of derived categories.

In Chapter IV we deal with hereditary torsion theories generated by compact objects in a triangulated category with arbitrary small coproducts. Following the fundamental work of Keller [**69**] on derived categories of DG-algebras, we describe the torsion class of the hereditary torsion theory generated by a set of compact objects in terms of derived categories of appropriate DG-algebras. These results provide simple torsion theoretic proofs of basic results of Happel [**57**], Rickard [**91**] and Keller [**69**]. We end with the relationship between homological conjectures in the representation theory of Artin algebras and (hereditary) torsion theories in the unbounded derived category of the algebra.

In Chapters V and VI we investigate torsion theories in pretriangulated categories of the form $\mathcal{C}/\omega$ where $\mathcal{C}$ is abelian and ω is a functorially finite subcategory of $\mathcal{C}$. We give sufficient conditions for the existence of torsion theories, and give the relationship with cotorsion pairs in $\mathcal{C}$. We also give applications to tilting theory. In Chapter VI we discuss the problem of when a torsion or torsion free class in a stable category is triangulated. We show that this happens if and only if the functorially finite subcategory ω is the category of projective or injective modules. In this case the torsion subcategory is related to the subcategory of Cohen-Macaulay objects and the torsion free subcategory is related to the subcategory of objects with finite projective dimension. We show that natural sources for such torsion theories are cotorsion triples induced by resolving and coresolving functorially finite subcategories of $\mathcal{C}$.

In Chapter VII we introduce and investigate in detail Gorenstein and Cohen-Macaulay abelian categories, which appear to be the proper generalizations of the category of (finitely generated modules) over a (commutative Noetherian) Gorenstein and Cohen-Macaulay ring. We give structure results for them in the context of (co)torsion theories in connection with the finiteness of various interesting homological dimensions. In particular we show that, under mild conditions, the trivial extension, in the sense of [**48**], of a Cohen-Macaulay category is Gorenstein. We also study minimal Cohen-Macaulay approximations and we give applications to Gorenstein and Cohen-Macaulay rings which admit a Morita self-duality.

In Chapter VIII we investigate homotopy theoretic properties of torsion and cotorsion theories by studying the connections with closed model structures, thus giving a homotopy theoretic interpretation of the results of the previous chapters. In particular we give classifications of cotorsion pairs and cotorsion triples in an abelian category and torsion pairs in a stable category in terms of closed model structures. As an application we give a closed model theoretic classification of (co)tilting modules over an Artin algebra.

In the last two chapters we apply the results of the previous chapters to construct and investigate universal (co)homology extension functors on an abelian category $\mathcal{C}$, when the stable category $\mathcal{C}/\omega$ admits a (co)hereditary torsion theory. We show that these universal (co)homology extension functors are natural generalizations of the Tate-Vogel (co)homology functors studied in homological group theory, commutative algebra and representation theory. More importantly we show that the generalized Tate-Vogel (co)homology functors fit nicely in long exact sequences involving the relative extension functors induced by the torsion or torsion-free class. Working in a Nakayama abelian category, which is a natural generalization of the module category of an Artin algebra, we show that there are well-behaved (co)hereditary torsion pairs which are intimately related to Cohen-Macaulay and CoCohen-Macaulay objects. We close the paper by studying the resulting universal (co)homology extension functors induced by the (Co)Cohen-Macaulay objects.

Convention. From now on and following the current increasingly strong trend, we use throughout the paper the terminology **torsion pair** instead of torsion theory and **TTF-triple** instead of TTF-theory.

Throughout the paper we compose morphisms in the diagrammatic order, i.e. the composition of morphisms $f : A \to B$ and $g : B \to C$ in a given category is denoted by $f \circ g$. Our additive categories admit finite direct sums.

Acknowledgement. The authors would like to thank the referees for their useful comments.

CHAPTER I

Torsion Pairs in Abelian and Triangulated Categories

In this chapter we give definitions and useful properties of torsion pairs in abelian and triangulated categories. For triangulated categories we show that our definition is closely related to the notion of t-structure in the sense of [**18**] and to the notion of aisle in the sense of [**74**]. We also recall an interesting interplay between torsion pairs in abelian and triangulated categories related to tilting theory. This serves as background for more general results proved in later chapters.

1. Torsion Pairs in Abelian Categories

In this section we recall some basic results concerning torsion pairs in an abelian category.

We start by fixing some notation.

If $\mathcal{Z}$ is a class of objects in an additive category $\mathcal{C}$, we denote by

$$^{\perp}\mathcal{Z} := \{X \in \mathcal{C} \mid \mathcal{C}(X, \mathcal{Z}) = 0\}$$

the **left orthogonal subcategory** of $\mathcal{Z}$, and by

$$\mathcal{Z}^{\perp} := \{Y \in \mathcal{C} \mid \mathcal{C}(\mathcal{Z}, Y) = 0\}$$

the **right orthogonal subcategory** of $\mathcal{Z}$.

Assume now that $\mathcal{C}$ is an abelian category.

DEFINITION 1.1. A **torsion pair** in an abelian category $\mathcal{C}$ is a pair $(\mathcal{X}, \mathcal{Y})$ of strict (i.e. closed under isomorphisms) full subcategories of $\mathcal{C}$ satisfying the following conditions:

(i) $\mathcal{C}(X, Y) = 0$, $\forall X \in \mathcal{X}$, $\forall Y \in \mathcal{Y}$.

(ii) For any object $C \in \mathcal{C}$ there exists a short exact sequence:

$$0 \longrightarrow X_C \xrightarrow{f_C} C \xrightarrow{g^C} Y^C \longrightarrow 0 \tag{1}$$

in $\mathcal{C}$ such that $X_C \in \mathcal{X}$ and $Y^C \in \mathcal{Y}$.

If $(\mathcal{X}, \mathcal{Y})$ is a torsion pair, then $\mathcal{X}$ is called a **torsion class** and $\mathcal{Y}$ is called a **torsion-free class**.

It is well–known and easy to see that for a torsion pair $(\mathcal{X}, \mathcal{Y})$ in $\mathcal{C}$ we have that $\mathcal{X}$ is closed under factors, extensions and coproducts and $\mathcal{Y}$ is closed under extensions, subobjects and products, and moreover: $\mathcal{X}^{\perp} = \mathcal{Y}$ and $^{\perp}\mathcal{Y} = \mathcal{X}$. Conversely if $\mathcal{C}$ is a locally small complete and cocomplete abelian category, then any full subcategory of $\mathcal{C}$ which is closed under factors, extensions and coproducts, is a torsion class. Dually any full subcategory of $\mathcal{C}$ which is closed under subobjects, extensions and products, is a torsion-free class.

For a torsion pair $(\mathcal{X}, \mathcal{Y})$ in $\mathcal{C}$ the assignment $\mathbf{R}(C) = X_C$ extends to an additive functor $\mathbf{R} : \mathcal{C} \to \mathcal{X}$ which is a right adjoint of the inclusion $\mathbf{i} : \mathcal{X} \hookrightarrow \mathcal{C}$, that is $\mathcal{X}$ is a coreflective subcategory of $\mathcal{C}$. Moreover the morphism f_C serves as the counit of the adjoint pair $(\mathbf{i}, \mathbf{R})$ evaluated at C. Dually the assignment $\mathbf{L}(C) = Y^C$ extends to an additive functor $\mathbf{L} : \mathcal{C} \to \mathcal{Y}$ which is a left adjoint of the inclusion $\mathbf{j} : \mathcal{Y} \hookrightarrow \mathcal{C}$, that is $\mathcal{Y}$ is a reflective subcategory of $\mathcal{C}$. Moreover the morphism g^C serves as the unit of the adjoint pair $(\mathbf{L}, \mathbf{j})$ evaluated at C.

To make comparison with our later results easier, we include the following characterization of torsion or torsion-free classes.

PROPOSITION 1.2. *Let $\mathcal{C}$ be an abelian category and let $\mathcal{X}$ and $\mathcal{Y}$ be full subcategories of $\mathcal{C}$ closed under isomorphisms. Then we have the following.*

(1) *$\mathcal{Y}$ is a torsion-free class in $\mathcal{C}$ if and only if the inclusion $\mathbf{j} : \mathcal{Y} \hookrightarrow \mathcal{C}$ admits a left adjoint $\mathbf{L} : \mathcal{C} \to \mathcal{Y}$, and $\mathcal{Y}$ is closed under extensions of left exact sequences, that is, if $0 \to Y_1 \to C \to Y_2$ is a left exact sequence with $Y_1, Y_2 \in \mathcal{Y}$, then C lies in $\mathcal{Y}$.*

(2) *$\mathcal{X}$ is a torsion class in $\mathcal{C}$ if and only if the inclusion $\mathbf{i} : \mathcal{X} \hookrightarrow \mathcal{C}$ admits a right adjoint $\mathbf{R} : \mathcal{C} \to \mathcal{X}$, and $\mathcal{X}$ is closed under extensions of right exact sequences, that is, if $X_1 \to C \to X_2 \to 0$ is a right exact sequence with $X_1, X_2 \in \mathcal{X}$, then C lies in $\mathcal{X}$.*

PROOF. (1) ($\Leftarrow$) If $\mathcal{Y}$ is closed under extensions of left exact sequences, then trivially $\mathcal{Y}$ is closed under extensions and subobjects. Let $\mathbf{L} : \mathcal{C} \to \mathcal{Y}$ be the left adjoint of the inclusion $\mathbf{j} : \mathcal{Y} \hookrightarrow \mathcal{C}$ and let $g^C : C \to \mathbf{L}(C)$ be the unit evaluated at C. Since $\mathcal{Y}$ is closed under subobjects, it is trivial to see that the inclusion $\mathrm{Im}(g^C) \hookrightarrow \mathbf{L}(C)$ is invertible, hence g^C is epic. Let $f_C : X_C \to C$ be the kernel of g^C. Then $X_C \in {}^{\perp}\mathcal{Y}$. Indeed if $\alpha : X_C \to Y$ is a morphism with $Y \in \mathcal{Y}$, then in the push-out $0 \to Y \to Y' \xrightarrow{h} \mathbf{L}(C) \to 0$ of $0 \to X_C \to C \to \mathbf{L}(C) \to 0$ along α, the object Y' is in $\mathcal{Y}$ since $\mathcal{Y}$ is closed under extensions. Since g^C is the reflection morphism of C in $\mathcal{Y}$ it follows directly that h splits, hence α factors through f_C, say via a morphism $\beta : C \to Y$. Since $Y \in \mathcal{Y}$, then β factors through g^C, and this implies trivially that $\beta = 0$, so $\alpha = 0$. Hence we have $X_C \in {}^{\perp}\mathcal{Y}$. Then by definition we have that $\mathcal{Y}$ is the torsion-free class of the torsion pair $({}^{\perp}\mathcal{Y}, \mathcal{Y})$ in $\mathcal{C}$.

($\Rightarrow$) Let $\mathcal{Y}$ be a torsion-free class and let $0 \to Y_1 \xrightarrow{\alpha} C \xrightarrow{\beta} Y_2$ be a left exact sequence with $Y_1, Y_2 \in \mathcal{Y}$. Since $\mathcal{Y}$ is closed under subobjects, $\mathrm{Im}(\beta) \in \mathcal{Y}$, and since $\mathcal{Y}$ is closed under extensions, we have $C \in \mathcal{Y}$. Hence $\mathcal{Y}$ is closed under extensions of left exact sequences and as noted before the inclusion $\mathcal{Y} \hookrightarrow \mathcal{C}$ admits a left adjoint.

(2) The proof is dual. □

Let $(\mathcal{X}, \mathcal{Y})$ be a torsion pair in $\mathcal{C}$, let $\mathbf{R}$ be the right adjoint of the inclusion $\mathbf{i} : \mathcal{X} \hookrightarrow \mathcal{C}$ and let $\mathbf{L}$ be the left adjoint of the inclusion $\mathbf{j} : \mathcal{Y} \hookrightarrow \mathcal{C}$. We recall that $(\mathcal{X}, \mathcal{Y})$ is said to be **hereditary**, resp. **cohereditary**, if the torsion class $\mathcal{X}$ is closed under subobjects, resp. the torsion–free class $\mathcal{Y}$ is closed under factors. Hereditary torsion pairs form a well-behaved class of torsion pairs which is important in ring theory in connection with the localization theory of rings and modules, since they correspond bijectively with Gabriel topologies, see the books of Golan [**56**] and

Stenström [**100**] for details and more information. It is well–known that $(\mathcal{X}, \mathcal{Y})$ is hereditary if and only if $\mathcal{X}$ is a Serre subcategory if and only if the (idempotent radical) functor $\mathbf{iR} : \mathcal{C} \to \mathcal{C}$ is left exact [**100**]. Dually $(\mathcal{X}, \mathcal{Y})$ is cohereditary if and only if $\mathcal{Y}$ is a Serre subcategory if and only if the (idempotent coradical) functor $\mathbf{jL} : \mathcal{C} \to \mathcal{C}$ is right exact.

A nice situation occurs when the torsion class $\mathcal{Y}$ of a hereditary torsion pair $(\mathcal{Y}, \mathcal{Z})$ is a torsion-free class. In this case $\mathcal{Y}$ induces a **torsion, torsion-free triple**, **TTF-triple** for short, in the abelian category $\mathcal{C}$, that is, a triple $(\mathcal{X}, \mathcal{Y}, \mathcal{Z})$ of strict full subcategories of $\mathcal{C}$ such that $(\mathcal{X}, \mathcal{Y})$ and $(\mathcal{Y}, \mathcal{Z})$ are torsion pairs. Then $\mathcal{Y}$ is called a **TTF-class**. TTF-triples occur frequently in practice in connection with recollement of abelian categories. For instance TTF-triples in generic representation theory of the finite general linear group are related to stratifications of categories which are of interest in representation theory, see [**80**]. In ring theory TTF-triples correspond bijectively to idempotent ideals, see [**100**].

If $(\mathcal{X}, \mathcal{Y}, \mathcal{Z})$ is a TTF-triple in $\mathcal{C}$, then obviously $(\mathcal{X}, \mathcal{Y})$ is cohereditary and $(\mathcal{Y}, \mathcal{Z})$ is hereditary. It follows that the TTF-class $\mathcal{Y}$ is a reflective and coreflective Serre subcategory of $\mathcal{C}$. If $\mathcal{V}$ is a Serre subcategory of $\mathcal{C}$, then we denote by $\mathcal{C}/\mathcal{V}$ the induced Gabriel quotient, see [**49**] for details and more information. The following result, proved in [**53**] in a different way, describes the Gabriel quotient $\mathcal{C}/\mathcal{Y}$ and presents an interesting connection between the involved subcategories in a TTF-triple. We include a simple proof for comparison with our results in later chapters.

PROPOSITION 1.3. [**53**] *If $(\mathcal{X}, \mathcal{Y}, \mathcal{Z})$ is a TTF-triple in $\mathcal{C}$ then there exists an equivalence: $\mathcal{X} \cap \mathcal{Z} \xrightarrow{\approx} \mathcal{C}/\mathcal{Y}$. In particular the category $\mathcal{X} \cap \mathcal{Z}$ is abelian.*

PROOF. Let $\pi : \mathcal{C} \to \mathcal{C}/\mathcal{Y}$ be the exact quotient functor. We show that the composition $F : \mathcal{X} \cap \mathcal{Z} \hookrightarrow \mathcal{C} \xrightarrow{\pi} \mathcal{C}/\mathcal{Y}$ is an equivalence. If $f : U_1 \to U_2$ is a morphism in $\mathcal{X} \cap \mathcal{Z}$ such that $\pi(f) = 0$, then by the construction of $\mathcal{C}/\mathcal{Y}$ we have that $\mathrm{Im}(f)$ lies in $\mathcal{Y}$. Then $\mathrm{Im}(f) = 0$ since U_1 is in $\mathcal{X}$. Hence $f = 0$ and consequently F is faithful. If $\alpha : \pi(U_1) \to \pi(U_2)$ is a morphism in $\mathcal{C}/\mathcal{Y}$, where the U_i are in $\mathcal{X} \cap \mathcal{Z}$, then α is represented by a diagram $U_1 \xleftarrow{s} U_3 \xrightarrow{f} U_2$ where s is a morphism with kernel and cokernel in $\mathcal{Y}$. Since $(\mathcal{X}, \mathcal{Y}) = 0 = (\mathcal{Y}, \mathcal{Z})$ we infer that s is invertible. Then it is easy to see that $\alpha = \pi(s^{-1} \circ f)$. Hence F is full. Finally let $\pi(C)$ be an object in $\mathcal{C}/\mathcal{Y}$. Consider the canonical exact sequences $0 \to X_C \to C \to Y^C \to 0$ and $0 \to Y_C \to C \to Z^C \to 0$. Applying the exact functor π to the above sequences, we have isomorphisms $\pi(X_C) \xrightarrow{\cong} \pi(C) \xrightarrow{\cong} \pi(Z^C)$. Let U be the image of the composition $X_C \to C \to Z^C$. Then U lies in $\mathcal{X} \cap \mathcal{Z}$ since $\mathcal{X}$ is closed under factors and $\mathcal{Z}$ is closed under subobjects. Since the composition $\pi(X_C) \to \pi(Z^C)$ is invertible, we have $\pi(C) \cong \pi(U)$. This shows that F is surjective on objects. □

2. Torsion Pairs in Triangulated Categories

Motivated by the notion of a torsion pair in the abelian case we study in this section the analogous concept of a (hereditary) torsion pair in triangulated categories, and give the relationship to t-structures.

Let $\mathcal{C}$ be a triangulated category with suspension functor Σ. A direct analogue of the definition of a torsion pair in an abelian category in the triangulated case is the following.

DEFINITION 2.1. A **torsion pair** in $\mathcal{C}$ is a pair of strict full subcategories $(\mathcal{X}, \mathcal{Y})$ of $\mathcal{C}$ satisfying the following conditions:

(i) $\mathcal{C}(\mathcal{X}, \mathcal{Y}) = 0$.
(ii) $\Sigma(\mathcal{X}) \subseteq \mathcal{X}$ and $\Sigma^{-1}(\mathcal{Y}) \subseteq \mathcal{Y}$.
(iii) For any $C \in \mathcal{C}$ there exists a triangle

$$X_C \xrightarrow{f_C} C \xrightarrow{g^C} Y^C \xrightarrow{h^C} \Sigma(X_C) \tag{2}$$

in $\mathcal{C}$ such that $X_C \in \mathcal{X}$ and $Y^C \in \mathcal{Y}$.

Then $\mathcal{X}$ is called a **torsion class** and $\mathcal{Y}$ is called a **torsion-free class**.

The following observation shows that if (i) and (iii) hold, then $\Sigma(\mathcal{X}) \subseteq \mathcal{X}$ if and only if $\Sigma^{-1}(\mathcal{Y}) \subseteq \mathcal{Y}$. Moreover if $(\mathcal{X}, \mathcal{Y})$ is a torsion pair in $\mathcal{C}$, then $(\mathcal{X}, \mathcal{Y})$ is complete with respect to the vanishing of Hom.

REMARK 2.2. Suppose that the pair $(\mathcal{X}, \mathcal{Y})$ satisfies (i), (iii) above.

(1) If $\Sigma(\mathcal{X}) \subseteq \mathcal{X}$, then: $\mathcal{X}^{\perp} = \mathcal{Y}$ and $\Sigma^{-1}(\mathcal{Y}) \subseteq \mathcal{Y}$. Indeed by (i), $\mathcal{Y} \subseteq \mathcal{X}^{\perp}$. If $C \in \mathcal{X}^{\perp}$, then by (iii) we have $f_C = 0$. This implies that $Y^C \cong C \oplus \Sigma(X_C)$. Since $\Sigma(X_C) \in \mathcal{X}$, we have trivially that $C \cong Y^C \in \mathcal{Y}$. For any $Y \in \mathcal{Y}$ we now have $\mathcal{C}(\mathcal{X}, \Sigma^{-1}(Y)) \cong \mathcal{C}(\Sigma(\mathcal{X}), Y) = 0$. Hence $\Sigma^{-1}(Y) \in \mathcal{X}^{\perp} = \mathcal{Y}$, i.e. $\Sigma^{-1}(\mathcal{Y}) \subseteq \mathcal{Y}$.

(2) Dually if $\Sigma^{-1}(\mathcal{Y}) \subseteq \mathcal{Y}$, then: ${}^{\perp}\mathcal{Y} = \mathcal{X}$ and $\Sigma(\mathcal{X}) \subseteq \mathcal{X}$.

(3) Hence if $(\mathcal{X}, \mathcal{Y})$ is a torsion pair in $\mathcal{C}$, then: ${}^{\perp}\mathcal{Y} = \mathcal{X}$ and $\mathcal{X}^{\perp} = \mathcal{Y}$.

(4) If $(\mathcal{X}, \mathcal{Y})$ is a torsion pair in $\mathcal{C}$, then $\mathcal{X}$, resp. $\mathcal{Y}$, is a right, resp. left, triangulated subcategory of $\mathcal{C}$ in the sense of [**19**].

Recall that a full subcategory $\mathcal{Z}$ of $\mathcal{C}$ is said to be *closed under extensions*, if for any triangle $Z_1 \to C \to Z_2 \to \Sigma(Z_1)$ in $\mathcal{C}$ with $Z_1, Z_2 \in \mathcal{Z}$, the object C lies in $\mathcal{Z}$. Using the above remark, it is easy to see that if $(\mathcal{X}, \mathcal{Y})$ is a torsion pair in $\mathcal{C}$, then $\mathcal{X}$ and $\mathcal{Y}$ are closed under extensions.

The following result, first observed by Keller-Vosieck in [**74**], is a triangulated analogue of Proposition 1.2. Note that our torsion, resp. torsion-free, classes, coincide with aisles, resp. coaisles, in the sense of [**74**].

PROPOSITION 2.3. *Let $\mathcal{C}$ be a triangulated category and let $\mathcal{X}$ and $\mathcal{Y}$ be full subcategories of $\mathcal{C}$ closed under isomorphisms. Then we have the following.*

(1) *$\mathcal{Y}$ is a torsion-free class if and only if $\mathcal{Y}$ is closed under extensions and Σ^{-1}, and the inclusion $\mathbf{j} : \mathcal{Y} \hookrightarrow \mathcal{C}$ admits a left adjoint $\mathbf{L} : \mathcal{C} \to \mathcal{Y}$.*

(2) *$\mathcal{X}$ is a torsion class if and only if $\mathcal{X}$ is closed under extensions and Σ, and the inclusion $\mathbf{i} : \mathcal{X} \hookrightarrow \mathcal{C}$ admits a right adjoint $\mathbf{R} : \mathcal{C} \to \mathcal{X}$.*

PROOF. (1) ($\Leftarrow$) Let $\mathbf{L} : \mathcal{C} \to \mathcal{Y}$ be the left adjoint of the inclusion $\mathbf{j} : \mathcal{Y} \hookrightarrow \mathcal{C}$. For $C \in \mathcal{C}$ let $X_C \xrightarrow{f_C} C \xrightarrow{g^C} \mathbf{L}(C) \xrightarrow{h^C} \Sigma(X_C)$ be a triangle in $\mathcal{C}$ where g^C is the counit of $(\mathbf{L}, \mathbf{j})$ evaluated at C. Applying the cohomological functor $\mathcal{C}(-, \mathcal{Y})$ to the above triangle we get the long exact sequence $\mathcal{C}(\Sigma\mathbf{L}(C), \mathcal{Y}) \xrightarrow{(\Sigma(g^C), \mathcal{Y})} \mathcal{C}(\Sigma(C), \mathcal{Y}) \to \mathcal{C}(\Sigma(X_C), \mathcal{Y}) \to \mathcal{C}(\mathbf{L}(C), \mathcal{Y}) \xrightarrow{\mathcal{C}(g^C, \mathcal{Y})} \mathcal{C}(C, \mathcal{Y})$ in which $(g^C, \mathcal{Y})$ is invertible. Since $\mathcal{Y}$ is closed under Σ^{-1}, then $(\Sigma(g^C), \mathcal{Y})$ which is isomorphic to $(g^C, \Sigma^{-1}(\mathcal{Y}))$, is also invertible. It follows that $\mathcal{C}(\Sigma(X_C), \mathcal{Y}) \cong \mathcal{C}(X_C, \Sigma^{-1}(\mathcal{Y})) = 0$. We set $\mathcal{X} := {}^{\perp}\mathcal{Y}$, and it suffices to show that $X_C \in \mathcal{X}$. Now let $\alpha : X_C \to Y$ be a morphism with $Y \in \mathcal{Y}$. Consider the following morphism of triangles:

$$\begin{array}{ccccccc} X_C & \xrightarrow{f_C} & C & \xrightarrow{g^C} & \mathbf{L}(C) & \xrightarrow{h^C} & \Sigma(X_C) \\ {\scriptstyle\alpha}\downarrow & & {\scriptstyle\exists\beta}\downarrow & & \| & & \downarrow{\scriptstyle\Sigma(\alpha)} \\ Y & \xrightarrow{f'} & Y' & \xrightarrow{g'} & \mathbf{L}(C) & \xrightarrow{h'} & \Sigma(Y) \end{array}$$

Then $Y' \in \mathcal{Y}$ since $\mathcal{Y}$ is closed under extensions. Since g^C is the coreflection of C in $\mathcal{Y}$, there exists a unique morphism $\rho : \mathbf{L}(C) \to Y'$ such that $g^C \circ \rho = \beta$. Then $\alpha \circ f' = f_C \circ \beta = f_C \circ g^C \circ \rho = 0$. Hence there exists a morphism $\tau : X_C \to \Sigma^{-1}\mathbf{L}(C)$ such that $\tau \circ \Sigma^{-1}(h') = \alpha$. Since $\mathcal{C}(X_C, \Sigma^{-1}(\mathcal{Y})) = 0$, we have $\tau = 0$ and then $\alpha = 0$. We infer that $X_C \in {}^{\perp}\mathcal{Y} = \mathcal{X}$.

($\Rightarrow$) If $\mathcal{Y}$ is a torsion-free class with corresponding torsion class $\mathcal{X}$, then by Remark 2.2, we have $\mathcal{X} = {}^{\perp}\mathcal{Y}$ which is closed under Σ and $\mathcal{X}^{\perp} = \mathcal{Y}$ which is closed under Σ^{-1}. This implies trivially that $\mathcal{Y}$ is closed under extensions. Consider the triangle $X_C \xrightarrow{f_C} C \xrightarrow{g^C} Y^C \xrightarrow{h^C} \Sigma(X_C)$ of Definition 2.1. Applying the functor $\mathcal{C}(-, \mathcal{Y})$ to this triangle and using that $\Sigma(\mathcal{X}) \subseteq \mathcal{X}$ we see directly that $\mathcal{C}(g^C, \mathcal{Y}) : \mathcal{C}(Y^C, \mathcal{Y}) \to \mathcal{C}(C, \mathcal{Y})$ is invertible, i.e. g^C is the coreflection morphism of C in $\mathcal{Y}$. Hence setting $\mathbf{L}(C) = Y^C$ we obtain a left adjoint $\mathbf{L} : \mathcal{C} \to \mathcal{Y}$ of the inclusion $\mathbf{j} : \mathcal{Y} \hookrightarrow \mathcal{C}$.

(2) The proof is dual. □

REMARK 2.4. If $(\mathcal{X}, \mathcal{Y})$ is a torsion pair in $\mathcal{C}$, then so is $(\Sigma^n(\mathcal{X}), \Sigma^n(\mathcal{Y}))$, $\forall n \in \mathbb{Z}$. Indeed obviously we have $\mathcal{C}(\Sigma^n(\mathcal{X}), \Sigma^n(\mathcal{Y})) = 0$, $\Sigma(\Sigma^n(\mathcal{X})) \subseteq \Sigma^n(\mathcal{X})$ and for any $C \in \mathcal{C}$ we have the triangle $\Sigma^n\mathbf{R}\Sigma^{-n}(C) \to C \to \Sigma^n\mathbf{L}\Sigma^{-n}(C) \to \Sigma(\Sigma^n\mathbf{R}\Sigma^{-n}(C))$.

Inspired by the characterization of hereditary torsion pairs in abelian categories, we make the following definition (note that "left" or "right" exact functors between triangulated categories are exact).

DEFINITION 2.5. A torsion pair $(\mathcal{X}, \mathcal{Y})$ in $\mathcal{C}$ is called **hereditary**, resp. **cohereditary**, if the idempotent functor $\mathbf{iR} : \mathcal{C} \to \mathcal{C}$ is exact, resp. the idempotent functor $\mathbf{jL} : \mathcal{C} \to \mathcal{C}$ is exact.

We recall that a **thick** subcategory of $\mathcal{C}$ is a full triangulated subcategory of $\mathcal{C}$ closed under direct summands. A thick subcategory $\mathcal{Z} \subseteq \mathcal{C}$ is called **localizing**, resp. **colocalizing**, if the inclusion $\mathcal{Z} \hookrightarrow \mathcal{C}$ admits a right, resp. left, adjoint. If $\mathcal{Z}$ is a thick subcategory of $\mathcal{C}$ then we denote by $\mathcal{C}/\mathcal{Z}$ the induced quotient [**102**]. Thick subcategories can be regarded as triangulated analogues of Serre subcategories. If $\mathcal{U}, \mathcal{V}$ are full subcategories of $\mathcal{C}$, then we denote by $\mathcal{U} \star \mathcal{V}$ the category of extensions of $\mathcal{V}$ by $\mathcal{U}$, that is, the full subcategory of $\mathcal{C}$ consisting of objects C which may be included in a triangle $U \to C \to V \to \Sigma(U)$ in $\mathcal{C}$, with $U \in \mathcal{U}$ and $V \in \mathcal{V}$. The next result, which has no direct analogue in the abelian case, shows that hereditary torsion pairs coincide with cohereditary torsion pairs and includes a result of Bondal and Kapranov [**33**].

PROPOSITION 2.6. *If $\mathcal{X}$ and $\mathcal{Y}$ are full subcategories of $\mathcal{C}$, then the following conditions are equivalent.*

(i) $(\mathcal{X}, \mathcal{Y})$ *is a hereditary torsion pair in* $\mathcal{C}$.
(ii) $(\mathcal{X}, \mathcal{Y})$ *is a cohereditary torsion pair in* $\mathcal{C}$.
(iii) $(\mathcal{X}, \mathcal{Y})$ *is a torsion pair in* $\mathcal{C}$ *and* $\Sigma^{-1}(\mathcal{X}) \subseteq \mathcal{X}$.
(iv) $(\mathcal{X}, \mathcal{Y})$ *is a torsion pair in* $\mathcal{C}$ *and* $\Sigma(\mathcal{Y}) \subseteq \mathcal{Y}$.
(v) $\mathcal{C}(\mathcal{X}, \mathcal{Y}) = 0$, $\mathcal{X}$ *and* $\mathcal{Y}$ *are thick subcategories of* $\mathcal{C}$, *and* $\mathcal{C}$ *is generated as a triangulated category by* $\mathcal{X}$ *and* $\mathcal{Y}$.
(vi) $\mathcal{X}$ *is a localizing subcategory of* $\mathcal{C}$ *and there exists a short exact sequence:*

$$0 \longrightarrow \mathcal{X} \xrightarrow{\mathbf{i}} \mathcal{C} \xrightarrow{\mathbf{L}} \mathcal{Y} \longrightarrow 0$$

(vii) $\mathcal{Y}$ *is a colocalizing subcategory of* $\mathcal{C}$ *and there exists a short exact sequence:*

$$0 \longrightarrow \mathcal{Y} \xrightarrow{\mathbf{j}} \mathcal{C} \xrightarrow{\mathbf{R}} \mathcal{X} \longrightarrow 0$$

PROOF. (i) $\Leftrightarrow$ (iii) If $\Sigma^{-1}(\mathcal{X}) \subseteq \mathcal{X}$, then $\mathcal{X}$ is a full triangulated subcategory of $\mathcal{C}$. Hence $\mathbf{i} : \mathcal{X} \hookrightarrow \mathcal{C}$ is exact. It is well–known that left or right adjoints of exact functors between triangulated categories are exact, see [**102**]. Hence $\mathbf{R}$ is exact and then so is $\mathbf{iR} : \mathcal{C} \to \mathcal{C}$, i.e. $(\mathcal{X}, \mathcal{Y})$ is hereditary. If $(\mathcal{X}, \mathcal{Y})$ is hereditary, then $\mathbf{iR}$ is exact, in particular $\mathbf{iR}$ commutes functorially with Σ^{-1}. This implies trivially that $\Sigma^{-1}(\mathcal{X}) \subseteq \mathcal{X}$.

The proof of (ii) $\Leftrightarrow$ (iv) is dual and the proof of (iii) $\Leftrightarrow$ (iv) follows from the fact that if $\mathcal{X}^{\perp} = \mathcal{Y}$ and ${}^{\perp}\mathcal{Y} = \mathcal{X}$, then $\Sigma^{-1}(\mathcal{X}) \subseteq \mathcal{X}$ if and only if $\Sigma(\mathcal{Y}) \subseteq \mathcal{Y}$.

(i) $\Leftrightarrow$ (v) Obviously (i) implies (v). If (v) holds, then let $\mathcal{D} = \mathcal{X} \star \mathcal{Y}$ and it suffices to show that $\mathcal{D}$ is a triangulated subcategory of $\mathcal{C}$. Obviously $\mathcal{D}$ is closed under Σ, Σ^{-1}. Let $\alpha : D_1 \to D_2$ be a morphism in $\mathcal{D}$, so that there are triangles $X_i \xrightarrow{f_i} D_i \xrightarrow{g_i} Y_i \xrightarrow{h_i} \Sigma(X_i)$ with $X_i \in \mathcal{X}$ and $Y_i \in \mathcal{Y}$, $i = 1, 2$. Since $\mathcal{C}(\mathcal{X}, \mathcal{Y}) = 0$ we have $f_1 \circ \alpha \circ g_2 = 0$. Hence there exists a diagram

$$\begin{array}{ccccccc}
X_1 & \xrightarrow{f_1} & D_1 & \xrightarrow{g_1} & Y_1 & \xrightarrow{h_1} & \Sigma(X_1) \\
\beta \downarrow & & \alpha \downarrow & & \exists \gamma \downarrow & & \Sigma(\beta) \downarrow \\
X_2 & \xrightarrow{f_2} & D_2 & \xrightarrow{g_2} & Y_2 & \xrightarrow{h_2} & \Sigma(X_2)
\end{array}$$

By a result of Verdier [**104**], this can be completed to a 3×3–diagram of triangles which implies in particular that a cone D_3 of α is included in a triangle $X_3 \to D_3 \to Y_3 \to \Sigma(X_3)$ where X_3 is a cone of β and Y_3 is a cone of a morphism γ making the above diagram commutative. Since $\mathcal{X}, \mathcal{Y}$ are triangulated, it follows that $D_3 \in \mathcal{D}$. Hence $\mathcal{D}$ is triangulated category.

(i) $\Leftrightarrow$ (vi) Assume that (i) holds. Since $\mathcal{X} = {}^{\perp}\mathcal{Y}$, it follows trivially that $\mathcal{X}$ is thick and by Proposition 2.3 we know that $\mathcal{X}$ is coreflective and $\mathcal{Y}$ is reflective with coreflection functor $\mathbf{L}$. Consider the quotient $\mathcal{C}/\mathcal{X}$ which is defined as the category of fractions $\mathcal{C}[\mathfrak{J}^{-1}]$, where $\mathfrak{J}$ is the class of morphisms which admit a cone in $\mathcal{X}$. This implies that $\mathbf{L} : \mathcal{C} \to \mathcal{Y}$ sends $\mathfrak{J}$ to isomorphisms in $\mathcal{Y}$. If $F : \mathcal{C} \to \mathcal{E}$ is an exact functor to a triangulated category with this property, then F factorizes uniquely through $\mathbf{L}$ by defining $F^* : \mathcal{Y} \to \mathcal{E}$, $F^* = F\mathbf{j}$. Indeed $F^*\mathbf{L} = F\mathbf{jL}$ is isomorphic to F, since the unit $\mathrm{Id}_{\mathcal{C}} \to \mathbf{jL}$ lies in $\mathfrak{J}$ and F inverts $\mathfrak{J}$. Hence $F^*\mathbf{L} = F$. If $G : \mathcal{C} \to \mathcal{E}$ is another exact functor such that $G\mathbf{L} = F$, then $G = G\mathbf{Lj} = F\mathbf{j} = F^*$. It follows that $\mathbf{L} : \mathcal{C} \to \mathcal{Y}$ represents the quotient functor $\mathcal{C} \to \mathcal{C}[\mathfrak{J}^{-1}] = \mathcal{C}/\mathcal{X}$, i.e. it induces a triangle equivalence $\mathcal{C}/\mathcal{X} \to \mathcal{Y}$. Conversely since $\mathcal{X}$ is coreflective, by Proposition

2.3 we have a hereditary torsion pair $(\mathcal{X}, \mathcal{X}^{\perp})$ in $\mathcal{C}$. Then by a well-known result of Verdier, see [**102**], we have triangle equivalences $\mathcal{C}/\mathcal{X} \approx \mathcal{Y} = \mathcal{X}^{\perp}$.

(i) $\Leftrightarrow$ (vii) The proof is dual and is left to the reader. □

We recall that a Grothendieck category is an abelian category with a generator and exact filtered colimits. The triangulated analogue of a Grothendieck category is usually considered to be a **compactly generated** triangulated category. We recall that a set $\mathcal{U}$ of objects of $\mathcal{C}$ is called a **generating set**, if $\mathcal{U}^{\perp} = 0$ and $\mathcal{U}$ is closed under Σ and Σ^{-1}. An object $T \in \mathcal{C}$ is called **compact**, if the functor $\mathcal{C}(T, -) : \mathcal{C} \to \mathcal{A}b$ preserves coproducts. In what follows we denote by $\mathcal{C}^b$ the full subcategory of $\mathcal{C}$ consisting of all compact objects. Then $\mathcal{C}$ is compactly generated if $\mathcal{C}$ has coproducts and admits a set of compact generators. This is equivalent to saying that $\mathcal{C}$ has coproducts and coincides with the smallest full triangulated subcategory which is closed under coproducts and contains the compact objects. Important examples of compactly generated triangulated categories include the following:

- The unbounded derived category $\mathbf{D}(\mathrm{Mod}(\Lambda))$ of an associative ring Λ; the compact objects are the perfect complexes.
- The stable homotopy category of spectra, see [**81**]; the compact objects are the finite CW-complexes.
- The unbounded derived category of quasi-coherent sheaves over a quasi-compact separated scheme; the compact objects are the complexes of the thick subcategory generated by the suspensions of powers of an ample line bundle, see [**85**].
- The stable module category $\underline{\mathrm{Mod}}(kG)$ of the group algebra of a finite group over a field k; the compact objects are the objects induced by the finitely generated kG-modules.

The corresponding characterization of torsion or torsion-free classes in a compactly generated triangulated category is analogous to the characterization in Grot–hendieck categories, modulo some set-theoretic restrictions which are necessary for the existence of (left or right) triangulated quotients.

COROLLARY 2.7. *Let $\mathcal{C}$ be a compactly generated triangulated category.*

(1) *For a strict full subcategory $\mathcal{X}$ of $\mathcal{C}$, the following are equivalent.*

(i) *$\mathcal{X}$ is the torsion class of a (hereditary) torsion pair in $\mathcal{C}$.*

(ii) *$\mathcal{X}$ is closed under Σ (and Σ^{-1}), extensions, coproducts and the right triangulated quotient $\mathcal{C}/\mathcal{X}$ has small* Hom *sets.*

(2) *For a strict full subcategory $\mathcal{Y}$ of $\mathcal{C}$, the following are equivalent.*

(i) *$\mathcal{Y}$ is the torsion-free class of a (cohereditary) torsion pair in $\mathcal{C}$.*

(ii) *$\mathcal{Y}$ is closed under Σ^{-1} (and Σ), extensions, products and the left triangulated quotient $\mathcal{C}/\mathcal{Y}$ has small* Hom *sets.*

PROOF. (1) (ii) $\Rightarrow$ (i) By hypothesis $\mathcal{X}$ is a full right triangulated subcategory of $\mathcal{C}$ closed under coproducts. Then by Proposition 2.3 it suffices to show that the inclusion $\mathcal{X} \hookrightarrow \mathcal{C}$ admits a right adjoint, and this follows from [**24**]. The converse follows trivially from Proposition 2.6.

(2) The proof is similar to (1) using Neeman's Theorem [**86**] that in a compactly generated triangulated category $\mathcal{C}$, a product preserving homological functor $\mathcal{C} \to \mathcal{A}b$ is representable. $\square$

Motivated by the definition of a TTF-triple in an abelian category, we introduce the notion of a TTF-triple in a triangulated category.

DEFINITION 2.8. A **torsion, torsion-free triple**, **TTF-triple** for short, in $\mathcal{C}$ is a triple $(\mathcal{X}, \mathcal{Y}, \mathcal{Z})$ of full subcategories of $\mathcal{C}$ such that the pairs $(\mathcal{X}, \mathcal{Y})$ and $(\mathcal{Y}, \mathcal{Z})$ are torsion pairs. In this case $\mathcal{Y}$ is called a **TTF-class**.

If $(\mathcal{X}, \mathcal{Y}, \mathcal{Z})$ is a TTF-triple in $\mathcal{C}$, then we have that $(\mathcal{X}, \mathcal{Y})$ is cohereditary and $(\mathcal{Y}, \mathcal{Z})$ is hereditary. Hence by Proposition 2.6, $\mathcal{X}$ is a thick localizing subcategory of $\mathcal{C}$, $\mathcal{Z}$ is a thick colocalizing subcategory of $\mathcal{C}$, and $\mathcal{Y}$ is a thick localizing and colocalizing subcategory of $\mathcal{C}$. As usual we denote by $\mathbf{R} : \mathcal{C} \to \mathcal{X}$ the right adjoint of the inclusion $\mathbf{i} : \mathcal{X} \hookrightarrow \mathcal{C}$, by $\mathbf{T} : \mathcal{C} \to \mathcal{Z}$ the left adjoint of the inclusion $\mathbf{k} : \mathcal{Z} \hookrightarrow \mathcal{C}$, and by $\mathbf{L} : \mathcal{C} \to \mathcal{Y}$, resp. $\mathbf{S} : \mathcal{C} \to \mathcal{Y}$ the left, resp. right, adjoint of the inclusion $\mathbf{j} : \mathcal{Y} \hookrightarrow \mathcal{C}$. Then we have adjoint pairs:

$$(\mathbf{i},\mathbf{R}) \ : \ \mathcal{X} \ \underset{\mathbf{R}}{\overset{\mathbf{i}}{\rightleftarrows}} \ \mathcal{C} \ \underset{\mathbf{j}}{\overset{\mathbf{L}}{\rightleftarrows}} \ \mathcal{Y} \ : \ (\mathbf{L},\mathbf{j}) \qquad \text{and} \qquad (\mathbf{j},\mathbf{S}) \ : \ \mathcal{Y} \ \underset{\mathbf{S}}{\overset{\mathbf{j}}{\rightleftarrows}} \ \mathcal{C} \ \underset{\mathbf{k}}{\overset{\mathbf{T}}{\rightleftarrows}} \ \mathcal{Z} \ : \ (\mathbf{T},\mathbf{k})$$

and $\forall C \in \mathcal{C}$ we have the following functorial exact commutative diagram of triangles, where the involved morphisms are the adjunctions:

$$\begin{array}{ccccccc}
 & & \mathbf{iR}(C) & = & \mathbf{iR}(C) & & \\
 & & \downarrow & & \downarrow & & \\
\mathbf{jS}(C) & \longrightarrow & C & \longrightarrow & \mathbf{kT}(C) & \longrightarrow & \Sigma\mathbf{jS}(C) \\
\| & & \downarrow & & \downarrow & & \| \\
\mathbf{jS}(C) & \longrightarrow & \mathbf{jL}(C) & \longrightarrow & \mathbf{jLkT}(C) & \longrightarrow & \Sigma\mathbf{jS}(C) \\
 & & \downarrow & & \downarrow & & \\
 & & \Sigma\mathbf{iR}(C) & = & \Sigma\mathbf{iR}(C) & &
\end{array}$$

The second vertical triangle in the above diagram shows that the above reflection/coreflection functors are connected with a natural isomorphism:

$$\mathbf{jS}\,\mathbf{iR} \ \xrightarrow{\cong} \ \Sigma^{-1}\mathbf{jL}\,\mathbf{kT} \ : \ \mathcal{C} \longrightarrow \mathcal{C}$$

In other words for any object C in $\mathcal{C}$ we have: $Y_{X_C} \xrightarrow{\cong} \Sigma^{-1}(Y^{Z^C})$. Observe that for any two objects A, B in $\mathcal{C}$ we have natural isomorphisms:

$$\mathcal{C}\big(\mathbf{L}(A), B\big) \ \xleftarrow{\cong} \ \mathcal{Y}(\mathbf{L}(A), \mathbf{S}(B)) \ \xrightarrow{\cong} \ \mathcal{C}\big(A, \mathbf{S}(B)\big).$$

The proof of the following analogue of Proposition 1.3 is a direct consequence of Proposition 2.6 and the above diagram.

COROLLARY 2.9. *If $(\mathcal{X}, \mathcal{Y}, \mathcal{Z})$ is a TTF-triple in $\mathcal{C}$, then there are triangle equivalences:* $\mathcal{X} \xleftarrow{\approx} \mathcal{C}/\mathcal{Y} \xrightarrow{\approx} \mathcal{Z}$.

Note that the equivalence $\mathcal{X} \xrightarrow{\approx} \mathcal{Z}$ above is given explicitly by the functor $\mathbf{Ti} : \mathcal{X} \to \mathcal{Z}$ which is given by $X \mapsto Z^X$, with quasi-inverse the functor $\mathbf{Rk} : \mathcal{Z} \to \mathcal{X}$ which is given by $Z \mapsto X_Z$.

REMARK 2.10. If $(\mathcal{X}, \mathcal{Y}, \mathcal{Z})$ is a TTF-triple in an abelian category $\mathcal{C}$, then in general it is not true that $\mathcal{X}$ and $\mathcal{Z}$ are equivalent. However if $\mathcal{C}$ has enough projectives and enough injectives, then it is not difficult to see that the analogous functors as above induce an equivalence $\mathcal{A} \xrightarrow{\approx} \mathcal{B}$, where $\mathcal{B}$ is the Giraud subcategory of $\mathcal{C}$ corresponding to the hereditary torsion pair $(\mathcal{Y}, \mathcal{Z})$, and $\mathcal{A}$ is the co-Giraud subcategory of $\mathcal{C}$ corresponding to the cohereditary torsion pair $(\mathcal{X}, \mathcal{Y})$, see [**100**] for the concept of a (co-)Giraud subcategory corresponding to a (co-)hereditary torsion pair. Since there are equivalences $\mathcal{A} \xleftarrow{\approx} \mathcal{C}/\mathcal{Y} \xrightarrow{\approx} \mathcal{B}$, by Proposition 1.3 we have equivalences $\mathcal{A} \xleftarrow{\approx} \mathcal{X} \cap \mathcal{Y} \xrightarrow{\approx} \mathcal{B}$.

With a proper definition of TTF-triple $(\mathcal{X}, \mathcal{Y}, \mathcal{Z})$ in a stable category of an abelian category, we shall see in Chapter VI that the subcategories $\mathcal{X}$ and $\mathcal{Z}$ are equivalent up to projective or injective summands.

TTF-triples in triangulated categories occur frequently in practice and play an important role in connection with stratifications of derived categories of stratified spaces and highest weight categories, see [**18**], [**39**]. In the setting of compactly generated triangulated categories we have the following existence result of TTF-triples which will be useful later in connection with tilting theory in derived categories.

PROPOSITION 2.11. *Let $\mathcal{C}$ and $\mathcal{D}$ be compactly generated triangulated categories, and let $F : \mathcal{D} \to \mathcal{C}$ be a fully faithful exact functor which preserves coproducts and compact objects. Then F admits a right adgoint $G : \mathcal{C} \to \mathcal{D}$ which preserves coproducts, and if $\mathcal{X} = \mathrm{Im}(F)$ is the essential image of F, then there exists a TTF-triple $(\mathcal{X}, \mathcal{Y}, \mathcal{Z})$ in $\mathcal{C}$, where $\mathcal{Y} := \mathrm{Ker}(G)$, and $\mathcal{Z} := \mathcal{Y}^{\perp}$.*

PROOF. Since the functor F preserves coproducts, the existence of G follows by [**85**]. Note that by [**85**] the functor $G : \mathcal{C} \to \mathcal{D}$ preserves coproducts, since F preserves compact objects. Since G preserves coproducts, it follows that G admits a right adjoint $H : \mathcal{D} \to \mathcal{C}$. Hence we have an adjoint triple (F, G, H). Now it is easy to see that in such an adjoint triple, F is fully faithful if and only if H is fully faithful [**34**]. Hence H is fully faithful. Let $\mathcal{Z} := \mathrm{Im}(H)$ be the essential image of H, and let $\mathcal{Y} := \mathrm{Ker}(G)$. Observe that $\mathcal{C}(\mathcal{X}, \mathcal{Y}) = 0$ and $\mathcal{C}(\mathcal{Y}, \mathcal{Z}) = 0$. Indeed if $X \in \mathcal{X}$, $Y \in \mathcal{Y}$, and $Z \in \mathcal{Z}$, then $X = F(A)$ for some object $A \in \mathcal{D}$, $G(Y) = 0$, and $Z = H(B)$ for some object B in $\mathcal{D}$. Then $\mathcal{C}(X, Y) = \mathcal{C}(F(A), Y) \xrightarrow{\cong} \mathcal{D}(A, G(Y)) = 0$. Similarly $\mathcal{C}(Y, Z) = \mathcal{C}(Y, H(B)) \xrightarrow{\cong} \mathcal{D}(G(Y), B) = 0$. Now let $\varepsilon : FG \to \mathrm{Id}_{\mathcal{C}}$ be the counit of the adjoint pair (F, G), and let $\delta : \mathrm{Id}_{\mathcal{C}} \to HG$ be the unit of the adjoint pair (G, H). For any object C in $\mathcal{C}$ consider the functorial triangles $FG(C) \xrightarrow{\varepsilon_C} C \to Y^C \to \Sigma(FG(C))$ and $Y_C \to C \xrightarrow{\delta_C} HG(C) \to \Sigma(Y_C)$ in $\mathcal{C}$. Using standard properties of adjoint functors we have $G(Y_C) = 0 = G(Y^C)$. Hence $Y_C, Y^C \in \mathcal{Y}$. Since by definition $FG(C) \in \mathcal{X}$ and $HG(C) \in \mathcal{Z}$, we infer that $(\mathcal{X}, \mathcal{Y})$ and $(\mathcal{Y}, \mathcal{Z})$ are torsion pairs. Hence $(\mathcal{X}, \mathcal{Y}, \mathcal{Z})$ is a TTF-triple in $\mathcal{C}$. □

REMARK 2.12. In the situation of the above Proposition it is easy to see that we have functorial triangles $FG \xrightarrow{\varepsilon} \mathrm{Id}_{\mathcal{C}} \to \mathbf{jL} \to \Sigma FG$ and $\Sigma^{-1} HG \to \mathbf{jS} \to \mathrm{Id}_{\mathcal{C}} \xrightarrow{\delta}$

HG, where ε is the counit of the adjoint pair (F, G), and δ is the unit of the adjoint pair (G, H). Moreover we have identifications: $\mathbf{iR} = FG$, $\mathbf{kT} = HG$.

Torsion pairs in triangulated categories are used in the literature mainly in the form of *t*-**structures**. Recall from [**18**] that a *t*-structure in a triangulated category $\mathcal{C}$ is a pair $(\mathcal{T}^{\leq 0}, \mathcal{T}^{\geq 0})$ of full subcategories such that setting $\mathcal{T}^{\leq n} = \Sigma^{-n}(\mathcal{T}^{\leq 0})$ and $\mathcal{T}^{\geq n} = \Sigma^{-n}(\mathcal{T}^{\geq 0})$, $\forall n \in \mathbb{Z}$, the following are satisfied. First $\mathcal{C}(\mathcal{T}^{\leq 0}, \mathcal{T}^{\geq 1}) = 0$. Second $\mathcal{T}^{\leq 0} \subseteq \mathcal{T}^{\leq 1}$ and $\mathcal{T}^{\geq 1} \subseteq \mathcal{T}^{\geq 0}$. Thirdly any object $C \in \mathcal{C}$ is included in a triangle $C^{\leq 0} \to C \to C^{\geq 1} \to \Sigma(C^{\leq 0})$, where $C^{\leq 0} \in \mathcal{T}^{\leq 0}$ and $C^{\geq 1} \in \mathcal{T}^{\geq 1}$.

The following result shows that torsion pairs and *t*-structures essentially coincide.

PROPOSITION 2.13. *The maps*

$$\Phi : (\mathcal{X}, \mathcal{Y}) \longmapsto (\mathcal{X}, \Sigma\mathcal{Y}) \quad \textit{and} \quad \Psi : (\mathcal{T}^{\leq 0}, \mathcal{T}^{\geq 0}) \longmapsto (\mathcal{T}^{\leq 0}, \mathcal{T}^{\geq 1})$$

are mutually inverse bijections between torsion pairs and t-structures in $\mathcal{C}$.

PROOF. If $(\mathcal{X}, \mathcal{Y})$ is a torsion pair in $\mathcal{C}$, then set $\mathcal{T}^{\leq 0} = \mathcal{X}$ and $\mathcal{T}^{\geq 0} = \Sigma(\mathcal{Y})$ and further $\mathcal{T}^{\leq n} = \Sigma^{-n}(\mathcal{X})$ and $\mathcal{T}^{\geq n} = \Sigma^{-n+1}(\mathcal{Y})$. Then $\mathcal{C}(\mathcal{T}^{\leq 0}, \mathcal{T}^{\geq 1}) = \mathcal{C}(\mathcal{X}, \mathcal{Y}) = 0$. Since $\Sigma(\mathcal{X}) \subseteq \mathcal{X}$, we have $\mathcal{T}^{\leq 0} = \mathcal{X} \subseteq \Sigma^{-1}(\mathcal{X}) = \mathcal{T}^{\leq 1}$. Since $\Sigma^{-1}(\mathcal{Y}) \subseteq \mathcal{Y}$, we have $\mathcal{T}^{\geq 1} = \mathcal{Y} \subseteq \Sigma(\mathcal{Y}) = \mathcal{T}^{\geq 0}$. Finally since $(\mathcal{X}, \mathcal{Y})$ is a torsion pair, $\forall C \in \mathcal{C}$, there exists a triangle $X_C \to C \to Y^C \to \Sigma(X_C)$ with $X_C \in \mathcal{T}^{\leq 0}$ and $Y^C \in \mathcal{T}^{\geq 1}$. So $(\mathcal{X}, \Sigma(\mathcal{Y}))$ is a *t*-structure in $\mathcal{C}$. Conversely if $(\mathcal{T}^{\leq 0}, \mathcal{T}^{\geq 0})$ is a *t*-structure, then set $\mathcal{X} = \mathcal{T}^{\leq 0}$ and $\mathcal{Y} = \mathcal{T}^{\geq 1}$. By definition $\mathcal{C}(\mathcal{X}, \mathcal{Y}) = 0$ and $\Sigma(\mathcal{X}) = \mathcal{T}^{\leq -1} \subseteq \mathcal{T}^{\leq 0} = \mathcal{X}$. Since $\forall C \in \mathcal{C}$ there exists a triangle $X_C \to C \to Y^C \to \Sigma(\mathcal{X}_C)$ with $X_C \in \mathcal{T}^{\leq 0}$ and $Y^C \in \mathcal{T}^{\geq 1}$, we infer that $(\mathcal{T}^{\leq 0}, \mathcal{T}^{\geq 1})$ is a torsion pair. □

Let $(\mathcal{X}, \mathcal{Y})$ be a torsion pair in $\mathcal{C}$. As usual we denote by $\mathbf{R}$ the right adjoint of the inclusion $\mathbf{i} : \mathcal{X} \hookrightarrow \mathcal{C}$ and by $\mathbf{L}$ the left adjoint of the inclusion $\mathbf{j} : \mathcal{Y} \hookrightarrow \mathcal{C}$. Let $\mathcal{T}^{\leq n} = \Sigma^{-n}(\mathcal{X})$ and $\mathcal{T}^{\geq n} = \Sigma^{-n+1}(\mathcal{Y})$, $\forall n \in \mathbb{Z}$. Then we have the functor $\tau^{\leq n} := \Sigma^{-n}\mathbf{R}\Sigma^{n} : \mathcal{C} \to \mathcal{T}^{\leq n}$ which is a right adjoint of the inclusion $\mathcal{T}^{\leq n} \hookrightarrow \mathcal{C}$ and the functor $\tau^{\geq n} := \Sigma^{-n}\mathbf{L}\Sigma^{n} : \mathcal{C} \to \mathcal{T}^{\leq n}$ which is a left adjoint of the inclusion $\mathcal{T}^{\geq n} \hookrightarrow \mathcal{C}$. In other words the functors $\tau^{\leq n}$ and $\tau^{\geq n}$ are the truncation functors in the sense of [**18**] associated to the *t*-structure $(\mathcal{X}, \Sigma(\mathcal{Y}))$.

We recall from [**18**] that the **heart** of a *t*-structure $(\mathcal{T}^{\leq 0}, \mathcal{T}^{\geq 0})$ in $\mathcal{C}$ is defined to be the full subcategory $\mathcal{H} = \mathcal{T}^{\leq 0} \cap \mathcal{T}^{\geq 0}$. We also define the heart of a torsion pair $(\mathcal{X}, \mathcal{Y})$ in $\mathcal{C}$ to be the full subcategory $\mathcal{H} = \mathcal{X} \cap \Sigma(\mathcal{Y})$. As in [**18**], we define homology functors by $\mathrm{H}^n := \Sigma\mathbf{L}\Sigma^{-1}\mathbf{R}\Sigma^{n} = \mathbf{R}\Sigma\mathbf{L}\Sigma^{n-1} : \mathcal{C} \to \mathcal{H}$, $\forall n \in \mathbb{Z}$. By [**18**], the heart $\mathcal{H}$ of the torsion pair $(\mathcal{X}, \mathcal{Y})$ is an abelian category and the functors H^n are indeed homological, i.e. they send triangles in $\mathcal{C}$ to long exact sequences in $\mathcal{H}$.

REMARK 2.14. Proposition 2.13 has an analogue for TTF-triples. In fact, it is not difficult to see that TTF-triples are in bijective correspondence with recollement situations in the sense of [**18**].

There are connections between torsion pairs in $\mathcal{C}$ and in $\mathcal{C}/\mathcal{V}$ for a thick subcategory $\mathcal{V}$ of $\mathcal{C}$, and between torsion pairs and the corresponding hearts. This is also interesting since it provides new ways of constructing torsion pairs.

We have the following result, where we denote by $\mathsf{Q} : \mathcal{C} \to \mathcal{C}/\mathcal{V}$ the localization functor of the quotient $\mathcal{C}/\mathcal{V}$ with respect to the thick subcategory $\mathcal{V}$.

PROPOSITION 2.15. *Let $\mathcal{V}$ be a thick subcategory of $\mathcal{C}$ and $(\mathcal{X}, \mathcal{Y})$ a (hereditary) torsion pair in $\mathcal{C}$. Then the following statements are equivalent:*

(i) $\mathbf{L}(\mathcal{V}) \subseteq \mathcal{V}$.
(ii) $\mathbf{R}(\mathcal{V}) \subseteq \mathcal{V}$.
(iii) *The pair $(\mathsf{Q}(\mathcal{X}), \mathsf{Q}(\mathcal{Y}))$ is a (hereditary) torsion pair in $\mathcal{C}/\mathcal{V}$.*
(iv) *For any object A in $\mathcal{C}$ which is isomorphic in $\mathcal{C}/\mathcal{V}$ to an object from $\mathcal{X}$, the object $\mathbf{L}(A)$ is in $\mathcal{V}$.*
(v) *For any object A in $\mathcal{C}$ which is isomorphic in $\mathcal{C}/\mathcal{V}$ to an object from $\mathcal{Y}$, the object $\mathbf{R}(A)$ is in $\mathcal{V}$.*

If (iii) *holds and $\mathcal{H} = \mathcal{X} \cap \Sigma(\mathcal{Y})$ is the heart of $(\mathcal{X}, \mathcal{Y})$ in $\mathcal{C}$ and $\mathcal{K} = \mathsf{Q}(\mathcal{X}) \cap \Sigma(\mathsf{Q}(\mathcal{Y}))$ is the heart of $(\mathsf{Q}(\mathcal{X}), \mathsf{Q}(\mathcal{Y}))$ in $\mathcal{C}/\mathcal{V}$, then $\mathcal{U} := \mathcal{X} \cap \Sigma(\mathcal{Y}) \cap \mathcal{V}$ is a Serre subcategory of $\mathcal{H}$ and there exists a short exact sequence of abelian categories:*

$$0 \longrightarrow \mathcal{U} \longrightarrow \mathcal{H} \longrightarrow \mathcal{K} \longrightarrow 0$$

Moreover if $\mathcal{V}$ is a localizing, resp. colocalizing, subcategory of $\mathcal{C}$, then the functor $\mathcal{H} \to \mathcal{K}$ admits a right, resp. left, adjoint.

PROOF. (i) $\Rightarrow$ (ii) Let V be in $\mathcal{V}$ and consider the standard triangle $X_V \to V \to Y^V \to \Sigma(X_V)$ with $X_V \in \mathcal{X}$ and $Y^V \in \mathcal{Y}$. By hypothesis we have $Y^V \in \mathcal{V}$. Since $\mathcal{V}$ is thick, we infer that X_V lies in $\mathcal{V}$ and this implies that $\mathbf{R}(\mathcal{V}) \subseteq \mathcal{V}$.

(ii) $\Rightarrow$ (iii) It is clear that $(\mathsf{Q}(\mathcal{X}), \mathsf{Q}(\mathcal{Y}))$ is a torsion pair in $\mathcal{C}/\mathcal{V}$ if and only if $(\mathsf{Q}(\mathcal{X}), \mathsf{Q}(\mathcal{Y})) = 0$. Let $\alpha : \mathsf{Q}(X) \to \mathsf{Q}(Y)$ be a morphism in $\mathcal{C}/\mathcal{V}$. Then α is represented by a fraction $X \xleftarrow{s} A \xrightarrow{\rho} Y$, where in the triangle $V \xrightarrow{\kappa} A \xrightarrow{s} X \to \Sigma(V)$ the object $V \in \mathcal{V}$. Consider the triangle $\mathbf{R}(A) \xrightarrow{f_A} A \xrightarrow{g^A} \mathbf{L}(A) \to \Sigma\mathbf{R}(A)$. Applying the Octahedral Axiom to the composition $f_A \circ s$ we see easily that there is a triangle $\mathbf{L}(A) \to X' \to V \to \Sigma\mathbf{L}(A)$ with $X' \in \mathcal{X}$ and $V \in \mathcal{V}$. From this triangle it is clear that $X' = \mathbf{R}(V)$ and $\Sigma\mathbf{L}(A) = \mathbf{L}(V)$. By hypothesis we have that X' lies in $\mathcal{V}$. Hence $f_A \circ s : \mathbf{R}(A) \to X$ is invertible in $\mathcal{C}/\mathcal{V}$, and then so is f_A. This implies that the fraction $X \xleftarrow{s} A \xrightarrow{\rho} Y$ is equal to the fraction $X \xleftarrow{f_A \circ s} \mathbf{R}(A) \xrightarrow{0} Y$ which represents the zero morphism in $\mathcal{C}/\mathcal{V}$. Hence $\alpha = 0$ in $\mathcal{C}/\mathcal{V}$.

(iii) $\Rightarrow$ (iv) Let A be an object in $\mathcal{C}$ such that $\mathsf{Q}(A) \xrightarrow{\cong} \mathsf{Q}(X)$ where X is in $\mathcal{X}$. Consider the triangle $\mathbf{R}(A) \to A \to \mathbf{L}(A) \to \Sigma\mathbf{R}(A)$ in $\mathcal{C}$. Then we have the triangle $\mathsf{Q}\mathbf{R}(A) \to \mathsf{Q}(A) \to \mathsf{Q}\mathbf{L}(A) \to \mathsf{Q}\Sigma\mathbf{R}(A)$ in $\mathcal{C}/\mathcal{V}$. Since $(\mathsf{Q}(\mathcal{X}), \mathsf{Q}(\mathcal{Y}))$ is a torsion pair we have $\mathsf{Q}\mathbf{L}(A) \in \mathsf{Q}(\mathcal{X}) \cap \mathsf{Q}(\mathcal{Y}) = 0$. This implies that $\mathbf{L}(A)$ lies in $\mathcal{V}$.

(iv) $\Rightarrow$ (i) Let V be in $\mathcal{V}$ and consider the triangle $X_V \to V \to Y^V \to \Sigma(X_V)$ with $X_V \in \mathcal{X}$ and $Y^V \in \mathcal{Y}$. Then obviously $\Sigma^{-1}(Y^V)$ is isomorphic in $\mathcal{C}/\mathcal{V}$ to $X_V \in \mathcal{X}$. By hypothesis we have that $\Sigma^{-1}(Y^V)$, or equivalently Y^V, lies in $\mathcal{V}$. Hence $\mathbf{L}(\mathcal{V}) \subseteq \mathcal{V}$.

(v) $\Leftrightarrow$ (iii) Follows by dual arguments.

Now let $A_1 \rightarrowtail A_2 \twoheadrightarrow A_3$ be a short exact sequence in $\mathcal{H}$. Then there exists a morphism $C \to \Sigma(A)$ such that $A_1 \to A_2 \to A_3 \to \Sigma(A_1)$ is a triangle in $\mathcal{C}$. Applying Q we have a triangle $\mathsf{Q}(A_1) \to \mathsf{Q}(A_2) \to \mathsf{Q}(A_3) \to \Sigma\mathsf{Q}(A_1)$ in $\mathcal{C}/\mathcal{V}$ with the $\mathsf{Q}(A_i)$ in $\mathcal{K}$. Then the sequence $\mathsf{Q}(A_1) \rightarrowtail \mathsf{Q}(A_2) \twoheadrightarrow \mathsf{Q}(A_3)$ is short exact in $\mathcal{K}$ and we infer that the induced functor $\mathsf{Q} : \mathcal{H} \to \mathcal{K}$ is exact. By definition Q is surjective on objects and $\operatorname{Ker}(\mathsf{Q}) := \mathcal{U} = \mathcal{X} \cap \Sigma(\mathcal{Y}) \cap \mathcal{V}$. Using the universal property of $\mathsf{Q} : \mathcal{C} \to \mathcal{C}/\mathcal{V}$ we infer that $\mathsf{Q} : \mathcal{H} \to \mathcal{K}$ is the localization functor $\mathcal{H} \to \mathcal{H}/\mathcal{U}$. We leave the easy proof of the last assertion to the reader. □

A thick subcategory $\mathcal{V}$ of $\mathcal{C}$ is called **$(\mathcal{X}, \mathcal{Y})$-stable**, where $(\mathcal{X}, \mathcal{Y})$ is a torsion pair in $\mathcal{C}$, if the equivalent conditions of Proposition 2.15 hold. If $\mathcal{V}$ is a thick $(\mathcal{X}, \mathcal{Y})$-stable subcategory of $\mathcal{C}$, then by [**18**] we get an induced torsion pair $(\mathcal{X} \cap \mathcal{V}, \mathcal{Y} \cap \mathcal{V})$ in $\mathcal{V}$ with heart precisely the category $\mathcal{U}$. In this case the inclusion functor $\mathcal{V} \hookrightarrow \mathcal{C}$ and the quotient functor $\mathcal{C} \to \mathcal{C}/\mathcal{V}$ are t-exact in the sense of [**18**].

The above result can be used to construct interesting torsion theories:

EXAMPLE. Let $\mathcal{C}$ be an abelian category and let $\mathcal{H}(\mathcal{C})$ be the homotopy category of, say unbounded, complexes over $\mathcal{C}$. If $\mathcal{X}$, resp. $\mathcal{Y}$, denotes the full subcategory of complexes which are acyclic in degrees > 0, resp. < 1, then it is easy to see that $(\mathcal{X}, \mathcal{Y})$ is a torsion pair in $\mathcal{H}(\mathcal{C})$ and the subcategory $\mathcal{V}$ of acyclic complexes is $(\mathcal{X}, \mathcal{Y})$-stable. Then the quotient $\mathcal{H}(\mathcal{C})/\mathcal{V}$ is the derived category $\mathbf{D}(\mathcal{C})$ of $\mathcal{C}$ and the induced torsion pair $(\mathsf{Q}(\mathcal{X}), \mathsf{Q}(\mathcal{Y}))$ is the natural torsion pair (= t-structure) $(\mathbf{D}^{\leq 0}(\mathcal{C}), \mathbf{D}^{\geq 0}(\mathcal{C}))$ of $\mathbf{D}(\mathcal{C})$, see [**18**]. In our case the exact sequence of the hearts in Proposition 2.15 reduces to the following exact sequence of abelian categories

$$0 \longrightarrow \operatorname{mod}_e(\mathcal{C}) \longrightarrow \operatorname{mod}(\mathcal{C}) \longrightarrow \mathcal{C} \longrightarrow 0$$

where $\operatorname{mod}(\mathcal{C})$ is the category of contravariant coherent functors over $\mathcal{C}$ and $\operatorname{mod}_e(\mathcal{C})$ is the full subcategory consisting of all coherent functors F which admit a presentation $\mathcal{C}(-, A) \to \mathcal{C}(-, B) \to F \to 0$ where $A \to B$ is an epimorphism. Note that the above exact sequence of abelian categories, which is due to Auslander [**4**], plays an important role in representation theory.

EXAMPLE. Let Λ be a self-injective Artin algebra and let $\underline{\operatorname{mod}}(\Lambda)$ be its stable module category. Let $\mathbf{D}^b(\operatorname{mod}(\Lambda))$ be the bounded derived category and let $(\mathcal{X}, \mathcal{Y})$ be a torsion pair in $\mathbf{D}^b(\operatorname{mod}(\Lambda))$. Setting $\mathcal{V} = \mathcal{H}^b(\mathcal{P}_\Lambda)$ to be the full subcategory of perfect complexes and assuming that $\mathcal{V}$ is $(\mathcal{X}, \mathcal{Y})$-stable, we get a torsion pair in $\underline{\operatorname{mod}}(\Lambda)$, since it is well-known that the quotient $\mathbf{D}^b(\operatorname{mod}(\Lambda))/\mathcal{H}^b(\mathcal{P}_\Lambda)$ is triangle equivalent to $\underline{\operatorname{mod}}(\Lambda)$, see [**74**], [**37**], [**92**], [**20**].

3. Tilting Torsion Pairs

In this section we discuss an interesting interplay between torsion pairs in abelian and triangulated categories, related to tilting theory.

In classical tilting theory, the torsion pair associated with a tilting module of projective dimension less than or equal to one plays an important role in connection with homological or representation theoretic questions. Actually the tilting process can be made starting directly with a *tilting* torsion pair $(\mathcal{T}, \mathcal{F})$ in an abelian category $\mathcal{C}$, in the sense that any object of $\mathcal{C}$ embeds in an object from $\mathcal{T}$. This tilting process, which was initiated by Happel, Reiten and Smalø in [**60**], takes place in the bounded derived category $\mathbf{D}^b(\mathcal{C})$ and presents an interesting interplay between torsion pairs in an abelian category and torsion pairs in the bounded derived category, thus generalizing the classical tilting situation. More precisely let $\mathcal{T}$ and $\mathcal{F}$ be full subcategories of $\mathcal{C}$ closed under isomorphisms, and define full subcategories in $\mathbf{D}^b(\mathcal{C})$ as follows:

$$\mathcal{X}(\mathcal{T}) := \{C^\bullet \in \mathbf{D}^b(\mathcal{C}) \mid \mathrm{H}^n(C^\bullet) = 0, \forall n > 0,\ \mathrm{H}^0(C^\bullet) \in \mathcal{T}\}$$

$$\mathcal{Y}(\mathcal{F}) := \{C^\bullet \in \mathbf{D}^b(\mathcal{C}) \mid \mathrm{H}^n(C^\bullet) = 0, \forall n < 0, \ \mathrm{H}^0(C^\bullet) \in \mathcal{F}\}$$

Then we have the following relationship.

THEOREM 3.1. [**60**] *The following conditions are equivalent.*

(i) *$(\mathcal{T}, \mathcal{F})$ is a torsion pair in $\mathcal{C}$.*

(ii) *$(\mathcal{X}(\mathcal{T}), \mathcal{Y}(\mathcal{F}))$ is a torsion pair in $\mathbf{D}^b(\mathcal{C})$.*

PROOF. For a proof that (i) $\Rightarrow$ (ii) we refer to [**60**]. Assume that (ii) holds. We view $\mathcal{C}$ as the full subcategory of $\mathbf{D}^b(\mathcal{C})$ consisting of complexes concentrated in degree zero. If $T \in \mathcal{T}$ and $F \in \mathcal{F}$, then $T \in \mathcal{X}(\mathcal{T})$ and $F \in \mathcal{Y}(\mathcal{F})$. Hence $\mathcal{C}(T, F) \xrightarrow{\cong} \mathbf{D}^b(\mathcal{C})(T, F) = 0$. Now let C be in $\mathcal{C}$, and let $X_C \to C \to Y^C \to \Sigma(X_C)$ be the standard triangle associated with the torsion pair $(\mathcal{X}(\mathcal{T}), \mathcal{Y}(\mathcal{F}))$ in $\mathbf{D}^b(\mathcal{C})$. Applying the usual homological functor $\mathrm{H}^0 : \mathbf{D}^b(\mathcal{C}) \to \mathcal{C}$ to this triangle, we infer an exact sequence $0 \to \mathrm{H}^0(X_C) \to C \to \mathrm{H}^0(Y^C) \to 0$ in $\mathcal{C}$ with $\mathrm{H}^0(X_C) \in \mathcal{T}$ and $\mathrm{H}^0(Y^C) \in \mathcal{F}$. Hence $(\mathcal{T}, \mathcal{F})$ is a torsion pair in $\mathcal{C}$. □

If $(\mathcal{T}, \mathcal{F})$ is a torsion pair in $\mathcal{C}$ and if $\mathcal{H}$ is the heart of the torsion pair $(\mathcal{X}(\mathcal{T}), \mathcal{Y}(\mathcal{F}))$ in $\mathbf{D}^b(\mathcal{C})$, then $(\Sigma(\mathcal{F}), \mathcal{T})$ is a torsion pair in $\mathcal{H}$ [**60**]. Moreover if $(\mathcal{T}, \mathcal{F})$ is a tilting torsion pair then another application of the construction to $\mathbf{D}^b(\mathcal{H})$ recovers the original torsion pair. In addition if $\mathcal{C}$ has enough projectives or $\mathcal{H}$ has enough injectives, then there exists a derived equivalence $\mathbf{D}^b(\mathcal{H}) \xrightarrow{\approx} \mathbf{D}^b(\mathcal{C})$, which extends the inclusion $\mathcal{H} \hookrightarrow \mathcal{C}$. The case of a classical tilting module gives rise to a special case of this construction. The map $(\mathcal{T}, \mathcal{F}) \mapsto (\mathcal{X}(\mathcal{T}), \mathcal{Y}(\mathcal{F}))$, which establishes an injective function between the poset of torsion pairs in $\mathcal{C}$ and the poset of torsion pairs in $\mathbf{D}^b(\mathcal{C})$, seems to be a very useful tool for interchanging information between module categories and derived categories. For instance this generalized tilting process was used in an essential way in the classification of hereditary Noetherian abelian categories with Serre duality by Reiten-Van den Bergh in [**89**] and the classification of hereditary abelian categories with tilting object by Happel in [**59**].

It would be interesting to know under what conditions a torsion pair in $\mathbf{D}^b(\mathcal{C})$ can be recovered in some way from a torsion pair in $\mathcal{C}$.

EXAMPLE. Let $(\mathcal{T}, \mathcal{F})$ be the torsion pair generated by the injective envelope $E(\Lambda)$ of Λ, where Λ is any ring, see [**100**]. Then $\mathcal{T}$ is a tilting torsion class, since it contains all the injectives. It follows that there is a triangle equivalence $\mathbf{D}^b(\mathcal{H}) \xrightarrow{\approx} \mathbf{D}^b(\mathrm{Mod}(\Lambda))$, where $\mathcal{H}$ is the heart of the induced torsion pair in $\mathbf{D}^b(\mathrm{Mod}(\Lambda))$.

As noted above an important source of examples illustrating the above considerations emerges from tilting modules.

Let Λ be an Artin algebra and T a finitely generated Λ-module. We denote by $\mathrm{add}(T)$ the full subcategory of $\mathrm{mod}(\Lambda)$ consisting of all direct summands of finite direct sums of copies of T. We recall that T is called a *tilting module*, if:

(i) $\mathrm{Ext}^n_\Lambda(T, T) = 0, \forall n \geq 1$,

(ii) $\mathrm{pd}_\Lambda T < \infty$.

(iii) There exists an exact sequence $0 \to \Lambda \to T^0 \to T^1 \to \cdots \to T^r \to 0$ with $T^i \in \mathrm{add}(T)$ for all $i = 0, 1, ..., r$.

Dually T is called a *cotilting module*, if:

(i) $\mathrm{Ext}^n_\Lambda(T,T) = 0$, $\forall n \geq 1$,
(ii) $\mathrm{id}_\Lambda T < \infty$.
(iii) There exists an exact sequence $0 \to T_s \to \cdots \to T_1 \to T_0 \to \mathrm{D}(\Lambda) \to 0$ with $T_i \in \mathrm{add}(T)$ for all $i = 0, 1, ..., s$, where $\mathrm{D}(\Lambda)$ is the minimal injective cogenerator of $\mathrm{mod}(\Lambda)$.

Let T be a tilting module with $\mathrm{pd}_\Lambda T = 1$. Then $(\mathcal{T}, \mathcal{F})$ is a torsion pair in the category $\mathrm{mod}(\Lambda)$ of finitely generated Λ-modules, where:

$$\mathcal{T} = \{C \in \mathrm{mod}(\Lambda) \mid \mathrm{Ext}^1_\Lambda(T, C) = 0\} \text{ and } \mathcal{F} = \{C \in \mathrm{mod}(\Lambda) \mid \mathrm{Hom}_\Lambda(T, C) = 0\}$$

The heart of the corresponding torsion pair $(\mathcal{X}(\mathcal{T}), \mathcal{Y}(\mathcal{F}))$ in $\mathbf{D}^b(\mathrm{mod}(\Lambda))$ is equivalent to $\mathrm{mod}(\mathrm{End}_\Lambda(T))$ and Λ and $\mathrm{End}_\Lambda(T)$ are derived equivalent. See [**60**] for a generalization of this situation, which led to the concept of a quasi-tilted algebra.

In general a tilting module T with $\mathrm{pd}_\Lambda T > 1$ does not define a reasonable torsion pair in $\mathrm{mod}(\Lambda)$. However T induces important torsion pairs in the derived category and in an appropriate stable category, see Chapters III, V.

CHAPTER II

Torsion Pairs in Pretriangulated Categories

Motivated by the results of the first chapter, we introduce and investigate in this chapter the concept of a torsion pair in the fairly general setting of a pretriangulated category. Pretriangulated categories provide a common generalization of abelian, triangulated and stable categories, so it is interesting to develop our theory in this more general setting. There are natural triangulated categories associated with a pretriangulated category, and we investigate connections between torsion pairs in this situation.

1. Pretriangulated Categories

In this section we discuss the concept of a pretriangulated category which gives the general setting for the rest of the paper. We give many examples, illustrating the wide variety of contexts that are covered by the notion of a pretriangulated category.

Left and Right Triangulated Categories. First we recall basic definitions and results about left, right and pretriangulated categories. For details and more information on one-sided triangulated categories we refer to [**19**], [**24**], [**74**]. Let $\mathcal{C}$ be an additive category equipped with an additive endofunctor $\Omega : \mathcal{C} \to \mathcal{C}$. Consider the category $\mathcal{LT}(\mathcal{C}, \Omega)$ whose objects are diagrams of the form $\Omega(C) \xrightarrow{h} A \xrightarrow{g} B \xrightarrow{f} C$ and where the morphisms are indicated by the following diagram:

$$\begin{array}{ccccccc}
\Omega(C) & \xrightarrow{\ h\ } & A & \xrightarrow{\ g\ } & B & \xrightarrow{\ f\ } & C \\
{\scriptstyle \Omega(\gamma)}\downarrow & & {\scriptstyle \alpha}\downarrow & & {\scriptstyle \beta}\downarrow & & {\scriptstyle \gamma}\downarrow \\
\Omega(C') & \xrightarrow{\ h'\ } & A' & \xrightarrow{\ g'\ } & B' & \xrightarrow{\ f'\ } & C'
\end{array}$$

A **left triangulation** of the pair $(\mathcal{C}, \Omega)$ is a full subcategory Δ of $\mathcal{LT}(\mathcal{C}, \Omega)$ which satisfies all the axioms of a triangulated category, except that Ω is not necessarily an equivalence. Then the triple $(\mathcal{C}, \Omega, \Delta)$ is called a **left triangulated category**, the functor Ω is the **loop functor** and the diagrams in Δ are the **left triangles**.

Dually if $\mathcal{C}$ is an additive category equipped with an additive endofunctor $\Sigma : \mathcal{C} \to \mathcal{C}$, consider in $\mathcal{C}$ the category $\mathcal{RT}(\mathcal{C}, \Sigma)$ with objects diagrams of the form

$A \xrightarrow{f} B \xrightarrow{g} C \xrightarrow{h} \Sigma(A)$ and morphisms indicated by the following diagram:

$$\begin{array}{ccccccc} A & \xrightarrow{f} & B & \xrightarrow{g} & C & \xrightarrow{h} & \Sigma(A) \\ {\scriptstyle\alpha}\downarrow & & {\scriptstyle\beta}\downarrow & & {\scriptstyle\gamma}\downarrow & & {\scriptstyle\Sigma(\alpha)}\downarrow \\ A' & \xrightarrow{f'} & B' & \xrightarrow{g'} & C' & \xrightarrow{h'} & \Sigma(C') \end{array}$$

A **right triangulation** of the pair $(\mathcal{C}, \Sigma)$ is a full subcategory ∇ of $\mathcal{RT}(\mathcal{C}, \Sigma)$ which satisfies all the axioms of a triangulated category, except that Σ is not necessarily an equivalence. Then the triple $(\mathcal{C}, \Sigma, \nabla)$ is called a **right triangulated category**, Σ is the **suspension functor** and the diagrams in ∇ are the **right triangles**.

Exact and Homological Functors. The most important functors defined on a left or right triangulated category are the homological functors and the (left or right exact) functors. We recall here the basic definitions.

If $(\mathcal{C}, \Omega, \Delta)$ is a left triangulated category, then an additive functor $F : \mathcal{C} \to \mathcal{A}b$ is called **homological**, if for any triangle $\Omega(C) \to A \to B \to C$ in Δ, the sequence $\cdots \to F\Omega(C) \to F(A) \to F(B) \to F(C)$ is exact. It is easy to see that $\forall E \in \mathcal{C}$, the functor $\mathcal{C}(E, -) : \mathcal{C} \to \mathcal{A}b$ is homological. We recall that a morphism $g : A \to B$ is a *weak kernel* of $f : B \to C$, if the sequence of functors $\mathcal{C}(-, A) \xrightarrow{(-,g)} \mathcal{C}(-, B) \xrightarrow{(-,f)} \mathcal{C}(-, C)$ is exact. It follows that any morphism in a left triangle is a weak kernel of the next one. Similarly an additive functor $F : \mathcal{C}^{\mathrm{op}} \to \mathcal{A}b$ is called **cohomological**, if for any triangle $\Omega(C) \to A \to B \to C$ in Δ, the sequence $F(C) \to F(B) \to F(A) \to F\Omega(C) \to \cdots$ is exact.

Dually if $(\mathcal{C}, \Sigma, \nabla)$ is a right triangulated category, then an additive functor $F : \mathcal{C}^{\mathrm{op}} \to \mathcal{A}b$ is called **cohomological**, if for any triangle $A \to B \to C \to \Sigma(A)$ in ∇, the sequence $\cdots \to F\Sigma(A) \to F(C) \to F(B) \to F(A)$ is exact. It is easy to see that $\forall E \in \mathcal{C}$, the functor $\mathcal{C}(-, E) : \mathcal{C}^{\mathrm{op}} \to \mathcal{A}b$ is cohomological. We recall that a morphism $g : B \to C$ is a *weak cokernel* of $f : A \to B$ if the sequence of functors $\mathcal{C}(C, -) \xrightarrow{(g,-)} \mathcal{C}(B, -) \xrightarrow{(f,-)} \mathcal{C}(A, -)$ is exact. It follows that any morphism in a right triangle is a weak cokernel of the previous one. Similarly an additive functor $F : \mathcal{C} \to \mathcal{A}b$ is called **homological**, if for any triangle $A \to B \to C \to \Sigma(A)$ in ∇, the sequence $F(A) \to F(B) \to F(C) \to F\Sigma(C) \to \cdots$ is exact.

If $(\mathcal{C}_1, \Sigma_1, \nabla_1), (\mathcal{C}_2, \Sigma_2, \nabla_2)$ are right triangulated categories, then a functor $F : \mathcal{C}_1 \to \mathcal{C}_2$ is called **right exact** if there exists a natural isomorphism $\xi : F\Sigma_1 \xrightarrow{\cong} \Sigma_2 F$ such that for any triangle $A \xrightarrow{f} B \xrightarrow{g} C \xrightarrow{h} \Sigma_1(A)$ in $\mathcal{C}_1$, the diagram $F(A) \xrightarrow{F(f)} F(B) \xrightarrow{F(g)} F(C) \xrightarrow{F(h)\circ\xi_C} \Sigma_2(F(A))$ is a triangle in $\mathcal{C}_2$. Similarly one defines left exact functors between left triangulated categories.

Pretriangulated Categories. If an additive category is left and right triangulated, then usually the left and right structures are compatible in a nice way. Following [**24**], [**64**] we formalize this situation in the following definition.

DEFINITION 1.1. Let $\mathcal{C}$ be an additive category. A **pre-triangulation** of $\mathcal{C}$ consists of the following data:

(i) An adjoint pair (Σ, Ω) of additive endofunctors $\Sigma, \Omega : \mathcal{C} \to \mathcal{C}$. Let $\varepsilon : \Sigma\Omega \to \mathrm{Id}_{\mathcal{C}}$ be the counit and let $\delta : \mathrm{Id}_{\mathcal{C}} \to \Omega\Sigma$ be the unit of the adjoint pair.

(ii) A collection of diagrams Δ in $\mathcal{C}$ of the form $\Omega(C) \to A \to B \to C$, such that the triple $(\mathcal{C}, \Omega, \Delta)$ is a left triangulated category.

(iii) A collection of diagrams ∇ in $\mathcal{C}$ of the form $A \to B \to C \to \Sigma(A)$, such that the triple $(\mathcal{C}, \Sigma, \nabla)$ is a right triangulated category.

(iv) For any diagram in $\mathcal{C}$ with commutative left square:

$$\begin{array}{ccccccc} A & \xrightarrow{f} & B & \xrightarrow{g} & C & \xrightarrow{h} & \Sigma(A) \\ {\scriptstyle\alpha}\downarrow & & {\scriptstyle\beta}\downarrow & & {\scriptstyle\exists\gamma}\downarrow & & \downarrow{\scriptstyle\Sigma(\alpha)\circ\varepsilon_{C'}} \\ \Omega(C') & \xrightarrow{f'} & A' & \xrightarrow{g'} & B' & \xrightarrow{h'} & C' \end{array}$$

where the upper row is in ∇ and the lower row is in Δ, there exists a morphism $\gamma : C \to B'$ making the diagram commutative.

(v) For any diagram in $\mathcal{C}$ with commutative right square:

$$\begin{array}{ccccccc} A & \xrightarrow{f} & B & \xrightarrow{g} & C & \xrightarrow{h} & \Sigma(A) \\ {\scriptstyle\delta_A\circ\Omega(\alpha)}\downarrow & & {\scriptstyle\exists\gamma}\downarrow & & {\scriptstyle\beta}\downarrow & & {\scriptstyle\alpha}\downarrow \\ \Omega(C') & \xrightarrow{f'} & A' & \xrightarrow{g'} & B' & \xrightarrow{h'} & C' \end{array}$$

where the upper row is in ∇ and the lower row is in Δ, there exists a morphism $\gamma : B \to A'$ making the diagram commutative.

A **pretriangulated category** is an additive category together with a pre-triangulation, and is denoted by $(\mathcal{C}, \Sigma, \Omega, \nabla, \Delta, \epsilon, \delta)$.

If $\mathcal{C}$ is a pretriangulated category and if $F : \mathcal{C} \to \mathcal{D}$ is a functor to a right, resp. left, triangulated category $\mathcal{D}$, then F is called right, resp. left, exact, if F is a right, resp. left, exact functor of right, resp. left, triangulated categories. Let $H : \mathcal{C}_1 \to \mathcal{C}_2$ be a functor between pretriangulated categories. Then H is called **exact** if H is left and right exact. More generally we define a *morphism of pretriangulated categories* to be an adjoint pair (F, G) of additive functors $F : \mathcal{C}_1 \to \mathcal{C}_2$, $G : \mathcal{C}_2 \to \mathcal{C}_1$, such that F is right exact and G is left exact. Note that if (F, G) is an adjoint pair of functors between pretriangulated categories, then it is not a formal consequence of the adjointness that, in our terminology, F is right exact and G is left exact. The reason is that left or right exactness depends on the pretriangulations.

We now give some first examples of pretriangulated categories; later we shall see more examples.

EXAMPLE. (1) Triangulated categories are pretriangulated. Here $\nabla = \Delta$ and $\Omega = \Sigma^{-1}$ and the notions of right or left exactness coincide (with the usual notions).

(2) Any additive category with kernels and cokernels (in particular any abelian category) is pretriangulated, with $\Omega = \Sigma = 0$ and Δ the class of left exact sequences and ∇ the class of right exact sequences. The notions of right or left exactness have their usual meaning.

(3) There are abelian pretriangulated categories for which Ω and Σ are not zero. An example is the category of $\mathbb{Z}$-graded vector spaces, where Σ is the usual shift functor and Ω its right adjoint.

(4) The homotopy category of an additive closed model category in the sense of Quillen [**88**] is pretriangulated.

Homologically Finite Subcategories and Triangulations. One important source of examples of pretriangulated categories arises from stable categories, as we now explain.

Let $\mathcal{C}$ be an additive category with split idempotents and let $\mathcal{X}$ be a full additive subcategory of $\mathcal{C}$ closed under direct summands and isomorphisms.

The induced stable category is denoted by $\mathcal{C}/\mathcal{X}$. We recall that the objects of $\mathcal{C}/\mathcal{X}$ are the objects of $\mathcal{C}$. If $A, B \in \mathcal{C}$, then $\mathcal{C}/\mathcal{X}(A, B)$ is defined to be the quotient $\mathcal{C}(A, B)/\mathcal{I}_{\mathcal{X}}(A, B)$, where $\mathcal{I}_{\mathcal{X}}(A, B)$ is the subgroup of $\mathcal{C}(A, B)$ consisting of all morphisms factorizing through an object of $\mathcal{X}$. We denote by $\pi : \mathcal{C} \to \mathcal{C}/\mathcal{X}$ the natural projection functor and we set $\pi(A) = \underline{A}$ and $\pi(f) = \underline{f}$.

We recall from [**13**] that a morphism $f_A : X_A \to A$ in $\mathcal{C}$ is called a *right $\mathcal{X}$-approximation* of A if X_A is in $\mathcal{X}$ and any morphism $X \to A$ with $X \in \mathcal{X}$ factors through f_A. The subcategory $\mathcal{X}$ is called *contravariantly finite* [**13**] if any object of $\mathcal{C}$ admits a right $\mathcal{X}$-approximation. The dual notions are *left $\mathcal{X}$-approximation* and *covariantly finite*. The subcategory $\mathcal{X}$ is called *functorially finite* if it is both contravariantly finite and covariantly finite. Finally the subcategory $\mathcal{X}$ is called *homologically finite* if it is contravariantly, covariantly or functorially finite.

Assume now that $\mathcal{C}$ is abelian and $\mathcal{X}$ is contravariantly finite in $\mathcal{C}$. By [**19**] the subcategory $\mathcal{X}$ induces on the stable category $\mathcal{C}/\mathcal{X}$ a left triangulated structure $(\Omega_{\mathcal{X}}, \Delta_{\mathcal{X}})$, where $\Omega_{\mathcal{X}} : \mathcal{C}/\mathcal{X} \to \mathcal{C}/\mathcal{X}$ is the loop functor and $\Delta_{\mathcal{X}}$ is the triangulation. The loop functor $\Omega_{\mathcal{X}} : \mathcal{C}/\mathcal{X} \to \mathcal{C}/\mathcal{X}$ is defined as follows. Let $A \in \mathcal{C}$ and let $0 \to K_A \xrightarrow{\kappa_A} X_A \xrightarrow{\chi_A} A$ be a sequence in $\mathcal{C}$, where χ_A is a right $\mathcal{X}$-approximation of A and κ_A is the kernel of χ_A. Then in $\mathcal{C}/\mathcal{X} : \Omega_{\mathcal{X}}(\underline{A}) = \underline{K}_A$. If $f : A \to B$ is a morphism in $\mathcal{C}$, then we have the following commutative diagram:

$$\begin{array}{ccccc}
K_A & \xrightarrow{\kappa_A} & X_A & \xrightarrow{\chi_A} & A \\
{\scriptstyle \kappa_f}\downarrow & & {\scriptstyle \chi_f}\downarrow & & {\scriptstyle f}\downarrow \\
K_B & \xrightarrow{\kappa_B} & X_B & \xrightarrow{\chi_B} & B
\end{array}$$

It is easy to see that setting $\Omega(\underline{f}) = \underline{\kappa}_f$, we obtain an additive functor $\Omega_{\mathcal{X}}$. The triangulation $\Delta_{\mathcal{X}}$ is defined as follows. Let $0 \to C \xrightarrow{g} B \xrightarrow{f} A$ be a sequence in $\mathcal{C}$, where f has the property that $\mathcal{C}(\mathcal{X}, f) : \mathcal{C}(\mathcal{X}, A) \to \mathcal{C}(\mathcal{X}, B) \to 0$ is surjective and $g = ker(f)$. Then we have a commutative diagram:

$$\begin{array}{ccccc}
K_A & \xrightarrow{\kappa_A} & X_A & \xrightarrow{A} & A \\
{\scriptstyle \kappa_f}\downarrow & & {\scriptstyle \chi_f}\downarrow & & \Vert \\
C & \xrightarrow{g} & B & \xrightarrow{f} & A
\end{array}$$

Hence in $\mathcal{C}/\mathcal{X}$ we have a diagram $\Omega_{\mathcal{X}}(\underline{A}) \xrightarrow{h} \underline{C} \xrightarrow{g} \underline{B} \xrightarrow{f} \underline{A}$. The triangulation $\Delta_{\mathcal{X}}$ consists of all diagrams $\Omega_{\mathcal{X}}(\underline{H}) \to \underline{F} \to \underline{G} \to \underline{H}$ which are isomorphic in $\mathcal{C}/\mathcal{X}$ to diagrams of the above form.

REMARK 1.2. A morphism $f : A \to B$ is said to be an $\mathcal{X}$-**epic** if the morphism $\mathcal{C}(\mathcal{X}, f) : \mathcal{C}(\mathcal{X}, A) \to \mathcal{C}(\mathcal{X}, B)$ is surjective. Then the above construction works in the more general situation in which any $\mathcal{X}$-epic in $\mathcal{C}$ has a kernel [**19**]. For instance let $\mathcal{A}$ be a full subcategory of an abelian category $\mathcal{C}$ with enough projectives, and assume that $\mathcal{X}$ contains the projectives and is closed under extensions and kernels of epimorphisms. If $\mathcal{P}$ is the full subcategory of projectives of $\mathcal{C}$, then the stable category $\mathcal{X}/\mathcal{P}$ is a left triangulated category, in fact a left triangulated subcategory of $\mathcal{C}/\mathcal{P}$.

Dually a morphism $f : A \to B$ is said to be an $\mathcal{X}$-**monic** if the morphism $\mathcal{C}(f, \mathcal{X}) : \mathcal{C}(B, \mathcal{X}) \to \mathcal{C}(A, \mathcal{X})$ is surjective. If $\mathcal{Y}$ is covariantly finite in $\mathcal{C}$ and $\mathcal{C}$ is abelian, or more generally if any $\mathcal{Y}$-monic has a cokernel in $\mathcal{C}$, then the stable category $\mathcal{C}/\mathcal{Y}$ admits a right triangulated structure $(\mathcal{C}/\mathcal{Y}, \Sigma_{\mathcal{Y}}, \nabla_{\mathcal{Y}})$, where $\Sigma_{\mathcal{Y}}$ is the suspension functor and $\nabla_{\mathcal{Y}}$ is the right triangulation. The construction is completely dual using left $\mathcal{Y}$-approximations. For instance if $\mathcal{C}$ has enough injectives and $\mathcal{Y}$ contains the injectives and is closed under extensions and cokernels of monomorphisms, then the stable category $\mathcal{Y}/\mathcal{I}$ is a right triangulated category, in fact a right triangulated subcategory of $\mathcal{C}/\mathcal{I}$, where $\mathcal{I}$ is the full subcategory of injective objects of $\mathcal{C}$.

Our basic source of examples of pretriangulated categories is the following.

EXAMPLE. Let $\mathcal{X}$ be a functorially finite subcategory of an additive category $\mathcal{C}$ and assume that $\mathcal{C}$ is abelian, or more generally that any $\mathcal{X}$-epic has a kernel and any $\mathcal{X}$-monic has a cokernel in $\mathcal{C}$. Then $\mathcal{C}/\mathcal{X}$ admits a right triangulated structure $(\Sigma_{\mathcal{X}}, \nabla_{\mathcal{X}})$ and a left triangulated structure $(\Omega_{\mathcal{X}}, \Delta_{\mathcal{X}})$. Moreover by [**20**] we have an adjoint pair $(\Sigma_{\mathcal{X}}, \Omega_{\mathcal{X}})$ and the compatibility conditions (iv), (v) in the above definition hold. Hence the stable category $\mathcal{C}/\mathcal{X}$ is a pretriangulated category. Note that by [**24**], $\mathcal{C}/\mathcal{X}$ is the homotopy category of a naturally defined closed model structure in $\mathcal{C}$. We refer to Chapter VII for more about this class of pretriangulated categories in connection with torsion pairs and closed model structures.

Actually the above example also covers abelian categories (just take $\mathcal{X} = 0$) and unbounded derived categories of module categories (take $\mathcal{C}$ to be the category of homotopically projective complexes [**69**] and $\mathcal{X}$ to be the full subcategory of contractible complexes).

An Example from Artin Algebras. We now give a class of examples of pretriangulated stable categories for which the loop and the suspension functor can be computed rather explicitly.

Let Λ be an Artin algebra and let $\operatorname{mod}(\Lambda)$ be the category of finitely generated right Λ-modules. Let $\mathcal{P}_\Lambda$ be the full subcategory of finitely generated projective right Λ-modules and $\mathcal{I}_\Lambda$ the full subcategory of finitely generated injective right Λ-modules. Since $\mathcal{P}_\Lambda$ is functorially finite, the stable category $\operatorname{mod}(\Lambda)/\mathcal{P}_\Lambda := \underline{\operatorname{mod}}(\Lambda)$ is pretriangulated with loop functor the usual first syzygy functor Ω and suspension functor $\Sigma_{\mathbf{P}} = \operatorname{Tr}\Omega\operatorname{Tr}$, where $\operatorname{Tr} : \underline{\operatorname{mod}}(\Lambda) \to \underline{\operatorname{mod}}(\Lambda^{\mathrm{op}})$ is the transpose duality functor [**6**]. The left $\mathcal{P}_\Lambda$-approximation $p^A : A \to P^A$ of A is computed as follows. Let $\mathrm{d} = \mathrm{Hom}_\Lambda(-,\Lambda)$ denote both Λ-dual functors. If $Q \to \mathrm{d}(A)$ is a projective cover, then $P^A = \mathrm{d}(Q)$ and p^A is composition $A \to \mathrm{d}^2(A) \to \mathrm{d}(Q)$ where the first morphism is the natural one, and there is an exact sequence

$$0 \to \mathrm{Ext}^1_\Lambda(\mathrm{Tr}(A),\Lambda) \to A \xrightarrow{p^A} P^A \to \Sigma_{\mathbf{P}}(A) \to 0$$

Similarly, since $\mathcal{I}_\Lambda$ is functorially finite, the stable category $\mathrm{mod}(\Lambda)/\mathcal{I}_\Lambda = \overline{\mathrm{mod}}(\Lambda)$ is pretriangulated with suspension functor the usual first cosyzygy functor Σ, and loop functor $\Omega_{\mathbf{I}} = \mathrm{DTr}\Omega\mathrm{TrD}$, where D is the usual duality for Artin algebras. The right $\mathcal{I}_\Lambda$-approximation i_A of A is included in an exact sequence

$$0 \to \Omega_{\mathbf{I}}(A) \to \mathrm{N}^+(P) \xrightarrow{i_A} A \to \mathrm{DExt}^1_\Lambda(\mathrm{TrD}(A),\Lambda) \to 0$$

where $\mathrm{N}^+ = \mathrm{Dd}$ is the Nakayama functor, $\mathrm{N}^- = \mathrm{dD}$ its right adjoint, $P \to \mathrm{N}^-(A)$ is a projective cover and i_A is the composition $\mathrm{N}^-(P) \to \mathrm{N}^+\mathrm{N}^-(A) \to A$ where the last morphism is the counit of the adjoint pair $(\mathrm{N}^+,\mathrm{N}^-)$.

It is easy to see that the adjoint pair (d,d) of Λ-dual functors induces a morphism of pretriangulated categories $\mathrm{d} : \underline{\mathrm{mod}}(\Lambda) \leftrightarrows \underline{\mathrm{mod}}(\Lambda^{\mathrm{op}}) : \mathrm{d}$. It follows that the adjoint pair $(\mathrm{N}^+,\mathrm{N}^-)$ of Nakayama functors induces a morphism of pretriangulated categories $\mathrm{N}^+ : \underline{\mathrm{mod}}(\Lambda) \leftrightarrows \overline{\mathrm{mod}}(\Lambda) : \mathrm{N}^-$.

More generally let $T \in \mathrm{mod}(\Lambda)$ and let $\omega = \mathrm{add}(T)$ be the full subcategory consisting of all direct summands of finite direct sums of copies of T. It is well-known that ω is functorially finite, so the stable category $\underline{\mathrm{mod}}_T(\Lambda) := \mathrm{mod}(\Lambda)/\omega$ is pretriangulated.

Other Examples from Ring Theory. The previous example can be generalized to the stable module category (of not necessarily finitely generated modules) modulo projectives or injectives, for certain classes of rings that we discuss below.

Let Λ be a ring and let $\mathrm{Mod}(\Lambda)$ be the category of right Λ-modules. Let $\mathbf{P}_\Lambda$ be the category of projective right Λ-modules and $\mathbf{I}_\Lambda$ the category of injective right Λ-modules. By [**24**] we have the following.

- If Λ is left coherent and right perfect, then $\mathbf{P}_\Lambda$ is functorially finite, hence the stable category $\underline{\mathrm{Mod}}(\Lambda)$ modulo projectives is pretriangulated.
- If Λ is right Noetherian, then $\mathbf{I}_\Lambda$ is functorially finite, hence the stable category $\overline{\mathrm{Mod}}(\Lambda)$ modulo injectives is pretriangulated.

More generally let $T \in \mathrm{Mod}(\Lambda)$ and let $\omega = \mathrm{Add}(T)$ be the full subcategory consisting of all direct summands of arbitrary direct sums of copies of T. We denote the stable category $\mathrm{Mod}(\Lambda)/\omega$ by $\underline{\mathrm{Mod}}_T(\Lambda)$. In a given category with coproducts, we denote by $C^{(I)}$ the coproduct of copies of the object C indexed by the set I. Assume that T has the property that the canonical map $(T,T)^{(I)} \to (T,T^{(I)})$ is invertible for any index set I, and that the endomorphism ring $\mathrm{End}(T)$ is left coherent and right perfect. It is not difficult to see that ω is functorially finite, so the stable category $\underline{\mathrm{Mod}}_T(\Lambda)$ is pretriangulated. Also the category $\underline{\mathrm{Mod}}_T(\Lambda)$ is pretriangulated if the module T is endofinite in the sense that T has finite length as a module over its endomorphism ring. More generally $\underline{\mathrm{Mod}}_T(\Lambda)$ is pretriangulated if T is product-complete in the sense that any product of copies of T is a direct summand of a coproduct of copies of T. This is equivalent to saying that $\mathrm{Add}(T) = \mathrm{Prod}(T)$, where $\mathrm{Prod}(T)$ is the full subcategory of $\mathrm{Mod}(\Lambda)$ consisting of all direct summands of arbitrary direct products of copies of T. We refer to [**78**] for more information on endofinite or product-complete modules.

For more examples we refer to Chapter V.

2. Adjoints and Orthogonal Subcategories

In this section we examine more closely the connections between subcategories of a pretriangulated category which are orthogonal with respect to Hom, and the existence of left or right adjoints for the corresponding inclusion functors. These connections will be useful when discussing torsion pairs.

Throughout this section we fix a pretriangulated category $\mathcal{C} = (\mathcal{C}, \Sigma, \Omega, \Delta, \nabla)$.

Let $\mathcal{X}, \mathcal{Y}$ be full additive subcategories of $\mathcal{C}$. We assume tacitly that $\mathcal{X}, \mathcal{Y}$ are closed under direct summands and isomorphisms. We consider the left orthogonal subcategory ${}^{\perp}\mathcal{Y} = \{A \in \mathcal{C} \mid \mathcal{C}(A, \mathcal{Y}) = 0\}$ of $\mathcal{Y}$ and the right orthogonal subcategory $\mathcal{X}^{\perp} = \{B \in \mathcal{C} \mid \mathcal{C}(\mathcal{X}, B) = 0\}$ of $\mathcal{X}$.

We begin with some elementary necessary conditions for a subcategory to be a left or right orthogonal subcategory.

LEMMA 2.1. (1) *Assume that* ${}^{\perp}\mathcal{Y} = \mathcal{X}$. *If* $\mathcal{Y}$ *is closed under the loop functor* Ω *(for instance if* $\mathcal{Y}$ *is a left triangulated subcategory of* $\mathcal{C}$*), then* $\mathcal{X}$ *is a right triangulated subcategory of* $\mathcal{C}$*, closed under extensions of right triangles.*

(2) *Assume that* $\mathcal{X}^{\perp} = \mathcal{Y}$. *If* $\mathcal{X}$ *is closed under the suspension functor* Σ *(for instance if* $\mathcal{X}$ *is a right triangulated subcategory of* $\mathcal{C}$*), then* $\mathcal{Y}$ *is a left triangulated subcategory of* $\mathcal{C}$*, closed under extensions of left triangles.*

PROOF. (1) If $\Omega(\mathcal{Y}) \subseteq \mathcal{Y}$, then $\forall X \in \mathcal{C}$, $\mathcal{C}(\Sigma(X), \mathcal{Y}) \cong \mathcal{C}(X, \Omega(\mathcal{Y})) = 0$. Hence $\Sigma(X) \in \mathcal{X}$ and $\mathcal{X}$ is closed under suspension. Let $X_1 \to C \to X_2 \to \Sigma(X_1)$ be a right triangle with $X_1, X_2 \in \mathcal{X}$. Since the functor $\mathcal{C}(-, \mathcal{Y})$ is cohomological with respect to right triangles, the sequence $\cdots \to \mathcal{C}(\Sigma(X_1), \mathcal{Y}) \to \mathcal{C}(X_2, \mathcal{Y}) \to \mathcal{C}(C, \mathcal{Y}) \to \mathcal{C}(X_1, \mathcal{Y})$ is exact. It follows that $C \in {}^{\perp}\mathcal{Y} = \mathcal{X}$, so $\mathcal{X}$ is closed under extensions of right triangles. This implies that $\mathcal{X}$ is a right triangulated subcategory of $\mathcal{C}$. The proof of (2) is dual. □

COROLLARY 2.2. *Assume that* $\mathcal{X}^{\perp} = \mathcal{Y}$ *and* ${}^{\perp}\mathcal{Y} = \mathcal{X}$. *Then* $\mathcal{X}$ *is a right triangulated subcategory of* $\mathcal{C}$ *if and only if* $\mathcal{Y}$ *is a left triangulated subcategory of* $\mathcal{C}$.

It is trivial to see that in an arbitrary category, any coreflective, resp. reflective, subcategory is contravariantly, resp. covariantly, finite. The coreflection, resp. reflection, of an object gives the right, resp. left, approximation. Reflections and coreflections are examples of minimal approximations. We recall that a morphism $\alpha : X \to C$ is called *right minimal*, if any endomorphism ρ of X such that $\rho \circ \alpha = \alpha$, is invertible. A *minimal right* $\mathcal{X}$*-approximation* is a right minimal, right $\mathcal{X}$-approximation. *Left minimal* morphisms and *minimal left approximations* are defined dually. In an abelian category Wakamatsu's Lemma, which we state below for later use, gives connections between existence of minimal approximations and vanishing of the extension functor.

WAKAMATSU'S LEMMA. [**10**] *Let* $\mathcal{X}$ *be a full extension closed subcategory of an abelian category.*

(i) *If* $X_C \xrightarrow{f_C} C$ *is a minimal right* $\mathcal{X}$*-approximation, then*

$$\mathrm{Ext}^1\big(\mathcal{X}, \mathrm{Ker}(f_C)\big) = 0.$$

(ii) *If* $C \xrightarrow{g^C} X^C$ *is a minimal left* $\mathcal{X}$*-approximation, then*

$$\mathrm{Ext}^1\big(\mathrm{Coker}(g^C), \mathcal{X}\big) = 0.$$

In the setting of pretriangulated categories we have the following connection between coreflective/reflective subcategories, contravariantly/covariantly finite subcategories and right/left orthogonal subcategories with respect to the vanishing of Hom. This connection can be regarded as a stable analogue of Wakamatsu's Lemma.

LEMMA 2.3. *Let* $\mathcal{X}$, $\mathcal{Y}$ *be full subcategories of a pretriangulated category* $\mathcal{C}$.

(i) *If* $\Sigma(\mathcal{X}) \subseteq \mathcal{X}$, *then the following are equivalent.*
  (a) *The inclusion* $\mathbf{i} : \mathcal{X} \hookrightarrow \mathcal{C}$ *admits a right adjoint* $\mathbf{R}$.
  (b) $\mathcal{X}$ *is contravariantly finite and* $\forall C \in \mathcal{C}$, *there exists a left triangle* $\Omega(C) \to Y_C \to X_C \xrightarrow{f_C} C$, *where* f_C *is a right* $\mathcal{X}$*-approximation and* $Y_C \in \mathcal{X}^{\perp}$.

(ii) *If* $\Omega(\mathcal{Y}) \subseteq \mathcal{Y}$, *then the following are equivalent.*
  (a) *The inclusion* $\mathbf{j} : \mathcal{Y} \hookrightarrow \mathcal{C}$ *admits a left adjoint* $\mathbf{L}$.
  (b) $\mathcal{Y}$ *is covariantly finite and* $\forall C \in \mathcal{C}$, *there exists a right triangle* $C \xrightarrow{g^C} Y^C \to X^C \to \Sigma(C)$, *where* g^C *is a left* $\mathcal{Y}$*-approximation and* $X^C \in {}^{\perp}\mathcal{Y}$.

PROOF. We prove only (i), since the proof of (i) is dual.

(a) $\Rightarrow$ (b) Let $C \in \mathcal{C}$ and let f_C be the coreflection of C. Then trivially f_C is a right $\mathcal{X}$-approximation, hence $\mathcal{X}$ is contravariantly finite. Consider a left triangle $\Omega(C) \to Y_C \to \mathbf{R}(C) \xrightarrow{f_C} C$ in $\mathcal{C}$. Then applying the homological functor $\mathcal{C}(\mathcal{X}, -)$, we have the long exact sequence

$$\cdots \to \mathcal{C}(\mathcal{X}, \Omega\mathbf{R}(C)) \to \mathcal{C}(\mathcal{X}, \Omega(C)) \to \mathcal{C}(\mathcal{X}, Y_C) \to \mathcal{C}(\mathcal{X}, \mathbf{R}(C)) \to \mathcal{C}(\mathcal{X}, C).$$

Since the morphism $\mathcal{C}(\mathcal{X}, f_C) : \mathcal{C}(\mathcal{X}, \mathbf{R}(C)) \to \mathcal{C}(\mathcal{X}, C)$ is invertible, the sequence $\mathcal{C}(\mathcal{X}, \Omega\mathbf{R}(C)) \to \mathcal{C}(\mathcal{X}, \Omega(C)) \to \mathcal{C}(\mathcal{X}, Y_C) \to 0$ is exact. Using the adjoint pair (Σ, Ω) it follows that the morphism $\mathcal{C}(\mathcal{X}, \Omega(f_C)) : \mathcal{C}(\mathcal{X}, \Omega\mathbf{R}(C)) \to \mathcal{C}(\mathcal{X}, \Omega(C))$ is isomorphic to $\mathcal{C}(\Sigma(\mathcal{X}), f_C) : \mathcal{C}(\Sigma(\mathcal{X}), \mathbf{R}(C)) \to \mathcal{C}(\Sigma(\mathcal{X}), C)$. Since $\Sigma(\mathcal{X}) \subseteq \mathcal{X}$, the above isomorphism shows that the morphism $\mathcal{C}(\mathcal{X}, \Omega(f_C))$ is also invertible. From the long exact sequence above we infer that $\mathcal{C}(\mathcal{X}, Y_C) = 0$. Hence $Y_C \in \mathcal{X}^{\perp}$.

(b) $\Rightarrow$ (a) It suffices to show that f_C is the coreflection of C in $\mathcal{X}$. If $\alpha : X \to C$ is a morphism with $X \in \mathcal{X}$ which admits two liftings $\beta, \gamma : X \to X_C$ through f_C, then $\beta - \gamma$ factors through the weak kernel $Y_C \to X_C$ of f_C. Since $Y_C \in \mathcal{X}^{\perp}$, it follows that $\beta = \gamma$ and this implies that $X_C \to C$ is the coreflection of C in $\mathcal{X}$. □

The following result shows that in some cases contravariant/covariant finiteness implies coreflectivity/reflectivity. First we recall that an additive category is called a *Krull-Schmidt category* if any of its objects is a finite coproduct of objects with local endomorphism ring. If $\mathcal{X}$ is a skeletally small additive category, then $\mathrm{Mod}(\mathcal{X})$ denotes the category of additive functors $\mathcal{X}^{\mathrm{op}} \to \mathcal{A}b$.

PROPOSITION 2.4. *Assume that $\mathcal{X}$ is a skeletally small Krull-Schmidt right, resp. left, triangulated subcategory of $\mathcal{C}$. Then the following are equivalent.*

(i) *The inclusion functor $\mathcal{X} \hookrightarrow \mathcal{C}$ has a right, resp. left, adjoint.*
(ii) *$\mathcal{X}$ is contravariantly, resp. covariantly, finite in $\mathcal{C}$.*

PROOF. We prove only that (ii) $\Rightarrow$ (i) for the contravariantly finite case. Consider the functor $\mathcal{S} : \mathcal{C} \to \mathrm{Mod}(\mathcal{X})$ defined by $\mathcal{S}(C) = \mathcal{C}(-, C)|_{\mathcal{X}}$. Then $\mathcal{S}$ is a homological functor with respect to the left triangulation Δ of $\mathcal{C}$. Let $f_C : X_C \to C$ be a right $\mathcal{X}$-approximation of $C \in \mathcal{C}$ and let $\Omega(C) \to Y_C \to X_C \xrightarrow{f_C} C$ be a left triangle in $\mathcal{C}$. Then applying $\mathcal{S}$ to this triangle and using that $\mathcal{X}$ is closed under Σ, we have a short exact sequence $0 \to \mathcal{S}(Y_C) \to \mathcal{S}(X_C) \to \mathcal{S}(C) \to 0$ in $\mathrm{Mod}(\mathcal{X})$, which shows that $\mathcal{S}(C)$ is finitely generated functor, since $\mathcal{S}(X_C) = \mathcal{X}(-, X_C)$ is a finitely generated projective functor. Since $\mathcal{X}$ is a Krull-Schmidt category, it is well-known that any finitely generated functor over $\mathcal{X}$ has a projective cover. Let $r_C : \mathcal{X}(-, \mathbf{R}(C)) \to \mathcal{S}(C)$ be the projective cover. Now we claim that $\mathcal{S}(C)$ is a flat functor over $\mathcal{X}$. Indeed since the functor $\mathrm{Tor}_*^{\mathcal{X}}(\mathcal{S}(C), -)$ commutes with filtered colimits, it suffices to show that $\mathrm{Tor}_1^{\mathcal{X}}(\mathcal{S}(C), F) = 0$, for any finitely presented functor $F : \mathcal{X} \to \mathcal{A}b$. Let $\mathcal{X}(X_1, -) \to \mathcal{X}(X_0, -) \to F \to 0$ be a finite presentation of F and let (T): $X_0 \to X_1 \to X_2 \to \Sigma(X_0)$ be a right triangle in $\mathcal{X}$. Since $\mathcal{X}$ is a right triangulated subcategory of $\mathcal{C}$, (T) is also a right triangle in $\mathcal{C}$. Then the sequence $\cdots \to \mathcal{X}(\Sigma(X_0), -) \to \mathcal{X}(X_2, -) \to \mathcal{X}(X_1, -) \to \mathcal{X}(X_0, -) \to F \to 0$ is a projective resolution of F, hence $\mathrm{Tor}_1^{\mathcal{X}}(\mathcal{S}(C), F)$ is the homology of the complex $\mathcal{S}(C) \otimes_{\mathcal{X}} \mathcal{X}(X_2, -) \to \mathcal{S}(C) \otimes_{\mathcal{X}} \mathcal{X}(X_1, -) \to \mathcal{S}(C) \otimes_{\mathcal{X}} \mathcal{X}(X_0, -)$ which is isomorphic to $\mathcal{C}(X_2, C) \to \mathcal{C}(X_1, C) \to \mathcal{C}(X_0, C)$. However the last complex is exact since it is the result of the application of the cohomological functor $\mathcal{C}(-, C)$ to the right triangle $(T) \in \nabla$. We infer that $\mathcal{S}(C)$ is flat. By a well-known result of Bass, the only flat functors admitting a projective cover are the projective ones. Hence $\mathcal{S}(C)$ is projective, i.e. $r_C : \mathcal{X}(-, \mathbf{R}(C)) \to \mathcal{S}(C) = \mathcal{C}(-, C)|_{\mathcal{X}}$ is invertible. We infer that the morphism $r_C(1_{\mathbf{R}(C)}) : \mathbf{R}(C) \to C$ is the coreflection of C in $\mathcal{X}$. □

EXAMPLE. Let $\mathbf{D}^b(\mathrm{mod}(\Lambda))$ be the bounded derived category of finitely generated modules over an Artin algebra Λ. Then $\mathrm{gl.dim}\Lambda < \infty$ if and only if $\mathcal{H}^b(\mathcal{P}_\Lambda)$ is contravariantly finite in $\mathbf{D}^b(\mathrm{mod}(\Lambda))$ if and only if $\mathcal{H}^b(\mathcal{I}_\Lambda)$ is covariantly finite in $\mathbf{D}^b(\mathrm{mod}(\Lambda))$. We prove only the assertion for $\mathcal{H}^b(\mathcal{I}_\Lambda)$. Indeed if $\mathrm{gl.dim}\Lambda < \infty$, then $\mathcal{H}^b(\mathcal{I}_\Lambda) = \mathbf{D}^b(\mathrm{mod}(\Lambda)$, hence the assertion is trivial. If $\mathcal{H}^b(\mathcal{I}_\Lambda)$ is covariantly finite, then by Proposition 2.4, $\mathcal{H}^b(\mathcal{I}_\Lambda)$ is reflective, so by Proposition I.2.3 it is a torsion-free class in $\mathbf{D}^b(\mathrm{mod}(\Lambda))$. Since trivially ${}^{\perp}\mathcal{H}^b(\mathcal{I}_\Lambda) = 0$, we infer that $\mathcal{H}^b(\mathcal{I}_\Lambda) = \mathbf{D}^b(\mathrm{mod}(\Lambda))$, and this obviously implies that $\mathrm{gl.dim}\Lambda < \infty$.

Viewing an abelian category as a pretriangulated category (with $\Omega = 0 = \Sigma$), we have the following well-known observation.

EXAMPLE. Let $\mathcal{C}$ be abelian and $\mathcal{X}$ a skeletally small full additive Krull-Schmidt subcategory of $\mathcal{C}$, closed under quotient objects, resp. subobjects. Then $\mathcal{X}$ is a right, resp. left, triangulated subcategory of $\mathcal{C}$. It follows that $\mathcal{X}$ is coreflective, resp. reflective, if and only if $\mathcal{X}$ is contravariantly, resp. covariantly, finite.

Now we give conditions such that a full coreflective subcategory $\mathcal{X}$ of $\mathcal{C}$ is pretriangulated, and such that the right adjoint of the inclusion is left exact. These conditions will be useful later in connection with hereditary torsion pairs.

LEMMA 2.5. (1) *Let $\mathcal{X}$ be a right triangulated subcategory of $\mathcal{C}$, closed under the loop functor Ω and assume that the inclusion $\mathcal{X} \hookrightarrow \mathcal{C}$ admits a right adjoint* $\mathbf{R}$. *Then $\mathcal{X}$ is a pretriangulated subcategory of $\mathcal{C}$.*

(2) *Let $\mathcal{Y}$ be a left triangulated subcategory of $\mathcal{C}$, closed under the suspension functor Σ and assume that the inclusion $\mathcal{Y} \hookrightarrow \mathcal{C}$ admits a left adjoint* $\mathbf{L}$. *Then $\mathcal{Y}$ is a pretriangulated subcategory of $\mathcal{C}$.*

PROOF. We prove only the assertion for $\mathcal{X}$. First by definition $\mathcal{X}$ is closed under Ω. Let $X_2 \to X_1$ be a morphism in $\mathcal{X}$ and let $\Omega(X_1) \to A \to X_2 \to X_1$ be a left triangle in $\mathcal{C}$. It is easy to see that if (F, G) is an adjoint pair of functors between pretriangulated categories, then F preserves weak cokernels and G preserves weak kernels. It follows that $\mathbf{R}$ preserves weak kernels, hence we have a complex $\cdots \to \mathbf{R}\Omega(X_2) \to \mathbf{R}\Omega(X_1) \to \mathbf{R}(A) \to \mathbf{R}(X_2) \to \mathbf{R}(X_1)$ in $\mathcal{C}$ in which any morphism is a weak kernel of the next one, and a commutative diagram

$$\begin{array}{ccccccccc}
\mathbf{R}\Omega(X_2) & \longrightarrow & \mathbf{R}\Omega(X_1) & \longrightarrow & \mathbf{R}(A) & \longrightarrow & \mathbf{R}(X_2) & \longrightarrow & \mathbf{R}(X_1) \\
{\scriptstyle f_{\Omega(X_2)}}\downarrow & & {\scriptstyle f_{\Omega(X_1)}}\downarrow & & {\scriptstyle f_A}\downarrow & & {\scriptstyle f_{X_2}}\downarrow & & {\scriptstyle f_{X_1}}\downarrow \\
\Omega(X_2) & \longrightarrow & \Omega(X_1) & \longrightarrow & A & \longrightarrow & X_2 & \longrightarrow & X_1
\end{array}$$

in which the coreflection morphisms $f_{\Omega(X_2)}, f_{\Omega(X_1)}, f_{X_2}, f_{X_1}$ are invertible. Let $\operatorname{mod}(\mathcal{C})$ be the category of finitely presented contravariant additive functors over $\mathcal{C}$ and let $\mathbb{Y} : \mathcal{C} \hookrightarrow \operatorname{mod}(\mathcal{C})$ be the Yoneda embedding. Since $\mathcal{C}$ has weak kernels, $\operatorname{mod}(\mathcal{C})$ is abelian and the image under $\mathbb{Y}$ of the above diagram in $\operatorname{mod}(\mathcal{C})$ is exact. By the 5—Lemma, it follows that the map $\mathbb{Y}(f_A) : \mathbb{Y}(\mathbf{R}(A)) \to \mathbb{Y}(A)$ is invertible. Since $\mathbb{Y}$ is fully faithful, f_A is invertible, so $A \in \mathcal{X}$. It follows that $\mathcal{X}$ is a left triangulated subcategory of $\mathcal{C}$. It is easy to see that the compatibility conditions of definition 1.1 hold for $\mathcal{X}$, hence $\mathcal{X}$ is a pretriangulated subcategory of $\mathcal{C}$. □

Note that if $\mathbf{R}$ exists in the above Lemma, then the suspension functor $\Sigma : \mathcal{X} \to \mathcal{X}$ admits the functor $\mathbf{R}\Omega : \mathcal{X} \to \mathcal{X}$ as a right adjoint, even if $\mathcal{X}$ is not closed under Ω. This follows from the natural isomorphisms $\mathcal{X}(\Sigma(X), X') \cong \mathcal{C}(X, \Omega(X')) \cong \mathcal{X}(X, \mathbf{R}\Omega(X'))$ for all $X, X' \in \mathcal{X}$. This suggests that $\mathcal{X}$ is closed under Ω if and only if the functors $\mathbf{R}$, Ω commute. The following result which will be useful later in connection with hereditary torsion pairs, shows that this is the case.

PROPOSITION 2.6. *Assume that the inclusion functor* $\mathbf{i} : \mathcal{X} \hookrightarrow \mathcal{C}$ *has a right adjoint* $\mathbf{R}$ *and $\mathcal{X}$ is closed under the suspension functor Σ. Then there exists a natural morphism $\alpha : \mathbf{R}\Omega \to \Omega\mathbf{R}$ and the following are equivalent:*

(i) $\alpha : \mathbf{R}\Omega \to \Omega\mathbf{R}$ *is an isomorphism.*

(ii) $\Omega(\mathcal{X}) \subseteq \mathcal{X}$.

If (ii) *holds and in addition $\mathcal{X}$ is closed under extensions of left triangles, then $\mathbf{R} : \mathcal{C} \to \mathcal{X}$ and the idempotent functor $\mathbf{iR} : \mathcal{C} \to \mathcal{C}$ are left exact.*

PROOF. Set $\mathcal{X}^{\perp} := \mathcal{Y}$ and consider the counit $f_C : \mathbf{R}(C) \to C$ of the adjoint pair $(\mathbf{i}, \mathbf{R})$. Let $\Omega(C) \xrightarrow{h_C} Y_C \xrightarrow{g_C} \mathbf{R}(C) \xrightarrow{f_C} C$ be a triangle in Δ. By Lemma 2.3, we have that $Y_C \in \mathcal{Y}$. Consider the rotation $\Omega\mathbf{R}(C) \xrightarrow{\Omega(f_C)} \Omega(C) \xrightarrow{h_C} Y_C \xrightarrow{g_C} \mathbf{R}(C)$ of the above triangle, and let $f_{\Omega(C)} : \mathbf{R}\Omega(C) \to \Omega(C)$ be the counit evaluated at $\Omega(C)$. Since $Y_C \in \mathcal{Y}$, the composition $f_{\Omega(C)} \circ h_C = 0$. Hence there exists a morphism $\alpha_C : \mathbf{R}\Omega(C) \to \Omega\mathbf{R}(C)$ such that $\alpha_C \circ \Omega(f_C) = f_{\Omega(C)}$. The morphism α_C is unique with this property, since if $\alpha' : \mathbf{R}\Omega(C) \to \Omega\mathbf{R}(C)$ is another morphism with $\alpha' \circ \Omega(f_C) = f_{\Omega(C)}$, then the morphism $\alpha_C - \alpha'$ factors through $\Omega(Y_C)$. By Lemma 2.1, $\Omega(Y_C) \in \mathcal{Y}$ and then obviously $\alpha_C = \alpha'$, since $(\mathcal{X}, \mathcal{Y}) = 0$. It is not difficult to see that the morphism α_C is the component of a natural morphism $\alpha : \mathbf{R}\Omega \to \Omega\mathbf{R}$. (i) $\Rightarrow$ (ii) If the natural morphism $\alpha : \mathbf{R}\Omega \to \Omega\mathbf{R}$ is invertible, then $\forall C \in \mathcal{C}$ we have that $\Omega\mathbf{R}(C) \in \mathcal{X}$. This implies trivially that $\Omega(\mathcal{X}) \subseteq \mathcal{X}$.

(ii) $\Rightarrow$ (i) Assume that $\Omega(\mathcal{X}) \subseteq \mathcal{X}$. Then $\Omega\mathbf{R}(C) \in \mathcal{X}$. Hence there exists a unique morphism $\beta : \Omega\mathbf{R}(C) \to \mathbf{R}\Omega(C)$ such that $\beta \circ f_{\Omega(C)} = \Omega(f_C)$. Then we have: $\alpha_C \circ \beta \circ f_{\Omega(C)} = f_{\Omega(C)}$ and $\beta \circ \alpha_C \circ \Omega(f_C) = \Omega(f_C)$. By the uniqueness properties of the involved morphisms, we conclude that $\alpha_C \circ \beta = 1_{\mathbf{R}\Omega(C)}$ and $\beta \circ \alpha_C = 1_{\Omega\mathbf{R}(C)}$. Hence the natural morphism $\alpha : \mathbf{R}\Omega \to \Omega\mathbf{R}$ is invertible.

The easy proof of the last assertion is left to the reader. $\square$

3. Torsion Pairs

In this section we introduce the concept of torsion pair in a pretriangulated category and we give some basic properties.

As in the previous section, throughout $\mathcal{C}$ denotes a pretriangulated category. Let $\mathcal{X}, \mathcal{Y}$ be two full additive subcategories of $\mathcal{C}$, closed under direct summands and isomorphisms. Inspired by the definition of a torsion pair in an abelian or triangulated category, we introduce the following concept of torsion pair in a pretriangulated category.

DEFINITION 3.1. The pair $(\mathcal{X}, \mathcal{Y})$ is called a **torsion pair** in $\mathcal{C}$, if:

(i) $\mathcal{C}(\mathcal{X}, \mathcal{Y}) = 0$.

(ii) $\Sigma(\mathcal{X}) \subseteq \mathcal{X}$ and $\Omega(\mathcal{Y}) \subseteq \mathcal{Y}$.

(†) **[The glueing condition]**: $\forall C \in \mathcal{C}$, there are triangles

$$\begin{cases} \Delta(C) : \ \Omega(Y^C) \xrightarrow{g_C} X_C \xrightarrow{f_C} C \xrightarrow{g^C} Y^C & \in \Delta \\ \nabla(C) : \ X_C \xrightarrow{f_C} C \xrightarrow{g^C} Y^C \xrightarrow{f^C} \Sigma(X_C) & \in \nabla \end{cases} \qquad \text{(Glueing Triangles)}$$

with $X_C \in \mathcal{X}, Y^C \in \mathcal{Y}$.

If $(\mathcal{X}, \mathcal{Y})$ is a torsion pair in $\mathcal{C}$, then $\mathcal{X}$, resp. $\mathcal{Y}$, is called a **torsion class**, resp. **torsion-free class**. A **TTF-triple** is an ordered triple $(\mathcal{X}, \mathcal{Y}, \mathcal{Z})$, where $(\mathcal{X}, \mathcal{Y})$ and $(\mathcal{Y}, \mathcal{Z})$ are torsion pairs in $\mathcal{C}$. In this case $\mathcal{Y}$ is called a **TTF-class**.

The following remark shows that condition (ii) in the definition can be relaxed.

REMARK 3.2. If the pair $(\mathcal{X}, \mathcal{Y})$ satisfies (i) and the glueing condition, then $\Sigma(\mathcal{X}) \subseteq \mathcal{X}$ if and only if $\Omega(\mathcal{Y}) \subseteq \mathcal{Y}$. Indeed if $\Sigma(\mathcal{X}) \subseteq \mathcal{X}$ and $C \in \mathcal{X}^{\perp}$, then $f_C = 0$. Then from the right triangle $X_C \xrightarrow{f_C} C \xrightarrow{g^C} Y^C \xrightarrow{f^C} \Sigma(X_C)$, we have that g^C is split monic, so $C \in \mathcal{Y}$. It follows that $\mathcal{X}^{\perp} = \mathcal{Y}$. Then by Lemma 2.1, we have that $\Omega(\mathcal{Y}) \subseteq \mathcal{Y}$. Conversely $\Omega(\mathcal{Y}) \subseteq \mathcal{Y}$ implies that $\Sigma(\mathcal{X}) \subseteq \mathcal{X}$.

As in the case of torsion pairs in an abelian category we say that a torsion pair $(\mathcal{X}, \mathcal{Y})$ in a pretriangulated category $\mathcal{C}$ *splits* if any object C in $\mathcal{C}$ admits a direct sum decomposition $C = X \oplus Y$ where $X \in \mathcal{X}$ and $Y \in \mathcal{Y}$. We record the following immediate consequence of the above definition.

PROPOSITION 3.3. *Assume that the pair $(\mathcal{X}, \mathcal{Y})$ is a torsion pair in $\mathcal{C}$.*

(1) *$\mathcal{X} = {}^{\perp}\mathcal{Y}$ is a right triangulated subcategory of $\mathcal{C}$ closed under extensions of right triangles and the inclusion $\mathbf{i} : \mathcal{X} \hookrightarrow \mathcal{C}$ admits a right adjoint $\mathbf{R} : \mathcal{C} \to \mathcal{X}$.*

(2) *$\mathcal{Y} = \mathcal{X}^{\perp}$ is a left triangulated subcategory of $\mathcal{C}$ closed under extensions of left triangles, and the inclusion $\mathbf{j} : \mathcal{Y} \hookrightarrow \mathcal{C}$ admits a left adjoint $\mathbf{L} : \mathcal{C} \to \mathcal{Y}$.*

(3) *$\mathcal{C}$ is generated as a left or right triangulated category by $\mathcal{X}, \mathcal{Y}$.*

(4) *We have a "direct sum decomposition" $\mathcal{C} = \mathcal{X} \oplus \mathcal{Y}$ of $\mathcal{C}$, in the sense that* $\operatorname{Ker}\mathbf{R} := \{C \in \mathcal{C} \mid \mathbf{R}(C) = 0\} = \mathcal{Y}$ *and* $\operatorname{Ker}\mathbf{L} := \{C \in \mathcal{C} \mid \mathbf{L}(C) = 0\} = \mathcal{X}$ *and there exist sequences*

$$0 \longrightarrow \mathcal{Y} \xrightarrow{\mathbf{j}} \mathcal{C} \xrightarrow{\mathbf{R}} \mathcal{X} \longrightarrow 0 \quad \text{and} \quad 0 \longrightarrow \mathcal{X} \xrightarrow{\mathbf{i}} \mathcal{C} \xrightarrow{\mathbf{L}} \mathcal{Y} \longrightarrow 0$$

such that $\mathbf{Rj} = 0, \mathbf{Li} = 0, \mathbf{Ri} = \mathrm{Id}_{\mathcal{X}}, \mathbf{Lj} = \mathrm{Id}_{\mathcal{Y}}$.

(5) *If $(\mathcal{X}, \mathcal{Y}, \mathcal{Z})$ is a TTF-triple in $\mathcal{C}$, then $\mathcal{Y}$ is a pretriangulated subcategory of $\mathcal{C}$, the inclusion $\mathcal{Y} \hookrightarrow \mathcal{C}$ admits a left and a right adjoint and finally the following relations are true: $\mathcal{X} = {}^{\perp}\mathcal{Y} = {}^{\perp\perp}\mathcal{Z}$, $\mathcal{Z} = \mathcal{Y}^{\perp} = \mathcal{X}^{\perp\perp}$, $\mathcal{Y} = \mathcal{X}^{\perp} = {}^{\perp}\mathcal{Z}$.*

(6) *If $(\mathcal{X}, \mathcal{Y}, \mathcal{Z})$ is a TTF-triple in $\mathcal{C}$, then $\mathcal{X} = \mathcal{Z}$ if and only if the torsion pairs $(\mathcal{X}, \mathcal{Y})$ and $(\mathcal{Y}, \mathcal{Z})$ split.*

PROOF. Obviously $\mathcal{X} \subseteq {}^{\perp}\mathcal{Y}$. If $C \in {}^{\perp}\mathcal{Y}$, then from the triangle $\Delta(C)$ it follows that $C \cong X_C$, so $C \in \mathcal{X}$. Hence $\mathcal{X} = {}^{\perp}\mathcal{Y}$. Now the remaining assertions of (1) follow from Lemmas 2.1, 2.3. The proof of (2) is dual and the proof of the remaining parts follow directly from (1), (2) and the definitions. □

In analogy with the definitions of Chapter I, we define a torsion pair $(\mathcal{X}, \mathcal{Y})$ in $\mathcal{C}$ to be **hereditary**, resp. **cohereditary**, if the idempotent functor $\mathbf{iR} : \mathcal{C} \to \mathcal{C}$ is left exact, resp. the idempotent functor $\mathbf{jL} : \mathcal{C} \to \mathcal{C}$ is right exact. If $(\mathcal{X}, \mathcal{Y})$ is hereditary then by Lemma 2.5 and Proposition 2.6, $\mathcal{X}$ is a pretriangulated subcategory of $\mathcal{C}$ and dually if $(\mathcal{X}, \mathcal{Y})$ is cohereditary, then $\mathcal{Y}$ is a pretriangulated subcategory of $\mathcal{C}$.

The class $\mathbb{T}(\mathcal{C})$ of (hereditary or cohereditary) torsion pairs on $\mathcal{C}$ is partially ordered if we define: $(\mathcal{X}, \mathcal{Y}) \prec (\mathcal{X}', \mathcal{Y}')$ if and only if $\mathcal{X} \subseteq \mathcal{X}'$ (or equivalently $\mathcal{Y}' \subseteq \mathcal{Y}$). The torsion pair $(0, \mathcal{C})$ is the minimum element and the torsion pair $(\mathcal{C}, 0)$ is the maximum element of $\mathbb{T}(\mathcal{C})$.

The following examples show that a torsion pair in a pretriangulated category is a common generalization of a torsion pair in an abelian category and in a triangulated category.

EXAMPLE. (1) Let $\mathcal{C}$ be abelian considered as a pretriangulated category. Then the concept of a torsion pair in $\mathcal{C}$ in the sense of the above definition, coincides with the concept of a torsion pair in the (usual) sense of Chapter I. Also the concept of a TTF-triple reduces to the well–known concept studied in abelian categories and in particular in ring theory, see [**100**].

(2) Let $\mathcal{C}$ be a triangulated category considered as a pretriangulated category. Then the concept of a torsion pair in $\mathcal{C}$ in the sense of the above definition, coincides

with the concept of a torsion pair (= t-structure) in the sense of Chapter I and the TTF-triples reduce to recollement situations [**18**].

Motivated by the results of Chapter I we introduce the concept of the heart of a torsion pair in $\mathcal{C}$. There are actually two, not necessarily equivalent, hearts, reflecting the fact that the suspension and the loop functor of $\mathcal{C}$ are not necessarily quasi-inverse equivalences.

DEFINITION 3.4. Let $(\mathcal{X}, \mathcal{Y})$ be a torsion pair in a pretriangulated category $\mathcal{C}$. Then its **left heart** is $\mathcal{H}_l := \mathcal{X} \cap \Sigma\mathcal{Y}$ and its **right heart** is $\mathcal{H}_r := \Omega\mathcal{X} \cap \mathcal{Y}$.

REMARK 3.5. (1) Let $(\mathcal{X}, \mathcal{Y})$ be a torsion pair in an abelian category $\mathcal{C}$ viewed as a pretriangulated category with $\Omega = 0 = \Sigma$. Then the hearts of a torsion pair $(\mathcal{X}, \mathcal{Y})$ in $\mathcal{C}$ are trivial: $\mathcal{H}_l = 0 = \mathcal{H}_r$.

(2) If $\mathcal{C}$ is triangulated, then by [**18**] the hearts of a torsion pair $(\mathcal{X}, \mathcal{Y})$ (= t-structure), are abelian and $\mathcal{H}_r = \Sigma^{-1}\mathcal{H}_l$. Observe that if $(\mathcal{X}, \mathcal{Y})$ is hereditary, then: $\mathcal{H}_r = 0 = \mathcal{H}_l$.

We refer to Chapter III for a study of the heart in the triangulated case and to Chapter V for an example of a torsion pair in a pretriangulated category which is not abelian or triangulated such that the left heart is non-zero abelian and the right heart is zero.

We close this section with a characterization of torsion pairs. First we need a simple lemma. We recall that in a pretriangulated category any morphism has a weak kernel and a weak cokernel.

LEMMA 3.6. *Let f be a morphism in $\mathcal{C}$. If f is a weak kernel (weak cokernel), then f is a weak kernel (weak cokernel) of its weak cokernel (weak kernel).*

PROOF. Let $A \xrightarrow{f} B \xrightarrow{g} C \to \Sigma(A)$ be a triangle in $\mathcal{C}$. Then g is a weak cokernel of f. Let $\alpha : B \to D$ be a morphism such that f is a weak kernel of α. Then $f \circ \alpha = 0$, hence there exists a morphism $\beta : C \to D$ such that $g \circ \beta = \alpha$. Now let $\kappa : M \to B$ be a morphism such that $\kappa \circ g = 0$. Then $\kappa \circ g \circ \beta = \kappa \circ \beta = 0$. Since f is a weak kernel of α, there exists a morphism $\lambda : M \to A$ such that $\lambda \circ f = \kappa$. Hence f is a weak kernel of g. The parenthetical case is dual. □

The following characterizes when a pair of subcategories forms a torsion pair.

PROPOSITION 3.7. *Let $(\mathcal{X}, \mathcal{Y})$ be a pair of subcategories of a pretriangulated category $\mathcal{C}$. If $\mathcal{C}(\mathcal{X}, \mathcal{Y}) = 0$, then the following are equivalent:*

(i) (α) *The inclusion $\mathbf{i} : \mathcal{X} \hookrightarrow \mathcal{C}$ has a right adjoint $\mathbf{R}$ and the counit $f_C : \mathbf{R}(C) \to C$ is a weak kernel, $\forall C \in \mathcal{C}$.*

(β) *If $\mathbf{R}(C) \xrightarrow{f_C} C \xrightarrow{g^C} Y^C \xrightarrow{h^C} \Sigma\mathbf{R}(C)$ is a triangle in ∇, then $Y^C \in \mathcal{Y}$.*

(γ) *$\Omega(\mathcal{Y}) \subseteq \mathcal{Y}$ and if $\Omega(Y^C) \to A \to C \xrightarrow{g^C} Y^C$ is a triangle in Δ, then $\mathcal{C}(\Sigma(A), Y^C) = 0$.*

(ii) (α) *The inclusion $\mathbf{j} : \mathcal{Y} \hookrightarrow \mathcal{C}$ has a left adjoint $\mathbf{L}$ and the unit $g^C : C \to \mathbf{L}(C)$ is a weak cokernel, $\forall C \in \mathcal{C}$.*

(β) *If $\Omega\mathbf{L}(C) \xrightarrow{h_C} X_C \xrightarrow{f_C} C \xrightarrow{g^C} \mathbf{L}(C)$ is a triangle in Δ, then $X_C \in \mathcal{X}$.*

(γ) *$\Sigma(\mathcal{X}) \subseteq \mathcal{X}$ and if $X_C \xrightarrow{f_C} C \to B \to \Sigma(X_C)$ is a triangle in ∇, then $\mathcal{C}(X_C, \Omega(B)) = 0$.*

(iii) *The pair $(\mathcal{X}, \mathcal{Y})$ is a torsion pair in $\mathcal{C}$.*

PROOF. We prove that (i) $\Leftrightarrow$ (iii). The proof of (ii) $\Leftrightarrow$ (iii) is similar. Since (iii) $\Rightarrow$ (i) follows directly from Proposition 3.3, it remains to show that (i) $\Rightarrow$ (iii). For any $C \in \mathcal{C}$, let $f_C : \mathbf{R}(C) \to C$ be the counit of the adjoint pair $(\mathbf{i}, \mathbf{R})$, and let $(*)$: $\mathbf{R}(C) \xrightarrow{f_C} C \xrightarrow{g^C} Y^C \xrightarrow{h^C} \Sigma\mathbf{R}(C)$ be a triangle in ∇. If $\Omega(Y^C) \xrightarrow{\beta} A \xrightarrow{\alpha} C \xrightarrow{g^C} Y^C$ is a triangle in Δ, then since $f_C \circ g^C = 0$, there exists a morphism $\kappa : \mathbf{R}(C) \to A$ such that: $\kappa \circ \alpha = f_C$. By Lemma 3.6, f_C is a weak kernel of g^C, hence since $\alpha \circ g^C = 0$, there exists a morphism $\lambda : A \to \mathbf{R}(C)$ such that $\lambda \circ f_C = \alpha$.

Consider now the left triangle $\Omega(C) \xrightarrow{h_C} Y_C \xrightarrow{g_C} \mathbf{R}(C) \xrightarrow{f_C} C$ in Δ. By the axioms of a left triangulated category, there are morphisms $\phi : Y_C \to \Omega(Y^C)$ and $\psi : \Omega(Y^C) \to Y_C$ such that the following diagram commutes:

$$\begin{array}{ccccccc}
\Omega(C) & \xrightarrow{h_C} & Y_C & \xrightarrow{g_C} & \mathbf{R}(C) & \xrightarrow{f_C} & C \\
\| & & \exists\,\phi\downarrow & & \kappa\downarrow & & \| \\
\Omega(C) & \xrightarrow{\Omega(g^C)} & \Omega(Y^C) & \xrightarrow{\beta} & A & \xrightarrow{\alpha} & C \\
\| & & \exists\,\psi\downarrow & & \lambda\downarrow & & \| \\
\Omega(C) & \xrightarrow{h_C} & Y_C & \xrightarrow{g_C} & \mathbf{R}(C) & \xrightarrow{f_C} & C \\
\| & & \exists\,\phi\downarrow & & \kappa\downarrow & & \| \\
\Omega(C) & \xrightarrow{\Omega(g^C)} & \Omega(Y^C) & \xrightarrow{\beta} & A & \xrightarrow{\alpha} & C
\end{array}$$

Since $\kappa \circ \lambda \circ f_C = f_C$, we have that $\kappa \circ \lambda = 1_{\mathbf{R}(C)}$. By the axioms of a left triangulated category, this implies that $\phi \circ \psi$ is invertible. Further we have that $\lambda \circ \kappa \circ \alpha = \alpha$, so $(1_A - \lambda \circ \kappa) \circ \alpha = 0$. Hence there exists a morphism $\sigma : A \to \Omega(Y^C)$ such that $\sigma \circ \beta = 1_A - \lambda \circ \kappa$. By condition ($\gamma$), $\mathcal{C}(\Sigma(A), Y^C) \cong \mathcal{C}(A, \Omega(Y^C)) = 0$. Hence $\sigma = 0$ and then $1_A = \lambda \circ \kappa$. It follows that $A \cong \mathbf{R}(C)$. But then also $Y_C \cong \Omega(Y^C)$. Since by hypothesis $Y^C \in \mathcal{Y}$ and $\Omega(\mathcal{Y}) \subseteq \mathcal{Y}$, we have that $\Omega(Y^C) \in \mathcal{Y}$. Now setting $X_C := \mathbf{R}(C)$ we have that in the following diagram

$$\Omega(Y^C) \xrightarrow{h_C} X_C \xrightarrow{f_C} C \xrightarrow{g^C} Y^C \xrightarrow{h^C} \Sigma(X_C)$$

the three morphisms on the right constitute a right triangle and the three morphisms on the left constitute a left triangle. Hence the pair $(\mathcal{X}, \mathcal{Y})$ is a torsion pair. □

4. Torsion Pairs and Localization Sequences

Associated to a (co)hereditary torsion pair in an abelian or triangulated category there are localization sequences connecting the ambient category with the torsion and the torsion-free subcategory. In this section we discuss an analogous situation in the pretriangulated case, which will be useful in the investigation of torsion pairs in a stable category in Chapter VI.

Let $(\mathcal{X},\mathcal{Y})$ be a torsion pair in the pretriangulated category $\mathcal{C}$. Our aim here is to construct localization sequences relating the categories $\mathcal{X},\mathcal{Y}$ and $\mathcal{C}$ in analogy with the localization sequences in the triangulated case, cf. Proposition I.2.6.

Since $\mathcal{X}$ is a right triangulated subcategory of $\mathcal{C}$, closed under direct summands and extensions of right triangles, we can construct the localization category $\mathcal{C}/\mathcal{X}$ as in the triangulated case, see [**50**], [**102**] for details. Recall that the category $\mathcal{C}/\mathcal{X}$ is defined as the category of fractions $\mathcal{C}[\mathcal{R}_{\mathcal{X}}^{-1}]$, where the class $\mathcal{R}_{\mathcal{X}}$ consists of all morphisms $f : A \to B$ such that in any right triangle $A \xrightarrow{f} B \to X \to \Sigma(A)$, the object X lies in $\mathcal{X}$. It is not difficult to see that the class of morphisms $\mathcal{R}_{\mathcal{X}}$ admits a calculus of left fractions in the sense of [**50**]. Then the right triangulated structures of $\mathcal{X}$ and $\mathcal{C}$ induce a right triangulated structure in the localization category $\mathcal{C}/\mathcal{X}$ in such a way that the canonical functor $\mathcal{C} \to \mathcal{C}/\mathcal{X}$ is right exact and we have an exact sequence of right triangulated categories $0 \to \mathcal{X} \to \mathcal{C} \to \mathcal{C}/\mathcal{X} \to 0$.

Dually since $\mathcal{Y}$ is a left triangulated subcategory of $\mathcal{C}$, closed under direct summands and extensions of left triangles, we can construct the localization category $\mathcal{C}/\mathcal{Y}$ which is the category of fractions $\mathcal{C}[\mathcal{L}_{\mathcal{Y}}^{-1}]$, where the class $\mathcal{L}_{\mathcal{Y}}$ consists of all morphisms $f : A \to B$ such that in any left triangle $\Omega(B) \to Y \to A \xrightarrow{f} B$, the object Y lies in $\mathcal{Y}$. Then the class of morphisms $\mathcal{L}_{\mathcal{Y}}$ admits a calculus of right fractions and the left triangulated structures of $\mathcal{Y}$ and $\mathcal{C}$ induce a left triangulated structure in the localization category $\mathcal{C}/\mathcal{Y}$ in such a way that the canonical functor $\mathcal{C} \to \mathcal{C}/\mathcal{Y}$ is left exact. In this case we have an exact sequence of left triangulated categories $0 \to \mathcal{Y} \to \mathcal{C} \to \mathcal{C}/\mathcal{Y} \to 0$.

We have the following analogue of Proposition I.2.6 and Corollary I.2.9.

PROPOSITION 4.1. *Let $(\mathcal{X},\mathcal{Y})$ be a torsion pair in a pretriangulated category $\mathcal{C}$.*

(i) *If $\Omega(\mathcal{Y}) = \mathcal{Y}$, then $\mathcal{Y}$ is right triangulated and the functor $\mathbf{L} : \mathcal{C} \to \mathcal{Y}$ induces an equivalence $\mathcal{C}/\mathcal{X} \xrightarrow{\approx} \mathcal{Y}$. In other words we have an exact sequence of right triangulated categories: $0 \longrightarrow \mathcal{X} \xrightarrow{\mathbf{i}} \mathcal{C} \xrightarrow{\mathbf{L}} \mathcal{Y} \longrightarrow 0$.*

(ii) *If $\Sigma(\mathcal{X}) = \mathcal{X}$, then $\mathcal{X}$ is left triangulated and the functor $\mathbf{R} : \mathcal{C} \to \mathcal{Y}$ induces an equivalence $\mathcal{C}/\mathcal{Y} \xrightarrow{\approx} \mathcal{X}$. In other words we have an exact sequence of left triangulated categories: $0 \longrightarrow \mathcal{Y} \xrightarrow{\mathbf{j}} \mathcal{C} \xrightarrow{\mathbf{R}} \mathcal{X} \longrightarrow 0$.*

(iii) *If $(\mathcal{X},\mathcal{Y},\mathcal{Z})$ is a TTF-triple in $\mathcal{C}$ where $\Sigma(\mathcal{X}) = \mathcal{X}$ and $\Omega(\mathcal{Z}) = \mathcal{Z}$, then there are equivalences of pretriangulated categories: $\mathcal{X} \xleftarrow{\approx} \mathcal{C}/\mathcal{Y} \xrightarrow{\approx} \mathcal{Z}$.*

PROOF. We prove only part (i), since part (ii) is dual and part (iii) follows from parts (i) and (ii).

Consider the class of morphisms $\mathcal{R}_{\mathcal{X}}$ defined above. We call an object $M \in \mathcal{C}$ left $\mathcal{R}_{\mathcal{X}}$–closed if for any $A \xrightarrow{f} B \in \mathcal{R}_{\mathcal{X}}$, the morphism $\mathcal{C}(f, M) : \mathcal{C}(B, M) \to \mathcal{C}(A, M)$ is invertible. Let $A \xrightarrow{f} B$ be in $\mathcal{R}_{\mathcal{X}}$ and let $A \xrightarrow{f} B \to X \to \Sigma(A)$ be a right triangle in $\mathcal{C}$ with $X \in \mathcal{X}$. Then for any $Y \in \mathcal{Y}$, we have the long exact sequence $\cdots \to \mathcal{C}(\Sigma(X), Y) \to \mathcal{C}(\Sigma(B), Y) \to \mathcal{C}(\Sigma(A), Y) \to \mathcal{C}(X, Y) \to \mathcal{C}(B, Y) \to \mathcal{C}(A, Y)$, which shows that $\mathcal{C}(\Sigma(f), Y) : \mathcal{C}(\Sigma(B), Y) \to \mathcal{C}(\Sigma(A), Y)$ or equivalently $\mathcal{C}(f, \Omega(Y)) : \mathcal{C}(B, \Omega(Y)) \to \mathcal{C}(A, \Omega(Y))$, is invertible. Hence for any $Y \in \mathcal{Y}$, the object $\Omega(Y)$ is left $\mathcal{R}_{\mathcal{X}}$–closed. Since $\Omega(\mathcal{Y}) = \mathcal{Y}$, it follows that any object of $\mathcal{Y}$ is left $\mathcal{R}_{\mathcal{X}}$–closed. Conversely if M is left $\mathcal{R}_{\mathcal{X}}$–closed, then M lies in $\mathcal{Y}$. Indeed let $C \in \mathcal{C}$ and consider the right triangle $\mathbf{R}(C) \xrightarrow{f_C} C \xrightarrow{g^C} Y^C \xrightarrow{h^C} \Sigma\mathbf{R}(C)$, where f_C is the coreflection of C in $\mathcal{X}$. Since $\Sigma\mathbf{R}(C)$ is in $\mathcal{X}$, it follows that the morphism g^C is in $\mathcal{R}_{\mathcal{X}}$. Hence for any left $\mathcal{R}_{\mathcal{X}}$–closed object M we have $\mathcal{C}(g^C, M) : \mathcal{C}(Y^C, M) \xrightarrow{\cong} \mathcal{C}(C, M)$. Now if $C \in \mathcal{X}$, the morphism f_C is invertible, hence $g^C = 0$. It follows that $\mathcal{C}(X, M) = 0$, for any $X \in \mathcal{X}$, i.e. $M \in \mathcal{X}^{\perp} = \mathcal{Y}$. We infer that $\mathcal{Y}$ coincides with the class of left $\mathcal{R}_{\mathcal{X}}$–closed objects. Now let $C \in \mathcal{C}$ and consider the right triangle $\mathbf{R}(C) \xrightarrow{f_C} C \xrightarrow{g^C} \mathbf{L}(C) \xrightarrow{h^C} \Sigma\mathbf{R}(C)$, in which the morphism $g^C : C \to \mathbf{L}(C)$ is in $\mathcal{R}_{\mathcal{X}}$ and the object $\mathbf{L}(C)$ is left $\mathcal{R}_{\mathcal{X}}$–closed. By [**98**], the localization functor $\mathcal{C} \to \mathcal{C}/\mathcal{X}$ admits a fully faithful right adjoint $\mathcal{C}/\mathcal{X} \hookrightarrow \mathcal{C}$ and the composite functor $\mathcal{Y} \xrightarrow{\mathbf{j}} \mathcal{C} \to \mathcal{C}/\mathcal{X}$ is an equivalence with inverse the composite $\mathcal{C}/\mathcal{X} \to \mathcal{C} \xrightarrow{\mathbf{L}} \mathcal{Y}$. Since $\mathcal{C}/\mathcal{X}$ is right triangulated, so is $\mathcal{Y}$. □

The above result can be applied when the torsion class $\mathcal{X}$ is triangulated or the torsion-free class $\mathcal{Y}$ is triangulated. For concrete examples of this situation we refer to Chapter VI.

5. Lifting Torsion Pairs

To any left or right triangulated category there is associated in a universal way a triangulated category called the *stabilization* and which keeps track of the complexity of its left or right triangulation. Our aim in this section is to study the relationship between the torsion or torsion-free class of a torsion pair defined in a pretriangulated category and their stabilizations via suitable stabilization functors. Indeed we shall show that if the torsion pair is (co)hereditary, then it can be lifted to a torsion pair in the stabilization. This lifting construction will be used in the investigation of universal cohomology theories with respect to (co)hereditary torsion pairs in Chapter VIII.

Throughout this section we fix a pretriangulated category $\mathcal{C}$ and a torsion pair $(\mathcal{X}, \mathcal{Y})$ in $\mathcal{C}$. As usual we denote by $\mathbf{R} : \mathcal{C} \to \mathcal{X}$ the right adjoint of the inclusion $\mathbf{i} : \mathcal{X} \hookrightarrow \mathcal{C}$ and by $\mathbf{L} : \mathcal{C} \to \mathcal{Y}$ the left adjoint of the inclusion $\mathbf{j} : \mathcal{Y} \hookrightarrow \mathcal{C}$.

Stabilizations. Before we study the lifting of torsion pairs in the left and right stabilization of $\mathcal{C}$, we need to recall some facts about the stabilization process. We also investigate when the left or right stabilization of $\mathcal{C}$ can be realized by the torsion or torsion-free subcategory, via the (co)reflection functor. This realization has interesting consequences for the structure of the category.

Recall from [**20**], [**74**] that with a given left triangulated category $\mathcal{C}$, we can associate in a universal way a triangulated category $\mathcal{T}(\mathcal{C})$ which reflects many important homological properties of $\mathcal{C}$. More precisely there exists a left exact functor $\mathsf{P} : \mathcal{C} \to \mathcal{T}(\mathcal{C})$ such that for any left exact functor $F : \mathcal{C} \to \mathcal{T}$ to a triangulated category $\mathcal{T}$, there exists a unique up to isomorphism exact functor $F^* : \mathcal{T}(\mathcal{C}) \to \mathcal{T}$ such that $F^*\mathsf{P} \cong F$. The category $\mathcal{T}(\mathcal{C})$ is called the **stabilization** of $\mathcal{C}$ and the functor P is called the **stabilization functor**. We refer to [**20**], [**74**], for details. Here we need only the following facts.

The objects of $\mathcal{T}(\mathcal{C})$ are pairs (C,n), where C is an object in $\mathcal{C}$ and $n \in \mathbb{Z}$. The space of morphisms $\mathcal{T}(\mathcal{C})[(A,n),(B,m)]$ is identified with the direct limit:

$$\mathcal{T}(\mathcal{C})[(A,n),(B,m)] = \varinjlim_{k\geq n, k\geq m} \mathcal{C}[\Omega^{k-n}(A), \Omega^{k-m}(B)].$$

The loop functor $\Omega : \mathcal{T}(\mathcal{C}) \to \mathcal{T}(\mathcal{C})$ is defined by $\Omega(C,n) = (C, n-1)$ and the stabilization functor P is defined by $\mathsf{P}(C) = (C,0)$. Finally the extension of the left exact functor F above is defined by $F^*(C,n) = \Omega^{-n}F(C)$. Note that $\mathcal{T}(\mathcal{C}) = 0$ if and only if Ω is locally nilpotent, that is for any $C \in \mathcal{C}$, $\Omega^n(C) = 0$ for some $n \geq 0$. Dually any right triangulated category admits its stabilization which has a dual description and dual properties.

The stabilizations are very natural objects of study:

Example. Let $(\mathcal{X},\mathcal{Y})$ be the natural torsion pair (= t-structure) in the unbounded derived category $\mathbf{D}(\mathcal{C})$ of an abelian category. Then $\mathcal{T}_r(\mathcal{X}) = \mathbf{D}^-(\mathcal{C})$ is the derived category of right bounded complexes and $\mathcal{T}_l(\mathcal{Y}) = \mathbf{D}^+(\mathcal{C})$ is the derived category of left bounded complexes.

The pretriangulated category $\mathcal{C}$ as a left and right triangulated category admits a stabilization $\mathsf{P}_l : \mathcal{C} \to \mathcal{T}_l(\mathcal{C})$ when considered as a left triangulated category and a stabilization $\mathsf{P}_r : \mathcal{C} \to \mathcal{T}_r(\mathcal{C})$ when considered as a right triangulated category. In general the stabilizations $\mathcal{T}_l(\mathcal{C})$ and $\mathcal{T}_r(\mathcal{C})$ are not connected in a nice way, in particular they are rarely equivalent. It is important to have computable descriptions of the left and right stabilizations of $\mathcal{C}$. We now study this problem in case $\mathcal{C}$ admits a torsion pair.

Consider a torsion pair $(\mathcal{X},\mathcal{Y})$ in $\mathcal{C}$. Since $\mathcal{X}$ is right triangulated and $\mathcal{Y}$ is left triangulated, it is natural to ask what is the relationship between the stabilizations of $\mathcal{X},\mathcal{Y}$ and $\mathcal{C}$. In this respect the following subcategories are useful:

$$\widehat{\mathcal{X}} = \{C \in \mathcal{C} \mid \exists n \geq 0 : \ \Omega^n(C) \in \mathcal{X}\},$$

$$\widetilde{\mathcal{Y}} = \{C \in \mathcal{C} \mid \exists n \geq 0 : \ \Sigma^n(C) \in \mathcal{Y}\}$$

Similarly we introduce the following full subcategories of $\mathcal{C}$:

$$\mathcal{P}^{<\infty}(\mathcal{C}) := \{C \in \mathcal{C} \mid \exists n \geq 0 : \ \Omega^n(C) = 0\},$$

$$\mathcal{I}^{<\infty}(\mathcal{C}) := \{C \in \mathcal{C} \mid \exists n \geq 0 : \ \Sigma^n(C) = 0\}$$

It follows from [**20**] that $\mathrm{Ker}\mathsf{P} = \mathcal{P}^{<\infty}(\mathcal{C})$. The category $\mathcal{P}^{<\infty}(\mathcal{C})$ is the prototype of the stable category of modules of finite projective dimension modulo projectives, see Chapter V. It is clear that $\mathcal{P}^{<\infty}(\mathcal{C})$ is a left triangulated subcategory of $\mathcal{C}$ and its stabilization is trivial: $\mathcal{T}_l(\mathcal{P}^{<\infty}(\mathcal{C})) = 0$. Moreover any left triangulated subcategory of $\mathcal{C}$ with trivial stabilization is contained in $\mathcal{P}^{<\infty}(\mathcal{C})$. Dual remarks hold for the category $\mathcal{I}^{<\infty}(\mathcal{C})$.

The following gives a useful connection between the above subcategories.

Lemma 5.1. (i) *If $\mathcal{X}$ is closed under Ω, then: $\mathcal{Y} \subseteq \mathcal{P}^{<\infty}(\mathcal{C})$ if and only if $\widehat{\mathcal{X}} = \mathcal{C}$.*

(ii) *If $\mathcal{Y}$ is closed under Σ, then: $\mathcal{X} \subseteq \mathcal{I}^{<\infty}(\mathcal{C})$ if and only if $\widetilde{\mathcal{Y}} = \mathcal{C}$.*

PROOF. (i) Assume that $\mathcal{Y} \subseteq \mathcal{P}^{<\infty}(\mathcal{C})$. If $C \in \mathcal{C}$, then consider the standard triangle $\Delta(C): \ \Omega(C) \xrightarrow{h_C} Y_C \xrightarrow{g_C} \mathbf{R}(C) \xrightarrow{f_C} C$. Since $Y_C \in \mathcal{Y}$, there exists $n \geq 0$ such that $\Omega^n(Y_C) = 0$. Consider the $n-$th rotation $\Omega^{n+1}(C) \rightarrow \Omega^n Y_C \rightarrow \Omega^n\mathbf{R}(C) \rightarrow \Omega^n(C)$ of the above triangle. Then $\Omega^{n+1}(f_C) : \Omega^{n+1}\mathbf{R}(C) \xrightarrow{\cong} \Omega^{n+1}(C)$ is an isomorphism, since $\Omega^n(Y_C) = 0$. Since $\mathbf{R}(C) \in \mathcal{X}$ and $\mathcal{X}$ is closed under Ω, it follows that $\Omega^{n+1}(C) \in \mathcal{X}$. Hence $C \in \widehat{\mathcal{X}}$ and then $\mathcal{C} = \widehat{\mathcal{X}}$. Conversely if this equality holds, then fix an object $C \in \mathcal{C}$ and let $n \geq 0$ be such that $\Omega^n(C) \in \mathcal{X}$. Since by Proposition 2.6, $\mathbf{R}$ commutes with Ω, it follows easily that $\Omega^n\mathbf{R}(C) \cong \mathbf{R}\Omega^n(C)$. Then the morphism $\Omega^n(f_C) : \Omega^n\mathbf{R}(C) \rightarrow \Omega^n(C)$ is invertible. From the triangle $\Delta(C)$ above we infer that $\Omega^{n+1}(Y_C) = 0$. Hence for any object C in $\mathcal{C}$ there exists $n \geq 0$, such that $\Omega^n(Y_C) = 0$. Let now $Y \in \mathcal{Y}$ be an arbitrary object. Then in the standard triangle $\Omega(Y) \xrightarrow{h_Y} Y_Y \xrightarrow{g_Y} \mathbf{R}(Y) \xrightarrow{f_Y} Y$ we have that $f_Y = 0$. This implies trivially that $Y_Y = \mathbf{R}(Y) \oplus \Omega(Y)$. Since $\mathbf{R}(Y) \in \mathcal{X}$ and $Y_Y \in \mathcal{Y}$ we have $\mathbf{R}(Y) = 0$. Hence $Y_Y = \Omega(Y)$. By the above argument there exists $n \geq 0$, such that $\Omega^n(Y_Y) = 0$. Hence $\Omega^{n+1}(Y) = 0$, i.e. $\mathcal{Y} \subseteq \mathcal{P}^{<\infty}(\mathcal{C})$. □

The following result characterizes when the coreflection functor $\mathbf{R}$ is the stabilization of $\mathcal{C}$ with respect to its left triangulation, and when the reflection functor $\mathbf{L}$ is the stabilization of $\mathcal{C}$ with respect to its right triangulation, in terms of properties of the subcategories introduced above.

PROPOSITION 5.2. *Using the above notation, we have the following.*

(i) *The following are equivalent:*
  (a) $\mathbf{R} : \mathcal{C} \rightarrow \mathcal{X}$ *is the stabilization functor with respect to the left triangulation of* $\mathcal{C}$.
  (b) $\mathcal{X}$ *is triangulated and* $\widehat{\mathcal{X}} = \mathcal{C}$ *(or equivalently* $\mathcal{Y} = \mathcal{P}^{<\infty}(\mathcal{C})$*).*

(ii) *The following are equivalent:*
  (a) $\mathbf{L} : \mathcal{C} \rightarrow \mathcal{Y}$ *is the stabilization functor with respect to the right triangulation of* $\mathcal{C}$.
  (b) $\mathcal{Y}$ *is triangulated and* $\widetilde{\mathcal{Y}} = \mathcal{C}$ *(or equivalently* $\mathcal{X} = \mathcal{I}^{<\infty}(\mathcal{C})$*).*

PROOF. We prove only part (i). Assume that $\mathcal{X}$ is triangulated and $\widehat{\mathcal{X}} = \mathcal{C}$. Then the coreflection functor $\mathbf{R}$ and the inclusion functor $\mathbf{i}$ are left exact. If $F : \mathcal{C} \rightarrow \mathcal{D}$ is a left exact functor to a triangulated category $\mathcal{D}$, then $F(\mathcal{Y}) = 0$. Indeed, since $\widehat{\mathcal{X}} = \mathcal{C}$, by Lemma 5.1 we have $\mathcal{Y} \subseteq \mathcal{P}^{<\infty}(\mathcal{C})$, hence for any $Y \in \mathcal{Y}$, there exists $n \geq 0$ such that $\Omega^n(Y) = 0$. Since F commutes with Ω, we have $\Omega^n F(Y) = 0$ where Ω is the loop functor in $\mathcal{D}$. Then $F(Y) = 0$, since Ω is invertible in $\mathcal{D}$. Now consider the functorial left triangle $\Omega\mathbf{L}(C) \xrightarrow{h_C} \mathbf{R}(C) \xrightarrow{f_C} C \xrightarrow{g^C} \mathbf{L}(C)$ in $\mathcal{C}$, where $f_C : \mathbf{iR}(C) \rightarrow C$ is the coreflection of C in $\mathcal{X}$. Then we have a triangle $\Omega F\mathbf{L}(C) \rightarrow F\mathbf{R}(C) \xrightarrow{F(f_C)} F(C) \rightarrow F\mathbf{L}(C)$ in $\mathcal{D}$. Since $F(\mathcal{Y}) = 0$, it follows that $F(f_C) : F\mathbf{R}(C) \rightarrow F(C)$ is invertible. Define a functor $F^* : \mathcal{X} \rightarrow \mathcal{D}$ by $F^* = F\mathbf{i}$. Then F^* is obviously left exact and the above isomorphism shows that $F(f) : F^*\mathbf{R} \xrightarrow{\cong} F$. If $G : \mathcal{X} \rightarrow \mathcal{D}$ is another exact functor endowed with a natural isomorphism $\xi : G\mathbf{R} \xrightarrow{\cong} F$, then $\xi\mathbf{i} : G \cong G\mathbf{Ri} \xrightarrow{\cong} F\mathbf{i} = F^*$. Hence F^* is the unique up to isomorphism left exact functor which extends F. Then $\mathbf{R} : \mathcal{C} \rightarrow \mathcal{X}$ represents $\mathcal{X}$ as the stabilization of $\mathcal{C}$. Conversely assume that $\mathbf{R}$ represents $\mathcal{X}$ as the stabilization of $\mathcal{C}$ with respect to its left triangulation. By [**20**], we know that the kernel of the stabilization functor is $\mathcal{P}^{<\infty}(\mathcal{C})$. Since $\mathrm{Ker}\mathbf{R} = \mathcal{Y}$, we infer that $\mathcal{P}^{<\infty}(\mathcal{C}) = \mathcal{Y}$ or equivalently by Lemma 5.1, $\widehat{\mathcal{X}} = \mathcal{C}$. □

Lifting Torsion Pairs. Our aim now is to show that in some cases the exact localization sequences of left or right triangulated categories in Proposition 4.1 can be lifted to exact sequences in the stabilizations, thus providing a lifting of the torsion pair from the pretriangulated category to the triangulated stabilization. Assume throughout that the torsion pair $(\mathcal{X}, \mathcal{Y})$ is hereditary, the cohereditary case being treated in a dual way.

Since $(\mathcal{X}, \mathcal{Y})$ is hereditary we have that $\mathcal{X}$ is pretriangulated, and the inclusion functor $\mathbf{i} : \mathcal{X} \hookrightarrow \mathcal{C}$ and the coreflection functor $\mathbf{R} : \mathcal{C} \to \mathcal{X}$ are left exact. Consider the stabilization functors $\mathsf{P} : \mathcal{C} \to \mathcal{T}_l(\mathcal{C})$ and $\mathsf{P} : \mathcal{X} \to \mathcal{T}_l(\mathcal{X})$ of $\mathcal{C}$ and $\mathcal{X}$ when these are considered as **left** triangulated categories. Since $\mathbf{i}$ and $\mathbf{R}$ are left exact, by the universal property of the stabilization functors $\mathsf{P} : \mathcal{X} \to \mathcal{T}_l(\mathcal{X})$ and $\mathsf{P} : \mathcal{C} \to \mathcal{T}_l(\mathcal{C})$, there exist unique exact functors $\mathbf{i}^* : \mathcal{T}_l(\mathcal{X}) \to \mathcal{T}_l(\mathcal{C})$ and $\mathbf{R}^* : \mathcal{T}_l(\mathcal{C}) \to \mathcal{T}_l(\mathcal{X})$, such that the following diagrams commute:

$$\begin{array}{ccc} \mathcal{X} & \xrightarrow{\mathbf{i}} & \mathcal{C} \\ {\scriptstyle \mathsf{P}}\downarrow & & \downarrow{\scriptstyle \mathsf{P}} \\ \mathcal{T}_l(\mathcal{X}) & \xrightarrow{\mathbf{i}^*} & \mathcal{T}_l(\mathcal{C}) \end{array} \qquad\qquad \begin{array}{ccc} \mathcal{C} & \xrightarrow{\mathbf{R}} & \mathcal{X} \\ {\scriptstyle \mathsf{P}}\downarrow & & \downarrow{\scriptstyle \mathsf{P}} \\ \mathcal{T}_l(\mathcal{C}) & \xrightarrow{\mathbf{R}^*} & \mathcal{T}_l(\mathcal{X}). \end{array}$$

PROPOSITION 5.3. *Let $(\mathcal{X}, \mathcal{Y})$ be a hereditary torsion pair in a pretriangulated category $\mathcal{C}$. Then $\big(\mathcal{T}_l(\mathcal{X}), \mathcal{T}_l(\mathcal{Y})\big)$ is a hereditary torsion pair in $\mathcal{T}_l(\mathcal{C})$, that is, there exists a short exact sequence of triangulated categories*

$$0 \longrightarrow \mathcal{T}_l(\mathcal{Y}) \xrightarrow{\mathbf{j}^*} \mathcal{T}_l(\mathcal{C}) \xrightarrow{\mathbf{R}^*} \mathcal{T}_l(\mathcal{X}) \longrightarrow 0$$

The functor $\mathbf{i}^*$ is a left adjoint of $\mathbf{R}^*$ and the functor $\mathbf{j}^*$ admits a left adjoint.

PROOF. It suffices to show that the functor $\mathbf{R}^*$ admits a fully faithful left adjoint and $\mathrm{Ker}\mathbf{R}^* = \mathcal{T}_l(\mathcal{Y})$. By Proposition 3.4 of [**20**] we have that the left exact functor $\mathbf{j} : \mathcal{Y} \hookrightarrow \mathcal{C}$ induces a fully faithful exact functor $\mathbf{j}^* : \mathcal{T}_l(\mathcal{Y}) \hookrightarrow \mathcal{T}_l(\mathcal{C})$. Let (C, n) be in $\mathcal{T}_l(\mathcal{C})$ such that $\mathbf{R}^*(C, n) = 0$. Then $\Omega^{-n}\mathsf{P}(\mathbf{R}(C)) = (\mathbf{R}(C), n) = 0$ and this means that there exists k with $k + n \geq 0$ such that $\Omega^{k+n}\mathbf{R}(C) = 0$. Since $\mathbf{R}$ commutes with the loop functor Ω we have $\mathbf{R}\Omega^{k+n}(C) = 0$. Hence $\Omega^{k+n}(C)$ lies in $\mathcal{Y}$ and then $\mathsf{P}(\Omega^{k+n}(C)) = \Omega^{n+k}\mathsf{P}(C)$ lies in $\mathcal{T}_l(\mathcal{Y})$. Since the latter is a triangulated subcategory of $\mathcal{T}_l(\mathcal{C})$ we have that $\Omega^{-2n-k}\Omega^{n+k}\mathsf{P}(C) = \Omega^{-n}\mathsf{P}(C) = (C, n)$ lies in $\mathcal{Y}$. On the other hand if (Y, n) is in $\mathcal{T}_l(\mathcal{Y})$ then $\mathbf{R}^*(Y, n) = (\mathbf{R}(Y), n) = 0$ since $\mathbf{R}(\mathcal{Y}) = 0$. We infer that $\mathrm{Ker}\mathbf{R}^* = \mathcal{T}_l(\mathcal{Y})$. Since $\mathbf{Ri} = \mathrm{Id}_{\mathcal{X}}$ it follows that $\mathbf{R}^*\mathbf{i}^* = \mathrm{Id}_{\mathcal{T}_l(\mathcal{X})}$. Consider now the coreflection morphism $f : \mathbf{iR} \to \mathrm{Id}_{\mathcal{X}}$. It is not difficult to see that the lifting $\mathbf{i}^*\mathbf{R}^* \to \mathrm{Id}_{\mathcal{T}_l(\mathcal{X})}$ of f in $\mathcal{T}_l(\mathcal{C})$ serves as the coreflection morphism for the adjoint pair $(\mathbf{i}^*, \mathbf{R}^*)$ and of course the functor $\mathbf{i}^*$ is fully faithful, since $\mathbf{i}$ is so. Now standard arguments [**102**] show that $\mathbf{j}^*$ admits a left adjoint. □

Dually when the torsion pair is cohereditary, then $\mathcal{Y}$ is pretriangulated, and the inclusion functor $\mathbf{j} : \mathcal{Y} \hookrightarrow \mathcal{C}$ and the reflection functor $\mathbf{L} : \mathcal{C} \to \mathcal{Y}$ are right exact. Consider the stabilizations $\mathsf{Q} : \mathcal{C} \to \mathcal{T}_r(\mathcal{C})$ and $\mathsf{Q} : \mathcal{Y} \to \mathcal{T}_r(\mathcal{Y})$ of $\mathcal{C}$ and $\mathcal{Y}$ when these are considered as **right** triangulated categories. Since $\mathbf{j}$ and $\mathbf{L}$ are right exact, there exist unique exact functors $\mathbf{j}^* : \mathcal{T}_r(\mathcal{X}) \to \mathcal{T}_r(\mathcal{C})$ and $\mathbf{L}^* : \mathcal{T}_r(\mathcal{C}) \to \mathcal{T}_r(\mathcal{Y})$, such that the following diagrams commute:

$$\begin{array}{ccc} \mathcal{Y} & \xrightarrow{\mathbf{j}} & \mathcal{C} \\ {\scriptstyle Q}\downarrow & & {\scriptstyle Q}\downarrow \\ \mathcal{T}_r(\mathcal{Y}) & \xrightarrow{\mathbf{j}^*} & \mathcal{T}_r(\mathcal{C}) \end{array} \qquad \begin{array}{ccc} \mathcal{C} & \xrightarrow{\mathbf{L}} & \mathcal{Y} \\ {\scriptstyle Q}\downarrow & & {\scriptstyle Q}\downarrow \\ \mathcal{T}_r(\mathcal{C}) & \xrightarrow{\mathbf{L}^*} & \mathcal{T}_r(\mathcal{Y}). \end{array}$$

The following is a dual version of Proposition 5.3.

PROPOSITION 5.4. *Let $(\mathcal{X}, \mathcal{Y})$ be a cohereditary torsion pair in a pretriangulated category $\mathcal{C}$. Then $\big(\mathcal{T}_r(\mathcal{X}), \mathcal{T}_r(\mathcal{Y})\big)$ is a hereditary torsion pair in $\mathcal{T}_r(\mathcal{C})$, that is, there exists a short exact sequence of triangulated categories*

$$0 \longrightarrow \mathcal{T}_r(\mathcal{X}) \xrightarrow{\mathbf{i}^*} \mathcal{T}_r(\mathcal{C}) \xrightarrow{\mathbf{L}^*} \mathcal{T}_r(\mathcal{Y}) \longrightarrow 0$$

The functor $\mathbf{j}^*$ is a right adjoint of $\mathbf{L}^*$ and the functor $\mathbf{i}^*$ admits a right adjoint.

We have the following direct consequence.

COROLLARY 5.5. (1) *If $(\mathcal{X}, \mathcal{Y})$ is hereditary, then the functor $\mathbf{R}^* : \mathcal{T}_l(\mathcal{C}) \to \mathcal{T}_l(\mathcal{X})$ is a triangle equivalence if and only if $\mathcal{Y} \subseteq \mathcal{P}^{<\infty}(\mathcal{C})$. If this is the case and $\mathcal{X}$ is triangulated, then $\mathbf{R}^* : \mathcal{T}_l(\mathcal{C}) \to \mathcal{X}$ is a triangle equivalence and $\mathbf{R} : \mathcal{C} \to \mathcal{X}$ is the stabilization functor with respect to the left triangulation of $\mathcal{C}$.*

(2) *If $(\mathcal{X}, \mathcal{Y})$ is cohereditary, then the functor $\mathbf{L}^* : \mathcal{T}_r(\mathcal{C}) \to \mathcal{T}_r(\mathcal{Y})$ is a triangle equivalence if and only if $\mathcal{X} \subseteq \mathcal{I}^{<\infty}(\mathcal{C})$. If this is the case and $\mathcal{Y}$ is triangulated, then $\mathbf{L}^* : \mathcal{T}_r(\mathcal{C}) \to \mathcal{Y}$ is a triangle equivalence and $\mathbf{L} : \mathcal{C} \to \mathcal{Y}$ is the stabilization functor with respect to the right triangulation of $\mathcal{C}$.*

Combining Proposition 4.1 and Corollary 5.5 we have the following consequence which generalizes Corollary I.2.9.

COROLLARY 5.6. *Let $(\mathcal{X}, \mathcal{Y}, \mathcal{Z})$ be a TTF-triple in a pretriangulated category $\mathcal{C}$. If $\mathcal{X}$ and $\mathcal{Z}$ are triangulated subcategories of $\mathcal{C}$, then the following are equivalent:*

(i) *$\mathcal{X}$ is the left stabilization of $\mathcal{C}$ and $\mathcal{Z}$ is the right stabilization of $\mathcal{C}$.*
(ii) *$\mathcal{P}^{<\infty}(\mathcal{C}) \supseteq \mathcal{Y} \subseteq \mathcal{I}^{<\infty}(\mathcal{C})$.*
(iii) *$\widehat{\mathcal{X}} = \mathcal{C} = \widetilde{\mathcal{Z}}$.*

If (i) holds, then $\mathcal{P}^{<\infty}(\mathcal{C}) = \mathcal{Y} = \mathcal{I}^{<\infty}(\mathcal{C})$, the categories $\mathcal{X}$, $\mathcal{Z}$ and $\mathcal{C}/\mathcal{Y}$ are triangulated and there exist triangle equivalences:

$$\mathcal{X} \xleftarrow{\approx} \mathcal{C}/\mathcal{Y} \xrightarrow{\approx} \mathcal{Z}.$$

Grothendieck Groups. We close this chapter with an application of the lifting of torsion pairs to K-theory.

If $\mathcal{C}$ is a left triangulated category with loop functor Ω and (left) triangulation Δ, then its Grothendieck group $\mathrm{K}_0(\mathcal{C})$ is defined as the quotient of the free abelian group on the isoclasses of objects of $\mathcal{C}$, modulo the relations $(A) - (B) + (C)$, for each triangle $\Omega(C) \to A \to B \to C$ in Δ. The definition of the Grothendieck group of a right triangulated category is dual. If $\mathcal{C}$ is pretriangulated, then we denote by $\mathrm{K}_0^{\Delta}(\mathcal{C})$ the Grothendieck group when $\mathcal{C}$ is considered as a left triangulated category, and by $\mathrm{K}_0^{\nabla}(\mathcal{C})$ the Grothendieck group when $\mathcal{C}$ is considered as a right triangulated category. Of course if $\mathcal{C}$ is triangulated, then $\mathrm{K}_0^{\Delta}(\mathcal{C}) = \mathrm{K}_0^{\nabla}(\mathcal{C}) := \mathrm{K}_0(\mathcal{C})$ is the usual Grothendieck group of $\mathcal{C}$.

COROLLARY 5.7. (1) *Let* $(\mathcal{X}, \mathcal{Y})$ *be a hereditary torsion pair in* $\mathcal{C}$. *Then the coreflection functor* $\mathbf{R} : \mathcal{C} \to \mathcal{X}$ *induces an isomorphism*

$$\mathrm{K}_0^{\Delta}(\mathcal{C}) \xrightarrow{\cong} \mathrm{K}_0^{\Delta}(\mathcal{X}) \oplus \mathrm{K}_0^{\Delta}(\mathcal{Y})$$

In particular $\mathrm{K}_0^{\Delta}(\mathcal{C}) \cong \mathrm{K}_0^{\Delta}(\mathcal{X})$ *if* $\mathcal{Y} \subseteq \mathcal{P}^{<\infty}(\mathcal{C})$.

(2) *Let* $(\mathcal{X}, \mathcal{Y})$ *be a cohereditary torsion pair in* $\mathcal{C}$. *Then the reflection functor* $\mathbf{L} : \mathcal{C} \to \mathcal{Y}$ *induces an isomorphism*

$$\mathrm{K}_0^{\nabla}(\mathcal{C}) \xrightarrow{\cong} \mathrm{K}_0^{\nabla}(\mathcal{X}) \oplus \mathrm{K}_0^{\nabla}(\mathcal{Y})$$

In particular $\mathrm{K}_0^{\nabla}(\mathcal{C}) \cong \mathrm{K}_0^{\nabla}(\mathcal{Y})$ *if* $\mathcal{X} \subseteq \mathcal{I}^{<\infty}(\mathcal{C})$.

PROOF. We prove (1) since (2) is dual. By a well-known result of Grothendieck the exact sequence of triangulated categories in Proposition 5.3 induces an exact sequence $\mathrm{K}_0(\mathcal{T}_l(\mathcal{Y})) \to \mathrm{K}_0(\mathcal{T}_l(\mathcal{C})) \to \mathrm{K}_0(\mathcal{T}_l(\mathcal{X})) \to 0$ which is a split short exact sequence, since the functor $\mathbf{j}^*$ admits a left adjoint. Now the assertion follows from a result of [**20**] which says that the Grothendieck group is invariant under stabilization. □

The first part of the following consequence was first observed in [**89**].

COROLLARY 5.8. (1) *Let* $(\mathcal{X}, \mathcal{Y})$ *be a hereditary torsion pair in a triangulated category* $\mathcal{C}$. *Then the assignment* $[C] \longmapsto ([\mathbf{R}(C)], [\mathbf{L}(C)])$ *induces an isomorphism*

$$\mathrm{K}_0(\mathcal{C}) \xrightarrow{\cong} \mathrm{K}_0(\mathcal{X}) \oplus \mathrm{K}_0(\mathcal{Y}).$$

(2) *Let* $(\mathcal{X}, \mathcal{Y}, \mathcal{Z})$ *be a TTF-triple in a pretriangulated category* $\mathcal{C}$. *If* $\mathcal{X}$ *and* $\mathcal{Z}$ *are triangulated subcategories of* $\mathcal{C}$ *and* $\mathcal{P}^{<\infty}(\mathcal{C}) \supseteq \mathcal{Y} \subseteq \mathcal{I}^{<\infty}(\mathcal{C})$, *then we have isomorphisms:*

$$\mathrm{K}_0(\mathcal{X}) \xrightarrow{\cong} \mathrm{K}_0^{\Delta}(\mathcal{C}) \xrightarrow{\cong} \mathrm{K}_0^{\nabla}(\mathcal{C}) \xleftarrow{\cong} \mathrm{K}_0(\mathcal{Z}).$$

CHAPTER III

Compactly Generated Torsion Pairs in Triangulated Categories

Throughout this chapter we fix a triangulated category $\mathcal{C}$. Our purpose here is to develop a technique for constructing torsion pairs in $\mathcal{C}$ induced by a given set of objects, and to give some applications to tilting theory by investigating the heart. In particular we are interested in finding conditions such that the heart is a module category. This will be important in connection with tilting theory in derived categories and the construction of derived equivalences. As it happens in the case of abelian categories, for the construction of torsion theories generated by a set of objects, we have to assume that the category $\mathcal{C}$ is big enough to allow certain infinite constructions.

1. Torsion Pairs of Finite Type

It is well-known that an important and well-behaved class of torsion pairs in an abelian category $\mathcal{G}$ with exact filtered colimits, is the class of torsion pairs of finite type. We recall from [**78**] that a torsion pair $(\mathcal{T}, \mathcal{F})$ in $\mathcal{G}$ is said to be of finite type if the torsion free class $\mathcal{F}$ is closed under filtered colimits (in case $\mathcal{G}$ is locally coherent this is equivalent to the condition that the torsion subfunctor commutes with filtered colimits). These torsion pairs are important in connection with the Ziegler spectrum and the localization theory of abelian categories, see [**78**] for a detailed analysis from a representation theoretic point of view.

Filtered colimits rarely exist in a triangulated category. However all small coproducts, which are special filtered colimits, exist in many interesting triangulated categories, for instance the unbounded derived category of an abelian category with exact coproducts, the stable homotopy category of spectra, and the stable module category of a group algebra. So we are led to introduce and investigate the analogous concept of torsion theories of finite type in triangulated categories.

Throughout this section $\mathcal{C}$ denotes a triangulated category containing all small coproducts. In analogy with the abelian case we make the following definition.

DEFINITION 1.1. A torsion pair $(\mathcal{X}, \mathcal{Y})$ in $\mathcal{C}$ is said to be **of finite type** if the torsion free class $\mathcal{Y}$ is closed under all small coproducts.

Let $(\mathcal{X}, \mathcal{Y})$ be a torsion pair in $\mathcal{C}$, let $\mathbf{R} : \mathcal{C} \to \mathcal{X}$ be the right adjoint of the inclusion $\mathbf{i} : \mathcal{X} \hookrightarrow \mathcal{C}$, and let $\mathbf{L} : \mathcal{C} \to \mathcal{Y}$ be the left adjoint of the inclusion $\mathbf{j} : \mathcal{Y} \hookrightarrow \mathcal{C}$. The following criterion will be useful.

LEMMA 1.2. (1) *The torsion pair* $(\mathcal{X},\mathcal{Y})$ *is of finite type if and only if* $\mathbf{R}$ *preserves coproducts. In this case the functors* $\mathbf{i}:\mathcal{X}\hookrightarrow\mathcal{C}$ *and* $\mathbf{L}:\mathcal{C}\to\mathcal{Y}$ *preserve compact objects.*

(2) $\mathbf{L}$ *preserves products if and only if* $\mathcal{X}$ *is closed under products.*

PROOF. (1) If $\mathbf{R}$ preserves coproducts and $\{Y_i; i\in I\}$ is a set of objects of $\mathcal{Y}$, then: $\mathcal{C}(\mathcal{X},\oplus_{i\in I}Y_i)\cong\mathcal{C}(\mathcal{X},\mathbf{R}(\oplus_{i\in I}Y_i)) = \mathcal{C}(\mathcal{X},\oplus_{i\in I}\mathbf{R}(Y_i)) = 0$. Hence $\oplus_{i\in I}Y_i\in\mathcal{X}^{\perp}=\mathcal{Y}$ and $\mathcal{Y}$ is closed under coproducts. Conversely if $\mathcal{Y}$ is closed under coproducts and $\{C_i; i\in I\}$ is a set of objects of $\mathcal{C}$, then consider the triangle $\oplus_{i\in I}\mathbf{R}(C_i)\xrightarrow{\oplus f_i}\oplus_{i\in I}C_i\xrightarrow{\oplus g_i}\oplus_{i\in I}\mathbf{L}(C_i)\xrightarrow{\oplus h_i}\Sigma(\oplus_{i\in I}\mathbf{R}(C_i))$ which is a coproduct of the triangles $\mathbf{R}(C_i)\xrightarrow{f_i}C_i\xrightarrow{g_i}\mathbf{L}(C_i)\xrightarrow{h_i}\Sigma\mathbf{R}(C_i)$. Since $\mathcal{Y}$ is closed under coproducts, we have $\oplus_{i\in I}\mathbf{L}(C_i)\in\mathcal{Y}$. Then any morphism $\alpha: X\to\oplus_{i\in I}C_i$ with $X\in\mathcal{X}$ factors through $\oplus f_i$ since $\alpha\circ\oplus g_i=0$. Moreover the factorization is unique since for any two of them, their difference factorizes through $\Sigma^{-1}(\oplus_{i\in I}\mathbf{L}(C_i))\in\mathcal{Y}$, so it is zero. We infer that $\oplus_{i\in I}\mathbf{R}(C_i)$ is the coreflection of $\oplus_{i\in I}C_i$ in $\mathcal{X}$. This implies that the canonical morphism $\oplus_{i\in I}\mathbf{R}(C_i)\to\mathbf{R}(\oplus_{i\in I}C_i)$ is invertible.

If X is compact in $\mathcal{X}$, and $\{C_i \mid i\in I\}$ is a family of objects in $\mathcal{C}$, then the following isomorphisms show that X is compact in $\mathcal{C}$:

$$\oplus_{i\in I}\mathcal{C}(X,C_i)\xrightarrow{\cong}\oplus_{i\in I}\mathcal{X}(X,\mathbf{R}(C_i))\xrightarrow{\cong}\mathcal{X}(X,\oplus_{i\in I}\mathbf{R}(C_i))$$
$$\xrightarrow{\cong}\mathcal{X}(X,\mathbf{R}(\oplus_{i\in I}C_i))\xrightarrow{\cong}\mathcal{C}(X,\oplus_{i\in I}C_i).$$

Now since $\mathcal{Y}$ is closed under coproducts, the inclusion $\mathbf{j}:\mathcal{Y}\hookrightarrow\mathcal{C}$ preserves coproducts. Then working as above we have that the reflection functor $\mathbf{L}$ preserves compact objects.

The proof of (2) is dual and is left to the reader. □

NOTE. It is easy to see that the above proof works in any pretriangulated category with coproducts, resp. products. This more general result will be useful in the study of the torsion pair induced by Cohen-Macaulay objects in the stable category of a Nakayama category, see the last chapter.

EXAMPLE. Let $(\mathcal{T},\mathcal{F})$ be a torsion pair in a Grothendieck category $\mathcal{G}$ and assume that $(\mathcal{T},\mathcal{F})$ is of finite type, or more generally that $\mathcal{F}$ is closed under coproducts. The tilting construction of Section I.3 (which as easily seen works for unbounded complexes) produces a torsion pair of finite type $\big(\mathcal{X}(\mathcal{T}),\mathcal{Y}(\mathcal{F})\big)$ in the unbounded derived category $\mathbf{D}(\mathcal{G})$. For instance let G be a compact object in $\mathcal{G}$ and let $(\mathcal{T},\mathcal{F})$ be the torsion pair in $\mathcal{G}$ generated by G, i.e. $\mathcal{F}=G^{\perp}$ and $\mathcal{T}={}^{\perp}(G^{\perp})$. Then the torsion pair $\big(\mathcal{X}(\mathcal{T}),\mathcal{Y}(\mathcal{F})\big)$ in $\mathbf{D}(\mathcal{G})$ is of finite type.

It would be very interesting to have a classification of all torsion pairs of finite type in $\mathcal{C}$. This seems to be very difficult; see the remarks in the last section of the next chapter. But already the special class of torsion pairs generated by compact objects is very interesting, and has important applications. So we concentrate on this case.

2. Compactly Generated Torsion Pairs

Throughout this section we fix a triangulated category $\mathcal{C}$. Our purpose here is to construct a torsion pair in $\mathcal{C}$ generated by a set of objects, and to give a handy description of the torsion or the torsion-free class. This will be important in connection with having a nice description of the heart in the next section.

We fix a set of objects $\mathcal{P} \subseteq \mathcal{C}$. To define a torsion pair $(\mathcal{X}_{\mathcal{P}}, \mathcal{Y}_{\mathcal{P}})$ in $\mathcal{C}$ generated by $\mathcal{P}$ it is necessary by Proposition II.3.3 that $\mathcal{X}_{\mathcal{P}}$ is right triangulated and $\mathcal{Y}_{\mathcal{P}}$ is left triangulated. As in the case of torsion pairs in abelian categories it is natural to define the pair $(\mathcal{X}_{\mathcal{P}}, \mathcal{Y}_{\mathcal{P}})$ as follows (we use non-negative suspensions in the definition of $\mathcal{Y}_{\mathcal{P}}$ since a torsion-free class is closed under non-positive suspensions):

$$\mathcal{Y}_{\mathcal{P}} := \{C \in \mathcal{C} \mid \mathcal{C}(\Sigma^n(P), C) = 0, \forall P \in \mathcal{P}, \forall n \geq 0\} \quad \text{and} \quad \mathcal{X}_{\mathcal{P}} := {}^{\perp}\mathcal{Y}_{\mathcal{P}}.$$

The above definition is also similar to the definition of the natural torsion pair (= t-structure) in the derived category of a module category $\mathrm{Mod}(\Lambda)$, where the role of $\mathcal{P}$ is played by the stalk complex Λ concentrated in degree zero. We are interested in finding sufficient conditions for the pair $(\mathcal{X}_{\mathcal{P}}, \mathcal{Y}_{\mathcal{P}})$ to be a torsion pair in $\mathcal{C}$. First we record the following immediate consequence of the above definition.

LEMMA 2.1. *$\mathcal{X}_{\mathcal{P}}$ is a right triangulated subcategory of $\mathcal{C}$ closed under coproducts and extensions, and $\mathcal{Y}_{\mathcal{P}}$ is a left triangulated subcategory of $\mathcal{C}$ closed under products and extensions.*

From now on we assume that $\mathcal{C}$ contains all small coproducts and the set $\mathcal{P}$ consists of compact objects of $\mathcal{C}$. Throughout we use the following notation:

$$\mathcal{Q} := \{\Sigma^n(P) \mid P \in \mathcal{P}, \ n \geq 0\}$$

so $\mathcal{Q}$ remains a set of compact objects and by construction $\mathcal{Q}$ is closed under Σ. We denote by $\mathrm{Add}(\mathcal{Q})$ the full subcategory of $\mathcal{C}$ consisting of direct summands of arbitrary coproducts of objects of $\mathcal{Q}$. Observe that $\mathrm{Add}(\mathcal{Q})^{\perp} = \mathcal{Q}^{\perp}$.

The above setup and notation will be fixed throughout this chapter.

We begin with the following result which will be useful for the construction of coreflections of objects of $\mathcal{C}$ in $\mathcal{X}_{\mathcal{P}}$.

LEMMA 2.2. $\mathrm{Add}(\mathcal{Q})$ *is contravariantly finite in $\mathcal{C}$ and for any $C \in \mathcal{C}$, there exists a triangle $Q_0 \xrightarrow{f_0} C \xrightarrow{g_0} Y_0 \xrightarrow{h_0} \Sigma(Q_0)$ in $\mathcal{C}$ with the following properties:*

(i) *f_0 is a right $\mathrm{Add}(\mathcal{Q})$-approximation of C.*
(ii) *$0 \to \mathcal{C}(\mathcal{Q}, \Sigma^{-n-1}(Y_0)) \to \mathcal{C}(\mathcal{Q}, \Sigma^{-n}(Q_0)) \to \mathcal{C}(\mathcal{Q}, \Sigma^{-n}(C)) \to 0$ is a short exact sequence, $\forall n \geq 0$.*
(iii) *The morphism $\mathcal{C}(\Sigma^n(g_0), \mathcal{Y}_{\mathcal{P}}) : \mathcal{C}(\Sigma^n(Y_0), \mathcal{Y}_{\mathcal{P}}) \to \mathcal{C}(\Sigma^n(C), \mathcal{Y}_{\mathcal{P}})$ is invertible, $\forall n \geq 0$.*

PROOF. For any $C \in \mathcal{C}$, consider the set of morphisms $I_C := \{P \to C \mid P \in \mathcal{Q}\}$ and let $Q_0 = \oplus_{\lambda \in I_C} P_\lambda$. The set I_C induces a morphism $f_0 : Q_0 \to C$ which by construction has the property that $(\mathcal{Q}, f_0)$ is epic. This implies that (A, f_0) is epic, $\forall A \in \mathrm{Add}(\mathcal{Q})$, i.e. f_0 is a right $\mathrm{Add}(\mathcal{Q})$-approximation of C. If (T) : $Q_0 \xrightarrow{f_0} C \to Y_0 \to \Sigma(Q_0)$ is a triangle in $\mathcal{C}$, then the morphism $\mathcal{C}(\mathcal{Q}, \Sigma^{-n}(f_0))$: $\mathcal{C}(\mathcal{Q}, \Sigma^{-n}(Q_0)) \to \mathcal{C}(\mathcal{Q}, \Sigma^{-n}(C))$ is isomorphic to $\mathcal{C}(\Sigma^n(\mathcal{Q}), f_0) : \mathcal{C}(\Sigma^n(\mathcal{Q}), Q_0) \to \mathcal{C}(\Sigma^n(\mathcal{Q}), C)$, $\forall n \geq 0$. Hence assertion (ii) follows if we apply $\mathcal{C}(\mathcal{Q}, -)$ to (T), using that $\Sigma(\mathcal{Q}) \subseteq \mathcal{Q}$. Since $\mathcal{Y}_{\mathcal{P}} = \mathcal{Q}^{\perp}$, applying $\mathcal{C}(-, \mathcal{Y}_{\mathcal{P}})$ to the triangle (T), the last assertion follows. □

In the following we use a version of Bousfield's localization, called finite localization, popularized by Miller [**82**] in case $\mathcal{C}$ is the stable homotopy category of spectra, and the constructed torsion or torsion-free class is closed under Σ and Σ^{-1}. Our result is more general since in our case $\mathcal{C}$ is an arbitrary triangulated category with coproducts, and the torsion class we construct is closed only under Σ, equivalently the torsion-free class is closed only under Σ^{-1}.

THEOREM 2.3. *Let $\mathcal{C}$ be a triangulated category with coproducts and let $\mathcal{P} \subseteq \mathcal{C}$ be a set of compact objects. Then $(\mathcal{X}_{\mathcal{P}}, \mathcal{Y}_{\mathcal{P}})$ is a torsion pair of finite type in $\mathcal{C}$, where as before $\mathcal{Y}_{\mathcal{P}} := \{\Sigma^n(\mathcal{P}) \mid n \geq 0\}^{\perp}$ and $\mathcal{X}_{\mathcal{P}} := {}^{\perp}\mathcal{Y}_{\mathcal{P}}$.*

PROOF. By Lemma 2.1, $\mathcal{X}_{\mathcal{P}}$ is a full subcategory of $\mathcal{C}$, closed under Σ and extensions, and $\mathcal{Y}_{\mathcal{P}}$ is a full subcategory of $\mathcal{C}$, closed under Σ^{-1} and extensions. Let $Q_0 \xrightarrow{f_0} C \xrightarrow{g_0} Y_0 \xrightarrow{h_0} \Sigma(Q_0)$ be the triangle constructed in Lemma 2.2. Inductively $\forall n \geq 0$, we can construct triangles $Q_n \xrightarrow{f_n} Y_{n-1} \xrightarrow{g_n} Y_n \xrightarrow{h_n} \Sigma(Q_n)$ in $\mathcal{C}$, where $Y_{-1} = C$ and f_n is a right $\mathrm{Add}(\mathcal{Q})$−approximation of Y_{n-1} satisfying the properties of Lemma 2.2. Consider the tower of objects

$$C \xrightarrow{g_0} Y_0 \xrightarrow{g_1} Y_1 \xrightarrow{g_2} Y_2 \longrightarrow \cdots \longrightarrow Y_n \xrightarrow{g_{n+1}} Y_{n+1} \longrightarrow \cdots \tag{II}$$

and let $\underrightarrow{\mathrm{holim}} Y_n$ be its *homotopy colimit* which is defined by the triangle

$$\bigoplus_{n\geq 0} Y_n \xrightarrow{1-g} \bigoplus_{n\geq 0} Y_n \xrightarrow{\alpha} \underrightarrow{\mathrm{holim}} Y_n \xrightarrow{\beta} \Sigma \bigoplus_{n\geq 0} Y_n$$

where the morphism $1-g$ is induced by $Y_{n-1} \xrightarrow{(1_{Y_{n-1}}, -g_n)} Y_{n-1} \oplus Y_n \hookrightarrow \oplus_{n\geq 0} Y_n$. Since $\mathcal{Q}$ consists of compact objects, it is well-known that the canonical morphism $\underrightarrow{\lim}\, \mathcal{C}(\mathcal{Q}, Y_n) \to \mathcal{C}(\mathcal{Q}, \underrightarrow{\mathrm{holim}} Y_n)$ is invertible [**21**]. Since by construction $\mathcal{C}(\mathcal{Q}, g_n) = 0, \forall n \geq 0$, the short exact sequence

$$0 \longrightarrow \bigoplus_{n\geq 0} \mathcal{C}(\mathcal{Q}, Y_n) \xrightarrow{1-(\mathcal{Q},g)} \bigoplus_{n\geq 0} \mathcal{C}(\mathcal{Q}, Y_n) \longrightarrow \underrightarrow{\lim}\, \mathcal{C}(\mathcal{Q}, Y_n) \longrightarrow 0$$

shows that $0 = \underrightarrow{\lim}\, \mathcal{C}(\mathcal{Q}, Y_n) \cong \mathcal{C}(\mathcal{Q}, \underrightarrow{\mathrm{holim}} Y_n)$. Hence $\underrightarrow{\mathrm{holim}} Y_n \in \mathcal{Y}_{\mathcal{P}}$, and since $\mathcal{Y}_{\mathcal{P}}$ is closed under Σ^{-1}, we have that $\Sigma^{-1}\underrightarrow{\mathrm{holim}} Y_n \in \mathcal{Y}_{\mathcal{P}}$. The tower of objects (II) induces a morphism $g_C : C \to \underrightarrow{\mathrm{holim}} Y_n$. Let

$$\Sigma^{-1}\underrightarrow{\mathrm{holim}} Y_n \xrightarrow{h_C} X \xrightarrow{f_C} C \xrightarrow{g_C} \underrightarrow{\mathrm{holim}} Y_n \tag{$*$}$$

be a triangle in $\mathcal{C}$. We claim that g_C is the reflection of C in $\mathcal{Y}_{\mathcal{P}}$. By construction if we apply $\mathcal{C}(-, \mathcal{Y}_{\mathcal{P}})$ to the tower of objects (II), we get an inverse tower $\cdots \to \mathcal{C}(Y_2, \mathcal{Y}_{\mathcal{P}}) \xrightarrow{(g_2, \mathcal{Y}_{\mathcal{P}})} \mathcal{C}(Y_1, \mathcal{Y}_{\mathcal{P}}) \xrightarrow{(g_1, \mathcal{Y}_{\mathcal{P}})} \mathcal{C}(Y_0, \mathcal{Y}_{\mathcal{P}}) \xrightarrow{(g_0, \mathcal{Y}_{\mathcal{P}})} \mathcal{C}(C, \mathcal{Y}_{\mathcal{P}})$ where the involved morphisms are invertible. Hence $\underleftarrow{\lim}\, \mathcal{C}(Y_n, \mathcal{Y}_{\mathcal{P}}) \cong \mathcal{C}(C, \mathcal{Y}_{\mathcal{P}})$. Now by [**21**], we have a short exact sequence

$$0 \longrightarrow \underleftarrow{\lim}^{(1)} \mathcal{C}(\Sigma(Y_n), \mathcal{Y}_{\mathcal{P}}) \longrightarrow \mathcal{C}(\underrightarrow{\mathrm{holim}} Y_n, \mathcal{Y}_{\mathcal{P}}) \longrightarrow \underleftarrow{\lim}\, \mathcal{C}(Y_n, \mathcal{Y}_{\mathcal{P}}) \longrightarrow 0$$

where $\underleftarrow{\lim}^{(1)}$ denotes the first right derived functor of the inverse limit functor [**104**]. By part (iii) of Lemma 2.2, in the inverse system $\cdots \to \mathcal{C}(\Sigma(Y_2), \mathcal{Y}_{\mathcal{P}}) \xrightarrow{(\Sigma(g_2), \mathcal{Y}_{\mathcal{P}})}$

$\mathcal{C}(\Sigma(Y_1), \mathcal{Y}_\mathcal{P}) \xrightarrow{(\Sigma(g_1),\mathcal{Y}_\mathcal{P})} \mathcal{C}(\Sigma(Y_0), \mathcal{Y}_\mathcal{P})$ the involved morphisms are invertible. Then by the Mittag-Leffler condition [**104**], we have $\varprojlim^{(1)} \mathcal{C}(\Sigma(Y_n), \mathcal{Y}_\mathcal{P}) = 0$. Hence $\mathcal{C}(\underrightarrow{\text{holim}} Y_n, \mathcal{Y}_\mathcal{P}) \cong \varprojlim \mathcal{C}(Y_n, \mathcal{Y}_\mathcal{P}) \cong \mathcal{C}(C, \mathcal{Y}_\mathcal{P})$. We infer that $g_C : C \to \underrightarrow{\text{holim}} Y_n$ induces an isomorphism $\mathcal{C}(g_C, \mathcal{Y}_\mathcal{P}) : \mathcal{C}(\underrightarrow{\text{holim}} Y_n, \mathcal{Y}_\mathcal{P}) \xrightarrow{\cong} \mathcal{C}(C, \mathcal{Y}_\mathcal{P})$, i.e. g_C is the reflection of $C \in \mathcal{Y}_\mathcal{P}$. Hence by Proposition I.2.3, $f_C : X \to C$ is the coreflection of C in $\mathcal{X}_\mathcal{P}$ and then $(\mathcal{X}_\mathcal{P}, \mathcal{Y}_\mathcal{P})$ is a torsion pair in $\mathcal{C}$. Since $\mathcal{Y}_\mathcal{P} = \mathcal{Q}^\perp$ and $\mathcal{Q}$ consists of compact objects, it follows trivially that $\mathcal{Y}_\mathcal{P}$ is closed under coproducts, so $(\mathcal{X}_\mathcal{P}, \mathcal{Y}_\mathcal{P})$ is of finite type. □

Since compact objects in $\mathcal{C}$ remain compact in $\mathcal{X}_\mathcal{P}$, it follows that any object in $\mathcal{P}$ is compact in $\mathcal{X}_\mathcal{P}$. The following observation shows that it is necessary to start with a set $\mathcal{P}$ of compact objects in $\mathcal{C}$ in order to get a torsion pair of finite type $(\mathcal{X}_\mathcal{P}, \mathcal{Y}_\mathcal{P})$ in $\mathcal{C}$ such that $\mathcal{P}$ consists of compact objects in the torsion class.

NOTE. If $\mathcal{P}$ is a set of, not necessarily compact, objects in $\mathcal{C}$, then $\mathcal{P} \subseteq \mathcal{X}_\mathcal{P}$ and the following are equivalent for the pair $(\mathcal{X}_\mathcal{P}, \mathcal{Y}_\mathcal{P})$, where as before $\mathcal{Y}_\mathcal{P} := \{\Sigma^n(\mathcal{P}) \mid n \geq 0\}^\perp$ and $\mathcal{X}_\mathcal{P} := {}^\perp\mathcal{Y}_\mathcal{P}$.

(i) $\mathcal{P}$ is a set of compact objects in $\mathcal{C}$.
(ii) $(\mathcal{X}_\mathcal{P}, \mathcal{Y}_\mathcal{P})$ is a torsion pair of finite type in $\mathcal{C}$ and $\mathcal{P}$ is a set of compact objects in $\mathcal{X}_\mathcal{P}$.

That (i) implies (ii) follows from Theorem 2.3. If (ii) holds, then let P be in $\mathcal{P}$ and let $\{C_i \mid i \in I\}$ be a family of objects in $\mathcal{C}$. Since the torsion pair $(\mathcal{X}_\mathcal{P}, \mathcal{Y}_\mathcal{P})$ is of finite type, it follows that the torsion-free class $\mathcal{Y}_\mathcal{P}$ is closed under coproducts. This implies that in the standard triangle $X_{\oplus_{i\in I} C_i} \to \oplus_{i\in I} C_i \to Y^{\oplus_{i\in I} C_i} \to \Sigma(X_{\oplus_{i\in I} C_i})$, we have: $X_{\oplus_{i\in I} C_i} \xrightarrow{\cong} \oplus_{i\in I} X_{C_i}$ and $Y^{\oplus_{i\in I} C_i} \xrightarrow{\cong} \oplus_{i\in I} Y^{C_i}$. Hence the above triangle is isomorphic to the triangle $\oplus_{i\in I} X_{C_i} \to \oplus_{i\in I} C_i \to \oplus_{i\in I} Y^{C_i} \to \Sigma(\oplus_{i\in I} X_{C_i})$. Applying the functor $\mathcal{C}(P, -)$ to this triangle, we get the following commutative diagram

$$\begin{array}{ccc} \bigoplus_{i\in I} \mathcal{C}(P, X_{C_i}) & \longrightarrow & \bigoplus_{i\in I} \mathcal{C}(P, C_i) \\ \downarrow & & \downarrow \\ \mathcal{C}(P, \bigoplus_{i\in I} X_{C_i}) & \longrightarrow & \mathcal{C}(P, \bigoplus_{i\in I} C_i) \end{array}$$

where the horizontal maps are invertible. Since, by hypothesis, P is compact in $\mathcal{X}_\mathcal{P}$, it follows that the left vertical, hence the right vertical, map is invertible. Hence P is compact in $\mathcal{C}$.

If $\mathcal{C}$ is the unbounded derived category of right modules over a ring, and P is a set of, not necessarily perfect, complexes of modules which is closed under positive and negative suspensions, then it is shown in [**1**] that one gets a torsion pair $(\mathcal{X}_P, \mathcal{Y}_P)$ in $\mathcal{C}$. Of course the torsion pair $(\mathcal{X}_P, \mathcal{Y}_P)$ need not be of finite type. See also [**2**] for related results.

It is now natural to make the following definition.

DEFINITION 2.4. A torsion pair $(\mathcal{X}, \mathcal{Y})$ in $\mathcal{C}$ is **generated by a set of compact objects** or it is **compactly generated** if there exists a set of compact objects $\mathcal{P}$ in $\mathcal{C}$ such that $\mathcal{Y} = \{\Sigma^n(\mathcal{P}) \mid n \geq 0\}^\perp$ and $\mathcal{X} = {}^\perp\mathcal{Y}$. Such a torsion pair is always denoted by $(\mathcal{X}_\mathcal{P}, \mathcal{Y}_\mathcal{P})$.

By the above theorem it follows in particular that any compact object $T \in \mathcal{C}$ generates a torsion pair $(\mathcal{X}_T, \mathcal{Y}_T)$ of finite type in $\mathcal{C}$.

We are interested in having a convenient description of the torsion class $\mathcal{X}_\mathcal{P}$ in the torsion pair $(\mathcal{X}_\mathcal{P}, \mathcal{Y}_\mathcal{P})$ in $\mathcal{C}$ generated by a set compact objects $\mathcal{P}$. We denote by $\mathrm{Loc}^+(\mathcal{P})$ the smallest full subcategory of $\mathcal{C}$, containing $\mathcal{P}$, which is closed under Σ, extensions and coproducts. Then automatically $\mathrm{Loc}^+(\mathcal{P})$ is a right triangulated subcategory of $\mathcal{C}$. Clearly the torsion class $\mathcal{X}_\mathcal{P}$ contains $\mathrm{Loc}^+(\mathcal{P})$. It is useful to know under what conditions we have $\mathrm{Loc}^+(\mathcal{P}) = \mathcal{X}_\mathcal{P}$. We can prove that the equality holds under an additional assumption. For a full subcategory $\mathcal{U}$ of $\mathcal{C}$ we denote by $\mathcal{U}^{\star n} = \mathcal{U} \star \mathcal{U} \star \cdots \star \mathcal{U}$ ($n-$factors) the category of n-extensions of $\mathcal{U}$ by $\mathcal{U}$.

We need the following easy observation whose direct proof is left to the reader.

LEMMA 2.5. *If $\mathcal{C}(\mathcal{P}, \Sigma^t(\mathcal{P})) = 0, \forall t \geq 1$, then $\mathcal{C}(\mathcal{P}, \Sigma^t(C)) = 0, \forall t \geq 1$, for any $C \in \mathrm{Add}(\mathcal{Q})^{\star n}$ and $n \geq 1$.*

Using this lemma we get a description of the torsion class $\mathcal{X}_\mathcal{P}$ under an additional assumption. This assumption can be dropped whenever the triangulated category $\mathcal{C}$ admits a model, see Remark 2.7 below.

PROPOSITION 2.6. *If $\mathcal{C}(\mathcal{P}, \Sigma^n(\mathcal{P})) = 0, \forall n \geq 1$, then $\mathcal{X}_\mathcal{P} = \mathrm{Loc}^+(\mathcal{P})$.*

PROOF. Since $\mathcal{X}_\mathcal{P}$ is closed under coproducts, extensions, Σ and contains $\mathcal{P}$, it follows by construction that $\mathrm{Loc}^+(\mathcal{P}) \subseteq \mathcal{X}_\mathcal{P}$. Let $C \in \mathcal{C}$ and consider the triangles $Q_n \xrightarrow{f_n} Y_{n-1} \xrightarrow{g_n} Y_n \to \Sigma(Q_n)$, $n \geq 0$ constructed in Theorem 2.3, where $C = Y_{-1}$. Setting $\chi_0 = f_0$ and $Q_0 = T_0$, we can construct inductively a tower of objects $T_0 \xrightarrow{\tau_0} T_1 \xrightarrow{\tau_1} T_2 \to \cdots$ and the following tower of triangles:

$$
\begin{array}{ccccccc}
T_0 & \xrightarrow{\chi_0} & C & \xrightarrow{g_0} & Y_0 & \xrightarrow{h^C} & \Sigma(T_0) \\
{\scriptstyle \tau_0}\downarrow & & \| & & {\scriptstyle g_1}\downarrow & & {\scriptstyle \Sigma(\tau_0)}\downarrow \\
T_1 & \xrightarrow{\chi_1} & C & \xrightarrow{g_0 \circ g_1} & Y_1 & \longrightarrow & \Sigma(T_1) \\
{\scriptstyle \tau_1}\downarrow & & \| & & {\scriptstyle g_2}\downarrow & & {\scriptstyle \Sigma(\tau_1)}\downarrow \\
T_2 & \xrightarrow{\chi_2} & C & \xrightarrow{g_0 \circ g_1 \circ g_2} & Y_2 & \longrightarrow & \Sigma(T_2) \\
\downarrow & & \| & & \downarrow & & \downarrow \\
\vdots & & \vdots & & \vdots & & \vdots
\end{array}
$$

Using the Octahedral Axiom in each step it is not difficult to see that each object T_n lies in $\mathrm{Add}(\mathcal{Q})^{\star n}$ [**21**]. Consider the homotopy colimit $\underrightarrow{\mathrm{holim}} T_n$ of the left vertical tower, which is equipped with a morphism $\chi_C : \underrightarrow{\mathrm{holim}} T_n \to C$ induced by the above diagram, and let

$$\underrightarrow{\mathrm{holim}} T_n \xrightarrow{\chi_C} C \xrightarrow{g} \widehat{C} \xrightarrow{h} \Sigma(\underrightarrow{\mathrm{holim}} T_n) \tag{T}$$

be a triangle in $\mathcal{C}$. By construction $\underrightarrow{\mathrm{holim}} T_n$ lies in $\mathrm{Loc}^+(\mathcal{P})$. Using Lemma 2.2, we have short exact sequences $0 \to \mathcal{C}(\mathcal{Q}, \Sigma^{-1}(Y_n)) \to \mathcal{C}(\mathcal{Q}, T_n) \to \mathcal{C}(\mathcal{Q}, C) \to 0$, $\forall n \geq 0$. Taking direct limits we deduce a short exact sequence $0 \to \varinjlim \mathcal{C}(\mathcal{Q}, \Sigma^{-1}(Y_n)) \to \varinjlim \mathcal{C}(\mathcal{Q}, T_n) \to \mathcal{C}(\mathcal{Q}, C) \to 0$. Since $\mathcal{Q}$ consists of compact objects, we have

$\varinjlim \mathcal{C}(\mathcal{Q}, \Sigma^{-1}(Y_n)) \cong \mathcal{C}(\mathcal{Q}, \underrightarrow{\mathrm{holim}}\Sigma^{-1}(Y_n)) = 0$ and $\varinjlim \mathcal{C}(\mathcal{Q}, T_n) \cong \mathcal{C}(\mathcal{Q}, \underrightarrow{\mathrm{holim}}T_n)$. It follows that the morphism χ_C induces an isomorphism $\mathcal{C}(\mathcal{Q}, \chi_C) : \mathcal{C}(\mathcal{Q}, \underrightarrow{\mathrm{holim}}T_n) \to \mathcal{C}(\mathcal{Q}, C)$. Since $\Sigma(\mathcal{Q}) \subseteq \mathcal{Q}$, it follows that $\mathcal{C}(\mathcal{Q}, \Sigma^{-1}(\chi_C)) : \mathcal{C}(\mathcal{Q}, \Sigma^{-1}(\underrightarrow{\mathrm{holim}}T_n)) \to \mathcal{C}(\mathcal{Q}, \Sigma^{-1}(C))$ is invertible. This implies that $\mathcal{C}(\mathcal{Q}, \Sigma^{-1}(\widehat{C})) = 0$, i.e. $\widehat{C} \in \Sigma(\mathcal{Y}_{\mathcal{P}})$.

Assume now that $C \in \mathcal{X}_{\mathcal{P}}$. Then $\widehat{C} \in \mathcal{X}_{\mathcal{P}}$ since $\underrightarrow{\mathrm{holim}}T_n$ lies in $\mathcal{X}_{\mathcal{P}}$ and the latter is right triangulated. It follows that $\widehat{C} \in \mathcal{X}_{\mathcal{P}} \cap \Sigma(\mathcal{Y}_{\mathcal{P}})$. Applying the functor $\mathcal{C}(\mathcal{P}, -)$ to the triangle (T), we have an exact sequence $\mathcal{C}(\mathcal{P}, \underrightarrow{\mathrm{holim}}T_n) \to \mathcal{C}(\mathcal{P}, C) \to \mathcal{C}(\mathcal{P}, \widehat{C}) \to \mathcal{C}(\mathcal{P}, \Sigma(\underrightarrow{\mathrm{holim}}T_n))$. Since $\mathcal{P}$ consists of compact objects we have $\varinjlim \mathcal{C}(\mathcal{P}, \Sigma(T_n)) \cong \mathcal{C}(\mathcal{P}, \Sigma(\underrightarrow{\mathrm{holim}}T_n))$, and by Lemma 2.5, $\mathcal{C}(\mathcal{P}, \Sigma(\underrightarrow{\mathrm{holim}}T_n)) = 0$. Since $\mathcal{C}(\mathcal{P}, \chi_C)$ is invertible we infer that $\mathcal{C}(\mathcal{P}, \widehat{C}) = 0$, hence $\widehat{C} \in \mathcal{Y}_{\mathcal{P}}$. Then $\widehat{C} = 0$, since $\widehat{C}$ lies also in $\mathcal{X}_{\mathcal{P}}$. Hence $C = \underrightarrow{\mathrm{holim}}T_n \in \mathrm{Loc}^+(\mathcal{P})$, i.e. $\mathrm{Loc}^+(\mathcal{P}) = \mathcal{X}_{\mathcal{P}}$. □

REMARK 2.7. The assumption $\mathcal{C}(\mathcal{P}, \Sigma^n(\mathcal{P})) = 0, \forall n \geq 1$, in Proposition 2.6 can be removed, in case $\mathcal{C}$ admits a model, for example if $\mathcal{C}$ is the stable module category of a quasi-Frobenius ring or the stable homotopy category of spectra or the unbounded derived category of an AB4 category, that is an abelian category with exact coproducts. Indeed in any of the above cases, the tower of triangles constructed above induces a triangle $\underrightarrow{\mathrm{holim}}T_n \to C \to \underrightarrow{\mathrm{holim}}Y_n \to \Sigma(\underrightarrow{\mathrm{holim}}T_n)$ in $\mathcal{C}$ which is isomorphic to the triangle (T) above. In particular we have $\widehat{C} \cong \underrightarrow{\mathrm{holim}}Y_n$. Then as above we infer that if $C \in \mathcal{X}_{\mathcal{P}}$, then $C = \underrightarrow{\mathrm{holim}}T_n \in \mathrm{Loc}^+(\mathcal{P})$, since $\underrightarrow{\mathrm{holim}}Y_n \in \mathcal{Y}_{\mathcal{P}}$. It follows that $\mathrm{Loc}^+(\mathcal{P}) = \mathcal{X}_{\mathcal{P}}$. The problem is that in a general triangulated category it is not known if the homotopy colimit of a tower of triangles is a triangle, see [**21**].

Trivially the assumption can also be removed if $\mathcal{P}$ is closed under Σ^{-1}. In this case the torsion class $\mathcal{X}_{\mathcal{P}}$ coincides with the smallest triangulated subcategory of $\mathcal{C}$ which is closed under coproducts and contains $\mathcal{P}$.

Using the above result we get a description of the torsion class $\mathcal{X}_{\mathcal{P}}$, not only of the torsion-free class $\mathcal{Y}_{\mathcal{P}}$. This will be useful for a nice description of the heart of the torsion pair in the next section. First we recall from Chapter I that a set of objects $\mathcal{U}$ generates a triangulated category $\mathcal{C}$, or is a generating set, if $\{\Sigma^n(\mathcal{U}) \mid n \in \mathbb{Z}\}^{\perp} = 0$.

The following result gives an explicit description of the torsion class $\mathcal{X}_{\mathcal{P}}$, provided that $\mathcal{P}$ is a generating set and satisfies the assumption of Proposition 2.6. This will be useful in the next section for a handy description of the heart.

PROPOSITION 2.8. *If* $\mathcal{C}(\mathcal{P}, \Sigma^n(\mathcal{P})) = 0, \ \forall n \geq 1$, *then*

$$\mathcal{X}_{\mathcal{P}} \subseteq \{C \in \mathcal{C} \mid \mathcal{C}(\mathcal{P}, \Sigma^n(C)) = 0, \ \forall n \geq 1\}.$$

Moreover the following statements are equivalent:

(i) $\mathcal{P}$ *generates* $\mathcal{C}$ *and* $\mathcal{C}(\mathcal{P}, \Sigma^n(\mathcal{P})) = 0, \ \forall n \geq 1$.
(ii) $\mathcal{X}_{\mathcal{P}} = \{C \in \mathcal{C} \mid \mathcal{C}(\mathcal{P}, \Sigma^n(C)) = 0, \ \forall n \geq 1\}$.

PROOF. By Proposition 2.6, the hypothesis implies that $\mathcal{X}_{\mathcal{P}} = \mathrm{Loc}^+(\mathcal{P})$ is the smallest full right triangulated subcategory of $\mathcal{C}$ which is closed under coproducts and contains $\mathcal{P}$. Since obviously $\{C \in \mathcal{C} \mid \mathcal{C}(\mathcal{P}, \Sigma^n(C)) = 0, \ \forall n > 0\}$ is a right triangulated subcategory of $\mathcal{C}$, closed under coproducts and contains $\mathcal{P}$, it follows that $\mathcal{X}_{\mathcal{P}} \subseteq \{C \in \mathcal{C} \mid \mathcal{C}(\mathcal{P}, \Sigma^n(C)) = 0, \ \forall n \geq 1\}$.

(i) $\Rightarrow$ (ii) It suffices to show that $\{C \in \mathcal{C} \mid \mathcal{C}(\mathcal{P}, \Sigma^n(C)) = 0, \ \forall n \geq 1\} \subseteq \mathcal{X}_{\mathcal{P}}$. Let C be in $\mathcal{C}$ such that $\mathcal{C}(\mathcal{P}, \Sigma^t(C)) = 0, \forall t > 0$, and consider the glueing triangle $\mathbf{R}(C) \to C \to \mathbf{L}(C) \to \Sigma\mathbf{R}(C))$ in $\mathcal{C}$. By the proof of Proposition 2.6 we can write $\mathbf{R}(C) = \underrightarrow{\text{holim}} T_n$, where $T_n \in \text{Add}(\mathcal{Q})^{\star n}$. Applying $\mathcal{C}(\Sigma^{-t}(\mathcal{P}), -)$ to the above triangle and using Lemma 2.5, we see directly that $\mathcal{C}(\Sigma^{-t}(\mathcal{P}), \mathbf{L}(C)) = 0, \forall t > 0$. Since $\mathbf{L}(C) \in \mathcal{Y}_{\mathcal{P}}$, we infer that $\mathcal{C}(\mathcal{P}, \Sigma^t\mathbf{L}(C)) = 0, \forall t \in \mathbb{Z}$. Since $\mathcal{P}$ generates $\mathcal{C}$, we have $\mathbf{L}(C) = 0$. This implies that $C \cong \mathbf{R}(C) \in \mathcal{X}_{\mathcal{P}}$.

(ii) $\Rightarrow$ (i) Since $\mathcal{P} \subseteq \mathcal{X}_{\mathcal{P}}$, it suffices to show that $\mathcal{P}$ is a generating set. So let C be an object in $\mathcal{C}$ such that $\mathcal{C}(\mathcal{P}, \Sigma^t(C)) = 0, \ \forall t \in \mathbb{Z}$. Then obviously $C \in \mathcal{Y}_{\mathcal{P}}$. Moreover the description of $\mathcal{X}_{\mathcal{P}}$ implies that $C \in \mathcal{X}_{\mathcal{P}}$. Hence $C \in \mathcal{X}_{\mathcal{P}} \cap \mathcal{Y}_{\mathcal{P}} = 0$. We infer that $\mathcal{P}$ is a generating set. □

The following examples illustrate the above constructions.

Example. (1) Let $\mathcal{C} = \mathbf{D}(\text{Mod}(\Lambda))$ be the unbounded derived category of right Λ-modules over a ring Λ. Let $\mathcal{P} = \{\Lambda\}$ consist only of the regular module Λ concentrated in degree zero. Then $\mathcal{P}$ satisfies the conditions of Proposition 2.8, and the torsion pair generated by Λ is the natural torsion pair (= t-structure) in $\mathbf{D}(\text{Mod}(\Lambda))$. The torsion class coincides with the smallest right triangulated subcategory of $\mathbf{D}(\text{Mod}(\Lambda))$ which is closed under coproducts and contains Λ.

(2) Let $\text{Ho}(\mathcal{S}p)$ be the stable homotopy category of spectra and let $\mathcal{P} = \{S^0\}$ consist only of the sphere spectrum. It is well-known that S^0 is compact and $\{S^0\}$ satisfies the conditions of Proposition 2.8 [**81**]. Then the torsion pair generated by S^0 is the natural torsion pair (= t-structure) in $\text{Ho}(\mathcal{S}p)$. The torsion class $\mathcal{X}_{\mathcal{P}}$ is known in algebraic topology as the category of connective spectra.

3. The Heart of a Compactly Generated Torsion Pair

We continue to assume that $\mathcal{C}$ is a triangulated category with small coproducts. In this section we investigate the heart of the torsion pair $(\mathcal{X}_{\mathcal{P}}, \mathcal{Y}_{\mathcal{P}})$ in $\mathcal{C}$ compactly generated by $\mathcal{P}$. For our applications to tilting theory in the next section we are interested in finding sufficient conditions for the heart to be a module category. A similar approach was considered independently also in [**63**].

For the rest of this section we denote by $\mathcal{H}(\mathcal{P}) = \mathcal{X}_{\mathcal{P}} \cap \Sigma(\mathcal{Y}_{\mathcal{P}})$ the heart of the torsion pair $(\mathcal{X}_{\mathcal{P}}, \mathcal{Y}_{\mathcal{P}})$. As in Chapter I, $\mathcal{H}(\mathcal{P})$ is an abelian category and the functor $\text{H}^0 = \Sigma\mathbf{L}\Sigma^{-1}\mathbf{R} = \mathbf{R}\Sigma\mathbf{L}\Sigma^{-1} : \mathcal{C} \to \mathcal{H}(\mathcal{P})$ is homological. The higher homological functors are denoted by $\text{H}^n : \mathcal{C} \to \mathcal{H}(\mathcal{P})$ and are defined by $\text{H}^n := \text{H}^0\Sigma^n = \Sigma\mathbf{L}\Sigma^{-1}\mathbf{R}\Sigma^n = \mathbf{R}\Sigma\mathbf{L}\Sigma^{n-1}, \forall n \in \mathbb{Z}$.

If $\mathcal{A}$ is an abelian category then we call a set of objects $\mathcal{T} \subseteq \mathcal{A}$ a set of **weak generators** if $\mathcal{A}(T, A) = 0, \ \forall T \in \mathcal{T}$ implies that $A = 0$. We denote by $\text{H}^0(\mathcal{P})$ the full subcategory $\{\text{H}^0(P) \mid P \in \mathcal{P}\}$ of $\mathcal{C}$.

We have the following result which will be useful later.

Lemma 3.1. (i) (a) $\mathcal{C}(\mathcal{P}, H) \xrightarrow{\cong} \mathcal{H}(\mathcal{P})(\text{H}^0(\mathcal{P}), H), \ \forall H \in \mathcal{H}(\mathcal{P})$.
(b) $\text{H}^n(\mathcal{P}) = 0, \ \forall n \geq 1$.
(ii) *$\mathcal{H}(\mathcal{P})$ has exact coproducts, i.e. it is AB4.*
(iii) *$\text{H}^0(\mathcal{P})$ is a set of compact weak generators in $\mathcal{H}(\mathcal{P})$.*

PROOF. (i) Since $\mathcal{P} \subseteq \mathcal{X}_{\mathcal{P}}$, we have: $\mathrm{H}^0(\mathcal{P}) = \Sigma\mathbf{L}\Sigma^{-1}\mathbf{R}(\mathcal{P}) = \Sigma\mathbf{L}\Sigma^{-1}(\mathcal{P})$. Then for any object H in $\mathcal{H}(\mathcal{P})$ we have: $\mathcal{H}(\mathcal{P})(\mathrm{H}^0(\mathcal{P}), H) = \mathcal{H}(\mathcal{P})\big(\Sigma\mathbf{L}\Sigma^{-1}(\mathcal{P}), H\big)$. Since H lies in $\mathcal{H}(\mathcal{P})$, we have $H = \Sigma(Y)$ for some $Y \in \mathcal{Y}_{\mathcal{P}}$. Then $\mathcal{H}(\mathcal{P})(\mathrm{H}^0(\mathcal{P}), H)$ $= \mathcal{Y}_{\mathcal{P}}(\mathbf{L}(\Sigma^{-1}(\mathcal{P}), Y) \xrightarrow{\cong} \mathcal{C}(\Sigma^{-1}(\mathcal{P}), Y) = \mathcal{C}(\mathcal{P}, \Sigma(Y)) = \mathcal{C}(\mathcal{P}, H)$ and this proves (a). Since $\Sigma^n(\mathcal{P}) \in \mathcal{X}_{\mathcal{P}}$ for $n \geq 0$, we have $\mathrm{H}^n(\mathcal{P}) = \Sigma\mathbf{L}\Sigma^{n-1}(\mathcal{P})$. If $n \geq 1$, then $\mathbf{L}\Sigma^{n-1}(\mathcal{P}) = 0$, since $\Sigma^t(\mathcal{P})$ lies in $\mathcal{X}_{\mathcal{P}}$ for $t > 0$. This proves (b).

(ii) Since the torsion pair $(\mathcal{X}_{\mathcal{P}}, \mathcal{Y}_{\mathcal{P}})$ is of finite type, the heart $\mathcal{H}(\mathcal{P})$ has all small coproducts, so it is cocomplete. If $0 \to A_i \to B_i \to C_i \to 0$ is a set of short exact sequences in $\mathcal{H}(\mathcal{P})$ indexed by I, then there are triangles $A_i \to B_i \to C_i \to \Sigma(A_i)$ in $\mathcal{C}$. Since the coproduct of triangles is a triangle, we have a triangle $\oplus_{i\in I} A_i \to \oplus_{i\in I} B_i \to \oplus_{i\in I} C_i \to \Sigma(\oplus_{i\in I} A_i)$ in $\mathcal{C}$. Then the sequence $0 \to \oplus_{i\in I} A_i \to \oplus_{i\in I} B_i \to \oplus_{i\in I} C_i \to 0$ is short exact in $\mathcal{H}(\mathcal{P})$. Hence coproducts are exact and $\mathcal{H}(\mathcal{P})$ is an AB4 category.

(iii) Let H be an object in the heart such that $\mathcal{H}(\mathcal{P})(\mathrm{H}^0(\mathcal{P}), H) = 0$. Then by (i) we have $\mathcal{C}(\mathcal{P}, H) = 0$. This implies that $\mathcal{C}(\Sigma^n(P), H) = 0$, $\forall n \geq 0$ since H lies also in $\Sigma(\mathcal{Y}_{\mathcal{P}})$. Hence $H \in \mathcal{X}_{\mathcal{P}} \cap \mathcal{Y}_{\mathcal{P}} = 0$. It follows that $\mathrm{H}^0(\mathcal{P})$ is a set of weak generators of the heart. It remains to prove that the set $\mathrm{H}^0(\mathcal{P})$ consists of compact objects in $\mathcal{H}(\mathcal{P})$. However this follows directly from the isomorphism in (i)(a) and the fact that $\mathcal{P}$ consists of compact objects in $\mathcal{C}$. □

We would like to know when the set $\mathrm{H}^0(\mathcal{P})$ consists of compact projective generators in $\mathcal{H}(\mathcal{P})$. The following result gives a sufficient condition.

LEMMA 3.2. *$\mathrm{H}^0(\mathcal{P})$ is a set of projectives in $\mathcal{H}(\mathcal{P})$ if and only if $\mathbf{L}\Sigma^{-1}(\mathcal{P}) \subseteq {}^{\perp}\mathcal{H}(\mathcal{P})$. If this is the case then $\mathrm{H}^0(\mathcal{P})$ is a set of compact projective generators in $\mathcal{H}(\mathcal{P})$.*

PROOF. If $\mathrm{H}^0(P)$ is projective and $h : \mathrm{H}^0(P) \to \Sigma(A)$ is a morphism with $A \in \mathcal{H}(\mathcal{P})$, then it is easy to see that for any triangle $A \xrightarrow{f} B \xrightarrow{g} \mathrm{H}^0(P) \xrightarrow{h} \Sigma(A)$, the sequence $A \xrightarrow{f} B \xrightarrow{g} \mathrm{H}^0(P)$ is short exact in $\mathcal{H}(\mathcal{P})$. Since $\mathrm{H}^0(P)$ is projective, g splits and then $h = 0$. It follows that $\mathcal{C}(\mathrm{H}^0(P), \Sigma(A)) = \mathcal{C}(\Sigma\mathbf{L}\Sigma^{-1}\mathbf{R}(P), \Sigma(A)) = \mathcal{C}(\mathbf{L}\Sigma^{-1}(P), A) = 0$, $\forall P \in \mathcal{P}$, $\forall A \in \mathcal{H}(\mathcal{P})$. Hence $\mathbf{L}\Sigma^{-1}(\mathcal{P}) \subseteq {}^{\perp}\mathcal{H}(\mathcal{P})$. Conversely if this holds, let $A \rightarrowtail B \twoheadrightarrow C$ be an extension in $\mathcal{H}(\mathcal{P})$. Then there exists a morphism $C \to \Sigma(A)$ such that $A \to B \to C \to \Sigma(A)$ is a triangle in $\mathcal{C}$. Since any morphism $\mathrm{H}^0(P) \to \Sigma(A)$ is zero, any morphism $\mathrm{H}^0(P) \to C$ factors through $B \to C$ and this shows that $\mathrm{H}^0(P)$ is projective in $\mathcal{H}(\mathcal{P})$.

Now if the objects $\mathrm{H}^0(P)$, with $P \in \mathcal{P}$, are projective in $\mathcal{H}(\mathcal{P})$, the functors $\mathcal{H}(\mathcal{P})(\mathrm{H}^0(P), -)$, $P \in \mathcal{P}$, are faithful if they reflect zero objects, and this happens by Lemma 3.1. Hence $\mathrm{H}^0(\mathcal{P})$ consists of compact projective generators in $\mathcal{H}(\mathcal{P})$. □

We have the following connection between the generating set $\mathcal{P}$ of the torsion pair $(\mathcal{X}_{\mathcal{P}}, \mathcal{Y}_{\mathcal{P}})$ and the full subcategory $\mathrm{H}^0(\mathcal{P}) \subseteq \mathcal{H}(\mathcal{P})$, which will be useful for a nice description of the heart.

LEMMA 3.3. *If $\mathcal{C}(\mathcal{P}, \Sigma^n(\mathcal{P})) = 0$, $\forall n \geq 1$, then we have the following.*

(i) *For any $T \in \mathcal{X}_{\mathcal{P}}$ we have an isomorphism: $\mathcal{C}(\mathcal{P}, T) \xrightarrow{\cong} \mathcal{C}(\mathcal{P}, \mathrm{H}^0(T))$.*

(ii) *The composite functor $\mathcal{P} \hookrightarrow \mathcal{C} \xrightarrow{\mathrm{H}^0} \mathcal{H}(\mathcal{P})$ is fully faithful. Hence it induces an equivalence $\mathrm{H}^0|_{\mathcal{P}} : \mathcal{P} \xrightarrow{\approx} \mathrm{H}^0(\mathcal{P})$.*

PROOF. (i) Consider the glueing triangle for the object $\Sigma^{-1}(T)$:

$$\mathbf{R}\Sigma^{-1}(T) \xrightarrow{f_{\Sigma^{-1}(T)}} \Sigma^{-1}(T) \xrightarrow{g^{\Sigma^{-1}(T)}} \mathbf{L}\Sigma^{-1}(T) \xrightarrow{h_{\Sigma^{-1}(T)}} \Sigma\mathbf{R}\Sigma^{-1}(T).$$

Since $\mathrm{H}^0(T) = \Sigma\mathbf{L}\Sigma^{-1}(T)$, we have the following triangle in $\mathcal{C}$:

$$\Sigma\mathbf{R}\Sigma^{-1}(T) \xrightarrow{-\Sigma(f_{\Sigma^{-1}(T)})} T \xrightarrow{-\Sigma(g^{\Sigma^{-1}(T)})} \mathrm{H}^0(T) \xrightarrow{-\Sigma(h_{\Sigma^{-1}(T)})} \Sigma^2\mathbf{R}\Sigma^{-1}(T)$$

Setting $\mu_T := -\Sigma(g^{\Sigma^{-1}(T)}) : T \to \mathrm{H}^0(T)$, it is not difficult to see that $\mu : \mathrm{Id}_{\mathcal{X}_\mathcal{P}} \to \mathrm{H}^0\mathbf{i} = \mathrm{H}^0|_{\mathcal{X}_\mathcal{P}}$ is a natural morphism. By Proposition 2.8, the hypothesis implies that $\mathcal{C}(\mathcal{P}, \Sigma\mathbf{R}\Sigma^{-1}(T)) = 0 = \mathcal{C}(\mathcal{P}, \Sigma^2\mathbf{R}\Sigma^{-1}(T))$. It follows that $\mathcal{C}(\mathcal{P}, \mu_T) : \mathcal{C}(\mathcal{P}, T) \to \mathcal{C}(\mathcal{P}, \mathrm{H}^0(T))$ is invertible, for any $T \in \mathcal{X}_\mathcal{P}$.

(ii) Let $\alpha : P \to Q$ be a morphism in $\mathcal{P}$. Then we have a morphism of triangles:

$$\begin{array}{ccccccc}
\Sigma\mathbf{R}\Sigma^{-1}(P) & \xrightarrow{-\Sigma(f_{\Sigma^{-1}(P)})} & P & \xrightarrow{\mu_P} & \mathrm{H}^0(P) & \xrightarrow{-\Sigma(h_{\Sigma^{-1}(P)})} & \Sigma^2\mathbf{R}\Sigma^{-1}(P) \\
\downarrow{\scriptstyle \Sigma\mathbf{R}\Sigma^{-1}(\alpha)} & & \downarrow{\scriptstyle \alpha} & & \downarrow{\scriptstyle \mathrm{H}^0(\alpha)} & & \downarrow{\scriptstyle \Sigma^2\mathbf{R}\Sigma^{-1}(\alpha)} \\
\Sigma\mathbf{R}\Sigma^{-1}(Q) & \xrightarrow{-\Sigma(f_{\Sigma^{-1}(Q)})} & Q & \xrightarrow{\mu_Q} & \mathrm{H}^0(Q) & \xrightarrow{-\Sigma(h_{\Sigma^{-1}(Q)})} & \Sigma^2\mathbf{R}\Sigma^{-1}(Q)
\end{array}$$

If $\mathrm{H}^0(\alpha) = 0$, then α factors through $-\Sigma(f_{\Sigma^{-1}(Q)})$. Since by Proposition 2.8 we have $\mathcal{C}(P, \Sigma\mathbf{R}\Sigma^{-1}(Q)) = 0$, it follows that $\alpha = 0$. We infer that $\mathrm{H}^0|_\mathcal{P}$ is faithful. Now let $\beta : \mathrm{H}^0(P) \to \mathrm{H}^0(Q)$ be a morphism, where P and Q lie in $\mathcal{P}$. Since by Proposition 2.8 we have $\mathcal{C}\big(P, \Sigma^2\mathbf{R}\Sigma^{-1}(Q)\big) = 0$, the composition $\mu_P \circ \beta \circ (-\Sigma(h_{\Sigma^{-1}(Q)})) = 0$. Hence there exists a morphism $\alpha : P \to Q$ such that $\alpha \circ \mu_Q = \mu_P \circ \beta$. Then $\mu_P \circ \mathrm{H}^0(\alpha) = \mu_P \circ \beta$, hence $\mathrm{H}^0(\alpha) - \beta$ factors through $-\Sigma(h_{\Sigma^{-1}(P)})$. Since $\mathcal{C}\big(\Sigma^2\mathbf{R}\Sigma^{-1}(P), \mathrm{H}^0(Q)\big) = \mathcal{C}\big(\Sigma^2\mathbf{R}\Sigma^{-1}(P), \Sigma\mathbf{L}\Sigma^{-1}(Q)\big) = \mathcal{C}\big(\Sigma\mathbf{R}\Sigma^{-1}(P), \mathbf{L}\Sigma^{-1}(Q)\big) = 0$, we infer that $\mathrm{H}^0(\alpha) = \beta$. Hence $\mathrm{H}^0|_\mathcal{P}$ is full. □

Consider the restriction functor

$$\mathcal{R} : \mathcal{C} \longrightarrow \mathrm{Mod}(\mathcal{P}), \quad \mathcal{R}(C) := \mathcal{C}(-, C)|_\mathcal{P}$$

and let $\mathcal{R}^*$ be the restriction of $\mathcal{R}$ to the heart $\mathcal{H}(\mathcal{P})$:

$$\mathcal{R}^* : \mathcal{H}(\mathcal{P}) \longrightarrow \mathrm{Mod}(\mathcal{P}), \quad \mathcal{R}^*(H) := \mathcal{C}(-, H)|_\mathcal{P}$$

The following main result of this section gives a sufficient condition for the heart to be a module category.

THEOREM 3.4. *If* $\mathcal{C}(\mathcal{P}, \Sigma^n(\mathcal{P})) = 0$, $\forall n \geq 1$, *then the restriction functor*

$$\mathcal{R}^* : \mathcal{H}(\mathcal{P}) \xrightarrow{\approx} \mathrm{Mod}(\mathcal{P})$$

is an equivalence of categories.

PROOF. By Lemma 3.1 the heart $\mathcal{H}(\mathcal{P})$ is a cocomplete abelian category. We show that the set $\mathrm{H}^0(\mathcal{P})$ consists of projective objects of the heart. Let $P \in \mathcal{P}$, and consider the triangle $\mathbf{R}\Sigma^{-1}(P) \to \Sigma^{-1}(P) \to \mathbf{L}\Sigma^{-1}(P) \to \Sigma\mathbf{R}\Sigma^{-1}(P))$. Let $\alpha : \mathbf{L}\Sigma^{-1}(P) \to A$ be a morphism with $A \in \mathcal{H}(\mathcal{P})$. Since $A \in \mathcal{X}_\mathcal{P}$, by Proposition 2.8 we have $\mathcal{C}(\Sigma^{-1}(P), A) = 0$. Hence α factors through $\mathbf{L}\Sigma^{-1}(P) \to \Sigma\mathbf{R}\Sigma^{-1}(P))$ via a morphism $\beta \in \mathcal{C}(\Sigma\mathbf{R}\Sigma^{-1}(P), A) = \mathcal{C}(\mathbf{R}\Sigma^{-1}(P), \Sigma^{-1}(A))$. However the last group is zero since $\Sigma^{-1}(A) \in \mathcal{Y}_\mathcal{P}$. It follows that $\mathcal{C}(\mathbf{L}\Sigma^{-1}(P), A) = 0$, $\forall A \in \mathcal{H}(\mathcal{P})$. Then by Lemma 3.2 we infer that $\mathrm{H}^0(P)$ is projective in $\mathcal{H}(\mathcal{P})$. It follows that $\mathrm{H}^0(\mathcal{P})$

is a set of compact projective generators in $\mathcal{H}(\mathcal{P})$. Then by a classical result of Gabriel-Freyd, the restriction functor $\mathcal{S} : \mathcal{H}(\mathcal{P}) \to \mathrm{Mod}(\mathrm{H}^0(\mathcal{P}))$ defined by $\mathcal{S}(H) = \mathcal{H}(\mathcal{P})(-, H)|_{\mathrm{H}^0(\mathcal{P})}$, is an equivalence. By Lemma 3.3 we have an equivalence $\mathrm{H}^0|_{\mathcal{P}} : \mathcal{P} \xrightarrow{\approx} \mathrm{H}^0(\mathcal{P})$, and using this equivalence it is easy to see that the functor $\mathcal{S}$ is isomorphic to the functor $\mathcal{R}^*$. We conclude that $\mathcal{R}^*$ is an equivalence. □

We can get a more pleasant description of the projective objects of the heart $\mathcal{H}(\mathcal{P})$ and the equivalence of Theorem 3.4, if the generating set $\mathcal{P}$ of the torsion pair $(\mathcal{X}_{\mathcal{P}}, \mathcal{Y}_{\mathcal{P}})$ lies in the heart. We need the following result whose easy proof is left to the reader.

LEMMA 3.5. *$\mathcal{P} \subseteq \mathcal{H}(\mathcal{P})$ if and only if $\mathcal{C}(\mathcal{P}, \Sigma^n(\mathcal{P})) = 0, \ \forall n < 0$.*

Combining Theorem 3.4 and Lemma 3.5, we have the following.

THEOREM 3.6. *If $\mathcal{C}(\mathcal{P}, \Sigma^n(\mathcal{P})) = 0, \ \forall n < 0$, then the restriction functor*

$$\mathcal{R}^* : \mathcal{H}(\mathcal{P}) \xrightarrow{\approx} \mathrm{Mod}(\mathcal{P}), \qquad C \longmapsto \mathcal{R}^*(C) := \mathcal{C}(-, C)|_{\mathcal{P}}$$

is an equivalence of categories if and only if $\mathcal{C}(\mathcal{P}, \Sigma^n(\mathcal{P})) = 0, \ \forall n > 0$. If this is the case, then $\mathrm{Add}(\mathcal{P})$ is the full subcategory of projective objects of $\mathcal{H}(\mathcal{P})$.

We recall that, in the language of t-structures, a torsion pair $(\mathcal{X}, \mathcal{Y})$ in $\mathcal{C}$ is called **non-degenerate**, if $\bigcap_{n\in\mathbb{Z}} \Sigma^n(\mathcal{X}) = 0 = \bigcap_{n\in\mathbb{Z}} \Sigma^n(\mathcal{Y})$ [**18**]. It is well-known that the non-degenerate property is an important finiteness condition, for instance in this case the torsion and the torsion-free class can be described in terms of the vanishing of the homological functors $\mathrm{H}^n : \mathcal{C} \to \mathcal{H}$ where $\mathcal{H} = \mathcal{X} \cap \Sigma(\mathcal{Y})$ is the heart; in addition the functors $\{\mathrm{H}^n \mid n \in \mathbb{Z}\}$ reflect collectively isomorphisms [**18**].

In this connection we have the following result which gives a nice description of the heart under special assumptions.

COROLLARY 3.7. *If $\mathcal{C}(\mathcal{P}, \Sigma^n(\mathcal{P})) = 0, \forall n \geq 1$, then we have an inclusion $\mathcal{H}(\mathcal{P}) \subseteq \{C \in \mathcal{C} \mid \mathcal{C}(\mathcal{P}, \Sigma^n(C)) = 0, \ \forall n \neq 0\}$, which is an equality if and only if $\mathcal{P}$ generates $\mathcal{C}$. Moreover the following conditions are equivalent:*

(i) *$\mathcal{P}$ generates $\mathcal{C}$ and $\mathcal{C}(\mathcal{P}, \Sigma^n(\mathcal{P})) = 0, \forall n \geq 1$.*
(ii) *$(\mathcal{X}_{\mathcal{P}}, \mathcal{Y}_{\mathcal{P}})$ is non-degenerate and $\mathcal{C}(\mathcal{P}, \Sigma^n(\mathcal{P})) = 0, \forall n \geq 1$.*
(iii) *$\mathcal{H}(\mathcal{P}) = \{C \in \mathcal{C} \mid \mathcal{C}(\mathcal{P}, \Sigma^n(C)) = 0, \ \forall n \neq 0\}$.*

PROOF. For simplicity we set $\mathcal{X}^{\leq 0}_{\mathcal{P}} := \{C \in \mathcal{C} \mid \mathcal{C}(\mathcal{P}, \Sigma^n(C)) = 0, \ \forall n \geq 1\}$. Assume that $\mathcal{C}(\mathcal{P}, \Sigma^n(\mathcal{P})) = 0, \forall n \geq 1$. Then by Proposition 2.8 we have $\mathcal{X}_{\mathcal{P}} \subseteq \mathcal{X}^{\leq 0}_{\mathcal{P}}$. Hence $\mathcal{H}(\mathcal{P}) = \mathcal{X}_{\mathcal{P}} \cap \Sigma(\mathcal{Y}_{\mathcal{P}}) \subseteq \mathcal{X}^{\leq 0}_{\mathcal{P}} \cap \Sigma(\mathcal{Y}_{\mathcal{P}}) = \{C \in \mathcal{C} \mid \mathcal{C}(\mathcal{P}, \Sigma^n(C)) = 0, \ \forall n \neq 0\}$. If the last inclusion is an equality, then let C be in $\mathcal{X}^{\leq 0}_{\mathcal{P}}$. Consider the triangle $\underrightarrow{\mathrm{holim}} T_n \xrightarrow{\chi_C} C \xrightarrow{g} \widehat{C} \xrightarrow{h} \Sigma(\underrightarrow{\mathrm{holim}} T_n)$ constructed in Proposition 2.6. Then $\underrightarrow{\mathrm{holim}} T_n$ lies in $\mathcal{X}_{\mathcal{P}}$ and $\widehat{C}$ lies in $\Sigma(\mathcal{Y}_{\mathcal{P}})$. It follows that $\widehat{C}$ lies in $\mathcal{X}^{\leq 0}_{\mathcal{P}} \cap \Sigma(\mathcal{Y}_{\mathcal{P}}) = \mathcal{H}(\mathcal{P})$. In particular $\widehat{C}$ lies in $\mathcal{X}_{\mathcal{P}}$. Since $\mathcal{X}_{\mathcal{P}}$ is closed under extensions, we have that C lies in $\mathcal{X}_{\mathcal{P}}$. Hence $\mathcal{X}_{\mathcal{P}} = \mathcal{X}^{\leq 0}_{\mathcal{P}}$, and then by Proposition 2.8 we infer that $\mathcal{P}$ generates $\mathcal{C}$. If $\mathcal{P}$ generates $\mathcal{C}$, then the equality $\mathcal{H}(\mathcal{P}) = \{C \in \mathcal{C} \mid \mathcal{C}(\mathcal{P}, \Sigma^n(C)) = 0, \ \forall n \neq 0\}$ follows from Proposition 2.8. The equivalence (i) $\Leftrightarrow$ (iii) follows from Proposition 2.8 and the above arguments. Finally the equivalence (i) $\Leftrightarrow$ (ii) follows from the fact that $\bigcap_{n\in\mathbb{Z}} \Sigma^n(\mathcal{Y}_{\mathcal{P}}) = \{C \in \mathcal{C} \mid \mathcal{C}(\mathcal{P}, \Sigma^n(C)) = 0, \ \forall n \in \mathbb{Z}\}$. □

REMARK 3.8. We call a torsion pair $(\mathcal{X}, \mathcal{Y})$ in $\mathcal{C}$ **left**, resp. **right**, **non-degenerate** if $\bigcap_{n\in\mathbb{Z}} \Sigma^n(\mathcal{X}) = 0$, resp. $\bigcap_{n\in\mathbb{Z}} \Sigma^n(\mathcal{Y}) = 0$. Then $(\mathcal{X}, \mathcal{Y})$ is non-degenerate if and only if it is left and right non-degenerate. If the torsion pair is generated by a set of compact objects $\mathcal{P}$, then since $\bigcap_{n\in\mathbb{Z}} \Sigma^n(\mathcal{Y}_\mathcal{P}) = \{C \in \mathcal{C} \mid \mathcal{C}(\mathcal{P}, \Sigma^n(C)) = 0, \ \forall n \in \mathbb{Z}\}$, it follows that $(\mathcal{X}_\mathcal{P}, \mathcal{Y}_\mathcal{P})$ is right non-degenerate if and only if $\mathcal{P}$ generates $\mathcal{C}$. If $\mathcal{C}(\mathcal{P}, \Sigma^n(\mathcal{P})) = 0, \forall n \geq 1$, then it is easy to see that $\bigcap_{n\in\mathbb{Z}} \Sigma^n(\mathcal{X}_\mathcal{P}) = \bigcap_{n\in\mathbb{Z}} \Sigma^n(\mathcal{Y}_\mathcal{P})$. Hence in this case we have that $(\mathcal{X}_\mathcal{P}, \mathcal{Y}_\mathcal{P})$ is right non-degenerate if and only if $(\mathcal{X}_\mathcal{P}, \mathcal{Y}_\mathcal{P})$ is left non-degenerate if and only if $(\mathcal{X}_\mathcal{P}, \mathcal{Y}_\mathcal{P})$ is non-degenerate if and only if $\mathcal{P}$ generates $\mathcal{C}$.

EXAMPLE. Let $\mathrm{Ho}(\mathcal{S}p)$ be the stable homotopy category of spectra and let $(\mathcal{X}_{S^0}, \mathcal{Y}_{S^0})$ be the torsion pair in $\mathrm{Ho}(\mathcal{S}p)$ compactly generated by the sphere spectrum S^0. Since S^0 generates $\mathrm{Ho}(\mathcal{S}p)$, by Corollary 3.7 the torsion pair $(\mathcal{X}_{S^0}, \mathcal{Y}_{S^0})$ is non-degenerate and its heart consists of all spectra C with stable homotopy groups $\pi_n(C) = 0$, $\forall n \neq 0$. Since the endomorphism ring of S^0 is the ring $\mathbb{Z}$ of integers, by Theorem 3.4 it follows that the heart is equivalent to the category $\mathcal{A}b$ of abelian groups.

EXAMPLE. Let $(\mathcal{T}, \mathcal{F})$ be a torsion pair in an abelian category $\mathcal{A}$ with exact coproducts. Then the induced torsion pair $\big(\mathcal{X}(\mathcal{T}), \mathcal{Y}(\mathcal{F})\big)$ in $\mathbf{D}(\mathcal{A})$ is non-degenerate. It is interesting to know under what conditions the torsion pair $\big(\mathcal{X}(\mathcal{T}), \mathcal{Y}(\mathcal{F})\big)$ is compactly generated.

It is useful to know the structure of projective/injective objects of the heart as well as the relationship between the extension functors $\mathrm{Ext}^*_{\mathcal{H}(\mathcal{P})}$ of $\mathcal{H}(\mathcal{P})$ and the graded Hom functors $\mathrm{Hom}^*_\mathcal{C}$ of $\mathcal{C}$. For a description of the structure of the injective objects of $\mathcal{H}(\mathcal{P})$ we need the following construction.

If $\mathcal{P}$ generates $\mathcal{C}$ then by Brown's representability Theorem [**85**], for any $P \in \mathcal{P}$ there exists an object $\mathsf{D}_{\mathbb{Q}/\mathbb{Z}}(P)$ in $\mathcal{C}$, unique up to isomorphism, and a natural isomorphism of functors

$$\omega : [\mathcal{C}(P, -), \mathbb{Q}/\mathbb{Z}] \xrightarrow{\cong} \mathcal{C}(-, \mathsf{D}_{\mathbb{Q}/\mathbb{Z}}(P)).$$

The object $\mathsf{D}_{\mathbb{Q}/\mathbb{Z}}(P)$ is called the *dual* object of the compact object P.

The following result gives, under a special assumption, a description of the injective objects of the heart in terms of dual objects of compact objects from $\mathcal{P}$. In addition we get a handy description of the extension functors of the heart in terms of its morphism spaces. This will be useful in the next section for the construction of derived equivalences.

PROPOSITION 3.9. *Assume that $\mathcal{P}$ generates $\mathcal{C}$ and $\mathcal{C}(\mathcal{P}, \Sigma^n(\mathcal{P})) = 0, \forall n \neq 0$. Then we have the following.*

(α) *$\mathcal{P}$ is a set of projective generators of $\mathcal{H}(\mathcal{P})$.*

(β) *$\{\mathsf{D}_{\mathbb{Q}/\mathbb{Z}}(P) \mid P \in \mathcal{P}\}$ is a set of cogenerators in $\mathcal{C}$ and a set of injective cogenerators in $\mathcal{H}(\mathcal{P})$.*

(γ) *There exists a natural isomorphism:*

$$\mathrm{Ext}^n_{\mathcal{H}(\mathcal{P})}(-,-) \xrightarrow{\cong} \mathcal{C}(-, \Sigma^n(-)) \, : \, \mathcal{H}(\mathcal{P})^{\mathrm{op}} \times \mathcal{H}(\mathcal{P}) \to \mathcal{A}b, \quad \forall n \in \mathbb{Z}.$$

PROOF. (α) By Theorem 3.6 have $\mathrm{Proj}(\mathcal{H}(\mathcal{P})) = \mathrm{Add}(\mathcal{P})$. Hence $\mathcal{P}$ is a generating set of projectives in $\mathcal{H}(\mathcal{P})$.

(β) If $\mathcal{C}(\Sigma^n(C), \mathsf{D}_{\mathbb{Q}/\mathbb{Z}}(P)) = 0$, $\forall P \in \mathcal{P}$, $\forall n \in \mathbb{Z}$, then using the natural morphism ω we infer that $[\mathcal{C}(P, \Sigma^n(C)), \mathbb{Q}/\mathbb{Z}] = 0$, $\forall P \in \mathcal{P}$, $\forall n \in \mathbb{Z}$. This implies that $C = 0$ since $\mathcal{P}$ generates $\mathcal{C}$ and $\mathbb{Q}/\mathbb{Z}$ cogenerates $\mathcal{A}b$. Now $\mathcal{C}(\Sigma^n(P), \mathsf{D}_{\mathbb{Q}/\mathbb{Z}}(P)) \cong [\mathcal{C}(P, \Sigma^n(P)), \mathbb{Q}/\mathbb{Z}] = 0$, $\forall n \neq 0$ and $\forall P \in \mathcal{P}$. Hence the set $\mathsf{D}_{\mathbb{Q}/\mathbb{Z}}(\mathcal{P})$ lies in $\mathcal{H}(\mathcal{P})$. If $A \rightarrowtail B \twoheadrightarrow C$ is an extension in $\mathcal{H}(\mathcal{P})$, then there exists a triangle $A \to B \to C \to \Sigma(A)$ in $\mathcal{C}$. Applying the functor $\mathcal{C}(-, \mathsf{D}_{\mathbb{Q}/\mathbb{Z}}(\mathcal{P}))$ to the extension and using that $\mathcal{C}\big(\Sigma^{-1}(C), \mathsf{D}_{\mathbb{Q}/\mathbb{Z}}(\mathcal{P})\big) = [\mathcal{C}(\mathcal{P}, \Sigma^{-1}(C)), \mathbb{Q}/\mathbb{Z}] = 0$, we infer that $\mathsf{D}_{\mathbb{Q}/\mathbb{Z}}(\mathcal{P})$ is a set of injective objects which obviously cogenerates $\mathcal{H}(\mathcal{P})$.

(γ) The sequence $F_n := \mathcal{C}(\Sigma^{-n}(-), C) : \mathcal{H}(\mathcal{P})^{\mathrm{op}} \to \mathcal{A}b$ is obviously an exact connected sequence of functors and $F_0 = \mathrm{Ext}^0_{\mathcal{H}(\mathcal{P})}(-, C) = \mathcal{C}(-, C)$. By a well-known characterization of Ext^* in an abelian category it suffices to show that $F_n(Q) = 0$, $\forall Q \in \mathrm{Proj}(\mathcal{H}(\mathcal{P}))$, $\forall n \geq 1$. Indeed using that $\mathrm{Proj}(\mathcal{H}(\mathcal{P})) = \mathrm{Add}(\mathcal{P})$ it follows that for any projective Q we have $\mathcal{C}(\Sigma^{-n}(Q), C) = 0$, since $Q \in \mathrm{Add}(\mathcal{P})$ and $C \in \mathcal{H}(\mathcal{P})$. Interchanging projectives with injectives, $G_n := \mathcal{C}(A, \Sigma^n(-)) : \mathcal{H}(\mathcal{P}) \to \mathcal{A}b$ constitute an exact connected sequence of functors which kill injectives and $G_0 = \mathrm{Ext}^0_{\mathcal{H}(\mathcal{P})}(A, -) = \mathcal{C}(A, -)$. Hence $G_n \cong \mathrm{Ext}^n_{\mathcal{H}(\mathcal{P})}(A, -)$. □

4. Torsion Pairs Induced by Tilting Objects

We know from the results recalled in section 3 of Chapter I that if T is a tilting module over an Artin algebra of projective dimension at most one, then the category $\mathrm{mod}(\Gamma)$, where $\Gamma = \mathrm{End}_\Lambda(T)$, can be described as the heart of a torsion pair in $\mathbf{D}^b(\mathrm{mod}(\Lambda))$ constructed from T. In this section we use the results of the previous section to show that a similar result holds when we more generally start with a tilting module of finite projective dimension. We also give a new proof of Rickard's fundamental Morita theorem for derived categories, which describes explicitly when two rings have equivalent derived categories.

Throughout this section we assume that the triangulated category $\mathcal{C}$ has all small coproducts.

In analogy with Rickard's definition of a tilting complex in the derived category of a ring, see [**91**], we make the following definition.

DEFINITION 4.1. An object $T \in \mathcal{C}$ is called a **tilting object** if:

(i) T is compact.
(ii) $\mathcal{C}(T, \Sigma^n(T)) = 0$, $\forall n \neq 0$.
(iii) $\{T\}$ generates $\mathcal{C}$.

If conditions (i) and (ii) hold, then we call T a **partial tilting object**.

EXAMPLE. Let $\mathcal{A}$ be an abelian category with exact coproducts. If $T^\bullet$ is a bounded complex in $\mathcal{A}$ with components compact projective objects, and if $\mathbf{D}(\mathcal{A})(T^\bullet, \Sigma^n(T^\bullet)) = 0$, $\forall n \neq 0$, then $T^\bullet$ is a partial tilting object in $\mathbf{D}(\mathcal{A})$. $T^\bullet$ is a tilting complex, if in addition $T^\bullet$ is a generator of $\mathbf{D}(\mathcal{A})$.

We have the following direct consequence of the results of the previous section, which characterizes tilting objects in torsion-theoretic terms, and gives useful information for the structure of the heart of the torsion pair generated by a tilting object.

COROLLARY 4.2. *Let T be a partial tilting object in $\mathcal{C}$. Then T is a tilting object if and only if the torsion pair $(\mathcal{X}_T, \mathcal{Y}_T)$ in $\mathcal{C}$ generated by T is non-degenerate if and only if the heart $\mathcal{H}(T) = \mathcal{X}_T \cap \Sigma(\mathcal{Y}_T)$ is equal to $\{C \in \mathcal{C} \mid \mathcal{C}(T, \Sigma^n(C)) = 0, \ \forall n \neq 0\}$. If this is the case, then T is a compact projective generator of the heart $\mathcal{H}(T)$ and the functor $\mathcal{C}(T, -) : \mathcal{H}(T) \to \mathrm{Mod}(\mathrm{End}_{\mathcal{C}}(T))$ is an equivalence of categories.*

If $\mathcal{C}$ is compactly generated then by duality we can define an object $E \in \mathcal{C}$ to be a **cotilting object**, if: (1) E is pure-injective in $\mathcal{C}$ in the sense of [**21**], (2) $\mathcal{C}(\Sigma^n(E), E) = 0$, $\forall n \neq 0$ and (3) $\{E\}$ cogenerates $\mathcal{C}$, i.e. $\mathcal{C}(\Sigma^n(C), E) = 0$, $\forall n \in \mathbb{Z}$ $\Rightarrow C = 0$. By [**21**] any object of the form $\mathsf{D}_{\mathbb{Q}/\mathbb{Z}}(T)$ is pure-injective, for any compact object T. Then under the assumptions of the above corollary we have that $\mathsf{D}_{\mathbb{Q}/\mathbb{Z}}(T)$ is a cotilting object in $\mathcal{C}$ and an injective cogenerator of $\mathcal{H}(T) = \mathrm{Mod}(\mathrm{End}_{\mathcal{C}}(T))$.

The following result shows that a tilting object T in the unbounded derived category of a Grothendieck category $\mathcal{G}$ induces a triangle equivalence between its bounded derived category and the bounded derived category of the heart of the torsion pair generated by the tilting object T. This generalizes Rickard's theorem to Grothendieck categories and presents a torsion-theoretic approach via the heart to the construction of derived equivalences.

THEOREM 4.3. *Let $\mathcal{A}$ be an abelian category with exact coproducts, e.g. a Grothendieck category. If $T^{\bullet}$ is a tilting object in $\mathbf{D}(\mathcal{A})$, then there exists a triangle equivalence*

$$\mathbf{D}^b(\mathcal{A}) \xrightarrow{\approx} \mathbf{D}^b\big(\mathrm{Mod}(\mathrm{End}(T^{\bullet}))\big).$$

PROOF. Since $\mathbf{D}(\mathcal{A})$ admits a tilting object, it is compactly generated, and the heart of the torsion pair generated by $T^{\bullet}$ is equivalent to $\mathrm{Mod}(\mathrm{End}(T^{\bullet}))$. Since the latter has enough projective and injective objects, by [**60**], [**18**], there exists an exact functor $G : \mathbf{D}^b(\mathrm{Mod}(\mathrm{End}(T^{\bullet}))) \to \mathbf{D}^b(\mathcal{A})$ extending the identity functor of $\mathrm{Mod}(\mathrm{End}(T^{\bullet})))$. By the argument of Theorem 3.3 in [**60**], G is an equivalence if for any pair of $\mathrm{End}(T^{\bullet})$-modules B, B', the canonical morphism

$$\mathrm{Ext}^n_{\mathrm{End}(T^{\bullet})}[B, B'] \xrightarrow{\cong} \mathbf{D}^b(\mathrm{Mod}(\mathrm{End}(T)))[B, \Sigma^n(B')] \longrightarrow \mathbf{D}^b(\mathcal{A})[B, \Sigma^n(B')]$$

induced by G is bijective, $\forall n \in \mathbb{Z}$. However this holds, by Proposition 3.9(γ). □

It is well-known that the full subcategory of compact objects in the unbounded derived category $\mathbf{D}(\mathrm{Mod}(\Lambda))$ of a ring Λ coincides, up to equivalence, with the bounded homotopy category $\mathcal{H}^b(\mathcal{P}_{\Lambda})$ of the category of finitely generated projective modules. In this case the tilting objects of $\mathbf{D}(\mathrm{Mod}(\Lambda))$ are precisely the tilting complexes of Rickard [**91**].

The following consequence of Theorem 4.3 gives a new, torsion theoretic, proof of a part of Rickard's Theorem on the construction of derived equivalences [**91**].

COROLLARY 4.4. [**91**] *Let $T^{\bullet} \in \mathbf{D}(\mathrm{Mod}(\Lambda))$ be a tilting complex with endomorphism ring $\Gamma := \mathrm{End}(T^{\bullet})$. Then there exists a triangle equivalence:*

$$\mathbf{D}^b(\mathrm{Mod}(\Lambda)) \xrightarrow{\approx} \mathbf{D}^b(\mathrm{Mod}(\Gamma)).$$

When T is a tilting Λ-module with $\mathrm{pd}_{\Lambda} T \leq 1$, we have an associated torsion pair $(\mathcal{T}, \mathcal{F})$ in $\mathrm{mod}(\Lambda)$, and when we have a torsion pair in $\mathrm{mod}(\Lambda)$ there is an associated torsion pair $(\mathcal{X}, \mathcal{Y})$ in $\mathbf{D}^b(\mathrm{mod}(\Lambda))$, whose heart is $\mathrm{mod}(\Gamma)$ for $\Gamma =$

$\mathrm{End}_\Lambda(T)$. For a tilting module T of finite projective dimension there is not in general a natural associated torsion pair in $\mathrm{mod}(\Lambda)$ which plays a role in tilting theory. But we can pass directly to $\mathbf{D}^b(\mathrm{mod}(\Lambda))$ and consider the pair $(\widetilde{\mathcal{X}}_T, \widetilde{\mathcal{Y}}_T)$ in $\mathbf{D}^b(\mathrm{mod}(\Lambda))$ generated by T as defined in Section 2 of this chapter, that is: $\widetilde{\mathcal{Y}}_T = \{C^\bullet \in \mathbf{D}^b(\mathrm{mod}(\Lambda)) \mid (T, \Sigma^n(C^\bullet)) = 0, \ \forall n \leq 0\}$ and $\widetilde{\mathcal{X}}_T = {}^\perp\widetilde{\mathcal{Y}}_T$ where the left orthogonal is formed in $\mathbf{D}^b(\mathrm{mod}(\Lambda))$. Note that it is easy to see that when $\mathrm{pd}_\Lambda T \leq 1$, then T generates the torsion pair $(\mathcal{X}, \mathcal{Y})$ in this sense. We want to show that $(\widetilde{\mathcal{X}}_T, \widetilde{\mathcal{Y}}_T)$ is a torsion pair also for $\mathrm{pd}_\Lambda T < \infty$ and that $\mathrm{mod}(\Gamma)$ is equivalent to its heart.

In view of the results in Section 3 it is hence an important problem to have conditions ensuring that a given compactly generated torsion pair in the unbounded derived category restricts to a torsion pair in the bounded derived category. Here we deal only with the case that the torsion pair is induced by a (partial) tilting module. So let Λ be a ring (not necessarily an Artin algebra). Recall that a right Λ-module T is called a **partial tilting module** if $\mathrm{Ext}^n_\Lambda(T,T) = 0, \forall n \geq 1$, and T has a finite exact resolution by finitely generated projective modules.

Let T be a partial tilting right Λ-module, and let $\Gamma = \mathrm{End}_\Lambda(T)$ be its endomorphism ring. We view the module T as a stalk complex concentrated in degree zero in $\mathbf{D}(\mathrm{Mod}(\Lambda))$. Then it is easy to see that T is a compact object in $\mathbf{D}(\mathrm{Mod}(\Lambda))$ with endomorphism ring Γ. It is well-known that we have an adjoint pair $\big(-\otimes^{\mathbb{L}}_\Gamma T, \ \mathbb{R}\mathrm{Hom}_\Lambda(T,-)\big)$ of exact functors:

$$\mathbb{R}\mathrm{Hom}_\Lambda(T,-) \, : \, \mathbf{D}(\mathrm{Mod}(\Lambda)) \leftrightarrows \mathbf{D}(\mathrm{Mod}(\Gamma)) \, : \, -\otimes^{\mathbb{L}}_\Gamma T$$

Let $(\mathcal{X}_T, \mathcal{Y}_T)$ be the torsion pair in $\mathbf{D}(\mathrm{Mod}(\Lambda))$ generated by T. Then

$$\mathcal{X}_T \subseteq \mathcal{X}_T^{\leq 0} := \big\{C^\bullet \in \mathbf{D}(\mathrm{Mod}(\Lambda)) \mid \big(T, \Sigma^n(C^\bullet)\big) = 0, \, \forall n \geq 1\big\}$$

$$\mathcal{Y}_T = \big\{C^\bullet \in \mathbf{D}(\mathrm{Mod}(\Lambda)) \mid \big(T, \Sigma^n(C^\bullet)\big) = 0, \, \forall n \leq 0\big\}.$$

By Theorem 3.5 we have that the heart $\mathcal{H}(T)$ is equivalent to $\mathrm{Mod}(\Gamma)$. We consider also the natural torsion pair $(\mathcal{Z}_\Gamma, \mathcal{W}_\Gamma)$ in $\mathbf{D}(\mathrm{Mod}(\Gamma))$ generated by Γ, i.e. $\mathcal{Z}_\Gamma = \{C^\bullet \in \mathbf{D}(\mathrm{Mod}(\Gamma)) \mid \mathrm{H}^n(C^\bullet) = 0, \, \forall n > 0\}$ and $\mathcal{W}_\Gamma = \{C^\bullet \in \mathbf{D}(\mathrm{Mod}(\Gamma)) \mid \mathrm{H}^n(C^\bullet) = 0, \, \forall n \leq 0\}$, with heart $\mathrm{Mod}(\Gamma)$.

To proceed further we need to recall from [**18**] the concept of t-exactness of exact functors between triangulated categories each endowed with a t-structure. Let $F : \mathcal{C} \to \mathcal{D}$ be an exact functor between the triangulated categories $\mathcal{C}$, $\mathcal{D}$. We assume that $(\mathcal{X}, \mathcal{Y})$ is a torsion pair in $\mathcal{C}$ and $(\mathcal{Z}, \mathcal{W})$ is a torsion pair in $\mathcal{D}$. Then F is called **left $t-$exact**, resp. **right $t-$exact**, if $F(\mathcal{Y}) \subseteq \mathcal{W}$, resp. $F(\mathcal{X}) \subseteq \mathcal{Z}$. Finally the functor F is called **$t-$exact** if F is left and right t-exact.

We have the following result.

PROPOSITION 4.5. *Let T be a partial tilting Λ-module with endomorphism ring Γ. Then we have the following.*

(1) *The functor* $\mathbb{R}\mathrm{Hom}_\Lambda(T,-) : \mathbf{D}(\mathrm{Mod}(\Lambda)) \to \mathbf{D}(\mathrm{Mod}(\Gamma))$ *is t-exact.*
(2) *The functor* $-\otimes^{\mathbb{L}}_\Gamma T : \mathbf{D}(\mathrm{Mod}(\Gamma)) \to \mathbf{D}(\mathrm{Mod}(\Lambda))$ *is fully faithful t-exact.*

PROOF. (1) Let $X^\bullet \in \mathcal{X}_T$. By the above observations we have $(T, \Sigma^n(X^\bullet)) = 0$, $\forall n \geq 1$. Then $\mathrm{H}^n\mathbb{R}\mathrm{Hom}_\Lambda(T, X^\bullet) = (T, \Sigma^n(X^\bullet)) = 0$ for $n \geq 1$. It follows that $\mathbb{R}\mathrm{Hom}_\Lambda(T, X^\bullet)$ lies in $\mathcal{Z}_\Gamma$, hence $\mathbb{R}\mathrm{Hom}_\Lambda(T,-)$ is right t-exact. Now let $Y^\bullet \in \mathcal{Y}_T$, so $(T, \Sigma^n(Y^\bullet)) = 0$ for $n \leq 0$. Then $\mathrm{H}^n\mathbb{R}\mathrm{Hom}_\Lambda(T, Y^\bullet) = (T, \Sigma^n(Y^\bullet)) = 0$ for $n \leq 0$, hence $\mathbb{R}\mathrm{Hom}_\Lambda(T, Y^\bullet)$ lies in $\mathcal{W}_\Gamma$, so $\mathbb{R}\mathrm{Hom}_\Lambda(T, -)$ is left t-exact. We conclude that $\mathbb{R}\mathrm{Hom}_\Lambda(T, -)$ is t-exact.

(2) Let $Z^\bullet$ be a complex in $\mathcal{Z}_\Gamma$. By adjointness, for any complex $Y^\bullet$ in $\mathcal{Y}_T$ we have an isomorphism $\big(Z^\bullet \otimes^{\mathbb{L}}_\Gamma T, Y^\bullet\big) \xrightarrow{\cong} \big(Z^\bullet, \mathbb{R}\mathrm{Hom}_\Lambda(T, Y^\bullet)\big)$. Since $Y^\bullet$ lies in $\mathcal{Y}_T$, by (1) we have $\mathbb{R}\mathrm{Hom}_\Lambda(T, C^\bullet) \in \mathcal{W}_\Gamma$. Since $Z^\bullet$ lies in $\mathcal{Z}_\Gamma$ and the pair $(\mathcal{Z}_\Gamma, \mathcal{W}_\Gamma)$ is a torsion pair in $\mathbf{D}(\mathrm{Mod}(\Gamma))$, we infer that $\big(Z^\bullet \otimes^{\mathbb{L}}_\Gamma T, Y^\bullet\big) = 0$, $\forall Y^\bullet \in \mathcal{Y}_T$. Hence $Z^\bullet \otimes^{\mathbb{L}}_\Gamma T$ lies in $\mathcal{X}_T$, for any $Z^\bullet$ in $\mathcal{Z}_\Gamma$. In other words $- \otimes^{\mathbb{L}}_\Gamma T$ is right t-exact. Next we show that $- \otimes^{\mathbb{L}}_\Gamma T$ is fully faithful. First observe that $- \otimes^{\mathbb{L}}_\Gamma T$ preserves coproducts and sends Γ to T. Since $\mathrm{Ext}^n_\Lambda(T, T) = 0$, $\forall n \neq 0$, and $\mathrm{Hom}_\Lambda(T, T) = \Gamma$, the functor $- \otimes^{\mathbb{L}}_\Gamma T$ induces bijections $(\Gamma, \Sigma^n(\Gamma)) \xrightarrow{\cong} (T, \Sigma^n(T))$. Then by devissage [**69**] we have that $- \otimes^{\mathbb{L}}_\Gamma T$ is fully faithful.

Now let $W^\bullet$ be a complex in $\mathcal{W}_\Gamma$, and consider the standard triangle $X_{W^\bullet \otimes^{\mathbb{L}}_\Gamma T} \to W^\bullet \otimes^{\mathbb{L}}_\Gamma T \to Y^{W^\bullet \otimes^{\mathbb{L}}_\Gamma T} \to \Sigma(X_{W^\bullet \otimes^{\mathbb{L}}_\Gamma T})$ in $\mathbf{D}(\mathrm{Mod}(\Lambda))$ associated to the torsion pair $(\mathcal{X}_T, \mathcal{Y}_T)$. Applying the exact functor $\mathbb{R}\mathrm{Hom}_\Lambda(T, -)$ to this triangle, we have a triangle $\mathbb{R}\mathrm{Hom}_\Lambda(T, X_{W^\bullet \otimes^{\mathbb{L}}_\Gamma T}) \to \mathbb{R}\mathrm{Hom}_\Lambda(T, W^\bullet \otimes^{\mathbb{L}}_\Gamma T) \to \mathbb{R}\mathrm{Hom}_\Lambda(T, Y^{W^\bullet \otimes^{\mathbb{L}}_\Gamma T}) \to \Sigma\mathbb{R}\mathrm{Hom}_\Lambda(T, X_{W^\bullet \otimes^{\mathbb{L}}_\Gamma T})$ in $\mathbf{D}(\mathrm{Mod}(\Gamma))$. Since the functor $\mathbb{R}\mathrm{Hom}_\Lambda(T, -)$ is t-exact, we have that $\mathbb{R}\mathrm{Hom}_\Lambda(T, Y^{W^\bullet \otimes^{\mathbb{L}}_\Gamma T}) \in \mathcal{W}_\Gamma$ and $\mathbb{R}\mathrm{Hom}_\Lambda(T, X_{W^\bullet \otimes^{\mathbb{L}}_\Gamma T}) \in \mathcal{Z}_\Gamma$. Since the functor $- \otimes^{\mathbb{L}}_\Gamma T$ is fully faithful, the canonical morphism $W^\bullet \to \mathbb{R}\mathrm{Hom}_\Lambda(T, W^\bullet \otimes^{\mathbb{L}}_\Gamma T)$ is invertible. Since $W^\bullet$ lies in $\mathcal{W}_\Gamma$ and the latter is left triangulated, it follows that $\mathbb{R}\mathrm{Hom}_\Lambda(T, X_{W^\bullet \otimes^{\mathbb{L}}_\Gamma T})$ lies in $\mathcal{Z}_\Gamma \cap \mathcal{W}_\Gamma = 0$. Then $\mathrm{H}^n\mathbb{R}\mathrm{Hom}_\Lambda(T, X_{W^\bullet \otimes^{\mathbb{L}}_\Gamma T}) = (T, \Sigma^n(X_{W^\bullet \otimes^{\mathbb{L}}_\Gamma T})) = 0$, $\forall n \in \mathbb{Z}$. Hence $X_{W^\bullet \otimes^{\mathbb{L}}_\Gamma T}$ lies in $\mathcal{X}_T \cap \mathcal{Y}_T = 0$. It follows that $W^\bullet \otimes^{\mathbb{L}}_\Gamma T \xrightarrow{\cong} Y^{W^\bullet \otimes^{\mathbb{L}}_\Gamma T}$ lies in $\mathcal{Y}_T$, hence $- \otimes^{\mathbb{L}}_\Gamma T$ is left t-exact. □

COROLLARY 4.6. *T is a tilting module if and only if $\mathcal{X}_T = \mathcal{X}_T^{\leq 0}$. If T is a tilting module, then the t-exact quasi-inverse triangle equivalences*

$$\mathbb{R}\mathrm{Hom}_\Lambda(T, -) \,:\, \mathbf{D}(\mathrm{Mod}(\Lambda)) \leftrightarrows \mathbf{D}(\mathrm{Mod}(\Gamma)) \,:\, - \otimes^{\mathbb{L}}_\Gamma T$$

restrict to quasi-inverse equivalences:

$$\mathcal{X}_T \xrightarrow{\approx} \mathcal{Z}_\Gamma \quad \text{and} \quad \mathcal{Y}_T \xrightarrow{\approx} \mathcal{W}_\Gamma.$$

We assume now that Λ is right coherent and in addition the endomorphism ring $\Gamma = \mathrm{End}_\Lambda(T)$ of T is right coherent. For simplicity we set $F = - \otimes^{\mathbb{L}}_\Gamma T$, $G = \mathbb{R}\mathrm{Hom}_\Lambda(T, -)$. Consider the full subcategories:

$$\mathcal{X}_T^b = \mathcal{X}_T \cap \mathbf{D}^b(\mathrm{mod}(\Lambda)) \quad \text{and} \quad \mathcal{Y}_T^b = \mathcal{Y}_T \cap \mathbf{D}^b(\mathrm{mod}(\Lambda)).$$

PROPOSITION 4.7. *If T is a tilting module, then $(\mathcal{X}_T^b, \mathcal{X}_T^b)$ is a torsion pair in $\mathbf{D}^b(\mathrm{mod}(\Lambda))$ with heart equivalent to the abelian category* $\mathrm{mod}(\Gamma)$.

PROOF. By [**91**] the triangle equivalences $G : \mathbf{D}(\mathrm{Mod}(\Lambda)) \leftrightarrows \mathbf{D}(\mathrm{Mod}(\Gamma)) : F$ restrict to triangle equivalences $G : \mathbf{D}^b(\mathrm{mod}(\Lambda)) \leftrightarrows \mathbf{D}^b(\mathrm{mod}(\Gamma)) : F$. By [**18**], the torsion pair $(\mathcal{X}_T, \mathcal{Y}_T)$ restricts to a torsion pair in $\mathbf{D}^b(\mathrm{mod}(\Lambda))$ if and only if the coreflection functor $\mathbf{R} : \mathbf{D}(\mathrm{Mod}(\Lambda)) \to \mathcal{X}_T$ satisfies $\mathbf{R}(\mathbf{D}^b(\mathrm{mod}(\Lambda))) \subseteq \mathbf{D}^b(\mathrm{mod}(\Lambda))$. Hence to prove the assertion, it suffices to show that $\mathbf{R}$ preserves bounded complexes with finitely generated components. Let $C^\bullet \in \mathbf{D}^b(\mathrm{mod}(\Lambda))$. Then $G(C^\bullet)$ lies in $\mathbf{D}^b(\mathrm{mod}(\Gamma))$. Obviously the coreflection ($=$ truncation) $\mathbf{R}' : \mathbf{D}(\mathrm{Mod}(\Gamma)) \to \mathcal{Z}_\Gamma$

satisfies $\mathbf{R}'(\mathbf{D}^b(\mathrm{mod}(\Gamma))) \subseteq \mathbf{D}^b(\mathrm{mod}(\Gamma))$. Hence we have the standard triangle $\mathbf{R}'G(C^\bullet) \to G(C^\bullet) \to \mathbf{L}'G(C^\bullet) \to \Sigma\mathbf{R}'G(C^\bullet)$ in $\mathbf{D}^b(\mathrm{mod}(\Gamma))$. Applying F, we have a triangle $F\mathbf{R}'G(C^\bullet) \to C^\bullet \to F\mathbf{L}'G(C^\bullet) \to \Sigma F\mathbf{R}'G(C^\bullet)$ in $\mathbf{D}^b(\mathrm{mod}(\Lambda))$, where $F\mathbf{R}'G(C^\bullet) \in \mathcal{X}_T$ and $F\mathbf{L}'G(C^\bullet) \in \mathcal{Y}_T$. Hence the triangle is the standard one associated to the torsion pair $(\mathcal{X}_T, \mathcal{Y}_T)$, in particular $F\mathbf{R}'G(C^\bullet) \cong \mathbf{R}(C^\bullet)$. Hence $\mathbf{R}(C^\bullet) \in \mathbf{D}^b(\mathrm{mod}(\Lambda))$ and the assertion follows. $\square$

Since we obviously have $\mathcal{Y}_T^b = \widetilde{\mathcal{Y}}_T$ and by Proposition 4.7 and Proposition 2.8 we have that $(\mathcal{X}_T^b, \mathcal{Y}_T^b)$ is a torsion pair in $\mathbf{D}^b(\mathrm{mod}(\Lambda))$ with $\widetilde{\mathcal{X}}_T = {}^\perp\widetilde{\mathcal{Y}}_T = \mathcal{X}_T^b$, we infer the following consequence.

THEOREM 4.8. *Let Λ be a right coherent ring and T a tilting Λ-module with right coherent endomorphism ring Γ. Then T generates a torsion pair $(\widetilde{\mathcal{X}}_T, \widetilde{\mathcal{Y}}_T)$ in $\mathbf{D}^b(\mathrm{mod}(\Lambda))$ with heart* $\mathrm{mod}(\Gamma)$.

REMARK 4.9. (1) If the ring Λ is not right coherent, then the above arguments show that a tilting Λ-module T with endomorphism ring Γ, generates a torsion pair $(\widetilde{\mathcal{X}}_T, \widetilde{\mathcal{Y}}_T)$ in $\mathbf{D}^b(\mathrm{Mod}(\Lambda))$ with heart $\mathrm{Mod}(\Gamma)$.

(2) The above results can be extended to complexes, that is, we may replace the tilting module T with a tilting complex $T^\bullet$. More generally, replacing $\mathbf{D}(\mathrm{Mod}(\Gamma))$ with the derived category of the DG-algebra induced by $T^\bullet$ (see the arguments in the next chapter), we can replace the tilting complex $T^\bullet$ with a compact generator of $\mathbf{D}(\mathrm{Mod}(\Lambda))$.

CHAPTER IV

Hereditary Torsion Pairs in Triangulated Categories

In this chapter we investigate hereditary torsion pairs, usually generated by compact objects, in a triangulated category with all small coproducts. We show that compactly generated hereditary torsion pairs provide the proper setting for the Morita theory in derived and stable categories, and we give applications to tilting theory by presenting torsion theoretic proofs of some important results of Rickard, Keller and Happel which are central in the Morita theory of derived categories. Our torsion theoretic arguments allow us to give proofs of (generalizations of) these results which are considerably simpler and conceptual. We apply our results to the study of homological conjectures in the representation theory of Artin algebras by giving alternative formulations of the conjectures inside the unbounded derived category.

1. Hereditary Torsion Pairs

In this section we study hereditary torsion pairs generated by a set of compact objects, we give necessary conditions for the existence of TTF-triples, and we show that compactly generated hereditary torsion pairs provide the proper setting for various important constructions in derived categories. Finally we study the question of when a hereditary torsion pair induces a torsion pair on the full subcategory of compact objects. This will be useful later in connection with K-theory and the homological conjectures in the representation theory of Artin algebras.

Throughout this section we fix a triangulated category $\mathcal{C}$ and we assume that $\mathcal{C}$ contains all small coproducts. We recall from Section 2 of Chapter I that a torsion pair $(\mathcal{X}, \mathcal{Y})$ in $\mathcal{C}$ is hereditary if and only if the torsion class $\mathcal{X}$, or equivalently the torsion-free class $\mathcal{Y}$, is triangulated.

We begin with the following basic result which characterizes the hereditary torsion pairs of finite type and gives sufficient conditions for the existence of TTF-triples.

PROPOSITION 1.1. *Let $(\mathcal{X}, \mathcal{Y})$ be a hereditary torsion pair in $\mathcal{C}$.*

(i) *If $\mathcal{C}$ is compactly generated, then the following are equivalent:*
  (a) *$(\mathcal{X}, \mathcal{Y})$ is of finite type.*
  (b) *There exists a TTF-triple $(\mathcal{X}, \mathcal{Y}, \mathcal{Z})$ in $\mathcal{C}$.*

  If one of the above conditions holds, then $\mathcal{Y}$ is compactly generated.

(ii) *If $\mathcal{Y}$ is compactly generated and $(\mathcal{X}, \mathcal{Y})$ is of finite type, then there exists a TTF-triple $(\mathcal{X}, \mathcal{Y}, \mathcal{Z})$ in $\mathcal{C}$.*

In any of the above cases, there are triangle equivalences ${}^{\perp}\mathcal{Y} = \mathcal{X} \xrightarrow{\approx} \mathcal{Z} = \mathcal{Y}^{\perp}$.

PROOF. (i) Let $\mathcal{C}$ be compactly generated. The implication (b) $\Rightarrow$ (a) is trivial. (a) $\Rightarrow$ (b) Since $(\mathcal{X}, \mathcal{Y})$ is of finite type, $\mathcal{Y}$ is closed under coproducts and by Lemma III.1.2 the reflection functor $\mathbf{L} : \mathcal{C} \to \mathcal{Y}$ preserves compact objects. Let $\mathcal{T}$ be a generating set of compact objects in $\mathcal{C}$. If $Y \in \mathcal{Y}$ is such that $\mathcal{Y}(\mathbf{L}(\mathcal{T}), Y) = 0$ then $\mathcal{C}(T, Y) = 0$, $\forall T \in \mathcal{T}$, hence $Y = 0$. It follows that $\mathcal{Y}$ is compactly generated by the set $\mathbf{L}(\mathcal{T})$. Since $\mathcal{Y}$ is compactly generated and the inclusion $\mathbf{j} : \mathcal{Y} \hookrightarrow \mathcal{C}$ is exact and preserves coproducts, by [**85**], $\mathbf{j}$ admits a right adjoint. By the results of Section I.2, we infer that $\mathcal{Y}$ is a TTF-class with TTF-triple $(\mathcal{X}, \mathcal{Y}, \mathcal{Z})$, where $\mathcal{Z} = \mathcal{Y}^{\perp}$.

(ii) The same arguments as above show that if $\mathcal{Y}$ is compactly generated and closed under coproducts in $\mathcal{C}$, then there exists a TTF-triple $(\mathcal{X}, \mathcal{Y}, \mathcal{Z})$ in $\mathcal{C}$ even if the latter is not compactly generated.

The last assertion follows from Corollary I.2.9. □

EXAMPLE. Let $(\mathfrak{X}, \mathcal{O})$ be a ringed space and let $\mathcal{U}$ be an open subspace of $\mathfrak{X}$ with closed complement $\mathcal{V}$. Then let $\mathbf{D}(\mathfrak{X}, \mathcal{O})$, $\mathbf{D}(\mathcal{U}, \mathcal{O})$, $\mathbf{D}(\mathcal{V}, \mathcal{O})$ be the derived categories of sheaves of $\mathcal{O}$-modules over $\mathfrak{X}$, $\mathcal{U}$ and $\mathcal{V}$. Then $\mathbf{D}(\mathcal{V}, \mathcal{O})$ is a TTF-class in $\mathbf{D}(\mathfrak{X}, \mathcal{O})$, hence it induces a recollement in the derived category $\mathbf{D}(\mathfrak{X}, \mathcal{O})$.

REMARK 1.2. It is not difficult to see that if $(\mathcal{X}, \mathcal{Y})$ is a torsion pair in $\mathcal{C}$ and $\mathcal{X}$ is generated by the set $\mathcal{T}$ and $\mathcal{Y}$ is generated by the set $\mathcal{S}$, then $\mathcal{C}$ is generated by the set $\mathcal{T} \cup \mathcal{S}$. Indeed if C is in $\mathcal{C}$ and $\mathcal{C}(\mathcal{T}, C) = 0$, then by adjointness we have $\mathcal{X}(\mathcal{T}, \mathbf{R}(C)) = 0$ and then $\mathbf{R}(C) = 0$. Hence $C \in \mathcal{Y}$. Now if also $\mathcal{C}(\mathcal{S}, C) = 0$, then since $C \in \mathcal{Y}$ we have $C = 0$. It follows that $\mathcal{C}$ generated by the set $\mathcal{T} \cup \mathcal{S}$.

Let $\mathcal{P}$ be a set of compact objects in $\mathcal{C}$ and set $\mathcal{R} := \{\Sigma^n(P) \mid n \in \mathbb{Z},\ P \in \mathcal{P}\}$. Then as shown in Section III.2, by setting $\mathcal{Y}_{\mathcal{P}} = \mathcal{R}^{\perp}$ and $\mathcal{X}_{\mathcal{P}} = {}^{\perp}\mathcal{Y}_{\mathcal{P}}$, we obtain a torsion pair $(\mathcal{X}_{\mathcal{P}}, \mathcal{Y}_{\mathcal{P}})$ of finite type in $\mathcal{C}$ which obviously is hereditary. We call $(\mathcal{X}_{\mathcal{P}}, \mathcal{Y}_{\mathcal{P}})$ the **hereditary torsion pair generated by** $\mathcal{P}$. Since it will be clear from the context, we use the same notation as in the non-hereditary case to denote the hereditary torsion pair generated by $\mathcal{P}$. Note that if $(\widehat{\mathcal{X}}_{\mathcal{P}}, \widehat{\mathcal{Y}}_{\mathcal{P}})$ is the non-hereditary torsion pair generated by $\mathcal{P}$, then $\mathcal{Y}_{\mathcal{P}} = \bigcap_{n \in \mathbb{Z}} \Sigma^n(\widehat{\mathcal{Y}}_{\mathcal{P}})$. By Remark III.2.7 the torsion class $\mathcal{X}_{\mathcal{P}}$ coincides with $\operatorname{Loc}(\mathcal{P})$, the smallest thick subcategory of $\mathcal{C}$ which is closed under coproducts and contains $\mathcal{P}$. In particular $\mathcal{X}_{\mathcal{P}}$ is a compactly generated triangulated category with $\mathcal{P}$ as a set of compact generators.

REMARK 1.3. It is not difficult to see that if $\widehat{\mathcal{P}}$ is the thick subcategory of $\mathcal{C}$ generated by the set of compact objects $\mathcal{P}$, then $\widehat{\mathcal{P}}$ consists of compact objects and the hereditary torsion pair $(\mathcal{X}_{\widehat{\mathcal{P}}}, \mathcal{Y}_{\widehat{\mathcal{P}}})$ generated by $\widehat{\mathcal{P}}$ coincides with $(\mathcal{X}_{\mathcal{P}}, \mathcal{Y}_{\mathcal{P}})$.

Observe that if $\mathcal{C}$ is compactly generated, then by Proposition 1.1 $(\mathcal{X}_{\mathcal{P}}, \mathcal{Y}_{\mathcal{P}})$ is part of a TTF-triple $(\mathcal{X}_{\mathcal{P}}, \mathcal{Y}_{\mathcal{P}}, \mathcal{Z}_{\mathcal{P}})$, where $\mathcal{Z}_{\mathcal{P}} = \mathcal{Y}_{\mathcal{P}}^{\perp}$ and moreover the TTF-class $\mathcal{Y}_{\mathcal{P}}$ is compactly generated as a triangulated category. Observe also that $\mathcal{Y}_{\mathcal{P}} = 0$ if and only if $\mathcal{P}$ generates $\mathcal{C}$. Hence the size of $\mathcal{Y}_{\mathcal{P}}$ measures how far $\mathcal{P}$ is from being a compact generating set in $\mathcal{C}$.

The following result shows that a partial converse of Proposition 1.1 is true if the torsion pair is generated by compact objects.

COROLLARY 1.4. *Let $(\mathcal{X},\mathcal{Y})$ be a hereditary torsion pair of finite type in $\mathcal{C}$. If $\mathcal{X}$ is compactly generated by a set $\mathcal{T}$ and $\mathcal{Y}$ is compactly generated by a set of objects $\mathcal{S}$ which are compact in $\mathcal{C}$, then $\mathcal{C}$ is compactly generated by the set $\mathcal{T}\cup\mathcal{S}$.*

PROOF. By Lemma III.1.2, $\mathcal{T}$ is a set of compact objects in $\mathcal{C}$. Since $\mathcal{S}\subseteq\mathcal{C}^b$, Remark 1.2 implies that $\mathcal{T}\cup\mathcal{S}$ is a set of compact generators for $\mathcal{C}$. □

Assume now that $\mathcal{C}$ is compactly generated and let $(\mathcal{X},\mathcal{Y})$ be a hereditary torsion pair of finite type in $\mathcal{C}$. Then by Proposition 1.1 we have a TTF-triple $(\mathcal{X},\mathcal{Y},\mathcal{Z})$ in $\mathcal{C}$. Recall from Section III.3 that to any compact object P in $\mathcal{C}$ we can associate its dual object $\mathsf{D}_{\mathbb{Q}/\mathbb{Z}}(P)$, such that there exists a natural isomorphism:

$$\omega:\ [\mathcal{C}(P,-),\mathbb{Q}/\mathbb{Z}] \xrightarrow{\cong} \mathcal{C}(-,\mathsf{D}_{\mathbb{Q}/\mathbb{Z}}(P)).$$

The following result shows that taking duals of compact objects from the torsion class $\mathcal{X}$ we get objects in the torsion-free class $\mathcal{Z}$.

PROPOSITION 1.5. (1) *A compact object P lies in $\mathcal{X}$ if and only if its dual object $\mathsf{D}_{\mathbb{Q}/\mathbb{Z}}(P)$ lies in $\mathcal{Z}$. In particular $\mathcal{Y}=0$ if and only if $\mathsf{D}_{\mathbb{Q}/\mathbb{Z}}(P)$ lies in $\mathcal{Z}$ for any compact object P of $\mathcal{C}$.*

(2) *A compact object P lies in $\mathcal{Y}$ if and only if its dual object $\mathsf{D}_{\mathbb{Q}/\mathbb{Z}}(P)$ lies in $\mathcal{Z}^{\perp}$. In particular $\mathcal{X}=0$ (or equivalently $\mathcal{Z}=0$) if and only if all compact objects of $\mathcal{C}$ lie in $\mathcal{Y}$.*

PROOF. Using the natural isomorphism ω we have: $P\in\mathcal{X}$ if and only if $\mathcal{C}(P,\mathcal{Y})=0$ if and only if $[\mathcal{C}(P,\mathcal{Y}),\mathbb{Q}/\mathbb{Z}]=0$ if and only if $\mathcal{C}(\mathcal{Y},\mathsf{D}_{\mathbb{Q}/\mathbb{Z}}(P))=0$ if and only if $\mathsf{D}_{\mathbb{Q}/\mathbb{Z}}(P)\in\mathcal{Y}^{\perp}=\mathcal{Z}$. The last assertion of part (1) is trivial, and part (2) is similar. □

The next result, which is a consequence of Proposition 1.5, shows that the dual objects of a generating set of compact objects of $\mathcal{X}$ form a cogenerating set of $\mathcal{Z}$.

COROLLARY 1.6. *Let $\mathcal{C}$ be a compactly generated triangulated category and let $(\mathcal{X}_{\mathcal{P}},\mathcal{Y}_{\mathcal{P}},\mathcal{Z}_{\mathcal{P}})$ be the TTF-triple generated by a set $\mathcal{P}$ of compact objects from $\mathcal{C}$. Then the set $\{\mathsf{D}_{\mathbb{Q}/\mathbb{Z}}(P)\mid P\in\mathcal{P}\}$ cogenerates $\mathcal{Z}_{\mathcal{P}}$.*

The following three examples illustrate the importance of hereditary torsion pairs of finite type and TTF-triples in concrete situations.

A: The Realization of the Derived Category in the Homotopy Category. Let Λ be a ring and let $\mathcal{H}(\mathrm{Mod}(\Lambda))$ be the unbounded homotopy category of all right Λ-modules.

Let $(\mathcal{X}_\Lambda,\mathcal{Y}_\Lambda)$ be the hereditary torsion pair in $\mathcal{H}(\mathrm{Mod}(\Lambda))$ generated by the stalk complex Λ concentrated in degree zero with stalk Λ. Hence $\mathcal{Y}_\Lambda=\mathcal{P}^{\perp}$ and $\mathcal{X}_\Lambda={}^{\perp}\mathcal{Y}_\Lambda$, where $\mathcal{P}=\{\Sigma^n(\Lambda)\mid n\in\mathbb{Z}\}$. Obviously Λ is a compact object in $\mathcal{H}(\mathrm{Mod}(\Lambda))$, so $(\mathcal{X}_\Lambda,\mathcal{Y}_\Lambda)$ is of finite type. Since the space $\mathcal{H}(\mathrm{Mod}(\Lambda))[\Sigma^n(\Lambda),C^\bullet]$ is identified with the cohomology $\mathrm{H}^{-n}(C^\bullet)$ of the complex $C^\bullet$, it follows that $\mathcal{Y}_\Lambda$ is identified with the unbounded homotopy category $\mathcal{H}_{\mathsf{Ac}}(\mathrm{Mod}(\Lambda))$ of acyclic complexes and by Remark III.2.7, $\mathcal{X}_\Lambda$ is identified with the localizing subcategory of $\mathcal{H}(\mathrm{Mod}(\Lambda))$ generated by Λ, which we denote by $\mathcal{H}_{\mathsf{P}}(\mathrm{Mod}(\Lambda))$. Dually we denote by $\mathcal{H}^{\mathsf{I}}(\mathrm{Mod}(\Lambda))$ the colocalizing subcategory of $\mathcal{H}(\mathrm{Mod}(\Lambda))$ generated by an injective cogenerator of $\mathrm{Mod}(\Lambda)$. In the literature the complexes in $\mathcal{H}_{\mathsf{P}}(\mathrm{Mod}(\Lambda))$, resp.

$\mathcal{H}^{\mathsf{I}}(\mathrm{Mod}(\Lambda))$, are known as *homotopically projective*, resp. *injective*, complexes, see [**72**] for an explicit description of homotopically projective, resp. injective, complexes as direct, resp. inverse, limits of special complexes of projectives, resp. injectives.

The following well-known basic result shows that the unbounded derived category can be realized as a full subcategory of the homotopy category of complexes, hence it has small hom-sets, in particular $\mathbf{D}(\mathrm{Mod}(\Lambda))$ "exists".

COROLLARY 1.7. [**72**], [**21**], [**31**] *There exists a TTF-triple*

$$\Big(\mathcal{H}_{\mathsf{P}}(\mathrm{Mod}(\Lambda)),\ \mathcal{H}_{\mathsf{Ac}}(\mathrm{Mod}(\Lambda)),\ \mathcal{H}^{\mathsf{I}}(\mathrm{Mod}(\Lambda))\Big)$$

in $\mathcal{H}(\mathrm{Mod}(\Lambda))$ *which induces triangle equivalences:*

$$\mathcal{H}_{\mathsf{P}}(\mathrm{Mod}(\Lambda)) \xleftarrow{\approx} \mathbf{D}(\mathrm{Mod}(\Lambda)) \xrightarrow{\approx} \mathcal{H}^{\mathsf{I}}(\mathrm{Mod}(\Lambda)).$$

PROOF. By Proposition I.2.6, we have triangle equivalences $\mathbf{D}(\mathrm{Mod}(\Lambda)) = \mathcal{H}(\mathrm{Mod}(\Lambda))/\mathcal{H}_{\mathsf{Ac}}(\mathrm{Mod}(\Lambda)) \xrightarrow{\approx} {}^{\perp}\mathcal{H}_{\mathsf{Ac}}(\mathrm{Mod}(\Lambda)) = \mathcal{H}_{\mathsf{P}}(\mathrm{Mod}(\Lambda))$ which show the first part. By [**72**] the inclusion $\mathcal{H}_{\mathsf{Ac}}(\mathrm{Mod}(\Lambda)) \hookrightarrow \mathcal{H}(\mathrm{Mod}(\Lambda))$ admits a right adjoint and $\mathcal{H}_{\mathsf{Ac}}(\mathrm{Mod}(\Lambda))^{\perp} = \mathcal{H}^{\mathsf{I}}(\mathrm{Mod}(\Lambda))$, so we have the desired TTF-triple and in the same way as above, the right hand side equivalence. □

Observe that by Corollary 1.7 the torsion class $\mathcal{H}_{\mathsf{P}}(\mathrm{Mod}(\Lambda))$ and the torsion-free class $\mathcal{H}^{\mathsf{I}}(\mathrm{Mod}(\Lambda))$ are compactly generated. However in general the TTF-class $\mathcal{H}_{\mathsf{Ac}}(\mathrm{Mod}(\Lambda))$ is not compactly generated. Indeed if $\Lambda = \mathbb{Z}$ is the ring of integers, then by [**86**] it follows that $\mathcal{H}(\mathrm{Mod}(\mathbb{Z}))$ is not generated by a set, hence it is not compactly generated. Hence by Remark 1.2 the same is true for $\mathcal{H}_{\mathsf{Ac}}(\mathrm{Mod}(\mathbb{Z}))$.

B: The Projective Case. The above construction can be performed in the unbounded homotopy category $\mathcal{H}(\mathbf{P}_{\Lambda})$ of the category $\mathbf{P}_{\Lambda}$ of projective right Λ-modules, which obviously has all small coproducts. This leads to interesting connections with the stable module category which will be used later in Keller's Morita Theorem for stable categories.

Consider the hereditary torsion pair of finite type $\big({}^{\perp}\mathcal{H}_{\mathsf{Ac}}(\mathbf{P}_{\Lambda}), \mathcal{H}_{\mathsf{Ac}}(\mathbf{P}_{\Lambda})\big)$ in $\mathcal{H}(\mathbf{P}_{\Lambda})$ generated by the compact object $\Lambda \in \mathcal{H}(\mathbf{P}_{\Lambda})$. Observe that the torsion-free class $\mathcal{H}_{\mathsf{Ac}}(\mathbf{P}_{\Lambda})$ is the costabilization of the stable module category $\underline{\mathrm{Mod}}(\Lambda)$ modulo projectives via the exact functor $\mathcal{H}_{\mathsf{Ac}}(\mathbf{P}_{\Lambda}) \to \underline{\mathrm{Mod}}(\Lambda)$ which sends an acyclic complex of projectives $P^{\bullet}$ to the class of $\mathrm{Im}(d^{-1})$ in the stable category [**20**]. This means that the above functor is the universal exact functor from a triangulated category to $\underline{\mathrm{Mod}}(\Lambda)$. In particular if Λ is quasi-Frobenius, then the above functor is a triangle equivalence $\mathcal{H}_{\mathsf{Ac}}(\mathbf{P}_{\Lambda}) \xrightarrow{\approx} \underline{\mathrm{Mod}}(\Lambda)$.

PROPOSITION 1.8. *If Λ is left coherent and right perfect, or if Λ has finite right global dimension, then there exists a TTF-triple in* $\mathcal{H}(\mathbf{P}_{\Lambda})$*:*

$$\Big({}^{\perp}\mathcal{H}_{\mathsf{Ac}}(\mathbf{P}_{\Lambda}),\ \mathcal{H}_{\mathsf{Ac}}(\mathbf{P}_{\Lambda}),\ \mathcal{H}_{\mathsf{Ac}}(\mathbf{P}_{\Lambda})^{\perp}\Big).$$

PROOF. If Λ is left coherent and right perfect, then by [**67**] it follows that $\mathcal{H}_{\mathsf{Ac}}(\mathbf{P}_{\Lambda})$ is compactly generated. This also follows using that the stable category modulo projectives $\underline{\mathrm{Mod}}(\Lambda)$ is a compactly generated pretriangulated category [**24**] and $\mathcal{H}_{\mathsf{Ac}}(\mathbf{P}_\Lambda)$ is its costabilization [**20**]. Then Proposition 1.1 implies the existence of the desired TTF-triple.

If r.gl.dim$\Lambda < \infty$, then by [**99**] the homotopy category $\mathcal{H}(\mathbf{P}_\Lambda)$ is compactly generated. Hence the assertion follows from Proposition 1.1. □

Under the assumptions of Proposition 1.8, the torsion class ${}^\perp\mathcal{H}_{\mathsf{Ac}}(\mathbf{P}_\Lambda)$ in $\mathcal{H}(\mathbf{P}_\Lambda)$ coincides with the torsion class ${}^\perp\mathcal{H}_{\mathsf{Ac}}(\mathrm{Mod}(\Lambda))$ in $\mathcal{H}(\mathrm{Mod}(\Lambda))$, hence by Corollary 1.7 it is triangle equivalent to the unbounded derived category $\mathbf{D}(\mathrm{Mod}(\Lambda))$. This follows from the fact that ${}^\perp\mathcal{H}_{\mathsf{Ac}}(\mathrm{Mod}(\Lambda))$ is contained in $\mathcal{H}(\mathbf{P}_\Lambda)$, by the discussion preceding Corollary 1.7.

REMARK 1.9. Let Λ be a left coherent and right perfect ring. Since $\mathcal{H}_{\mathsf{Ac}}(\mathbf{P}_\Lambda)$ and ${}^\perp\mathcal{H}_{\mathsf{Ac}}(\mathbf{P}_\Lambda)$ are compactly generated, it follows by Remark 1.2 that the homotopy category $\mathcal{H}(\mathbf{P}_\Lambda)$ of projective modules is generated by a set.

We have seen that in general $\mathcal{H}(\mathrm{Mod}(\Lambda))$ is not generated by a set. However $\mathcal{H}(\mathrm{Mod}(\Lambda))$ is generated by a set if Λ is a ring of finite representation type. Indeed, this follows from the above observations since in this case the category $\mathrm{Mod}(\Lambda)$ is equivalent to the category of all projective modules over Γ, where Γ is the Auslander ring of Λ, which is left coherent and right perfect. Recall that the Auslander ring of a representation finite ring Λ is the endomorphism ring of a finitely presented Λ-module X such that $\mathrm{mod}(\Lambda) = \mathrm{add}(X)$. More generally one can show that $\mathcal{H}(\mathrm{Mod}(\Lambda))$ is generated by a set if Λ is a right pure semisimple ring.

C: Idempotent Functors. Let Λ be a QF-ring and let $\underline{\mathrm{Mod}}(\Lambda)$ be the stable category modulo projectives. It is triangulated with coproducts, and the stable category $\underline{\mathrm{mod}}(\Lambda)$ of the finitely presented modules is a skeletally small thick generating subcategory which is identified, up to equivalence, with the compact objects of $\underline{\mathrm{Mod}}(\Lambda)$. If $\mathcal{P}$ is a set of finitely presented modules, then the stable category $\underline{\mathcal{P}}$ is a set of compact objects in $\underline{\mathrm{Mod}}(\Lambda)$. Hence $\underline{\mathcal{P}}$ generates a torsion pair of finite type in $\underline{\mathrm{Mod}}(\Lambda)$. It follows that any thick subcategory $\mathcal{S} \subseteq \underline{\mathrm{mod}}(\Lambda)$ generates a hereditary torsion pair $(\mathcal{E}_\mathcal{S}, \mathcal{F}_\mathcal{S})$ in $\underline{\mathrm{Mod}}(\Lambda)$. Consequently any module C sits in a triangle $\mathsf{E}_\mathcal{S}(\underline{C}) \to \underline{C} \to \mathsf{F}_\mathcal{S}(\underline{C}) \to \Sigma\mathsf{E}_\mathcal{S}(\underline{C})$, with $\mathsf{E}_\mathcal{S}(\underline{C}) \in \mathcal{E}_\mathcal{S}$ and $\mathsf{F}_\mathcal{S}(\underline{C}) \in \mathcal{F}_\mathcal{S}$. If $\Lambda = kG$ is the group algebra over a field k of a finite group G, then the functors $\mathsf{E}_\mathcal{S}$, $\mathsf{F}_\mathcal{S}$ are the idempotent functors of Rickard [**94**], associated to the (tensor ideal) thick subcategory $\mathcal{S}$.

If $(\mathcal{X}, \mathcal{Y})$ is a hereditary torsion pair of finite type in $\mathcal{C}$, then it is natural to ask if $(\mathcal{X}, \mathcal{Y})$ restricts to a torsion pair in the full subcategory of compact objects of $\mathcal{C}$. This question has connections with K-theory and our results will be applied in the next section to give a proof of a result of Happel without using Auslander-Reiten theory.

From now on we assume that $\mathcal{C}$ is compactly generated and $(\mathcal{X}, \mathcal{Y})$ is a hereditary torsion pair of finite type in $\mathcal{C}$. As usual $\mathcal{C}^b$ denotes the full subcategory of compact objects of $\mathcal{C}$.

We fix a generating set $\mathcal{T}$ of compact objects in $\mathcal{C}$. By Proposition 1.1, we know that there exists a TTF-triple $(\mathcal{X}, \mathcal{Y}, \mathcal{Z})$ in $\mathcal{C}$ where the TTF-class $\mathcal{Y}$ is compactly generated as a triangulated category by the set $\mathbf{L}(\mathcal{T})$. Also by Lemma III.1.2 the functor $\mathbf{R} : \mathcal{C} \to \mathcal{X}$ preserves coproducts. Since $\mathcal{C}$ is compactly generated, by [**85**], $\mathbf{R}$ admits a right adjoint $\mathbf{h} : \mathcal{X} \to \mathcal{C}$. Using that $\mathbf{R}$ admits the fully faithful left adjoint $\mathbf{i}$, it is not difficult to see that $\mathbf{h}$ is fully faithful. Since $\mathcal{Y}$ is closed under coproducts in $\mathcal{C}$, by [**85**] the inclusion $\mathbf{j} : \mathcal{Y} \hookrightarrow \mathcal{C}$ admits a right adjoint $\mathbf{G} : \mathcal{C} \to \mathcal{Y}$.

LEMMA 1.10. (1) *The essential image of the functor* $\mathbf{h} : \mathcal{X} \hookrightarrow \mathcal{C}$ *is the torsion-free class* $\mathcal{Z}$. *Moreover the reflection functor* $\mathcal{C} \to \mathcal{Z}$ *is isomorphic to* $\mathbf{hR}$.

(2) *The kernel of the functor* $\mathbf{G} : \mathcal{C} \to \mathcal{Y}$ *is the torsion-free class* $\mathcal{Z}$. *Moreover* $\mathbf{G}$ *is the coreflection functor for* $\mathcal{Y}$.

PROOF. (1) Let X be in $\mathcal{X}$. Then $\mathcal{C}(\mathcal{Y}, \mathbf{h}(X)) \cong \mathcal{X}(\mathbf{R}(\mathcal{Y}), X) = 0$. Hence $\mathbf{h}(X) \in \mathcal{Y}^{\perp} = \mathcal{Z}$. Hence $\mathbf{h}$ has image in $\mathcal{Z}$. Now if C is an object in $\mathcal{C}$, let $A \to C \to \mathbf{hR}(C) \to \Sigma(A)$ be a triangle in $\mathcal{C}$ where $C \to \mathbf{hR}(C)$ is the unit of the adjoint pair $(\mathbf{R}, \mathbf{h})$ evaluated at C. Applying to this triangle the functor $\mathcal{C}(\mathcal{X}, -)$ we see easily that $\mathcal{C}(\mathcal{X}, A) = 0$, hence $A \in \mathcal{X}^{\perp} = \mathcal{Y}$. It follows that $\mathbf{hR}$ is the coreflection of C in $\mathcal{Z}$. In particular if Z is in $\mathcal{Z}$, then the morphism $Z \to \mathbf{hR}(Z)$ is invertible and this shows that the essential image of $\mathbf{h}$ coincides with $\mathcal{Z}$.

(2) The proof is similar. □

Keeping the above notation we have the following result which gives necessary and sufficient conditions for a hereditary torsion pair in $\mathcal{C}$ to induce a torsion pair in the full subcategory $\mathcal{C}^b$ of compact objects.

PROPOSITION 1.11. *Let* $\mathcal{C}$ *be a compactly generated triangulated category and let* $\mathcal{T}$ *be a set of compact generators of* $\mathcal{C}$. *If* $(\mathcal{X}, \mathcal{Y})$ *is a hereditary torsion pair of finite type in* $\mathcal{C}$, *then the following conditions are equivalent.*

(i) $\mathbf{R} : \mathcal{C} \to \mathcal{X}$ *preserves compact objects.*
(ii) $\mathbf{h} : \mathcal{X} \hookrightarrow \mathcal{C}$ *preserves coproducts.*
(iii) $\mathcal{Z}$ *is closed under coproducts, i.e. the torsion pair* $(\mathcal{Y}, \mathcal{Z})$ *is of finite type.*
(iv) $\mathbf{G} : \mathcal{C} \to \mathcal{Y}$ *preserves coproducts.*
(v) *Any compact object of* $\mathcal{Y}$ *remains compact in* $\mathcal{C}$.
(vi) $(\mathcal{X}^b, \mathcal{Y}^b)$ *is a hereditary torsion pair in* $\mathcal{C}^b$.
(vii) *The inclusion* $\mathcal{X}^b \hookrightarrow \mathcal{C}^b$ *admits a right adjoint.*

If one of the above equivalent conditions holds, then the torsion pair $(\mathcal{X}, \mathcal{Y})$ *is compactly generated by the set of objects* $\mathbf{R}(\mathcal{T})$, *and* $\mathcal{X}^b = \mathcal{X} \cap \mathcal{C}^b$.

PROOF. Since $\mathbf{h}$ is a right adjoint of $\mathbf{R}$, that (i) is equivalent to (ii) follows from [**85**]. Since $\mathcal{Y}$ is compactly generated, by [**85**] we have that the inclusion $\mathbf{j} : \mathcal{Y} \to \mathcal{C}$ preserves compact objects if and only if its right adjoint $\mathbf{G}$ preserves coproducts, hence (iv) is equivalent to (v). Since $\mathbf{G}$ is the coreflection functor for the hereditary torsion pair $(\mathcal{Y}, \mathcal{Z})$ by Lemma III.1.2, we have the equivalence of (iii) and (iv). Hence the first five conditions are equivalent. If one of these holds, then let T be a compact object of $\mathcal{C}$ and consider the glueing triangle $\mathbf{R}(T) \to T \to \mathbf{L}(T) \to \Sigma\mathbf{R}(T)$. By (i) we know that $\mathbf{R}(T) \in \mathcal{X}^b$ and by Lemma III.1.2 we know that $\mathbf{L}(T) \in \mathcal{Y}^b$. Since obviously $\mathcal{C}(\mathcal{X}^b, \mathcal{Y}^b) = 0$, we have that $(\mathcal{X}^b, \mathcal{Y}^b)$ is a hereditary torsion pair in $\mathcal{C}^b$. If this is true, then obviously $\mathbf{R}$ preserves compact objects. Finally it is easy to see that (vi) is equivalent to (vii). Hence all conditions are equivalent.

Finally we claim that if $\mathbf{R}$ preserves compactness, then the set $\mathbf{R}(\mathcal{T})$ generates $\mathcal{X}$. Indeed, if X is in $\mathcal{X}$ and $\mathcal{X}(\mathbf{R}(\mathcal{T}), X) = 0$, then $\mathcal{C}(\mathcal{T}, \mathbf{h}(X)) = 0$ and then

$\mathbf{h}(X) = 0$ since $\mathcal{T}$ is a generating set in $\mathcal{C}$. Since $\mathbf{h}$ is fully faithful, we have $X = 0$. Hence $\mathbf{R}(\mathcal{T})$ generates $\mathcal{X}$. □

If the torsion pair $(\mathcal{X}, \mathcal{Y})$ is generated by a set $\mathcal{P}$ of compact objects of $\mathcal{C}$, then $\mathcal{X}^b$ is the thick subcategory $\mathsf{thick}(\mathcal{P})$ of $\mathcal{C}$ generated by $\mathcal{P}$. The Localization Theorem of Neeman-Ravenel, see [**86**], asserts that the triangulated quotient $\mathcal{C}^b/\mathsf{thick}(\mathcal{P})$ is fully embedded in $\mathcal{Y}^b = (\mathcal{C}/\mathcal{X})^b$ and the latter is the closure of the former under direct summands. The following consequence of Proposition 1.11 gives a simple proof of Neeman-Ravenel's Theorem under an additional assumption.

COROLLARY 1.12. *Let $\mathcal{P}$ be a set of compact objects in $\mathcal{C}$ and let $(\mathcal{X}_\mathcal{P}, \mathcal{Y}_\mathcal{P})$ be the hereditary torsion pair in $\mathcal{C}$ generated by $\mathcal{P}$. Then $\mathbf{R} : \mathcal{C} \longrightarrow \mathcal{X}_\mathcal{P}$ preserves compact objects if and only if $(\mathcal{X}^b_\mathcal{P}, \mathcal{Y}^b_\mathcal{P})$ is a hereditary torsion pair in $\mathcal{C}^b$ if and only if the inclusion $\mathsf{thick}(\mathcal{P}) \hookrightarrow \mathcal{C}^b$ admits a right adjoint. In this case we have a triangle equivalence $\mathcal{C}^b/\mathsf{thick}(\mathcal{P}) \xrightarrow{\approx} \mathcal{Y}^b$ and an isomorphism $\mathrm{K}_0(\mathcal{C}^b) \xrightarrow{\cong} \mathrm{K}_0(\mathsf{thick}(\mathcal{P})) \oplus \mathrm{K}_0(\mathcal{Y}^b)$.*

2. Hereditary Torsion Pairs and Tilting

Working in a compactly generated triangulated category which is of algebraic origin [**73**] (for instance derived categories, stable categories) and using the techniques of differential graded algebras developed by Keller [**69**], we can describe in more explicit terms the torsion class of the hereditary torsion pair generated by a compact object. This description provides simple proofs and slight generalizations of the Morita Theorem for derived categories of Rickard [**91**], of the Morita Theorem for stable categories of Keller [**69**] and in particular of Happel's Theorem [**57**] on the description of the derived category of an Artin algebra in terms of the stable category of its repetitive algebra. Thus using torsion-theoretic methods we obtain simple proofs and at the same time a unified approach to the above mentioned results. A similar approach was considered independently by Dwyer and Greenlees in [**43**], although all the main ideas are due to Keller [**69**]. In contrast to the previous chapter where the investigation of the heart of a torsion pair was crucial in the construction of derived equivalences, we use in this section hereditary torsion pairs, so the heart is trivial. Therefore it is not possible to use the machinery of the last chapter. In a sense the following results are relative versions of Corollary 1.7.

Rickard's Theorem. Let Λ be a ring and let $T^\bullet$ be a fixed compact object in the unbounded derived category $\mathbf{D}(\mathrm{Mod}(\Lambda))$, i.e. $T^\bullet$ is a complex quasi-isomorphic to a bounded complex with components finitely generated projective modules. Let $(\mathcal{X}_{T^\bullet}, \mathcal{Y}_{T^\bullet})$ be the hereditary torsion pair in $\mathbf{D}(\mathrm{Mod}(\Lambda))$ generated by $T^\bullet$. Let $\Gamma = \mathsf{DGHom}_\Lambda(T^\bullet, T^\bullet)$ be the DG-algebra induced by the complex $T^\bullet$ and let $\mathbf{D}(\Gamma)$ be the derived category of Γ in the sense of [**69**]. By Proposition 1.1 we know that $\mathcal{Y}_{T^\bullet}$ is a TTF-class, so we have a TTF-triple $(\mathcal{X}_{T^\bullet}, \mathcal{Y}_{T^\bullet}, \mathcal{Z}_{T^\bullet})$ in $\mathbf{D}(\mathrm{Mod}(\Lambda))$.

The following extends the bounded version of the Morita Theorem proved in Corollary III.4.4 via the heart, and contains a version of Rickard's Morita Theorem for unbounded derived categories.

THEOREM 2.1. [**91**], [**93**], [**69**] *Using the above notation, we have the following:*

(i) *There are triangle equivalences:*

$$\mathbf{D}(\mathrm{Mod}(\Lambda)) \hookleftarrow \mathcal{X}_{T^\bullet} \xleftarrow{\approx} \mathbf{D}(\Gamma) \xrightarrow{\approx} \mathcal{Z}_{T^\bullet}$$

(ii) *The inclusion* $\mathbf{D}(\Gamma) \hookrightarrow \mathbf{D}(\mathrm{Mod}(\Lambda))$ *sends* Γ *to* $T^\bullet$ *and admits a right adjoint.*

(iii) $T^\bullet$ *is a generator of* $\mathbf{D}(\mathrm{Mod}(\Lambda))$ *if and only if the inclusion* $\mathbf{D}(\Gamma) \hookrightarrow \mathbf{D}(\mathrm{Mod}(\Lambda))$ *induces a triangle equivalence:*

$$\mathbf{D}(\Gamma) \xrightarrow{\approx} \mathbf{D}(\mathrm{Mod}(\Lambda)).$$

(iv) *If* $\mathrm{Hom}_{\mathbf{D}(\mathrm{Mod}(\Lambda))}(T^\bullet, \Sigma^n(T^\bullet)) = 0$, $\forall n \neq 0$, *then the DG-algebra* Γ *is the ordinary ring* $\mathrm{End}_{\mathbf{D}(\mathrm{Mod}(\Lambda))}(T^\bullet)$. *In particular if* $T^\bullet$ *is a tilting complex, then we have a triangle equivalence* $\mathbf{D}(\mathrm{Mod}(\Gamma)) \xrightarrow{\approx} \mathbf{D}(\mathrm{Mod}(\Lambda))$ *sending* Γ *to* $T^\bullet$.

PROOF. (i) As in [**69**], $T^\bullet$ can be considered as Γ-Λ-bimodule and then the total right derived functor $\mathbb{R}\mathrm{Hom}_\Lambda(T^\bullet, -) : \mathbf{D}(\mathrm{Mod}(\Lambda)) \to \mathbf{D}(\Gamma)$ is defined and admits as a left adjoint the total left derived functor $- \otimes^{\mathbb{L}}_\Gamma T^\bullet : \mathbf{D}(\Gamma) \to \mathbf{D}(\mathrm{Mod}(\Lambda))$. We first show that the image of $- \otimes^{\mathbb{L}}_\Gamma T^\bullet$ lies in the torsion class $\mathcal{X}_{T^\bullet}$. Indeed for any DG-module M and for any torsion-free object $Y^\bullet \in \mathcal{Y}_{T^\bullet}$ we have:

$$\mathrm{Hom}_{\mathbf{D}(\mathrm{Mod}(\Lambda))}\big(M \otimes^{\mathbb{L}}_\Gamma T^\bullet, Y^\bullet\big) \xrightarrow{\cong} \mathrm{Hom}_{\mathbf{D}(\Gamma)}\big(M, \mathbb{R}\mathrm{Hom}_\Lambda(T^\bullet, Y^\bullet)\big)$$

which is zero by the construction of the torsion-free class $\mathcal{Y}_{T^\bullet}$. It follows that we can consider the exact functor $- \otimes^{\mathbb{L}}_\Gamma T^\bullet : \mathbf{D}(\Gamma) \to \mathcal{X}_{T^\bullet}$ which obviously preserves all small coproducts and moreover we have obviously $\Gamma \otimes^{\mathbb{L}}_\Gamma T^\bullet = T^\bullet$. Now the torsion class $\mathcal{X}_{T^\bullet}$ is a compactly generated triangulated category with $T^\bullet$ as a compact generator. It is well-known that a coproduct preserving exact functor between compactly generated triangulated categories is a triangle equivalence if and only if it induces an equivalence between the full subcategories of compact objects, see Lemma 4.2 in [**69**]. Therefore we infer that $- \otimes^{\mathbb{L}}_\Gamma T^\bullet : \mathbf{D}(\Gamma) \to \mathcal{X}_{T^\bullet}$ is a triangle equivalence with quasi-inverse the restriction $\mathbb{R}\mathrm{Hom}_\Lambda(T^\bullet, -) : \mathcal{X}_{T^\bullet} \to \mathbf{D}(\Gamma)$. This completes the proof of (i) since by Corollary I.2.9, we always have a triangle equivalence $\mathcal{X}_{T^\bullet} \xrightarrow{\approx} \mathcal{Z}_{T^\bullet}$.

(ii) Since $\mathcal{X}_{T^\bullet}$ is a torsion class, the inclusion $\mathcal{X}_{T^\bullet} \hookrightarrow \mathbf{D}(\mathrm{Mod}(\Lambda))$ admits a right adjoint and the assertion follows from (i).

(iii) If $T^\bullet$ is a generator of $\mathbf{D}(\mathrm{Mod}(\Lambda))$ then by construction it follows trivially that the torsion-free class $\mathcal{Y}_{T^\bullet}$ is 0. Hence the torsion class $\mathcal{X}_{T^\bullet}$ coincides with the whole derived category $\mathbf{D}(\mathrm{Mod}(\Lambda))$ and then the assertion follows by (i). The converse is trivial, since Γ is a generator of $\mathbf{D}(\Gamma)$.

(iv) If $T^\bullet$ is a tilting object, then by (i), (ii), we have a triangle equivalence $\mathbf{D}(\Gamma) \xrightarrow{\approx} \mathbf{D}(\mathrm{Mod}(\Lambda))$. Since $\mathrm{Hom}_{\mathbf{D}(\mathrm{Mod}(\Lambda))}(T^\bullet, \Sigma^n(T^\bullet)) = 0, \forall n \neq 0$, the cohomology of the DG-algebra Γ is concentrated in degree zero. This implies that Γ is (quasi-)isomorphic to $\mathrm{End}_{\mathbf{D}(\mathrm{Mod}(\Lambda))}(T^\bullet)$ and then $\mathbf{D}(\Gamma) \xrightarrow{\approx} \mathbf{D}(\mathrm{Mod}(\Gamma))$. □

The above proof shows that the idempotent functor $\mathbf{iR} : \mathbf{D}(\mathrm{Mod}(\Lambda)) \to \mathbf{D}(\mathrm{Mod}(\Lambda))$ associated to the torsion class $\mathcal{X}_{T^\bullet}$ is given by $\mathbb{R}\mathrm{Hom}_\Lambda(T^\bullet, -) \otimes^{\mathbb{L}}_\Gamma T^\bullet : \mathbf{D}(\mathrm{Mod}(\Lambda)) \to \mathbf{D}(\mathrm{Mod}(\Lambda))$.

Keller's Theorem. Now let $\mathcal{F}$ be an exact category in the sense of Quillen, see [**57**]. Recall from [**57**] that $\mathcal{F}$ is called *Frobenius category* if $\mathcal{F}$ has enough projective and injective objects, and the projectives coincide with the injectives. Throughout we fix an exact Frobenius category $\mathcal{F}$ which admits all small coproducts, for instance $\mathrm{Mod}(\Lambda)$ where Λ is a quasi-Frobenius ring. Then trivially the stable category $\underline{\mathcal{F}}$ of $\mathcal{F}$ modulo projectives has all small coproducts. We fix a compact object T in $\underline{\mathcal{F}}$ and let $(\mathcal{X}_T, \mathcal{Y}_T)$ be the hereditary torsion pair in $\underline{\mathcal{F}}$ generated by T. If $\underline{\mathcal{F}}$ admits a set of compact generators, then, by Proposition 1.1, we have that $\mathcal{Y}_T$ is a TTF-class so we have a TTF-triple $(\mathcal{X}_T, \mathcal{Y}_T, \mathcal{Z}_T)$ in $\underline{\mathcal{F}}$. By [**69**], the stable category $\underline{\mathcal{F}}$ is triangle equivalent to the homotopy category $\mathcal{H}_{\mathsf{Ac}}(\mathcal{P})$ of acyclic complexes of projectives of $\mathcal{F}$. Let $T^\bullet$ be the object in $\mathcal{H}_{\mathsf{Ac}}(\mathcal{P})$ which corresponds to T via the above equivalence; this amounts to choosing a Tate resolution of T, i.e. an unbounded acyclic complex $\cdots \longrightarrow P^{-1} \xrightarrow{d^{-1}} P^0 \xrightarrow{d^0} P^1 \longrightarrow \cdots$ of projectives such that $\mathrm{Im}(d^{-1}) \cong T$ in $\underline{\mathcal{F}}$. Let Γ be the DG-algebra $\mathsf{DGHom}(T^\bullet, T^\bullet)$ induced by the complex $T^\bullet$. The following is essentially Keller's Morita Theorem for stable categories.

THEOREM 2.2. [**69**] *Using the above notation, we have the following:*

(i) *There are triangle equivalences:*

$$\underline{\mathcal{F}} \hookleftarrow \mathcal{X}_T \xleftarrow{\approx} \mathbf{D}(\Gamma) \xrightarrow{\approx} \mathcal{Z}_T.$$

(ii) *The inclusion* $\mathbf{D}(\Gamma) \hookrightarrow \underline{\mathcal{F}}$ *admits a right adjoint which sends* T *to* Γ.

(iii) T *is a generator of* $\underline{\mathcal{F}}$ *if and only if the above inclusion* $\mathbf{D}(\Gamma) \hookrightarrow \underline{\mathcal{F}}$ *induces a triangle equivalence:*

$$\mathbf{D}(\Gamma) \xrightarrow{\approx} \underline{\mathcal{F}}.$$

(iv) *If* T *is a tilting object, then the DG-algebra* Γ *is the ordinary ring* $\mathrm{End}_{\underline{\mathcal{F}}}(T^\bullet)$ *and then we have a triangle equivalence* $\underline{\mathcal{F}} \xrightarrow{\approx} \mathbf{D}(\mathrm{Mod}(\Gamma))$.

PROOF. (i) Identifying $\underline{\mathcal{F}}$ with $\mathcal{H}_{\mathsf{Ac}}(\mathcal{P})$, the object $T^\bullet$ induces an exact functor $G : \underline{\mathcal{F}} \to \mathbf{D}(\Gamma)$, given by $G(P^\bullet) = \mathrm{Hom}(T^\bullet, P^\bullet)$, which preserves coproducts. Then, by restriction, G induces an exact functor $G : \mathcal{X}_T \to \mathbf{D}(\Gamma)$ which preserves coproducts and sends the compact generator T of $\mathcal{X}_T$ to the compact generator Γ of $\mathbf{D}(\Gamma)$. Then, as in the proof of Theorem 2.1, we infer that G induces a triangle equivalence $G : \mathcal{X}_T \xrightarrow{\approx} \mathbf{D}(\Gamma)$. Parts (ii), (iii), (iv) follow as in Theorem 2.1. □

REMARK 2.3. In case we work with a set of compact objects, then the DG-algebra Γ above can be replaced by the induced DG-category in the sense of [**69**].

Happel's Theorem. Let Λ be an Artin algebra. For the definition of the repetitive algebra $\widehat{\Lambda}$ associated to Λ we refer to the book of Happel [**57**]. In particular by [**57**] we have that $\widehat{\Lambda}$ is self-injective, hence the stable category $\underline{\mathrm{Mod}}(\widehat{\Lambda})$ is a compactly generated triangulated category and moreover it is easy to see that $\underline{\mathrm{Mod}}(\widehat{\Lambda})^b = \underline{\mathrm{mod}}(\widehat{\Lambda})$. In addition $\mathrm{Mod}(\Lambda)$ is fully embedded in the stable category $\underline{\mathrm{Mod}}(\widehat{\Lambda})$ and then, via this embedding, Λ is a compact object in $\underline{\mathrm{Mod}}(\widehat{\Lambda})$. Consider the TTF-triple $(\mathcal{X}_\Lambda, \mathcal{Y}_\Lambda, \mathcal{Z}_\Lambda)$ in $\underline{\mathrm{Mod}}(\widehat{\Lambda})$ generated by Λ.

The following consequence of Theorem 2.2 is due to Happel. Later Keller provided a simpler proof and extended the result to the big module category. We give a torsion theoretic proof of Keller's version below.

THEOREM 2.4. [**57**], [**68**] *Let Λ be an Artin algebra and let $\widehat{\Lambda}$ be its repetitive algebra.*

(i) *There are triangle equivalences*

$$\underline{\mathrm{Mod}}(\widehat{\Lambda}) \hookleftarrow \mathcal{X}_\Lambda \xleftarrow{\approx} \mathbf{D}(\mathrm{Mod}(\Lambda)) \xrightarrow{\approx} \mathcal{Z}_\Lambda$$

which induce a full embedding $\mathcal{H}^b(\mathcal{P}_\Lambda) \hookrightarrow \underline{\mathrm{mod}}(\widehat{\Lambda})$.

(ii) *The inclusion $\mathsf{H} : \mathbf{D}(\mathrm{Mod}(\Lambda)) \hookrightarrow \underline{\mathrm{Mod}}(\widehat{\Lambda})$ admits a right adjoint $\mathbf{R}$: $\underline{\mathrm{Mod}}(\widehat{\Lambda}) \to \mathbf{D}(\mathrm{Mod}(\Lambda))$ which sends Λ to Λ.*

(iii) *The inclusion $\mathsf{H} : \mathbf{D}(\mathrm{Mod}(\Lambda)) \hookrightarrow \underline{\mathrm{Mod}}(\widehat{\Lambda})$ induces a triangle equivalence*

$$\mathbf{D}(\mathrm{Mod}(\Lambda)) \xrightarrow{\approx} \underline{\mathrm{Mod}}(\widehat{\Lambda})$$

if and only if Λ generates $\underline{\mathrm{Mod}}(\widehat{\Lambda})$ if and only if Λ is a tilting object in $\underline{\mathrm{Mod}}(\widehat{\Lambda})$ if and only if $\mathbf{R}$ preserves compact objects if and only if the inclusion $\mathcal{H}^b(\mathcal{P}_\Lambda) \hookrightarrow \underline{\mathrm{mod}}(\widehat{\Lambda})$ admits a right adjoint if and only if $\operatorname{gldim}\Lambda < \infty$.

PROOF. All the assertions, except for the last two in part (iii), are consequences of our previous results. If Λ is a tilting object, then H and consequently $\mathbf{R}$ are equivalences. Then trivially $\mathbf{R}$ preserves compactness. Conversely if the right adjoint $\mathbf{R}$ of H preserves compact objects, then by Proposition 1.9 any compact object of $\mathcal{Y}_\Lambda$ is a compact object in $\underline{\mathrm{Mod}}(\widehat{\Lambda})$, hence we have an inclusion $\mathcal{Y}^b_\Lambda \subseteq \underline{\mathrm{mod}}(\widehat{\Lambda})$. By the definition of $\mathcal{Y}_\Lambda$, inside $\underline{\mathrm{mod}}(\widehat{\Lambda})$ we have: $\mathcal{Y}^b_\Lambda \cap \mathrm{mod}(\Lambda) = 0$. Since $\mathrm{mod}(\Lambda)$ generates $\underline{\mathrm{mod}}(\widehat{\Lambda})$, we infer that $\mathcal{Y}^b_\Lambda = 0$. Since $\mathcal{Y}_\Lambda$ is compactly generated, this implies that $\mathcal{Y}_\Lambda = 0$, hence H and consequently $\mathbf{R}$ are equivalences. If this condition holds, then Λ generates $\underline{\mathrm{mod}}(\widehat{\Lambda})$. This implies easily by devissage that Λ has finite global dimension. Conversely if Λ has finite global dimension, then obviously Λ generates $\underline{\mathrm{mod}}(\widehat{\Lambda})$ as a thick subcategory and consequently Λ generates $\underline{\mathrm{Mod}}(\widehat{\Lambda})$ as category with infinite coproducts. Hence the latter coincides with $\mathcal{X}_\Lambda$ and consequently H is an equivalence. □

REMARK 2.5. Happel proved that $\mathbf{D}^b(\mathrm{mod}(\Lambda))$ is triangle equivalent to $\underline{\mathrm{mod}}(\widehat{\Lambda})$ if and only if Λ has finite global dimension, see [**57**], [**58**]. For the ($\Rightarrow$) direction he used the Auslander-Reiten theory on the existence of Auslander-Reiten sequences and triangles in the dualizing varieties $\mathrm{mod}(\widehat{\Lambda})$ and $\underline{\mathrm{mod}}(\widehat{\Lambda})$. The above proof avoids this.

REMARK 2.6. Using Neeman-Ravenel's Theorem [**86**], it follows that the Verdier quotient $\underline{\mathrm{mod}}(\widehat{\Lambda})/\mathcal{H}^b(\mathcal{P}_\Lambda)$ is fully embedded in $\mathcal{Y}^b$ and the latter is its closure under direct summands: $\mathrm{add}\big(\underline{\mathrm{mod}}(\widehat{\Lambda})/\mathcal{H}^b(\mathcal{P}_\Lambda)\big) = \mathcal{Y}^b$. Moreover $\mathcal{Y}^b$ is the thick subcategory generated by the isoclasses of the class of objects $\mathbf{L}\big(\underline{\mathrm{mod}}(\widehat{\Lambda})\big)$.

It follows by the proof of Theorem 2.4 that the full exact embedding H : $\mathbf{D}(\mathrm{Mod}(\Lambda)) \hookrightarrow \underline{\mathrm{Mod}}(\widehat{\Lambda})$ of (ii) above is Happel's functor constructed inductively in [**57**]. Hence the above result shows that it admits a right adjoint and identifies its image as the full subcategory $\mathcal{X}_\Lambda$ or $\mathcal{Z}_\Lambda$. Moreover it identifies the triangulated quotient $\underline{\mathrm{Mod}}(\widehat{\Lambda})/\mathbf{D}(\mathrm{Mod}(\Lambda))$ as the TTF-class $\mathcal{Y}_\Lambda$.

p-groups. The above results have an interesting application to the stable module category of a modular group algebra.

Let G be a finite group and k a field with characteristic $p > 0$. Then the stable module category $\underline{\mathrm{Mod}}(kG)$ is compactly generated and the trivial G-module k is a compact object. Let $(\mathcal{X}_k, \mathcal{Y}_k, \mathcal{Z}_k)$ be the TTF-triple of finite type in $\underline{\mathrm{Mod}}(kG)$ generated by k. The following gives a description of the localizing subcategory of $\underline{\mathrm{Mod}}(kG)$ generated by the trivial G-module k.

COROLLARY 2.7. (1) *There are triangle equivalences*

$$\underline{\mathrm{Mod}}(kG) \hookleftarrow \mathcal{X}_k \xleftarrow{\approx} \mathbf{D}(\Gamma) \xrightarrow{\approx} \mathcal{Z}_k$$

where Γ is the DG-algebra $\mathsf{DGHom}(k,k)$ of a Tate resolution of k, which is quasi-isomorphic to the Tate cohomology ring $\widehat{H}^*(G,k)$. Hence $\mathbf{D}(\Gamma)$ is triangle equivalent to $\mathbf{D}(\widehat{H}^*(G,k))$.

(2) *If G is a p-group, then the full exact embedding $\mathbf{D}(\Gamma) \hookrightarrow \underline{\mathrm{Mod}}(kG)$ of* (1) *induces a triangle equivalence*

$$\mathbf{D}\big(\widehat{H}^*(G,k)\big) \xrightarrow{\approx} \underline{\mathrm{Mod}}(kG).$$

PROOF. The first part follows by 2.2((i). Part (2) follows from 2.2((iii), since if G is a p-group, then the trivial G-module k is a generator of $\underline{\mathrm{Mod}}(kG)$. □

Note that if the centralizer of each element of order p in G is p-nilpotent in the sense of [**29**], then the full exact embedding $\mathbf{D}(\Gamma) \hookrightarrow \underline{\mathrm{Mod}}(kG)$ constructed in (1) above, induces a triangle equivalence

$$\mathbf{D}\big(\widehat{H}^*(G,k)\big) \xrightarrow{\approx} \underline{B_0}(kG)$$

where $\underline{B_0}(kG)$ denotes the full subcategory of $\underline{\mathrm{Mod}}(kG)$ consisting of modules in the principal block of G. This follows from the fact that, under the above assumption, the trivial G-module k generates the modules in the principal block of G, see [**29**].

3. Connections with the Homological Conjectures

In this section we use the previous results on hereditary torsion pairs to shed new light on some homological conjectures in the representation theory of Artin algebras. We show that an interesting problem in tilting theory, whose solution would imply several homological conjectures, is equivalent to a problem on hereditary torsion pairs in derived categories.

Throughout this section we work over an Artin R-algebra Λ where R is a commutative Artin ring. We fix a finitely generated right Λ-module T such that $\mathrm{Ext}^n_\Lambda(T,T) = 0, \forall n \geq 1$. Recall that T is called a *Wakamatsu tilting module* if the regular right Λ-module Λ admits an exact coresolution

$$0 \to \Lambda \to T^0 \xrightarrow{f^0} T^1 \to \cdots \to T^n \xrightarrow{f^n} T^{n+1} \to \cdots$$

where T^n lies in $\mathrm{add}(T)$ and $\mathrm{Ext}^1_\Lambda(\mathrm{Im}(f^n), T) = 0$, for any $n \geq 0$. Dually T is called a *Wakamatsu cotilting module* if the injective cogenerator $\mathrm{D}(\Lambda)$ admits an exact resolution

$$\cdots \to T^{-n-1} \xrightarrow{f_{-n}} T^{-n} \to \cdots \to T^{-1} \xrightarrow{f_0} T^0 \to \mathrm{D}(\Lambda) \to 0$$

where T^{-n} lies in $\mathrm{add}(T)$ and $\mathrm{Ext}^1_\Lambda(T, \mathrm{Im}(f_{-n})) = 0$, for any $n \geq 0$. A module is known to be a Wakamatsu tilting module if and only if it is a Wakamatsu cotilting module [**103**], [**36**]. A Wakamatsu (co)tilting module is not necessary (co)tilting, but it is not known whether the following, which we call (WT)-Conjecture, is true.

- **(WT)-Conjecture:** If T is a Wakamatsu tilting (cotilting) module of finite projective (injective) dimension, then T is a tilting (cotilting) module.

It would be interesting in itself to solve this problem, but even more because of the following consequence. First we recall that the generalized Nakayama conjecture as proposed by Auslander and Reiten [**8**] says the following. In the minimal injective resolution $0 \to \Lambda \to I^0 \to I^1 \to \cdots$ of Λ each indecomposable injective Λ-module occurs as a summand of some I^n. Obviously the generalized Nakayama conjecture implies the classical Nakayama Conjecture which says that if each I^n is projective then Λ is self-injective. Then we have the following result, part (i) of which was first observed by A. Buan, using [**36**]. First we recall from [**9**] that an Artin algebra Λ is called *Gorenstein*, if Λ has finite injective dimension both as a left and right Λ-module, equivalently Λ is a cotilting module.

PROPOSITION 3.1. *If the (WT)-Conjecture holds, then we have the following.*

(i) *The generalized Nakayama Conjecture holds.*

(ii) *The Gorenstein Symmetry Conjecture holds, that is: if* $\mathrm{id}_\Lambda \Lambda < \infty$, *then* Λ *is Gorenstein, i.e.* $\mathrm{id}\Lambda_\Lambda < \infty$.

PROOF. (i) Let $0 \to \Lambda_\Lambda \to I^0 \to I^1 \to \cdots$ be the minimal injective resolution of Λ, and let T be the direct sum of the indecomposable injectives occurring as summands of the terms I^n, $n \geq 0$. Then T is a Wakamatsu (co)tilting module, and hence a cotilting module if the (WT)-conjecture is assumed to hold. Hence the number of non-isomorphic indecomposable summands equals the number of non-isomorphic indecomposable injectives. In other words T contains all indecomposable injectives as summands, hence the generalized Nakayama conjecture holds.

(ii) Since $\mathrm{id}_\Lambda \Lambda < \infty$, the injective cogenerator $\mathrm{D}(\Lambda)$ has finite projective dimension as a right module. Since $\mathrm{D}(\Lambda)$ is obviously a Wakamatsu tilting module and the (WT)-Conjecture is assumed to hold, $\mathrm{D}(\Lambda)$ is a tilting module. Hence we have $\mathrm{pd}_\Lambda \mathrm{D}(\Lambda) < \infty$, so that $\mathrm{id}\, \Lambda_\Lambda < \infty$. □

Note that the (WT)-Conjecture also implies a related conjecture which says that if the nth term of the minimal injective resolution of Λ has projective dimension $\leq n$ for each $n \geq 0$, then the algebra is Gorenstein, see [**12**]. Indeed the direct sum of the indecomposable injectives occurring in the resolution form a Wakamatsu tilting module of finite projective dimension. Hence by the (WT)-conjecture this is a tilting module. It follows that ${}_\Lambda\Lambda$ has finite injective dimension, and then by [**12**] Λ is Gorenstein.

We now fix a Wakamatsu tilting module T and assume that T has finite projective dimension.

We view T as an object of the unbounded derived category $\mathbf{D}(\mathrm{Mod}(\Lambda))$ via the full embedding $\mathrm{Mod}(\Lambda) \hookrightarrow \mathbf{D}(\mathrm{Mod}(\Lambda))$, which sends a module A to the stalk complex concentrated in degree zero with stalk A. Since T is finitely presented and has finite projective dimension, it follows that T is a compact object in $\mathbf{D}(\mathrm{Mod}(\Lambda))$, that is, T lies in $\mathcal{H}^b(\mathcal{P}_\Lambda)$. Let $(\mathcal{X}_T, \mathcal{Y}_T, \mathcal{Z}_T)$ be the TTF-triple in $\mathbf{D}(\mathrm{Mod}(\Lambda))$ generated by T. Then the TTF-class $\mathcal{Y}_T$ is given by:

$$\mathcal{Y}_T = \Big\{ C^\bullet \in \mathbf{D}(\mathrm{Mod}(\Lambda)) \mid \big(T[n], C^\bullet\big) = 0, \ \ \forall n \in \mathbb{Z} \Big\}$$

By Proposition 1.1 we know that the categories $\mathcal{X}_T, \mathcal{Y}_T$ and $\mathcal{Z}_T$ are compactly generated, and $\mathcal{X}_T$ and $\mathcal{Z}_T$ are triangle equivalent. We denote as usual by $\mathbf{i} : \mathcal{X}_T \hookrightarrow \mathbf{D}(\mathrm{Mod}(\Lambda))$, $\mathbf{j} : \mathcal{Y}_T \hookrightarrow \mathbf{D}(\mathrm{Mod}(\Lambda))$, and $\mathbf{k} : \mathcal{Z}_T \hookrightarrow \mathbf{D}(\mathrm{Mod}(\Lambda))$ the inclusion functors, and let $(\mathbf{i}, \mathbf{R})$, $(\mathbf{L}, \mathbf{j})$, $(\mathbf{j}, \mathbf{S})$, $(\mathbf{k}, \mathbf{T})$, be the corresponding adjoint pairs. We shall now interpret the subcategories $\mathcal{X}$, $\mathcal{Y}$ and $\mathcal{Z}$ in different useful terms.

Let $\Gamma = \mathrm{End}(T)$ be the endomorphism ring of T. Then we have the adjoint pair of functors (F, G), where $F := - \otimes_\Gamma T_\Lambda : \mathrm{Mod}(\Gamma) \to \mathrm{Mod}(\Lambda)$ and $G := \mathrm{Hom}_\Lambda({}_\Gamma T_\Lambda, -) : \mathrm{Mod}(\Lambda) \to \mathrm{Mod}(\Gamma)$. Note that F preserves products since T is finitely presented. The adjoint pair (F, G) induces an adjoint pair of total derived functors $(\mathbb{F}, \mathbb{G})$ on the level of unbounded derived categories, where $\mathbb{F} := - \otimes^{\mathbb{L}}_\Gamma T_\Lambda : \mathbf{D}(\mathrm{Mod}(\Gamma)) \to \mathbf{D}(\mathrm{Mod}(\Lambda))$ and $\mathbb{G} := \mathbb{R}\mathrm{Hom}_\Lambda(T, -) : \mathbf{D}(\mathrm{Mod}(\Lambda)) \to \mathbf{D}(\mathrm{Mod}(\Gamma))$.

The following is essentially a consequence of Proposition I.2.11.

LEMMA 3.2. *Using the above notation, the functor* $\mathbb{F}$ *is fully faithful and the functor* $\mathbb{G}$ *admits a fully faithful right adjoint*

$$\mathbb{H} : \mathbf{D}(\mathrm{Mod}(\Gamma)) \longrightarrow \mathbf{D}(\mathrm{Mod}(\Lambda)), \quad \textit{given by} \quad \mathbb{H} := \mathbb{R}\mathrm{Hom}_\Gamma\big({}_\Lambda\mathbb{R}\mathrm{Hom}_\Lambda(T, \Lambda)_\Gamma, -\big).$$

Moreover we have the following identifications:

$$\mathrm{Im}(\mathbb{F}) = \mathcal{X}_T, \ \ \mathrm{Ker}(\mathbb{G}) = \mathcal{Y}_T, \ \ \mathrm{Im}(\mathbb{H}) = \mathcal{Z}_T, \ \ \textit{and} \ \ \mathbf{iR} = \mathbb{FG}, \ \ \mathbf{kT} = \mathbb{HG}.$$

PROOF. By Proposition III.4.5 we have that $\mathbb{F}$ is fully faithful. Since T is compact, it follows that $\mathbb{G}$ preserves coproducts. Hence by [**85**] the functor $\mathbb{F}$ preserves compact objects and the functor $\mathbb{G}$ admits a right adjoint $\mathbb{H} : \mathbf{D}(\mathrm{Mod}(\Gamma)) \to \mathbf{D}(\mathrm{Mod}(\Lambda))$. By Proposition I.2.11 we have identifications $\mathrm{Im}(\mathbb{F}) = \mathcal{X}_T$, $\mathrm{Im}(\mathbb{H}) = \mathcal{Z}_T$ and $\mathrm{Ker}(\mathbb{G}) = \mathcal{Y}_T$, and moreover $\mathbf{iR} = \mathbb{FG}$ and $\mathbf{kT} = \mathbb{HG}$. Now consider the complex ${}_\Lambda\mathbb{R}\mathrm{Hom}_\Lambda(T, \Lambda)_\Gamma$ and the canonical morphism $\phi : - \otimes^{\mathbb{L}}_\Lambda \mathbb{R}\mathrm{Hom}_\Lambda(T, \Lambda)_\Gamma \to \mathbb{R}\mathrm{Hom}_\Lambda(T, -) = \mathbb{G}$ of functors $\mathbf{D}(\mathrm{Mod}(\Lambda)) \to \mathbf{D}(\mathrm{Mod}(\Gamma))$, where $\phi_{C^\bullet}$ is the evaluation. Since both functors preserve coproducts and ϕ_Λ is invertible, by [**69**] it follows that ϕ is invertible. This implies that the right adjoint $\mathbb{H}$ of $\mathbb{G}$ is given by $\mathbb{H} = \mathbb{R}\mathrm{Hom}_\Gamma\big({}_\Lambda\mathbb{R}\mathrm{Hom}_\Lambda(T, \Lambda)_\Gamma, -\big)$. □

We have the following direct consequence.

COROLLARY 3.3. *The functor* $\mathbb{F} : \mathbf{D}(\mathrm{Mod}(\Gamma)) \to \mathbf{D}(\mathrm{Mod}(\Lambda))$ *admits a factorization* $\mathbb{F} = \mathbf{i}\,\mathbb{F}^*$, *where* $\mathbb{F}^* : \mathbf{D}(\mathrm{Mod}(\Gamma)) \xrightarrow{\approx} \mathcal{X}_T$ *is a triangle equivalence, and the functor* $\mathbb{H} : \mathbf{D}(\mathrm{Mod}(\Gamma)) \to \mathbf{D}(\mathrm{Mod}(\Lambda))$ *admits a factorization* $\mathbb{H} = \mathbf{k}\,\mathbb{H}^*$, *where* $\mathbb{H}^* : \mathbf{D}(\mathrm{Mod}(\Gamma)) \xrightarrow{\approx} \mathcal{Z}_T$ *is a triangle equivalence.*

It follows from Proposition I.2.11 and the above result that the triangle equivalence $\Phi : \mathcal{X}_T \xrightarrow{\approx} \mathcal{Z}_T$ and its quasi-inverse $\Psi : \mathcal{Z}_T \xrightarrow{\approx} \mathcal{X}_T$ are given as follows:

$$\Psi := - \otimes^{\mathbb{L}}_{\Lambda} \mathbb{R}\mathrm{Hom}_{\Lambda}(T, \Lambda) \otimes^{\mathbb{L}}_{\Gamma} T, \qquad \Phi := \mathbb{R}\mathrm{Hom}_{\Gamma}\big(\mathbb{R}\mathrm{Hom}_{\Lambda}(T, \Lambda), \mathbb{R}\mathrm{Hom}_{\Lambda}(T, -)\big) = \mathbb{R}\mathrm{Hom}_{\Lambda}\big(\mathbb{R}\mathrm{Hom}_{\Lambda}(T, \Lambda) \otimes^{\mathbb{L}}_{\Gamma} T, -\big)$$

The following result gives a variety of necessary and sufficient conditions for a Wakamatsu tilting module of finite projective dimension to be a tilting module.

THEOREM 3.4. *Let T be a Wakamatsu tilting right Λ-module of finite projective dimension with endomorphism ring Γ. Then the following are equivalent.*

(i) *T is a tilting module.*
(ii) *The torsion class $\mathcal{X}_T$ is closed under products in $\mathbf{D}(\mathrm{Mod}(\Lambda))$.*
(iii) *The total left derived functor $\mathbb{F} := - \otimes^{\mathbb{L}}_{\Gamma} T_{\Lambda} : \mathbf{D}(\mathrm{Mod}(\Gamma)) \to \mathbf{D}(\mathrm{Mod}(\Lambda))$ preserves products.*
(iv) *$\Lambda \in \mathcal{X}_T$.*
(v) *$\mathrm{D}(\Lambda) \in \mathcal{Z}_T$.*
(vi) *$\mathcal{X}_T \subseteq \mathcal{Z}_T$.*
(vii) *$\mathcal{Z}_T \subseteq \mathcal{X}_T$.*
(viii) *The canonical morphism $: \mathbb{R}\mathrm{Hom}_{\Lambda}(T, \Lambda) \otimes^{\mathbb{L}}_{\Gamma} T \to \Lambda$ is invertible.*
(ix) *The canonical morphism $: \mathrm{D}(\Lambda) \to \mathbb{R}\mathrm{Hom}_{\Lambda}\big(\mathbb{R}\mathrm{Hom}_{\Lambda}(T, \Lambda) \otimes^{\mathbb{L}}_{\Gamma} T, \mathrm{D}(\Lambda)\big)$ is invertible.*
(x) *The non-hereditary torsion pair in $\mathbf{D}(\mathrm{Mod}(\Lambda))$ generated by T is non-degenerate.*
(xi) *T has finite projective dimension as a left Γ-module.*

PROOF. Obviously (i) implies all the other conditions, since if T is a tilting module, then $\mathcal{X}_T = \mathcal{Z}_T = \mathbf{D}(\mathrm{Mod}(\Lambda))$, hence $\mathcal{Y}_T = 0$, and the functors $\mathbb{F}$ and $\mathbb{G}$ are quasi-inverse equivalences. By Lemma 3.2, we have that (iv) is equivalent to (viii) and (v) is equivalent to (ix). Since $\mathcal{X}_T$ is a localizing subcategory of $\mathbf{D}(\mathrm{Mod}(\Lambda))$, if $\mathcal{X}_T$ contains the generator Λ, then by infinite devissage we have $\mathcal{X}_T = \mathbf{D}(\mathrm{Mod}(\Lambda))$. Dually since $\mathcal{Z}_T$ is a colocalizing subcategory, (v) implies that $\mathcal{Z}_T = \mathbf{D}(\mathrm{Mod}(\Lambda))$. In both cases we have $\mathcal{Y}_T = 0$, hence T generates $\mathbf{D}(\mathrm{Mod}(\Lambda))$ and then T is a tilting module. By Proposition III.4.2 we have that (x) implies (i). If $\mathbb{F}$ preserves products, then so does the functor $\mathbb{F}\mathbb{G} = \mathbf{iR}$. This implies trivially that $\mathcal{X}_T$ is closed under products in $\mathbf{D}(\mathrm{Mod}(\Lambda))$. Conversely if this holds, then $\mathbb{F}$ preserves products since $\mathbb{F} = \mathbf{i}\,\mathbb{F}^*$ and $\mathbb{F}^*$ is an equivalence. Hence (iii) is equivalent to (ii). Since for a Wakamatsu tilting module T we have that $\mathrm{End}_{\Lambda}(T)$ is isomorphic to Γ, it is easy to see directly that (i) is equivalent to (xi). Using the glueing triangles for the torsion pairs $(\mathcal{X}_T, \mathcal{Y}_T)$ and $(\mathcal{Y}_T, \mathcal{Z}_T)$, it is easy to see that (vi) and (vii) are equivalent. It remains to show that (ii) $\Rightarrow$ (i) and (vi) $\Rightarrow$ (i).

We first show that the right Λ-module Λ, resp. $\mathrm{D}(\Lambda)$, as a stalk complex concentrated in degree zero is isomorphic in $\mathbf{D}(\mathrm{Mod}(\Lambda))$ to a (countable) homotopy limit, resp. homotopy colimit, of compact objects from $\mathcal{X}_T$. Indeed let $0 \to \Lambda_{\Lambda} \to T^0 \to T^1 \to T^2 \to \cdots$ be an exact coresolution of Λ_{Λ}, where T^n is in $\mathrm{add}(T)$, $\forall n \geq 0$. Then Λ_{Λ} is isomorphic to the complex $T^{\bullet} : \ \cdots \to 0 \to T^0 \to T^1 \to T^2 \to \cdots$ in the unbounded derived category. Filtering the complex $T^{\bullet}$ by using the stupid truncation complexes $T^{\bullet}_n : \ \cdots \to 0 \to T^0 \to T^1 \to T^2 \to \cdots \to T^n \to 0 \to \cdots$, we can form in the unbounded derived category the homotopy limit $\underleftarrow{\text{holim}} T_n^\bullet$ of the induced inverse tower $\cdots \to T_2^\bullet \to T_1^\bullet \to T_0^\bullet$. In other words we have a triangle

$$\underleftarrow{\text{holim}} T_n^\bullet \longrightarrow \prod_{n\geq 0} T_n^\bullet \longrightarrow \prod_{n\geq 0} T_n^\bullet \longrightarrow \underleftarrow{\text{holim}} T_n^\bullet[1] \qquad (\mathsf{T})$$

in $\mathbf{D}(\text{Mod}(\Lambda))$. By results of Boekstedt-Neeman [**31**], it follows that we have an isomorphism $T^\bullet \xrightarrow{\cong} \underleftarrow{\text{holim}} T_n^\bullet$. Hence we infer the following isomorphism in the derived category:

$$\Lambda_\Lambda \xrightarrow{\cong} \underleftarrow{\text{holim}} T_n^\bullet \qquad (\dagger)$$

and by construction all the complexes $T_n^\bullet$ lie in $\mathcal{H}^b(\text{add}(T)) = \mathcal{X}_T^b$. Since T is also a Wakamatsu cotilting module, there exists an exact sequence $\cdots \to T^{-3} \to T^{-2} \to T^{-1} \to \mathrm{D}(\Lambda) \to 0$, where $T^{-n} \in \text{add}(T)$, for $n \geq 1$. Working as above we infer the following isomorphism in the derived category:

$$\mathrm{D}(\Lambda)_\Lambda \xrightarrow{\cong} \underrightarrow{\text{holim}} T_{-n}^\bullet \qquad (\dagger\dagger)$$

where $T_{-n}^\bullet$ are bounded complexes with components in $\text{add}(T)$. In particular we have that $\mathrm{D}(\Lambda)$ lies in $\mathcal{X}_T$ since the latter is closed under coproducts, hence under homotopy colimits. This implies that ${}^\perp\mathcal{X}_T = 0$, since $\big(C^\bullet, \Sigma^n(\mathrm{D}(\Lambda))\big) \xrightarrow{\cong} \mathrm{H}^n(C^\bullet)$. Hence if (vi) holds, then $\mathcal{Y}_T = {}^\perp\mathcal{Z}_T \subseteq {}^\perp\mathcal{X}_T = 0$, and then T is a tilting module, i.e. (vi) implies (i). Now if $\mathcal{X}_T$ is closed under products, then it is closed under homotopy limits. We infer by the isomorphism ($\dagger$) that $\mathcal{X}_T$ contains Λ, hence by (iv) we have $\mathcal{X}_T = \mathbf{D}(\text{Mod}(\Lambda))$. Alternatively if $\mathcal{X}_T$ is closed under products, then $\mathcal{X}_T$ is a colocalizing subcategory of $\mathbf{D}(\text{Mod}(\Lambda))$. Since by the isomorphism ($\dagger\dagger$), $\mathcal{X}_T$ contains $\mathrm{D}(\Lambda)$, it follows by Corollary 1.7 that $\mathcal{X}_T = \mathbf{D}(\text{Mod}(\Lambda))$. This shows that (ii) implies (i). □

When T is a Wakamatsu tilting Λ-module of finite projective dimension, with endomorphism ring Γ, then, by Theorem 3.4 we have $\text{pd}_\Gamma T < \infty$, if and only if $- \otimes_\Gamma^{\mathbb{L}} T_\Lambda : \mathbf{D}(\text{Mod}(\Gamma)) \to \mathbf{D}(\text{Mod}(\Lambda))$ preserves products. In fact, there is more generally the following useful result due to Bernhard Keller. We thank him for allowing us to include his result here and for some valuable comments on these topics.

PROPOSITION 3.5. [B. Keller] *Let ${}_\Gamma T_\Lambda$ be a Γ-Λ-bimodule which as a left Γ-module is finitely presented. Then the total derived functor $-\otimes_\Gamma^{\mathbb{L}} T_\Lambda : \mathbf{D}(\text{Mod}(\Gamma)) \to \mathbf{D}(\text{Mod}(\Lambda))$ preserves products if and only if ${}_\Gamma T$ has finite projective dimension.*

Proposition 3.5 suggests the following problem which is of independent interest, since coherent functors play an important role in representation theory, see [**78**].

PROBLEM. Let $F : \text{Mod}(\Lambda) \to \text{Mod}(\Gamma)$ be a coherent functor, that is, F preserves filtered colimits and products. Under what conditions does the total left derived functor $\mathbb{L}F : \mathbf{D}(\text{Mod}(\Lambda)) \to \mathbf{D}(\text{Mod}(\Gamma))$ preserve products?

As an application of the above results we have the following characterization of Gorenstein algebras, where we denote by $- \otimes_\Lambda^{\mathbb{L}} \mathrm{D}(\Lambda) : \mathbf{D}(\text{Mod}(\Lambda)) \to \mathbf{D}(\text{Mod}(\Lambda))$ the total left derived functor of the Nakayama functor $- \otimes_\Lambda \mathrm{D}(\Lambda) : \text{Mod}(\Lambda)) \to \text{Mod}(\Lambda))$, and by $\mathbb{R}\text{Hom}_\Lambda(\mathrm{D}(\Lambda), -) : \mathbf{D}(\text{Mod}(\Lambda)) \to \mathbf{D}(\text{Mod}(\Lambda))$ its right adjoint.

COROLLARY 3.6. *The functor* $- \otimes^{\mathbb{L}}_{\Lambda} \mathrm{D}(\Lambda) : \mathbf{D}(\mathrm{Mod}(\Lambda)) \to \mathbf{D}(\mathrm{Mod}(\Lambda))$ *preserves products if and only if* $\mathrm{id}\Lambda_{\Lambda} < \infty$. *In particular the following are equivalent.*

(i) *The algebra* Λ *is Gorenstein.*
(ii) *The functor* $\mathbb{R}\mathrm{Hom}_{\Lambda}(\mathrm{D}(\Lambda), -)$ *preserves coproducts and the functor* $- \otimes^{\mathbb{L}}_{\Lambda} \mathrm{D}(\Lambda)$ *preserves products.*
(iii) *The functor* $- \otimes^{\mathbb{L}}_{\Lambda} \mathrm{D}(\Lambda) : \mathbf{D}(\mathrm{Mod}(\Lambda)) \xrightarrow{\approx} \mathbf{D}(\mathrm{Mod}(\Lambda))$ *is a triangle equivalence.*

PROOF. By Proposition 3.5 the functor $- \otimes^{\mathbb{L}}_{\Lambda} \mathrm{D}(\Lambda)$ preserves products if and only if ${}_{\Lambda}\mathrm{D}(\Lambda)$ has finite projective dimension, or equivalently Λ_{Λ} has finite injective dimension. On the other hand the functor $\mathbb{R}\mathrm{Hom}_{\Lambda}(\mathrm{D}(\Lambda), -)$ preserves coproducts if and only if $\mathrm{D}(\Lambda)_{\Lambda}$ is compact if and only if $\mathrm{D}(\Lambda)_{\Lambda}$ has finite projective dimension or equivalently ${}_{\Lambda}\Lambda$ has finite injective dimension. The remaining assertions follow from Theorem 3.4. □

The following observation shows that the behavior of the Nakayama functors plays a role in the investigation of when a Wakamatsu tilting module of finite projective dimension is a tilting module.

LEMMA 3.7. $\mathcal{Y}_T \subseteq \mathrm{Ker}\big(\mathbb{R}\mathrm{Hom}_{\Lambda}(\mathrm{D}(\Lambda), -)\big)$ *and* $\mathrm{Im}\big(- \otimes^{\mathbb{L}}_{\Lambda} \mathrm{D}(\Lambda)\big) \subseteq \mathcal{X}_T$.

PROOF. From the proof of Theorem 3.4 we have that $\mathrm{D}(\Lambda)$ lies in $\mathcal{X}_T$. Hence for any complex $Y^{\bullet} \in \mathcal{Y}_T$ we have $\mathbb{R}\mathrm{Hom}_{\Lambda}(\mathrm{D}(\Lambda), Y^{\bullet}) = 0$, and the first inclusion follows. From the isomorphism $\big(C^{\bullet} \otimes^{\mathbb{L}}_{\Lambda} \mathrm{D}(\Lambda), Y^{\bullet}\big) \xrightarrow{\cong} \big(C^{\bullet}, \mathbb{R}\mathrm{Hom}_{\Lambda}(\mathrm{D}(\Lambda), Y^{\bullet})\big) = 0$, it follows that for any complex $C^{\bullet} \in \mathbf{D}(\mathrm{Mod}(\Lambda))$, we have that $C^{\bullet} \otimes^{\mathbb{L}}_{\Lambda} \mathrm{D}(\Lambda)$ lies in $\mathcal{X}_T$, and the second inclusion follows. □

Now let $\mathfrak{F}$ denote the set of isoclasses of indecomposable compact objects of $\mathbf{D}(\mathrm{Mod}(\Lambda))$. Recall that the full subcategory of compact objects of $\mathbf{D}(\mathrm{Mod}(\Lambda))$ can be identified up to equivalence with the bounded homotopy category $\mathcal{H}^b(\mathcal{P}_{\Lambda})$, and it is well-known that the latter is a Krull-Schmidt category. By Brown representability, for any object $P^{\bullet} \in \mathfrak{F}$, there is a natural isomorphism

$$\big((P^{\bullet}, -), k\big) \xrightarrow{\cong} \big(-, \mathbb{D}_k(P^{\bullet})\big)$$

where k denotes the injective envelope of $R/\mathrm{rad}R$. As in Proposition III.3.9 it follows that the set $\mathbb{D}_k(\mathfrak{F}) := \{\mathbb{D}_k(P^{\bullet}) \mid P^{\bullet} \in \mathfrak{F}\}$ is a set of pure-injective cogenerators of $\mathbf{D}(\mathrm{Mod}(\Lambda))$ (see [**21**] and [**77**] for the concept of purity in triangulated categories). By results of Happel [**57**], it follows that for any object $P^{\bullet} \in \mathfrak{F}$: $\mathbb{D}_k(P^{\bullet}) \xrightarrow{\cong} P^{\bullet} \otimes^{\mathbb{L}}_{\Lambda} \mathrm{D}(\Lambda)$. Hence $\mathbb{D}_k(\mathfrak{F}) = \mathfrak{F} \otimes^{\mathbb{L}}_{\Lambda} \mathrm{D}(\Lambda) \subseteq \mathcal{X}_T$.

The following result, which possibly can be proved directly, shows that the (WT)-Conjecture holds if Λ has finite right self-injective dimension.

PROPOSITION 3.8. *Let* T_{Λ} *be a Wakamatsu tilting module of finite projective dimension. If* $\mathrm{id}\Lambda_{\Lambda} < \infty$, *then* T *is a tilting module.*

PROOF. Assume that $\mathrm{id}\Lambda_{\Lambda} < \infty$ or equivalently that $\mathrm{pd}_{\Lambda}\mathrm{D}(\Lambda) < \infty$. Let $E^{\bullet}$ be an indecomposable pure-injective complex which lies in $\mathcal{Y}_T$. Since the set $\mathbb{D}_k(\mathfrak{F})$ is a cogenerating set of pure-injectives, it follows that $E^{\bullet}$ is a direct summand of $\prod_{P^{\bullet} \in \mathfrak{F}} \mathbb{D}_k(P^{\bullet}) = \prod_{P^{\bullet} \in \mathfrak{F}} \big(P^{\bullet} \otimes^{\mathbb{L}}_{\Lambda} \mathrm{D}(\Lambda)\big)$. By Proposition 3.5 we infer that

$\prod_{P^\bullet \in \mathfrak{F}}(P^\bullet \otimes^{\mathbb{L}}_\Lambda \mathrm{D}(\Lambda)) \xrightarrow{\cong} (\prod_{P^\bullet \in \mathfrak{F}} P^\bullet) \otimes^{\mathbb{L}}_\Lambda \mathrm{D}(\Lambda)$ and this lies in $\mathcal{X}_T$ by Lemma 3.7. Hence $E^\bullet$ lies in $\mathcal{X}_T$ as a direct summand of $(\prod_{P^\bullet \in \mathfrak{F}} P^\bullet) \otimes^{\mathbb{L}}_\Lambda \mathrm{D}(\Lambda)$. Since $E^\bullet \in \mathcal{Y}_T$, we have $E^\bullet = 0$. Then by a result of Krause, see Theorem 4.4 in **[77]**, it follows that Λ lies in $\mathcal{X}_T$, and then by Theorem 3.4 we have that T is a tilting module. □

COROLLARY 3.9. *If $\mathcal{Z}_T$ is closed under coproducts and $\mathbb{R}\mathrm{Hom}_\Lambda(\mathrm{D}(\Lambda), Z^\bullet)$ lies in $\mathcal{Z}_T$ for any complex $Z^\bullet \in \mathcal{Z}_T$, then T is a tilting module.*

PROOF. By Proposition 1.11 we have that $\mathcal{Y}^b_T \subseteq \mathcal{H}^b(\mathcal{P}_\Lambda)$. Let $P^\bullet$ be an indecomposable compact object in $\mathcal{Y}_T$. Then $\mathbb{D}_k(P^\bullet) = P^\bullet \otimes^{\mathbb{L}}_\Lambda \mathrm{D}(\Lambda)$ lies in $\mathcal{X}_T$. For any $Z^\bullet$ in $\mathcal{Z}_T$ we have an isomorphism $(P^\bullet \otimes^{\mathbb{L}}_\Lambda \mathrm{D}(\Lambda), Z^\bullet) \xrightarrow{\cong} (P^\bullet, \mathbb{R}\mathrm{Hom}_\Lambda(\mathrm{D}(\Lambda), Z^\bullet))$, and the last space is zero by hypothesis. Hence $P^\bullet \otimes^{\mathbb{L}}_\Lambda \mathrm{D}(\Lambda) \in {}^\perp\mathcal{Z}_T = \mathcal{Y}_T$. Since $P^\bullet \otimes^{\mathbb{L}}_\Lambda \mathrm{D}(\Lambda)$ lies also in $\mathcal{X}_T$ we infer that $P^\bullet \otimes^{\mathbb{L}}_\Lambda \mathrm{D}(\Lambda) = 0$. By construction this is possible only if $P^\bullet = 0$. It follows that $\mathcal{Y}^b_T = 0$, hence $\mathcal{Y}_T = 0$ since $\mathcal{Y}_T$ is compactly generated. We infer that T is a tilting module. □

The (WT)-Conjecture is related to a problem of Grothendieck groups, which we formulate as the following conjecture.

CONJECTURE. Let T be a Wakamatsu tilting module of finite projective dimension, and assume that the Grothendieck group $\mathrm{K}_0(\mathcal{Y}^b_T)$ of the full subcategory of compact objects of $\mathcal{Y}_T$ is trivial. Then $\mathcal{Y}_T = 0$, i.e. T is a tilting module.

More generally the following problem is of interest in connection with the results of this section. In a sense the next problem measures how far a triangulated category is from being algebraic in the sense of Keller **[73]**.

PROBLEM. Let $\mathcal{T}$ be a triangulated category with coproducts and a compact generator. Is it true that $\mathrm{K}_0(\mathcal{T}^b) = 0$ implies $\mathcal{T}^b = 0$, or equivalently $\mathcal{T} = 0$?

We have the following connection between the above Conjecture and the (WT)-Conjecture, where for a finitely generated module M, $s(M)$ denotes the number of non-isomorphic indecomposable direct summands of M. Note that if T is a tilting module, then $s(\Lambda) = s(T)$.

PROPOSITION 3.10. *Let T_Λ be a Wakamatsu tilting module of finite projective dimension and assume that $s(\Lambda) = s(T)$. If the above Conjecture is true and $\mathcal{Z}_T$ is closed under coproducts in $\mathbf{D}(\mathrm{Mod}(\Lambda))$, then T is a tilting module.*

PROOF. Since $\mathcal{Z}_T$ is closed under coproducts, by Proposition 1.11 we have that $\mathbf{R} : \mathcal{X}_T \to \mathbf{D}(\mathrm{Mod}(\Lambda))$ preserves coproducts. Then by Corollary 1.12 we have an isomorphism $\mathrm{K}_0(\mathcal{H}^b(\mathcal{P}_\Lambda)) \xrightarrow{\cong} \mathrm{K}_0(\mathcal{H}^b(\mathrm{add}(T))) \oplus \mathrm{K}_0(\mathcal{Y}^b_T)$. Now it is easy to see that $\mathrm{K}_0(\mathcal{H}^b(\mathcal{P}_\Lambda)) \xrightarrow{\cong} \mathbb{Z}^{s(\Lambda)}$ and $\mathrm{K}_0(\mathcal{H}^b(\mathrm{add}(T))) \xrightarrow{\cong} \mathbb{Z}^{s(T)}$. Hence the hypothesis $s(\Lambda) = s(T)$ implies that $\mathrm{K}_0(\mathcal{H}^b(\mathcal{P}_\Lambda)) \xrightarrow{\cong} \mathrm{K}_0(\mathcal{H}^b(\mathrm{add}(T)))$. We infer that $\mathrm{K}_0(\mathcal{Y}^b_T) = 0$, and then $\mathcal{Y}_T = 0$ by our assumptions. This implies that T is a tilting module. □

It is not known if a partial tilting module with $s(T) = s(\Lambda)$ is always a tilting module. This is a well-known open problem in representation theory **[95]**. The above results suggest the following conjecture which lies between the (WT)-Conjecture and the Generalized Nakayama Conjecture.

CONJECTURE. Let T be a Wakamatsu tilting module of finite projective dimension. If $s(T) = s(\Lambda)$, then T is a tilting module.

REMARK 3.11. Observe that if $\mathcal{Z}_T$ is closed under coproducts, then the compact generator $\mathbf{L}(\Lambda)$ of $\mathcal{Y}_T$ is compact in $\mathbf{D}(\mathrm{Mod}(\Lambda))$. Hence by Theorem 2.1, there exists a triangle equivalence $\mathcal{Y}_T \xrightarrow{\cong} \mathbf{D}(\mathrm{Mod}(\Delta))$ where Δ is the DG-algebra of endomorphisms of $\mathbf{L}(\Lambda)$. In this case the TTF-triple $(\mathcal{X}_T, \mathcal{Y}_T, \mathcal{Z}_T)$ reduces, up to triangle equivalence, to a TTF-triple $\big(\mathbf{D}(\mathrm{Mod}(\Gamma)), \mathbf{D}(\mathrm{Mod}(\Delta)), \mathbf{D}(\mathrm{Mod}(\Gamma))\big)$, i.e. a recollement of derived categories, in $\mathbf{D}(\mathrm{Mod}(\Lambda))$.

We close this section with the following result which shows that the above conjecture has an affirmative answer if $\mathcal{Z}_T$ is closed under coproducts.

COROLLARY 3.12. *Let T be a Wakamatsu tilting module of finite projective dimension and assume that $s(T) = s(\Lambda)$. If the torsion pair $(\mathcal{Y}_T, \mathcal{Z}_T)$ is of finite type, that is $\mathcal{Z}_T$ is closed under coproducts, then T is a tilting module.*

PROOF. By Proposition 3.10 we have that $\mathrm{K}_0(\mathcal{Y}_T) = 0$. Then with the notation of the above remark we have $\mathrm{K}_0\big(\mathbf{D}(\mathrm{Mod}(\Delta))^b\big) = 0$. However this holds only if the DG-algebra Δ of endomorphisms of the compact object $\mathbf{L}(\Lambda)$ is (quasi-isomorphic to) 0. Hence $\mathbf{L}(\Lambda) = 0$, or equivalently Λ lies in $\mathcal{X}_T$. Then the assertion follows from Theorem 3.4. □

We refer to [**25**] for a discussion of the topics of this section using pure homological algebra and the structure of the Ziegler spectrum, in connection with the Auslander-Reiten theory in triangulated and derived categories.

4. Concluding Remarks and Comments

As noted in Chapter I, torsion pairs in a triangulated category are widely known as t-structures. They were introduced by Beilinson-Bernstein-Deligne in [**18**] as an important tool in the study of cohomology of sheaves over possibly singular spaces. In their language, recollement situations correspond to TTF-triples.

Hereditary torsion pairs are known in the recent literature as semi-orthogonal decompositions, a terminology introduced by Reiten-Van den Bergh, see [**90**], to a concept introduced and investigated in detail by Bondal and Kapranov, see [**33**]. It is known that semi-orthogonal decompositions provide a very powerful tool for the study of the derived category of coherent sheaves on manifolds and algebraic varieties.

On the other hand hereditary torsion classes of a torsion pair of finite type in a compactly generated triangulated category are known in the literature as *smashing* subcategories, generalizing a concept which has its origin in stable homotopy theory. We recall that a Grothendieck category $\mathcal{G}$ is called *locally coherent* if $\mathcal{G}$ admits a set of finitely presented generators and the full subcategory $\mathrm{f.p}(\mathcal{G})$ of finitely presented objects of $\mathcal{G}$ is abelian. It is known that hereditary torsion pairs of finite type in a locally coherent Grothendieck category $\mathcal{G}$ are in bijective correspondence with the Serre subcategories of the abelian category $\mathrm{f.p}(\mathcal{G})$ of finitely presented objects of $\mathcal{G}$ [**78**]. This result lies at the heart of the representation-theoretic investigation of $\mathcal{G}$ via the theory of the Ziegler spectrum of $\mathcal{G}$. In the triangulated case the substitute of a locally coherent Grothendieck category is a compactly generated triangulated category $\mathcal{C}$ and the substitute of the Serre subcategories of f.p($\mathcal{G}$) are the thick subcategories of the category of compact objects of $\mathcal{C}$.

Hence it is natural to ask if the above correspondence has a triangulated analogue. The validity of the triangulated analogue of the above correspondence is equivalent, in our language, to the following well-known conjecture which is connected with the classification of thick subcategories of compact objects.

Let $\mathcal{C}$ be a compactly generated triangulated category and let $\mathcal{C}^b$ be the full subcategory of compact objects of $\mathcal{C}$.

Ravenel's Telescope Conjecture: Any hereditary torsion pair of finite type in $\mathcal{C}$ is generated by a set of compact objects, i.e. is of the form $(\mathcal{X}_{\mathcal{P}}, \mathcal{Y}_{\mathcal{P}})$ for a set $\mathcal{P} \subseteq \mathcal{C}^b$ closed under Σ, Σ^{-1}. Equivalently the maps

$$\mathcal{P} \longmapsto (\mathcal{X}_{\mathcal{P}}, \mathcal{Y}_{\mathcal{P}}) \quad \text{and} \quad (\mathcal{X}, \mathcal{Y}) \longmapsto \mathcal{X} \cap \mathcal{C}^b$$

are mutually inverse bijections between thick subcategories of $\mathcal{C}^b$ and hereditary torsion pairs of finite type in $\mathcal{C}$.

Let $(\mathcal{X}, \mathcal{Y})$ be a hereditary torsion pair in $\mathcal{C}$. Then by Proposition 1.1 we have a TTF-triple $(\mathcal{X}, \mathcal{Y}, \mathcal{Z})$ in $\mathcal{C}$. Then by Proposition IV.1.11, Ravenel's Conjecture is implied by any one of the following equivalent statements:

- The torsion pair $(\mathcal{X}, \mathcal{Y})$ induces a torsion pair $(\mathcal{X}^b, \mathcal{Y}^b)$ in $\mathcal{C}^b$.
- The torsion pair $(\mathcal{Y}, \mathcal{Z})$ is of finite type.

In this generality the conjecture is false. In our language Keller [**70**] produced an example of a hereditary torsion pair $(\mathcal{X}, \mathcal{Y})$ of finite type in the unbounded derived category $\mathbf{D}(\mathrm{Mod}(\Lambda))$ of a non-Noetherian ring Λ such that the torsion class $\mathcal{X}$ contains no non-zero compact object from $\mathbf{D}(\mathrm{Mod}(\Lambda))$, so the torsion pair can not be generated by compact objects. By results of Hopkins-Neeman (see [**84**]), the conjecture holds in case $\mathcal{C}$ is the unbounded derived category of a commutative Noetherian ring. We refer to the work of Krause [**77**] for a detailed analysis and a partial solution. Another class of categories in which the conjecture is true is the class of compactly generated triangulated categories $\mathcal{C}$ which admit filtered colimits, that is: the pure-semisimple categories of [**21**]. In such a category any object is a coproduct of indecomposable compact objects with local endomorphism ring. This implies trivially that any hereditary torsion pair is generated by compact objects.

The class of pure-semisimple triangulated categories is very small but not quite trivial. Important examples include the unbounded derived category of a ring which is derived equivalent to a right pure-semisimple right hereditary ring, in particular the unbounded derived category of an iterated tilted algebra of Dynkin type, and the stable module category of a representation-finite quasi-Frobenius ring, in particular the stable module category of a self-injective Artin algebra of finite representation type, see [**21**].

CHAPTER V

Torsion Pairs in Stable Categories

Let throughout this chapter $\mathcal{C}$ be an abelian category with enough projective and enough injective objects. When ω is a functorially finite subcategory of $\mathcal{C}$ we know that the stable category $\mathcal{C}/\omega$ is pretriangulated. A pair $(\mathcal{X}, \mathcal{Y})$ of subcategories of $\mathcal{C}$ with $\omega \subseteq \mathcal{X} \cap \mathcal{Y}$ gives rise to a pair $(\mathcal{X}/\omega, \mathcal{Y}/\omega)$ of subcategories of $\mathcal{C}/\omega$. The aim of this chapter is to obtain descriptions of $(\mathcal{X}/\omega, \mathcal{Y}/\omega)$ being a torsion pair in terms of properties of the pair $(\mathcal{X}, \mathcal{Y})$. In particular we show that there is a strong relationship with pairs of contravariantly finite and covariantly finite subcategories, orthogonal with respect to Ext^1, investigated in [**9**],[**10**] and [**97**]. Important examples of such pairs of subcategories in the representation theory of Artin algebras emerge from tilting or cotilting modules. In particular we show that (co)tilting modules give rise to interesting torsion pairs, this time in stable categories induced by (co)tilting modules.

1. A Description of Torsion Pairs

Let $\mathcal{X}, \mathcal{Y}$ be full subcategories of $\mathcal{C}$ such that $\omega = \mathcal{X} \cap \mathcal{Y}$ is functorially finite in $\mathcal{C}$. Our aim in this section is to get a description of $(\mathcal{X}/\omega, \mathcal{Y}/\omega)$ being a torsion pair in $\mathcal{C}/\omega$ in terms of properties of $\mathcal{X}$ and $\mathcal{Y}$. This description is based on the comparison of pairs of subcategories of $\mathcal{C}$ and $\mathcal{C}/\omega$ with respect to being contravariantly or covariantly finite.

We start with the following useful observation.

LEMMA 1.1. (1) *Assume that $\omega \subseteq \mathcal{X}$ and ω is contravariantly finite in $\mathcal{C}$. Then $\mathcal{X}$ is contravariantly finite in $\mathcal{C}$ iff $\mathcal{X}/\omega$ is contravariantly finite in $\mathcal{C}/\omega$.*

(2) *Assume that $\omega \subseteq \mathcal{Y}$ and ω is covariantly finite in $\mathcal{C}$. Then $\mathcal{Y}$ is covariantly finite in $\mathcal{C}$ iff $\mathcal{Y}/\omega$ is covariantly finite in $\mathcal{C}/\omega$.*

PROOF. (1) Assume that $\mathcal{X}/\omega$ is contravariantly finite in $\mathcal{C}/\omega$ and let $C \in \mathcal{C}$. Let $\underline{\alpha} : \underline{X} \to \underline{C}$ be a right $\mathcal{X}/\omega$-approximation of $\underline{C}$. Choose an object X_C and a morphism $f_C : X_C \to C$ such that $\underline{X_C} = \underline{X}$ and $\underline{f_C} = \underline{\alpha}$. Without loss of generality we may assume that f_C is ω-epic (if it is not then $f'_C := {}^t(\mu, f_C) : X'_C := T \oplus X_C \to C$ is ω-epic, where $\mu : T \to C$ is a right ω-approximation of C and $\underline{X'_C} = \underline{X}$ and $\underline{f'_C} = \underline{\alpha}$). If $g : X' \to C$ is a morphism with $X' \in \mathcal{X}$, then there exists a morphism $\underline{h} : \underline{X'} \to \underline{X}$ such that $\underline{h} \circ \underline{\alpha} = \underline{g}$. Then $g - h \circ \alpha$ factors through a right ω-approximation $\phi_C : \omega_C \to C$ of C, i.e. there exists $\rho : X' \to \omega_C$ such that $\rho \circ \phi_C = g - h \circ f_C$. Since f_C is ω-epic and $\omega \subseteq \mathcal{X}$, there exists $\sigma : \omega_C \to X_C$ such that $\sigma \circ f_C = \phi_C$. Then $g = (h + \rho \circ \sigma) \circ f_C$. Hence g factors through f_C, and this shows that f_C is a right $\mathcal{X}$-approximation of C. Conversely if $f_C : X_C \to C$ is a right $\mathcal{X}$-approximation of C, then it is easy to see that $\underline{f_C} : \underline{X_C} \to \underline{C}$ is a right

$\mathcal{X}/\omega$-approximation of $\underline{C}$ in $\mathcal{C}/\omega$. It follows that if $\mathcal{X}$ is contravariantly finite in $\mathcal{C}$, then $\mathcal{X}/\omega$ is contravariantly finite in $\mathcal{C}/\omega$. Part (2) is dual. □

We now compare $\mathcal{X}$ being contravariantly finite in $\mathcal{C}$ with the inclusion $\mathcal{X}/\omega \hookrightarrow \mathcal{C}/\omega$ having a right adjoint. We have the following connection.

PROPOSITION 1.2. *Let ω be a functorially finite subcategory of $\mathcal{C}$.*
(α) *If $\omega \subseteq \mathcal{X}$, then the following are equivalent.*

(i) *$\mathcal{X}$ is closed under cokernels of left ω-approximations and the inclusion $\mathcal{X}/\omega \hookrightarrow \mathcal{C}/\omega$ admits a right adjoint.*

(ii) *$\mathcal{X}$ is contravariantly finite in $\mathcal{C}$ and for any object C in $\mathcal{C}$ there exists an exact sequence $0 \to Y_C \to X_C \xrightarrow{f_C} C$, where f_C is a right $\mathcal{X}$-approximation of C and $\underline{Y}_C \in (\mathcal{X}/\omega)^{\perp}$.*

(β) *If $\omega \subseteq \mathcal{Y}$, then the following are equivalent.*

(i) *$\mathcal{Y}$ is closed under kernels of right ω-approximations and the inclusion $\mathcal{Y}/\omega \hookrightarrow \mathcal{C}/\omega$ admits a left adjoint.*

(ii) *$\mathcal{Y}$ is covariantly finite in $\mathcal{C}$ and and for any object C in $\mathcal{C}$ there exists an exact sequence $C \xrightarrow{g^C} Y^C \to X^C \to 0$, where g^C is a left $\mathcal{Y}$-approximation $C \xrightarrow{g^C} Y^C \to X^C \to 0$ of C and $\underline{X}^C \in {}^{\perp}(\mathcal{Y}/\omega)$.*

PROOF. (α) (i) $\Rightarrow$ (ii) Since $\mathcal{X}$ is closed under cokernels of left ω-approximations it follows that $\Sigma(\mathcal{X}/\omega) \subseteq \mathcal{X}/\omega$. Let $\mathbf{R} : \mathcal{C}/\omega \to \mathcal{X}/\omega$ be the right adjoint of the inclusion $\mathbf{i} : \mathcal{X}/\omega \hookrightarrow \mathcal{C}/\omega$. Let $C \in \mathcal{C}$ and let $\alpha : \mathbf{R}(\underline{C}) \to \underline{C}$ be the coreflection of $\underline{C}$ in $\mathcal{X}/\omega$. If $(T) : \ \Omega(\underline{C}) \to \underline{Y}_C \to \mathbf{R}(\underline{C}) \xrightarrow{\alpha} \underline{C}$ is a left triangle in $\mathcal{C}/\omega$, then by Lemma II.2.3 we have $\underline{Y}_C \in (\mathcal{X}/\omega)^{\perp}$. As in Lemma 1.1, α induces a right $\mathcal{X}$-approximation $f_C : X_C \to C$ of C. Since f_C is ω-epic, it induces a left triangle $(T') : \ \Omega(\underline{C}) \to \underline{K}_C \to \underline{X}_C \to \underline{C}$ in $\mathcal{C}/\omega$, where $K_C = \operatorname{Ker}(f_C)$. By construction the left triangle (T') coincides with (T), so $\underline{K}_C \cong \underline{Y}_C \in (\mathcal{X}/\omega)^{\perp}$.

(ii) $\Rightarrow$ (i) Let $(E) : 0 \to Y_C \xrightarrow{g_C} X_C \xrightarrow{f_C} C$ be an exact sequence, where f_C is a right $\mathcal{X}$-approximation of C and the object $\underline{Y}^C$ lies in $(\mathcal{X}/\omega)^{\perp}$. Since $\omega \subseteq \mathcal{X}$, f_C is ω-epic, hence the sequence (E) induces a left triangle $\Omega(\underline{C}) \to \underline{Y}^C \to \underline{X}_C \xrightarrow{\underline{f}_C} \underline{C}$ in $\mathcal{C}/\omega$. Since $\underline{Y}^C \in (\mathcal{X}/\omega)^{\perp}$, by Lemma II.2.3, the morphism $\underline{f}_C$ is the coreflection of $\underline{C}$ in $\mathcal{X}/\omega$. Hence $\mathcal{X}/\omega$ is coreflective. It remains to show that $\mathcal{X}$ is closed under left ω-approximations. Let $X \xrightarrow{\kappa} \omega^X \xrightarrow{\lambda} C \to 0$ be an exact sequence, where κ is a left ω-approximation of $X \in \mathcal{X}$. Since $\omega \subseteq \mathcal{X}$, there exists $\beta : \omega^X \to X_C$ such that $\lambda = \beta \circ f_C$. Since λ is epic, so is f_C and we have an exact commutative diagram

$$\begin{array}{ccccccccc}
 & & X & \xrightarrow{\kappa} & \omega^X & \xrightarrow{\lambda} & C & \longrightarrow & 0 \\
 & & {\scriptstyle \exists\, \alpha}\downarrow & & {\scriptstyle \exists\, \beta}\downarrow & & \| & & \\
0 & \longrightarrow & Y_C & \xrightarrow{g_C} & X_C & \xrightarrow{f_C} & C & \longrightarrow & 0
\end{array}$$

Since $X \in \mathcal{X}$ and $\underline{Y}^C \in (\mathcal{X}/\omega)^{\perp}$, there exists a factorization $\alpha = \alpha_1 \circ \alpha_2 : X \xrightarrow{\alpha_1} T \xrightarrow{\alpha_2} Y_C$, where $T \in \omega$. Since κ is a left ω-approximation, there exists $\rho : \omega^X \to T$ such that $\alpha_1 = \kappa \circ \rho$. Setting $\tau := \rho \circ \alpha_2$ it follows that $\kappa \circ \tau = \alpha$. Then by standard arguments, there exists $\sigma : C \to X_C$ such that $\sigma \circ f_C = 1_C$. Hence $C \in \mathcal{X}$ as a direct summand of X_C.

(β) The proof is dual. □

For $\omega = 0$, we have the following special case, which is also easily seen directly (compare [**54**]).

COROLLARY 1.3. *(1) $\mathcal{X}$ is coreflective in $\mathcal{C}$ iff for any object $C \in \mathcal{C}$, there exists a right $\mathcal{X}$-approximation $f_C : X_C \to C$ with* $\mathrm{Ker}(f_C) \in \{C \in \mathcal{C} \mid \mathcal{C}(\mathcal{X}, C) = 0\}$.

(2) $\mathcal{Y}$ is reflective in $\mathcal{C}$ iff for any object $C \in \mathcal{C}$, there exists a left $\mathcal{Y}$-approximation $f^C : C \to Y^C$ with $\mathrm{Coker}(f^C) \in \{C \in \mathcal{C} \mid \mathcal{C}(C, \mathcal{Y}) = 0\}$.

Combining Lemma 1.1 and Proposition 1.2 we have the following consequence.

COROLLARY 1.4. *Let $(\mathcal{X}/\omega, \mathcal{Y}/\omega)$ be a torsion pair in $\mathcal{C}/\omega$ where ω is functorially finite in $\mathcal{C}$. Then we have the following.*

(1) Any object of $\mathcal{C}$ admits a right $\mathcal{X}$-approximation with kernel in $\mathcal{Y}$ and any object of $\mathcal{X}$ admits a left ω-approximation with cokernel in $\mathcal{X}$.

(2) Any object of $\mathcal{C}$ admits a left $\mathcal{Y}$-approximation with cokernel in $\mathcal{X}$ and any object of $\mathcal{Y}$ admits a right ω-approximation with kernel in $\mathcal{Y}$.

By Corollary 1.4 it follows that for C in $\mathcal{C}$, a torsion pair $(\mathcal{X}/\omega, \mathcal{Y}/\omega)$ in $\mathcal{C}/\omega$ gives a right $\mathcal{X}$-approximation $f_C : X_C \to C$ and a left $\mathcal{Y}$-approximation $g^C : C \to Y^C$. Also we get a left ω-approximation $\kappa : X_C \to T$ with cokernel in $\mathcal{X}$ and a left ω-approximation $\mu : T' \to Y^C$ with kernel in $\mathcal{Y}$. It turns out that, conversely, if we put this information together in an appropriate commutative diagram in $\mathcal{C}$, we obtain sufficient (and necessary) conditions to have a torsion pair.

THEOREM 1.5. *If ω is functorially finite in $\mathcal{C}$, then the following are equivalent.*

(i) The pair $(\mathcal{X}/\omega, \mathcal{Y}/\omega)$ is a torsion pair in $\mathcal{C}/\omega$.

(ii) $\forall C \in \mathcal{C}$, there exists a bicartesian (i.e. a pull-back and push-out) square

$$\begin{array}{ccc} X_C & \xrightarrow{f_C} & C \\ {\scriptstyle\kappa}\downarrow & & \downarrow{\scriptstyle g^C} \\ T & \xrightarrow{\mu} & Y^C \end{array} \qquad (\dagger)$$

with the following properties.

(α) $X_C \in \mathcal{X}$, $Y^C \in \mathcal{Y}$ and any morphism $X \to Y$ with $X \in \mathcal{X}$ and $Y \in \mathcal{Y}$ factors through an object in ω.

(β) κ is a left ω-approximation of X_C with cokernel in $\mathcal{X}$.

(γ) μ is a right ω-approximation of Y^C with kernel in $\mathcal{Y}$.

In this case $f_C : X_C \to C$ is a right $\mathcal{X}$-approximation of C with kernel in $\mathcal{Y}$ and $g^C : C \to Y^C$ is a left $\mathcal{Y}$-approximation of C with cokernel in $\mathcal{X}$.

PROOF. (ii) $\Rightarrow$ (i) The bicartesian square ($\dagger$) induces a short exact sequence

$$0 \longrightarrow X_C \xrightarrow{(\kappa, f_C)} T \oplus C \xrightarrow{{}^t(-\mu, g^C)} Y^C \longrightarrow 0 \qquad (1)$$

Since κ is ω-monic, it follows that (κ, f_C) is ω-monic. Hence we have a right triangle $\nabla(C)$: $\underline{X_C} \xrightarrow{\underline{f_C}} \underline{C} \xrightarrow{\underline{g^C}} \underline{Y^C} \xrightarrow{\underline{h^C}} \Sigma(\underline{X_C})$ in $\mathcal{C}/\omega$. Since μ is ω-epic, it follows that ${}^t(-\mu, g^C)$ is ω-epic. Hence we have a left triangle $\Delta(C):\ \Omega(\underline{Y}^C) \xrightarrow{\underline{h}_C} \underline{X}_C \xrightarrow{\underline{f}_C} \underline{C} \xrightarrow{\underline{g}^C} \underline{Y}^C$ in $\mathcal{C}/\omega$. By condition (α) we have $\underline{X}_C \in \mathcal{X}/\omega$, $\underline{Y}^C \in \mathcal{Y}/\omega$ and $(\mathcal{X}/\omega, \mathcal{Y}/\omega) = 0$. Next we show that g^C is ω-monic. Indeed if $\alpha : C \to T'$ is a morphism with $T' \in \omega$, then by (β), there exists a morphism $\alpha' : T \to T'$ such that $\kappa \circ \alpha' = \alpha$. Since ($\dagger$) is a push-out diagram, there exists $h : Y^C \to T'$ such that $g^C \circ h = \alpha'$, hence g^C is ω-monic. Dually f_C is ω-epic. It remains to show that $\Sigma(\mathcal{X}/\omega) \subseteq \mathcal{X}/\omega$ or equivalently that $\mathcal{X}$ is closed under left ω-approximations. Suppose that $C \in \mathcal{X}$ in the diagram ($\dagger$). Since $Y^C \in \mathcal{Y}$, the morphism g^C factors through ω, hence $\underline{g}^C = 0$ in $\mathcal{C}/\omega$. Then from the right triangle $\nabla(C)$ we infer that $\underline{Y}^C \cong \Sigma(\underline{X}_C)$, which by ($\beta$) lies in $\mathcal{X}/\omega$. Since $\underline{Y}^C \in \mathcal{Y}/\omega$, it follows that $\underline{Y}^C \in \mathcal{X}/\omega \cap \mathcal{Y}/\omega = 0$. Hence $Y^C \in \omega$. Set $X^C = \mathrm{Coker}(g^C)$. Then by the bicartesian square ($\dagger$) above, $X^C = \mathrm{Coker}(g^C) \cong \mathrm{Coker}(\kappa)$, so by definition $X^C \in \mathcal{X}$. Since g^C is ω-monic, it follows that $\Sigma(\underline{C}) \cong \underline{X}^C \cong \Sigma(\underline{X}_C) \in \mathcal{X}/\omega$.

(i) $\Rightarrow$ (ii) Assume now that the pair $(\mathcal{X}/\omega, \mathcal{Y}/\omega)$ is a torsion pair in $\mathcal{C}/\omega$. Then $(\mathcal{X}/\omega, \mathcal{Y}/\omega) = 0$, $\mathcal{X}$ is closed under left ω-approximations because $\Sigma(\mathcal{X}/\omega) \subseteq \mathcal{X}/\omega$, $\mathcal{Y}$ is closed under right ω-approximations because $\Omega(\mathcal{Y}/\omega) \subseteq \mathcal{Y}/\omega$ and finally for any $C \in \mathcal{C}$, there exist triangles:

$$\Omega(\underline{Y}^C) \to \underline{X}_C \xrightarrow{\alpha} \underline{C} \xrightarrow{\beta} \underline{Y}^C \ \in \Delta \quad \text{and} \quad \underline{X}_C \xrightarrow{\alpha} C \xrightarrow{\beta} \underline{Y}^C \to \Sigma(\underline{X}_C) \ \in \nabla$$

Let $\widetilde{g}^C : C \to Y^C$ be a morphism such that $\underline{\widetilde{g}}^C = \underline{\beta}$ and let $\mu : T_1 \to Y^C$ be a right ω-approximation of Y^C. Consider the left exact sequence $0 \to \widehat{X}_C \xrightarrow{(\lambda, f_C)} T_1 \oplus C \xrightarrow{{}^t(-\mu, \widetilde{g}^C)} Y^C$. It is clear that in $\mathcal{C}/\omega$ we have $\underline{\widehat{X}}_C \cong \underline{X}_C$ and $\underline{f}_C = \underline{\alpha}$. Let $\kappa : \widehat{X}_C \to T_2$ be a left ω-approximation of $\widehat{X}_C$ and consider the right exact sequence $\widehat{X}_C \xrightarrow{(\kappa, \lambda, f_C)} T_2 \oplus T_1 \oplus C \xrightarrow{{}^t(-\nu, -\xi, g^C)} \widetilde{Y}^C \to 0$. Since (λ, f_C) is monic, so is (κ, λ, f_C), hence we have a short exact sequence

$$0 \longrightarrow \widehat{X}_C \xrightarrow{(\kappa, \lambda, f_C)} T_2 \oplus T_1 \oplus C \xrightarrow{{}^t(-\nu, -\xi, g^C)} \widetilde{Y}^C \longrightarrow 0 \tag{2}$$

The above sequence induces a right triangle $\widehat{X}_C \xrightarrow{\underline{f}_C} \underline{C} \xrightarrow{\underline{g}^C} \underline{\widetilde{Y}}^C \to \Sigma(\widehat{X}_C)$ from which we infer that $\underline{\widetilde{Y}}^C \cong \underline{Y}^C$ and $\underline{\widetilde{g}}^C = \underline{\beta} = \underline{g}^C$. In particular $\widetilde{Y}^C \in \mathcal{Y}$. The above construction is included in the following exact commutative diagram:

$$\begin{array}{ccccccccc}
0 & \longrightarrow & \widehat{X}_C & \xrightarrow{(\kappa,\lambda,f_C)} & T_2 \oplus T_1 \oplus C & \xrightarrow{{}^t(-\nu,-\xi,g^C)} & \widetilde{Y}^C & \longrightarrow & 0 \\
 & & \| & & \downarrow {}^t(0,1_C) & & \downarrow \exists!\ \sigma & & \\
0 & \longrightarrow & \widehat{X}_C & \xrightarrow{(\lambda,f_C)} & T_1 \oplus C & \xrightarrow{{}^t(-\mu,\widetilde{g}^C)} & Y^C & &
\end{array}$$

Observe that the right square is a pull-back diagram. We claim that the morphism ${}^t(-\nu, -\xi) : T_2 \oplus T_1 \to \widetilde{Y}^C$ is a right ω-approximation. Indeed let $\phi : T \to \widetilde{Y}^C$ be a morphism with $T \in \omega$. Since by construction $-\mu : T_1 \to Y^C$ is a right ω-approximation, there exists a morphism $\theta : T \to T_1$ such that $\theta \circ \mu = \phi \circ \sigma$. Consider the morphisms $(\theta, 0) : T \to T_1 \oplus C$ and $\phi : T \to \widetilde{Y}^C$. By the pull-back property there exists a unique morphism $\zeta := (\zeta_1, \zeta_2, \zeta_3) : T \to T_2 \oplus T_1 \oplus C$ such that $(\zeta_1, \zeta_2, \zeta_3) \circ {}^t(-\nu, -\xi, g^C) = \phi$ and $(\zeta_1, \zeta_2, \zeta_3) \circ {}^t(0, 1_C) = (\theta, 0)$. It follows trivially that $\zeta_3 = 0$ and $\zeta_2 = \theta$. Then $(\zeta_1, \theta) : T \to T_2 \oplus T_1$ is a morphism such that $(\zeta_1, \theta) \circ {}^t(-\nu, -\xi) = \phi$. Hence ${}^t(-\nu, -\xi) : T_2 \oplus T_1 \to \widetilde{Y}^C$ is a right ω-approximation. Now since κ is a left ω-approximation, so is $(\kappa, \lambda) : \widehat{X}_C \to T_2 \oplus T_1$. Hence setting $T := T_2 \oplus T_1$, $\widehat{\kappa} := (\kappa, \lambda) : \widehat{X}_C \to T$ and $\widetilde{\mu} := {}^t(\nu, \xi) : T \to \widetilde{Y}^C$, we obtain a short exact sequence $0 \to \widehat{X}_C \xrightarrow{(\widehat{\kappa}, f_C)} T \oplus C \xrightarrow{{}^t(-\widetilde{\mu}, g^C)} \widetilde{Y}^C \to 0$, which is equivalent to the existence of a bicartesian square with the desired properties.

Assume now that (ii) holds. If $\tau : C \to Y$ is a morphism with $Y \in \mathcal{Y}$, then by (α), (β), the composition $f_C \circ \tau$ factors through κ. Since the square (†) is a push-out, this implies that τ factors through g^C. Hence g^C is a left $\mathcal{Y}$-approximation of C and then $\mathrm{Coker}(g^C) \cong \mathrm{Coker}(\kappa) \in \mathcal{X}$. The proof for f_C is dual. □

Now we deduce a consequence of Theorem 1.5 which will be useful later. First we recall that a subcategory ω of $\mathcal{X}$ is called a *cogenerator* for $\mathcal{X}$ if for any $X \in \mathcal{X}$ there exists a short exact sequence $0 \to X \to T \to X' \to 0$ with $T \in \omega$ and $X' \in \mathcal{X}$. Dually we have the notion of *generator* for a subcategory.

COROLLARY 1.6. *If $\omega = \mathcal{X} \cap \mathcal{Y}$ is functorially finite in $\mathcal{C}$ and $(\mathcal{X}/\omega, \mathcal{Y}/\omega)$ is a torsion pair in $\mathcal{C}/\omega$, then we have the following.*

(1) *$\mathcal{X}$ contains the projectives iff ω is a generator of $\mathcal{Y}$.*
(2) *$\mathcal{Y}$ contains the injectives iff ω is a cogenerator of $\mathcal{X}$.*

PROOF. (1) If $\mathcal{X}$ contains the projectives, the morphism f_C in the bicartesian diagram (†) is an epimorphism, and then μ is an epimorphism. If $C \in \mathcal{Y}$, then the morphism g^C is a split monomorphism. Chasing the diagram (†) we see easily that κ is split monomorphism, hence X_C is in ω as a direct summand of T. Conversely if ω is a generator of $\mathcal{Y}$ then μ is an epimorphism, since it is a right ω-approximation of Y^C. This implies that f_C is an epimorphism. Since f_C is a right $\mathcal{X}$-approximation of C it follows that $\mathcal{X}$ contains the projectives. Part (2) is dual. □

By the above theorem, for any torsion pair $(\mathcal{X}/\omega, \mathcal{Y}/\omega)$ in $\mathcal{C}/\omega$, where ω is functorially finite in $\mathcal{C}$, and any object C in $\mathcal{C}$, there exists an exact sequence

$$0 \longrightarrow X_C \xrightarrow{(f_C, -\kappa)} C \oplus T \xrightarrow{{}^t(g^C, \mu)} Y^C \longrightarrow 0$$

where:
(i) $f_C : X_C \to C$ is a right $\mathcal{X}$-approximation of C with kernel in $\mathcal{Y}$.
(ii) $g^C : C \to Y^C$ is a left $\mathcal{Y}$-approximation of C with cokernel in $\mathcal{X}$.
(iii) $\kappa : X_C \to T$ is a left ω-approximation of X_C with cokernel in $\mathcal{X}$.
(iv) $\mu : T \to Y^C$ is a right ω-approximation of Y^C with kernel in $\mathcal{Y}$.

DEFINITION 1.7. The short exact sequence above is called the **universal exact sequence** of C with respect to the torsion pair $(\mathcal{X}/\omega, \mathcal{Y}/\omega)$ in $\mathcal{C}/\omega$.

Note that for any object C in $\mathcal{C}$ the universal exact sequence of C remains exact after the application of the contravariant functors $\mathcal{C}(-, \mathcal{Y})$ and $\mathcal{C}(-, \omega)$, and the covariant functors $\mathcal{C}(\mathcal{X}, -)$ and $\mathcal{C}(\omega, -)$.

REMARK 1.8. Theorem 1.5 generalizes the description of a usual torsion pair. Indeed, if $(\mathcal{T}, \mathcal{F})$ is a torsion pair in the abelian category $\mathcal{C}$, and if t is the torsion subfunctor, then $\mathcal{T} \cap \mathcal{F} = 0$ is functorially finite and the following square

$$\begin{array}{ccc} t(C) & \longrightarrow & C \\ \downarrow & & \downarrow \\ 0 & \longrightarrow & C/t(C) \end{array}$$

is bicartesian and satisfies the conditions in part (ii) of Theorem 1.5. Moreover the induced short exact sequence $(S): \ 0 \to t(C) \to C \to C/t(C) \to 0$ is the universal exact sequence of C. Conversely if $\mathcal{C}(\mathcal{X}, \mathcal{Y}) = 0$ and we have a bicartesian square as above with $t(C) \in \mathcal{X}$, $C/t(C) \in \mathcal{Y}$, then the sequence (S) is exact, hence $(\mathcal{X}, \mathcal{Y})$ is a torsion pair in $\mathcal{C}$ in the usual sense. We infer that $(\mathcal{X}, \mathcal{Y})$ is a torsion pair in the usual sense iff $(\mathcal{X}, \mathcal{Y})$ is a torsion pair in our sense with $\mathcal{X} \cap \mathcal{Y} = 0$.

2. Comparison of Subcategories

Let ω be a functorially finite subcategory of $\mathcal{C}$. In the previous section we got a description of torsion pairs in $\mathcal{C}/\omega$ by comparing pairs of subcategories of $\mathcal{C}$ and $\mathcal{C}/\omega$ with respect to being contravariantly or covariantly finite. Our aim in this section is to compare pairs of subcategories of $\mathcal{C}$ and $\mathcal{C}/\omega$ with respect to being contravariantly or covariantly finite, and with respect to the vanishing of Ext^1 and Hom respectively (see [**5**] for a similar type of investigation). We start by recalling some of the main results on pairs $(\mathcal{X}, \mathcal{Y})$ investigated in [**9**]. These are proved in [**9**] when $\mathcal{C}$ is the category of of finitely generated modules over an Artin algebra, but the same proofs work when $\mathcal{C}$ is more generally a Krull–Schmidt category.

A subcategory $\mathcal{X}$ of $\mathcal{C}$ is *preresolving* if it is closed under extensions and contains the projective objects, and it is *resolving* if in addition it is closed under kernels of epimorphisms. Dually a subcategory $\mathcal{Y}$ of $\mathcal{C}$ is *precoresolving* if it is closed under extensions and contains the injective objects, and it is *coresolving* if in addition it is closed under cokernels of monomorphisms. Many of the results that follow work more generally, by dropping the assumption that $\mathcal{C}$ has enough projectives and injectives, and defining $\mathcal{X}$ in $\mathcal{C}$ to be preresolving if it is closed under extensions and whenever $f : A \to B$ is a morphism in $\mathcal{C}$ such that $\mathcal{C}(\mathcal{X}, f)$ is an epimorphism, then f is an epimorphism. For a subcategory $\mathcal{Z}$ of $\mathcal{C}$ we denote by $\mathcal{Z}^{\perp}$ the full subcategory of $\mathcal{C}$ whose objects are the Y with $\operatorname{Ext}^1(Z, Y) = 0$, for all $Z \in \mathcal{Z}$, and by ${}^{\perp}\mathcal{Z}$ the full subcategory of $\mathcal{C}$ whose objects are the X with $\operatorname{Ext}^1(X, Z) = 0$, for all $Z \in \mathcal{Z}$. When we use the symbol "$\perp$" in stable categories we mean vanishing with respect to Hom.

DEFINITION 2.1. A pair $(\mathcal{X}, \mathcal{Y})$ of full subcategories of $\mathcal{C}$ is called a **good pair** if $\mathcal{X}$ is contravariantly finite with $\mathcal{X}^{\perp} = \mathcal{Y}$ and $\mathcal{Y}$ is covariantly finite with $\mathcal{X} = {}^{\perp}\mathcal{Y}$.

We need the following result from [**9**] which shows how to construct good pairs in a Krull-Schmidt category.

THEOREM 2.2. *Let $\mathcal{C}$ be a Krull-Schmidt category.*

(i) *If $\mathcal{X}$ is preresolving and contravariantly finite in $\mathcal{C}$, then $(\mathcal{X}, \mathcal{X}^{\perp})$ is a good pair.*
(ii) *If $\mathcal{Y}$ is precoresolving and covariantly finite in $\mathcal{C}$, then $({}^{\perp}\mathcal{Y}, \mathcal{Y})$ is a good pair.*

A subcategory ω of $\mathcal{X}$ is called an *Ext-injective cogenerator* for $\mathcal{X}$ if it is a cogenerator $\omega \subseteq \mathcal{X}$ such that $\omega \subseteq \mathcal{X}^{\perp}$. Dually we have the notion of an *Ext-projective generator* for a subcategory. In what follows we shall need the following useful result from [**9**] which, combined with Theorem 2.2, shows how to construct Ext-injective cogenerators and Ext-projective generators in Krull-Schmidt categories.

PROPOSITION 2.3. *Let $(\mathcal{X}, \mathcal{Y})$ be a good pair in $\mathcal{C}$ with $\omega = \mathcal{X} \cap \mathcal{Y}$.*

(i) *$\mathcal{X}$ is resolving iff $\mathcal{Y}$ is coresolving.*
(ii) *If $\mathcal{C}$ is a Krull-Schmidt category, then ω is an Ext-injective cogenerator for $\mathcal{X}$ and an Ext-projective generator for $\mathcal{Y}$.*

We now make some comparisons between parallel properties for $(\mathcal{X}, \mathcal{Y})$ and $(\mathcal{X}/\omega, \mathcal{Y}/\omega)$. Note that if $\mathcal{X}$ is a subcategory of $\mathcal{C}$ and $\mathcal{Y} = \mathcal{X}^{\perp}$, then automatically $\mathcal{Y}$ is closed under extensions and contains the injectives. And if $\mathcal{X} = {}^{\perp}\mathcal{Y}$ for a subcategory $\mathcal{Y}$ of $\mathcal{C}$, then $\mathcal{X}$ is closed under extensions and contains the projectives.

LEMMA 2.4. *Let $\mathcal{X}$ and $\mathcal{Y}$ be subcategories of $\mathcal{C}$ with $\omega \subseteq \mathcal{X} \cap \mathcal{Y}$, and assume that ω is functorially finite and $\operatorname{Ext}^1(\mathcal{X}, \mathcal{Y}) = 0$.*

(i) *If $(\mathcal{X}/\omega)^{\perp} = \mathcal{Y}/\omega$, then $\mathcal{Y}$ is closed under extensions and cokernels of monics.*
(ii) *If $\mathcal{X}/\omega = {}^{\perp}(\mathcal{Y}/\omega)$, then $\mathcal{X}$ is closed under extensions and kernels of epics.*

PROOF. (i) Let $0 \to Y_1 \to C \to Y_2 \to 0$ be exact in $\mathcal{C}$ with $Y_1, Y_2 \in \mathcal{Y}$. Since $\operatorname{Ext}^1(\mathcal{X}, \mathcal{Y}) = 0$ and $\omega \subseteq \mathcal{X}$, it follows that the sequence $0 \to \mathcal{C}(\omega, Y_1) \to \mathcal{C}(\omega, C) \to \mathcal{C}(\omega, Y_2) \to 0$ is exact. Hence it induces a left triangle $\Omega(\underline{Y}_2) \to \underline{Y}_1 \to \underline{C} \to \underline{Y}_2$ in the pretriangulated category $\mathcal{C}/\omega$. Applying the homological functor $(\mathcal{X}/\omega, -)$ to the triangle we infer that $(\mathcal{X}/\omega, \underline{C}) = 0$, hence $\underline{C} \in \mathcal{Y}/\omega$. This means that C is in $\mathcal{Y}$ so that $\mathcal{Y}$ is closed under extensions.

Now let $0 \to Y_1 \to Y_2 \xrightarrow{\beta} C \to 0$ be an exact sequence in $\mathcal{C}$ with $Y_1, Y_2 \in \mathcal{Y}$. As above the sequence induces a left triangle $\Omega(\underline{C}) \to \underline{Y}_1 \to \underline{Y}_2 \xrightarrow{\underline{\beta}} \underline{C}$ in $\mathcal{C}/\omega$. It follows that the sequence $\cdots \to (\mathcal{X}/\omega, \underline{Y}_1) \to (\mathcal{X}/\omega, \underline{Y}_2) \xrightarrow{(\mathcal{X}/\omega, \underline{\beta})} (\mathcal{X}/\omega, \underline{C})$ is exact. Now observe that we have an exact commutative diagram

$$\begin{array}{ccccccc}
(\mathcal{X}, Y_1) & \longrightarrow & (\mathcal{X}, Y_2) & \xrightarrow{(\mathcal{X}, \beta)} & (\mathcal{X}, C) & \longrightarrow & 0 \\
{\scriptstyle \pi_1}\downarrow & & {\scriptstyle \pi_2}\downarrow & & {\scriptstyle \pi_3}\downarrow & & \\
(\mathcal{X}/\omega, \underline{Y}_1) & \longrightarrow & (\mathcal{X}/\omega, \underline{Y}_2) & \xrightarrow{(\mathcal{X}/\omega, \underline{\beta})} & (\mathcal{X}/\omega, \underline{C}) & &
\end{array}$$

where the π_i are the canonical epimorphisms. Since the composition $(\mathcal{X}, \beta) \circ \pi_3$ is epic, it follows that $\pi_2 \circ (\mathcal{X}/\omega, \underline{\beta})$ is also epic, and this implies that $(\mathcal{X}/\omega, \underline{\beta}) : (\mathcal{X}/\omega, \underline{Y}_2) \to (\mathcal{X}/\omega, \underline{C})$ is epic. Then $(\mathcal{X}/\omega, \underline{Y}_2) = 0$, since $(\mathcal{X}/\omega, \mathcal{Y}/\omega) = 0$. Hence $(\mathcal{X}/\omega, \underline{C}) = 0$, i.e. $\underline{C} \in (\mathcal{X}/\omega)^{\perp} = \mathcal{Y}/\omega$, so that $C \in \mathcal{Y}$.

The proof of part (ii) is dual. □

We have the following consequence of Lemma 2.4.

PROPOSITION 2.5. *Assume that $\operatorname{Ext}^1(\mathcal{X}, \mathcal{Y}) = 0$ and let ω be a functorially finite subcategory of $\mathcal{C}$ with $\omega \subseteq \mathcal{X} \cap \mathcal{Y}$. If $(\mathcal{X}/\omega, \mathcal{Y}/\omega)$ is a torsion pair in $\mathcal{C}/\omega$,* *then $\mathcal{X}$ is closed under extensions and kernels of epimorphisms, and $\mathcal{Y}$ is closed under extensions and cokernels of monomorphisms.*

The following example shows that the two orthogonality conditions are not always related.

EXAMPLE. Let Λ be a Nakayama algebra with admissible sequence $(2,2,1)$, and denote by S_1, S_2, S_3 the corresponding simple Λ-modules. We then have the Auslander-Reiten quiver

$$S_3 \nearrow \begin{pmatrix} S_2 \\ S_3 \end{pmatrix} \searrow S_2 \nearrow \begin{pmatrix} S_1 \\ S_2 \end{pmatrix} \searrow S_1.$$

Let $\mathcal{X} = \operatorname{add}\Big\{S_1, S_2, \begin{pmatrix} S_1 \\ S_2 \end{pmatrix}\Big\}$ and $\omega = \operatorname{add}\Big\{S_1, \begin{pmatrix} S_1 \\ S_2 \end{pmatrix}\Big\}$. Then we have $\mathcal{X}^{\perp} = \operatorname{add}\Big\{\begin{pmatrix} S_2 \\ S_3 \end{pmatrix}, \begin{pmatrix} S_1 \\ S_2 \end{pmatrix}, S_1\Big\}$ and $(\mathcal{X}/\omega)^{\perp} = \operatorname{add}\Big\{S_3, \begin{pmatrix} S_2 \\ S_3 \end{pmatrix}, S_1, \begin{pmatrix} S_1 \\ S_2 \end{pmatrix}\Big\}/\omega$.

The orthogonality condition $\operatorname{Ext}^1(\mathcal{X}, \mathcal{Y}) = 0$ is very useful. However if the pair $(\mathcal{X}/\omega, \mathcal{Y}/\omega)$ is a torsion pair in $\mathcal{C}/\omega$, so that $(\mathcal{X}/\omega, \mathcal{Y}/\omega) = 0$, then in general there is no reason to expect that $\operatorname{Ext}^1(\mathcal{X}, \mathcal{Y}) = 0$ (see Remark 2.10 and the example after Corollary 3.8 below). The following result, which will be useful later for constructing good and torsion pairs, gives a sufficient condition for this to happen.

LEMMA 2.6. *Let $\mathcal{X}$ be a contravariantly finite subcategory of $\mathcal{C}$ which is closed under kernels of epimorphisms, and let ω be a subcategory of $\mathcal{C}$ which consists of Ext-injective objects for $\mathcal{X}$. Then for any subcategory $\mathcal{Y}$ of $\mathcal{C}$ with the property $(\mathcal{X}/\omega, \mathcal{Y}/\omega) = 0$, we have $\operatorname{Ext}^1(\mathcal{X}, \mathcal{Y}) = 0$.*

PROOF. Let $0 \to Y \to C \to X \to 0$ be an extension with $Y \in \mathcal{Y}$ and $X \in \mathcal{X}$ and let $f_C : X_C \to C$ be a right $\mathcal{X}$-approximation of $\mathcal{C}$. Consider the exact commutative diagram:

$$\begin{array}{ccccccccc} 0 & \longrightarrow & X' & \xrightarrow{\alpha} & X_C & \xrightarrow{\beta} & X & \longrightarrow & 0 \\ & & {\scriptstyle\eta}\downarrow & & {\scriptstyle f_C}\downarrow & & \| & & \\ 0 & \longrightarrow & Y & \xrightarrow{g} & C & \xrightarrow{f} & X & \longrightarrow & 0 \end{array}$$

Since $\mathcal{X}$ is closed under kernels of epics, $X' \in \mathcal{X}$. Since Y is in $\mathcal{Y}$, the morphism η factors through an object in ω, i.e. there exists a factorization $\eta := \eta_1 \circ \eta_2 : X' \xrightarrow{\eta_1} T' \xrightarrow{\eta_2} Y$, with $T' \in \omega$. The push-out of the upper row along η_1 splits, since the objects of ω are Ext-injective in $\mathcal{X}$. It follows easily from this that η factors through α. This implies that f splits. We infer that $\operatorname{Ext}^1(\mathcal{X}, \mathcal{Y}) = 0$. $\square$

We have seen that the existence of a torsion pair $(\mathcal{X}/\omega, \mathcal{Y}/\omega)$ in $\mathcal{C}/\omega$ implies the existence of right $\mathcal{X}$-approximations with kernel in $\mathcal{Y}$ and left $\mathcal{Y}$-approximations with cokernel in $\mathcal{X}$. In order to have a better description of the pair $(\mathcal{X}, \mathcal{Y})$ we need to assume additional properties. It seems that the crucial properties are: (α) $\mathcal{X}$ and $\mathcal{Y}$ are Ext-orthogonal: $\operatorname{Ext}^1(\mathcal{X}, \mathcal{Y}) = 0$, and (β) $\mathcal{X}$ contains the projectives and $\mathcal{Y}$ contain the injectives, or equivalently, in view of (α) and Corollary 1.6, ω is an Ext-injective cogenerator of $\mathcal{X}$ and an Ext-projective generator of $\mathcal{Y}$.

To analyze the consequences of the above additional assumptions it is suggestive to use the notion of special approximations in the sense of the following definition.

DEFINITION 2.7. Let $\mathcal{X}$ and $\mathcal{Y}$ be full subcategories of $\mathcal{C}$.

(1) An epimorphism $f_C : X_C \to C$ is called a **special right $\mathcal{X}$-approximation** of C, if $X_C \in \mathcal{X}$ and $\mathrm{Ker}(f_C) \in \mathcal{X}^{\perp}$.

(2) A monomorphism $g^C : C \to Y^C$ is called a **special left $\mathcal{Y}$-approximation** of C, if $Y^C \in \mathcal{Y}$ and $\mathrm{Coker}(g^C) \in {}^{\perp}\mathcal{Y}$.

Observe that a special right $\mathcal{X}$-approximation $f_C : X_C \to C$ of C is a right $\mathcal{X}$-approximation of C and a special left $\mathcal{Y}$-approximation $g^C : C \to Y^C$ of C is a left $\mathcal{Y}$-approximation of C.

EXAMPLE. If $\mathcal{X} \subseteq \mathcal{C}$ is closed under extensions, then, by Wakamatsu's Lemma, any minimal right $\mathcal{X}$-approximation is special. The converse is false: any epimorphism $f : P \to C$ where P is projective, is a special $\mathcal{P}$-approximation. However f is right minimal iff f is a projective cover. Dually a minimal left $\mathcal{X}$-approximation is special, but in general a special left $\mathcal{X}$-approximation is not left minimal.

Using the terminology of special approximations we have the following consequence.

COROLLARY 2.8. *Let $(\mathcal{X}, \mathcal{Y})$ be a pair of subcategories of $\mathcal{C}$ with $\mathcal{X} \cap \mathcal{Y} = \omega$.*

(1) If $\mathcal{X}$ is extension closed with $\mathcal{X}^{\perp} = \mathcal{Y}$ and any object of $\mathcal{C}$ admits a special right $\mathcal{X}$-approximation, then ω is an Ext-injective cogenerator of $\mathcal{X}$.

(2) If $\mathcal{Y}$ is extension closed with ${}^{\perp}\mathcal{Y} = \mathcal{X}$ and any object of $\mathcal{C}$ admits a special left $\mathcal{Y}$-approximation, then ω is an Ext-projective generator of $\mathcal{Y}$.

PROOF. We only prove (1) since part (2) is dual. Let $X \in \mathcal{X}$ and let $0 \to X \to I \to A \to 0$ be exact with I injective. Then pulling back this sequence along the special right $\mathcal{X}$-approximation $f_A : X_A \to A$ of A, we obtain extensions $0 \to X \to T \to X_A \to 0$ and $0 \to Y_A \to T \to I \to 0$, where $Y_A = \mathrm{Ker}(f_A)$. Since X and X_A are in $\mathcal{X}$, the fact that $\mathcal{X}$ is extension closed implies that $T \in \mathcal{X}$. Since $\mathcal{X}^{\perp} = \mathcal{Y}$ it follows that $\mathcal{Y}$ is extension closed. Hence $T \in \mathcal{Y}$ since I and Y_A are in $\mathcal{Y}$. We infer that ω is a cogenerator of $\mathcal{X}$. Since $\omega \subseteq \mathcal{Y}$ and $\mathrm{Ext}^1(\mathcal{X}, \mathcal{Y}) = 0$, we conclude that ω is an Ext-injective cogenerator of $\mathcal{X}$. □

We close this section with the following useful corollary of the above results which summarizes the consequences in $\mathcal{C}$ of the existence of a torsion pair $(\mathcal{X}/\omega, \mathcal{Y}/\omega)$ in $\mathcal{C}/\omega$ under the assumptions that $\mathcal{X}$ contains the projectives, $\mathcal{Y}$ contains the injectives, and $\mathrm{Ext}^1(\mathcal{X}, \mathcal{Y}) = 0$.

COROLLARY 2.9. *Let ω be functorially finite in $\mathcal{C}$ and let $(\mathcal{X}/\omega, \mathcal{Y}/\omega)$ be a torsion pair in $\mathcal{C}/\omega$ where $\mathcal{X}$ contains the projectives and $\mathcal{Y}$ contains the injectives. If $\mathrm{Ext}^1(\mathcal{X}, \mathcal{Y}) = 0$, then $(\mathcal{X}, \mathcal{Y})$ is a good pair in $\mathcal{C}$, $\mathcal{X}$ is resolving and admits ω as an Ext-injective cogenerator, and $\mathcal{Y}$ is coresolving and admits ω as an Ext-projective generator.*

Note that under the assumptions of Corollary 2.9, Theorem 1.5 implies the existence of the following exact commutative diagram, for any $C \in \mathcal{C}$:

$$
\begin{array}{ccccccccc}
 & & & & 0 & & 0 & & \\
 & & & & \downarrow & & \downarrow & & \\
0 & \longrightarrow & Y_C & \xrightarrow{g_C} & X_C & \xrightarrow{f_C} & C & \longrightarrow & 0 \\
 & & \| & & {\scriptstyle\kappa}\downarrow & & {\scriptstyle g^C}\downarrow & & \\
0 & \longrightarrow & Y_C & \xrightarrow{\lambda} & T & \xrightarrow{\mu} & Y^C & \longrightarrow & 0 \\
 & & & & {\scriptstyle\nu}\downarrow & & {\scriptstyle f^C}\downarrow & & \\
 & & & & X^C & = & X^C & & \\
 & & & & \downarrow & & \downarrow & & \\
 & & & & 0 & & 0 & &
\end{array}
$$

where:

(i) f_C is a special right $\mathcal{X}$-approximation of C.
(ii) g^C is a special left $\mathcal{Y}$-approximation of C.
(iii) κ is a special left ω-approximation of X_C.
(iv) μ is a special right ω-approximation of Y^C.

REMARK 2.10. Since for a torsion pair $(\mathcal{X}, \mathcal{Y})$ in an abelian category $\mathcal{C}$ we have in general $\operatorname{Ext}^1(\mathcal{X}, \mathcal{Y}) \neq 0$, it follows that there exist torsion pairs $(\mathcal{X}/\omega, \mathcal{Y}/\omega)$ in $\mathcal{C}/\omega$ where $\omega = \mathcal{X} \cap \mathcal{Y}$ is functorially finite, such that $\operatorname{Ext}^1(\mathcal{X}, \mathcal{Y}) \neq 0$. Hence there exist torsion pairs in pretriangulated stable categories not arising from good pairs.

3. Torsion and Cotorsion pairs

Assume that ω is a functorially finite subcategory of $\mathcal{C}$ and let $\mathcal{X}$ and $\mathcal{Y}$ be full subcategories of $\mathcal{C}$ with $\omega = \mathcal{X} \cap \mathcal{Y}$. In this section we investigate more closely the relationship between the pair $(\mathcal{X}, \mathcal{Y})$ being good in $\mathcal{C}$ and the pair $(\mathcal{X}/\omega, \mathcal{Y}/\omega)$ being a torsion pair in $\mathcal{C}/\omega$.

We have seen in Corollary 2.9 that if $(\mathcal{X}/\omega, \mathcal{Y}/\omega)$ is a torsion pair in $\mathcal{C}/\omega$ and $\operatorname{Ext}^1(\mathcal{X}, \mathcal{Y}) = 0$ where $\mathcal{X}$ contains the projectives and $\mathcal{Y}$ contains the injectives, then the pair $(\mathcal{X}, \mathcal{Y})$ is good with some additional properties: any object of $\mathcal{C}$ admits a special right $\mathcal{X}$-approximation and a special left approximation. Using this observation and following the way we defined torsion pairs in abelian and (pre)triangulated categories with respect to the vanishing of Hom, it is suggestive to use the concept of a cotorsion pair, which can be thought of as a reasonable definition of a torsion pair in $\mathcal{C}$ with respect to the vanishing of Ext^1.

DEFINITION 3.1. Let $\mathcal{X}$ and $\mathcal{Y}$ be full subcategories of $\mathcal{C}$, closed under isomorphisms and direct summands. We call $(\mathcal{X}, \mathcal{Y})$ a **cotorsion pair** if the following conditions hold:

(i) $\operatorname{Ext}^1(\mathcal{X}, \mathcal{Y}) = 0$.

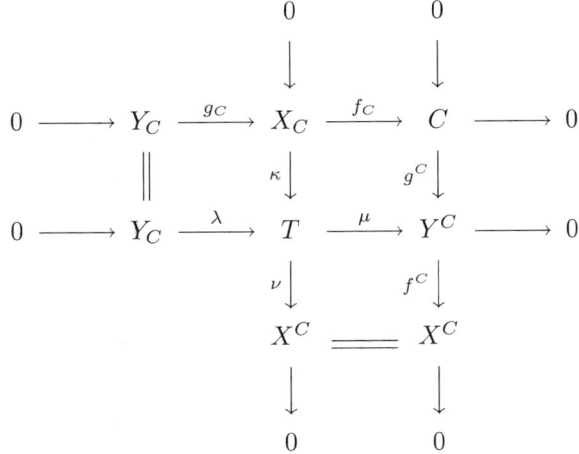

(ii) For any object C of $\mathcal{C}$ there exists a short exact sequence $0 \to Y_C \to X_C \xrightarrow{f_C} C \to 0$ where $Y_C \in \mathcal{Y}$ and $X_C \in \mathcal{X}$.

(iii) For any object C of $\mathcal{C}$ there exists a short exact sequence $0 \to C \xrightarrow{g^C} Y^C \to X^C \to 0$ where $Y^C \in \mathcal{Y}$ and $X^C \in \mathcal{X}$.

Then $\mathcal{X}$ is called a **cotorsion class** and $\mathcal{Y}$ is called a **cotorsion-free class**.

Note that there are slightly different definitions of cotorsion pairs which have appeared in the literature, see [**97**], [**79**]. The following remark shows that a cotorsion pair is a good pair with the additional property that the involved approximations are special.

REMARK 3.2. Let $(\mathcal{X}, \mathcal{Y})$ be a cotorsion pair in $\mathcal{C}$.

(1) The pair $(\mathcal{X}, \mathcal{Y})$ is complete with respect to the vanishing of Ext^1: $\mathcal{X}^{\perp} = \mathcal{Y}$ and ${}^{\perp}\mathcal{Y} = \mathcal{X}$. Indeed since $\mathrm{Ext}^1_{\mathcal{C}}(\mathcal{X}, \mathcal{Y}) = 0$, we have $\mathcal{X} \subseteq {}^{\perp}\mathcal{Y}$ and $\mathcal{Y} \subseteq \mathcal{X}^{\perp}$. Now let $C \in {}^{\perp}\mathcal{Y}$ and consider the short exact sequence $0 \to Y_C \to X_C \to C \to 0$ where $Y_C \in \mathcal{Y}$ and $X_C \in \mathcal{X}$. Since Y_C lies in $\mathcal{Y}$, the sequence splits and therefore C lies in $\mathcal{X}$ as a direct summand of X_C. Hence ${}^{\perp}\mathcal{Y} = \mathcal{X}$ and dually we have $\mathcal{X}^{\perp} = \mathcal{Y}$.

(2) It follows from (1) that the morphism f_C is a special right $\mathcal{X}$-approximation and the morphism g^C is a special left $\mathcal{Y}$-approximation.

(3) Conversely by a result of Auslander-Reiten [**9**] any good pair in a Krull-Schmidt category is a cotorsion pair.

The following result gives some equivalent conditions for a pair of subcategories $(\mathcal{X}, \mathcal{Y})$ in $\mathcal{C}$ to form a cotorsion pair.

LEMMA 3.3. *If $\mathcal{X}$, $\mathcal{Y}$ are subcategories of $\mathcal{C}$, then the following are equivalent.*

(i) *Any object admits a special right $\mathcal{X}$-approximation and $\mathcal{Y} = \mathcal{X}^{\perp}$.*
(ii) *Any object admits a special left $\mathcal{Y}$-approximation and ${}^{\perp}\mathcal{Y} = \mathcal{X}$.*
(iii) *$(\mathcal{X}, \mathcal{Y})$ is a cotorsion pair.*

PROOF. By Remark 3.2 we have that (iii) implies (i) and (ii). Hence it suffices to show that (i) and (ii) are equivalent. We prove only the direction (i) $\Rightarrow$ (ii) since the proof of the other direction is dual.

(i) $\Rightarrow$ (ii) First observe that $\mathcal{Y}$ is extension closed and contains the injectives. If C is an object in ${}^{\perp}\mathcal{Y}$ then the special right $\mathcal{X}$-approximation sequence $0 \to Y_C \to X_C \to C \to 0$ splits. Hence C lies in $\mathcal{X}$, and we infer that ${}^{\perp}\mathcal{Y} = \mathcal{X}$. Now let C be in $\mathcal{C}$ and let $0 \to C \xrightarrow{\alpha} E(C) \xrightarrow{\beta} \Sigma(C) \to 0$ be a short exact sequence in $\mathcal{C}$, where $E(C)$ is injective. Let $0 \to K_{\Sigma(C)} \xrightarrow{\psi} X_{\Sigma(C)} \xrightarrow{\chi} \Sigma(C) \to 0$ be a special right

$\mathcal{X}$-approximation of $\Sigma(C)$. Then we have the following exact commutative diagram

$$\begin{array}{ccccccccc}
 & & & & 0 & & 0 & & \\
 & & & & \downarrow & & \downarrow & & \\
 & & & & K_{\Sigma(C)} & \xrightarrow{=} & K_{\Sigma(C)} & & \\
 & & & & {\scriptstyle\lambda}\downarrow & & {\scriptstyle\psi}\downarrow & & \\
0 & \longrightarrow & C & \xrightarrow{g} & Y^C & \xrightarrow{f} & X_{\Sigma(C)} & \longrightarrow & 0 \\
 & & \| & & {\scriptstyle\mu}\downarrow & & {\scriptstyle\chi}\downarrow & & \\
0 & \longrightarrow & C & \xrightarrow{\alpha} & E(C) & \xrightarrow{\beta} & \Sigma(C) & \longrightarrow & 0 \\
 & & & & \downarrow & & \downarrow & & \\
 & & & & 0 & & 0 & &
\end{array}$$

Since $\mathcal{Y}$ is extension closed and $K_{\Sigma(C)}, E(C)$ are in $\mathcal{Y}$, we infer that Y^C lies in $\mathcal{Y}$. Since $X_{\Sigma(C)}$ lies in $\mathcal{X}$, the morphism g is a special left $\mathcal{Y}$-approximation of C. □

Now we are ready to characterize when a torsion pair in $\mathcal{C}/\omega$ comes from a cotorsion pair in $\mathcal{C}$.

PROPOSITION 3.4. *If $(\mathcal{X}/\omega, \mathcal{Y}/\omega)$ is a torsion pair in $\mathcal{C}/\omega$, then the following statements are equivalent:*

(i) *$\mathcal{X}$ is closed under kernels of epics, contains the projectives and admits ω as an Ext-injective cogenerator.*
(ii) *$\mathcal{Y}$ is closed under cokernels of monics, contains the injectives and admits ω as an Ext-projective generator.*
(iii) *$(\mathcal{X}, \mathcal{Y})$ is a cotorsion pair.*

PROOF. (i) $\Leftrightarrow$ (ii) Assume that (i) holds. Then by Lemma 2.6 we have $\mathrm{Ext}^1(\mathcal{X}, \mathcal{Y}) = 0$. In particular $\mathcal{Y} \subseteq \mathcal{X}^{\perp}$. Let C be in $\mathcal{X}^{\perp}$. Since ω is a cogenerator of $\mathcal{X}$, by Corollary 1.6, $\mathcal{Y}$ contains the injectives. Hence the left $\mathcal{Y}$-approximation g^C of C in Theorem 1.5 is a monomorphism. Since its cokernel X^C is in $\mathcal{X}$, it follows that g^C splits, hence $C \in \mathcal{Y}$. We infer that $\mathcal{Y} = \mathcal{X}^{\perp}$, in particular the objects of ω are Ext-projectives in $\mathcal{Y}$. Since $\mathcal{X}$ contains the projectives, by Corollary 1.6 we have that ω is a generator of $\mathcal{Y}$. It remains to show that $\mathcal{Y}$ is closed under cokernels of monics and this follows from Proposition 2.5. Part (ii) $\Rightarrow$ (i) is dual.

(iii) $\Leftrightarrow$ (i) If (i) holds, then by the equivalence of (i) with (ii) and Theorem 1.5 we have that $\mathcal{X}$ is contravariantly finite and $\mathcal{X}^{\perp} = \mathcal{Y}$, and $\mathcal{Y}$ is covariantly finite and ${}^{\perp}\mathcal{Y} = \mathcal{X}$. Hence $(\mathcal{X}, \mathcal{Y})$ is a good pair and then by Corollary 2.9 it follows that $(\mathcal{X}, \mathcal{Y})$ is a cotorsion pair. The converse follows from Corollary 2.9. □

We have seen in Lemma 2.4 that if $(\mathcal{X}/\omega, \mathcal{Y}/\omega)$ is a torsion pair in $\mathcal{C}/\omega$ where $\mathrm{Ext}^1(\mathcal{X}, \mathcal{Y}) = 0$, then $\mathcal{X}$ is closed under extensions and kernels of epics and $\mathcal{Y}$ is closed under extensions and cokernels of monics. This leads to the investigation of cotorsion pairs $(\mathcal{X}, \mathcal{Y})$ where $\mathcal{X}$ is resolving and $\mathcal{Y}$ is coresolving.

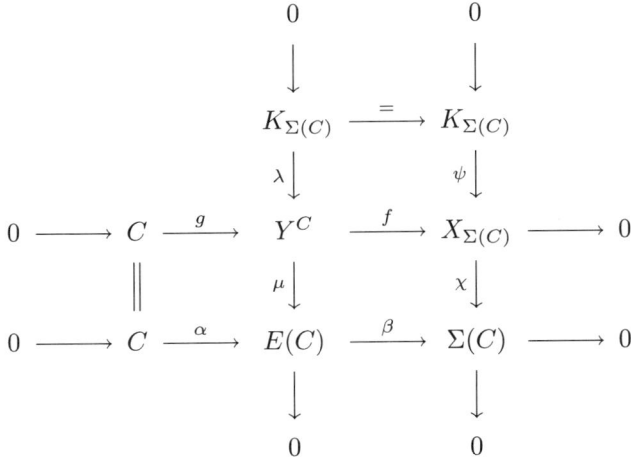

DEFINITION 3.5. A cotorsion pair $(\mathcal{X},\mathcal{Y})$ in $\mathcal{C}$ is called **resolving**, resp. **coresolving**, if $\mathcal{X}$ is resolving, resp. $\mathcal{Y}$ is coresolving.

The following result shows that a cotorsion pair is resolving if and only if it is coresolving.

PROPOSITION 3.6. *For a pair $(\mathcal{X},\mathcal{Y})$ of subcategories of $\mathcal{C}$, the following are equivalent:*

(i) *$\mathcal{Y} = \mathcal{X}^{\perp}$ is closed under cokernels of monics and any object of $\mathcal{C}$ admits a special right $\mathcal{X}$-approximation.*
(ii) *$\mathcal{X} = {}^{\perp}\mathcal{Y}$ is closed under kernels of epics and any object of $\mathcal{C}$ admits a special left $\mathcal{Y}$-approximation.*
(iii) *$(\mathcal{X},\mathcal{Y})$ is a resolving cotorsion pair in $\mathcal{C}$.*
(iv) *$(\mathcal{X},\mathcal{Y})$ is a coresolving cotorsion pair in $\mathcal{C}$.*

PROOF. Follows directly from Lemma 3.3 combined with Proposition 2.3. □

The following main result of this section gives the exact relationship between torsion pairs in a stable category and cotorsion pairs in the abelian category.

THEOREM 3.7. *Let $(\mathcal{X},\mathcal{Y})$ be a pair of subcategories of $\mathcal{C}$ such that $\omega = \mathcal{X} \cap \mathcal{Y}$ is functorially finite. Assume that $\mathcal{X}$ contains the projectives, $\mathcal{Y}$ contains the injectives, and one of the following vanishing conditions holds: (α) $\operatorname{Ext}^1(\mathcal{X},\mathcal{Y}) = 0$, (β) $\operatorname{Ext}^1(\mathcal{X},\omega) = 0$, or (γ) $\operatorname{Ext}^1(\omega,\mathcal{Y}) = 0$. Then the following are equivalent.*

(i) *$(\mathcal{X},\mathcal{Y})$ is a resolving cotorsion pair in $\mathcal{C}$.*
(ii) *$(\mathcal{X}/\omega,\mathcal{Y}/\omega)$ is a torsion pair in $\mathcal{C}/\omega$.*

PROOF. (ii) $\Rightarrow$ (i) Follows directly from Lemma 2.6 and Propositions 3.4, 3.6.

(i) $\Rightarrow$ (ii) Let C be in $\mathcal{C}$ and let $0 \to Y_C \xrightarrow{g_C} X_C \xrightarrow{f_C} C \to 0$ be a special right $\mathcal{X}$-approximation sequence of C. Since by Corollary 2.8 we have that ω is an Ext-injective cogenerator for $\mathcal{X}$, there exists an exact sequence $0 \to X_C \xrightarrow{\kappa} T \xrightarrow{\nu} X^C \to 0$ in $\mathcal{C}$ where T lies in ω and X^C lies in $\mathcal{X}$. Consider the following exact commutative diagram:

$$\begin{array}{ccccccccc}
 & & & & 0 & & 0 & & \\
 & & & & \downarrow & & \downarrow & & \\
0 & \longrightarrow & Y_C & \xrightarrow{g_C} & X_C & \xrightarrow{f_C} & C & \longrightarrow & 0 \\
 & & \| & & {\scriptstyle\kappa}\downarrow & & {\scriptstyle g^C}\downarrow & & \\
0 & \longrightarrow & Y_C & \xrightarrow{\lambda} & T & \xrightarrow{\mu} & Y^C & \longrightarrow & 0 \\
 & & & & {\scriptstyle\nu}\downarrow & & {\scriptstyle f^C}\downarrow & & \\
 & & & & X^C & = & X^C & & \\
 & & & & \downarrow & & \downarrow & & \\
 & & & & 0 & & 0 & &
\end{array}$$

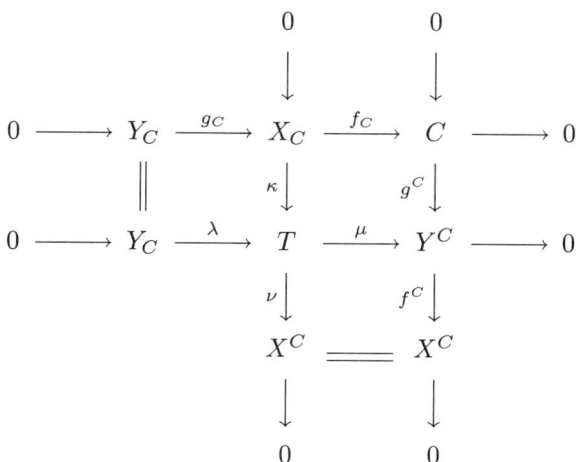

Since the cotorsion pair $(\mathcal{X}, \mathcal{Y})$ is resolving, it follows that $\mathcal{Y}$ is closed under cokernels of monomorphisms. In particular Y^C lies in $\mathcal{Y}$. Finally since the upper right square is bicartesian, by Theorem 1.5 we infer that $(\mathcal{X}/\omega, \mathcal{Y}/\omega)$ is a torsion pair in $\mathcal{C}/\omega$. □

COROLLARY 3.8. *If ω is a functorially finite subcategory of $\mathcal{C}$, then the map $(\mathcal{X}, \mathcal{Y}) \longmapsto (\mathcal{X}/\omega, \mathcal{Y}/\omega)$ gives a bijection between resolving cotorsion pairs $(\mathcal{X}, \mathcal{Y})$ in $\mathcal{C}$ such that $\mathcal{X} \cap \mathcal{Y} = \omega$, and torsion pairs $(\mathcal{X}/\omega, \mathcal{Y}/\omega)$ in $\mathcal{C}/\omega$ such that $\mathcal{X}$ contains the projectives, $\mathcal{Y}$ contains the injectives, and $\operatorname{Ext}^1(\mathcal{X}, \mathcal{Y}) = 0$.*

The following is another example of a torsion pair which is not induced by a cotorsion pair.

EXAMPLE. Assume that the full subcategory $\mathcal{P}$ of projective objects of $\mathcal{C}$ is functorially finite, so that the stable category $\mathcal{C}/\mathcal{P}$ is pretriangulated. Then $(\mathcal{C}/\mathcal{P}, 0)$ is obviously a torsion pair in $\mathcal{C}/\mathcal{P}$ but by Theorem 3.7 it is not induced by any good or cotorsion pair $(\mathcal{X}, \mathcal{Y})$, unless any projective object is injective.

The above results allow us to get further connections between a pair $(\mathcal{X}, \mathcal{Y})$ in $\mathcal{C}$ being cotorsion and the pair $(\mathcal{X}/\omega, \mathcal{Y}/\omega)$ being a torsion pair in $\mathcal{C}/\omega$, when $\omega = \mathcal{X} \cap \mathcal{Y}$ is functorially finite in $\mathcal{C}$. We also obtain connections with the conditions that the inclusion $\mathcal{X}/\omega \hookrightarrow \mathcal{C}/\omega$ has a right adjoint, together with $\mathcal{Y}/\omega$ being orthogonal to $\mathcal{X}/\omega$.

THEOREM 3.9. *Let $\mathcal{X}$ and $\mathcal{Y}$ be full subcategories of $\mathcal{C}$ such that $\omega := \mathcal{X} \cap \mathcal{Y}$ is functorially finite in $\mathcal{C}$. Assume that $\operatorname{Ext}^1(\mathcal{X}, \mathcal{Y}) = 0$, that $\mathcal{X}$ contains the projective objects and $\mathcal{Y}$ contains the injective objects, and that ω is a cogenerator for $\mathcal{X}$ and a generator for $\mathcal{Y}$. Then the following are equivalent:*

(i) *The inclusion $\mathbf{i} : \mathcal{X}/\omega \hookrightarrow \mathcal{C}/\omega$ admits a right adjoint and $\mathcal{Y}/\omega = (\mathcal{X}/\omega)^{\perp}$.*
(ii) *The inclusion $\mathbf{j} : \mathcal{Y}/\omega \hookrightarrow \mathcal{C}/\omega$ admits a left adjoint and $\mathcal{X}/\omega = {}^{\perp}(\mathcal{Y}/\omega)$.*
(iii) *The pair $(\mathcal{X}/\omega, \mathcal{Y}/\omega)$ is a torsion pair in $\mathcal{C}/\omega$.*

PROOF. Obviously (iii) $\Rightarrow$ (i), (ii). We show only that (i) $\Rightarrow$ (iii) since the proof of the implication (ii) $\Rightarrow$ (iii) is dual. Since ω is a cogenerator for $\mathcal{X}$ and $\mathcal{X}$ contains the projectives, by Proposition 1.2 there exists an exact sequence $0 \to Y_C \to X_C \to C \to 0$ where Y_C lies in $\mathcal{Y}$ and X_C lies in $\mathcal{X}$. Dually since ω is a generator for $\mathcal{Y}$ and $\mathcal{Y}$ contains the injectives, by Proposition 1.2 there exists an exact sequence $0 \to C \to Y^C \to X^C \to 0$ where X^C lies in $\mathcal{X}$ and Y^C lies in $\mathcal{Y}$. Finally since $\operatorname{Ext}^1(\mathcal{X}, \mathcal{Y}) = 0$, we infer that $(\mathcal{X}, \mathcal{Y})$ is a cotorsion pair in $\mathcal{C}$. Then by Lemma 2.4 we have that the cotorsion pair $(\mathcal{X}, \mathcal{Y})$ is resolving. Therefore by Theorem 3.7 we conclude that $(\mathcal{X}/\omega, \mathcal{Y}/\omega)$ is a torsion pair in $\mathcal{C}/\omega$. □

We have seen that if $(\mathcal{X}, \mathcal{Y})$ is a cotorsion pair in $\mathcal{C}$, then $\mathcal{X}$ is contravariantly finite and $\mathcal{Y}$ is covariantly finite. We close this section by investigating when $\mathcal{X}$ is covariantly finite or $\mathcal{Y}$ is contravariantly finite.

We begin with the following preliminary result.

LEMMA 3.10. *Let $\mathcal{X}$ be a full subcategory of $\mathcal{C}$ which admits an Ext-injective cogenerator ω. Then we have the following.*

(i) *If $\mathcal{X}$ is covariantly finite, then ω is covariantly finite.*

(ii) *If ω is contravariantly finite and $\Omega^d(\mathcal{C}/\omega) \subseteq \mathcal{X}/\omega$, then: ω is covariantly finite iff $\mathcal{X}$ is covariantly finite.*

PROOF. (i) Let C be in $\mathcal{C}$ and let $g^C : C \to X^C$ be a left $\mathcal{X}$-approximation of C. Since ω is an Ext-injective cogenerator of $\mathcal{X}$, there exists a short exact sequence $(E) : 0 \to X^C \xrightarrow{\mu} T \to X_C \to 0$ with $T \in \omega$ and X_C in $\mathcal{X}$. We claim that $g^C \circ \mu : C \to T$ is a left ω-approximation of C. Let $\alpha : C \to T'$ be a morphism with T' in ω. Since T' is in $\mathcal{X}$, there exists a morphism $\beta : X^C \to T'$ such that $g^C \circ \beta = \alpha$. Since the objects of ω are Ext-injective in $\mathcal{X}$, the push-out of the sequence (E) along β splits. Hence there exists a morphism $\gamma : T \to T'$ such that $\mu \circ \gamma = \beta$. Then $g^C \circ \mu \circ \gamma = g^C \circ \beta = \alpha$, hence α factors through $g^C \circ \mu$. We infer that $g^C \circ \mu$ is a left ω-approximation of C.

(ii) Assume that ω is functorially finite and there exists $d \geq 0$ such that $\Omega^d(\mathcal{C}/\omega) \subseteq \mathcal{X}/\omega$. Then the stable category $\mathcal{C}/\omega$ is pretriangulated and we have an adjoint pair of functors (Σ^n, Ω^n) in $\mathcal{C}/\omega$, for any $n \geq 0$. Since ω is an Ext-injective cogenerator of $\mathcal{X}$, for any $X \in \mathcal{X}$, there exists a long exact sequence $0 \to X \to T^0 \xrightarrow{g_0} T^1 \to \cdots \to T^{d-1} \xrightarrow{g_{d-1}} T^d \to \cdots$ with the T^i in ω, such that the sequence remains exact after the application of $\mathcal{C}(-, T)$, for any $T \in \omega$. This implies that for any $d \geq 0$, $\underline{X} = \Omega^d \underline{\mathrm{Im}}(g_d)$ in $\mathcal{C}/\omega$. Hence $\mathcal{X}/\omega \subseteq \Omega^d(\mathcal{C}/\omega)$, and then by hypothesis we infer that $\mathcal{X}/\omega = \Omega^d(\mathcal{C}/\omega)$. From the adjoint pair of functors (Σ^d, Ω^d) in $\mathcal{C}/\omega$, it follows that $\mathcal{X}/\omega$ is covariantly finite in $\mathcal{C}/\omega$, in fact reflective. Then Lemma 1.1 implies that $\mathcal{X}$ is covariantly finite in $\mathcal{C}$. □

Let $(\mathcal{X}, \mathcal{Y})$ be a cotorsion pair in $\mathcal{C}$. The following result, which will be useful in the next chapters in connection with Cohen-Macaulay objects, gives sufficient conditions for the subcategories $\mathcal{X}$, $\mathcal{Y}$ or ω to be functorially finite.

COROLLARY 3.11. *Let $(\mathcal{X}, \mathcal{Y})$ be a cotorsion pair in $\mathcal{C}$, and let $\omega = \mathcal{X} \cap \mathcal{Y}$.*

(i) *If $\mathcal{X}$ is covariantly finite, then ω is covariantly finite.*
(ii) *If $\mathcal{Y}$ is contravariantly finite, then ω is contravariantly finite.*
(iii) *If $\mathcal{X}$ is covariantly finite and $\mathcal{Y}$ is contravariantly finite, then ω is functorially finite.*
(iv) *If ω is contravariantly finite and $\Omega^d(\mathcal{C}/\omega) \subseteq \mathcal{X}/\omega$ for some $d \geq 0$, then ω is covariantly finite iff $\mathcal{X}$ is covariantly finite.*
(v) *If ω is covariantly finite and $\Sigma^d(\mathcal{C}/\omega) \subseteq \mathcal{Y}/\omega$ for some $d \geq 0$, then ω is contravariantly finite iff $\mathcal{Y}$ is contravariantly finite.*

PROOF. Follows from Lemma 3.10 and its dual, using that ω is an Ext-projective cogenerator of $\mathcal{X}$ and an Ext-projective generator of $\mathcal{Y}$. □

REMARK 3.12. Recently Krause and Solberg have shown that if $(\mathcal{X}, \mathcal{Y})$ is a (co)resolving cotorsion pair in $\mathrm{mod}(\Lambda)$, where Λ is an Artin algebra, then the subcategories $\mathcal{X}$, $\mathcal{Y}$, ω are functorially finite, see [**79**]. However there are (co)resolving cotorsion pairs $(\mathcal{X}, \mathcal{Y})$ in $\mathrm{Mod}(\Lambda)$, where Λ is a suitable ring, such that $\mathcal{X}$, hence ω, is not functorially finite (see the last example of the next section).

4. Torsion Classes and Cohen-Macaulay Objects

Let as before $\mathcal{C}$ be an abelian category with enough projectives and enough injective objects. In this section we are interested in finding when a full subcategory

$\mathcal{X}$ of $\mathcal{C}$ which admits a functorially finite Ext-injective cogenerator ω, induces a torsion class $\mathcal{X}/\omega$ in the stable category $\mathcal{C}/\omega$. And when such a torsion class exists, we are interested in having an explicit description of the corresponding torsion-free class. We also indicate some interesting examples of such subcategories $\mathcal{X}$, which arise naturally in practice in connection with Cohen-Macaulay approximations.

We start by stating some of the results of the previous section as equivalent conditions for the subcategory $\mathcal{X}$ of $\mathcal{C}$, leaving the formulation of the dual results concerning $\mathcal{Y}$ to the reader.

PROPOSITION 4.1. *Let $\mathcal{X}$ be a subcategory of $\mathcal{C}$, closed under kernels of epimorphisms and containing the projective objects of $\mathcal{C}$. If $\mathcal{X}$ has a functorially finite Ext-injective cogenerator ω, then the following are equivalent.*

(i) *$\mathcal{X}/\omega$ is a torsion class in $\mathcal{C}/\omega$.*
(ii) *For any $C \in \mathcal{C}$ there exists a right $\mathcal{X}$-approximation sequence $0 \to Y_C \to X_C \to C \to 0$ with $\underline{Y}_C \in (\mathcal{X}/\omega)^{\perp}$.*
(iii) *The inclusion $\mathcal{X}/\omega \hookrightarrow \mathcal{C}/\omega$ has a right adjoint.*

PROOF. Obviously (i) implies (iii). Since ω is an Ext-injective cogenerator of $\mathcal{X}$, we have that $\mathcal{X}$ is closed under cokernels of right ω-approximations. Then the equivalence (ii) $\Leftrightarrow$ (iii) follows by Proposition 1.2.

(iii) $\Rightarrow$ (i) Let $\mathcal{Y}$ be the subcategory of $\mathcal{C}$ defined by $\mathcal{Y}/\omega = (\mathcal{X}/\omega)^{\perp}$, so that $\omega = \mathcal{X} \cap \mathcal{Y}$. It suffices to show that $(\mathcal{X}, \mathcal{Y})$ satisfies the assumptions of Theorem 3.9. First observe that by Lemma 2.6 we have $\operatorname{Ext}^1(\mathcal{X}, \mathcal{Y}) = 0$. Then Lemma 2.4 ensures that $\mathcal{Y}$ is closed under extensions and cokernels of monics. Using that $\mathcal{X}$ contains the projectives and ω is an Ext-injective cogenerator of $\mathcal{X}$, Proposition 1.2 allows us to construct, for any $C \in \mathcal{C}$, the bicartesian square of the proof of Theorem 3.7. Using this we deduce easily that $\mathcal{Y}$ contains the injectives and has ω as an Ext-projective generator. Hence the assumptions of Theorem 3.9 hold for the pair $(\mathcal{X}, \mathcal{Y})$, and since the inclusion $\mathcal{X}/\omega \hookrightarrow \mathcal{C}/\omega$ admits a right adjoint, by Theorem 3.9 we infer that $(\mathcal{X}/\omega, \mathcal{Y}/\omega)$ is a torsion pair in $\mathcal{C}/\omega$. □

In case of Krull-Schmidt subcategories, we have the following characterizations of torsion classes.

COROLLARY 4.2. *Let $\mathcal{X}$ be a Krull-Schmidt subcategory of $\mathcal{C}$ closed under kernels of epimorphisms and containing the projective objects of $\mathcal{C}$. If $\mathcal{X}$ admits a functorially finite Ext-injective cogenerator ω, then the following are equivalent.*

(i) *$\mathcal{X}/\omega$ is a torsion class in $\mathcal{C}/\omega$.*
(ii) *$\mathcal{X}$ is contravariantly finite in $\mathcal{C}$.*

PROOF. By Lemma 1.1 we have that $\mathcal{X}$ is contravariantly finite in $\mathcal{C}$ iff $\mathcal{X}/\omega$ is contravariantly finite in $\mathcal{C}/\omega$. Since ω is an Ext-injective cogenerator for $\mathcal{X}$, we have that $\mathcal{X}/\omega$ is a right triangulated subcategory of the pretriangulated category $\mathcal{C}/\omega$. By Proposition II.2.4 we have that $\mathcal{X}/\omega$ is contravariantly finite in $\mathcal{C}/\omega$ iff $\mathcal{X}/\omega$ is coreflective in $\mathcal{C}/\omega$. Then the assertion follows from Proposition 4.1. □

COROLLARY 4.3. *Let $\mathcal{C}$ be a Krull-Schmidt category and $\mathcal{X}$ a preresolving contravariantly finite subcategory of $\mathcal{C}$ and assume that $\omega := \mathcal{X} \cap \mathcal{X}^{\perp}$ is functorially finite in $\mathcal{C}$. Then the following are equivalent.*

(i) $\mathcal{X}$ *is resolving.*
(ii) $\mathcal{X}/\omega$ *is a torsion class in* $\mathcal{C}/\omega$.

In view of the above results it is natural to consider the maximal subcategory of $\mathcal{C}$ containing ω as an Ext-injective cogenerator, which from now on we denote by $\mathrm{CM}(\omega)$. More explicitly if ω is a self-orthogonal subcategory $\omega \subseteq \mathcal{C}$, i.e. $\mathrm{Ext}^n(\omega,\omega) = 0, \forall n \geq 1$, then:

$$\mathrm{CM}(\omega) = \big\{ C \in \mathcal{C} \mid \exists \text{ exact sequence } 0 \to C \to T^0 \xrightarrow{f^0} T^1 \xrightarrow{f^1} T^2 \to \cdots,$$

$$T^s \in \omega,\ \forall s \geq 0: \ \mathrm{Ext}^t_{\mathcal{C}}(\mathrm{Ker}(f^n),\omega) = 0,\ \forall n \geq 0,\ \forall t \geq 1 \big\}.$$

We call the objects of $\mathrm{CM}(\omega)$ the **Cohen-Macaulay objects** of $\mathcal{C}$ with respect to ω. If $\omega = \mathrm{add}(T)$ for a self-orthogonal object T, then we use the notation $\mathrm{CM}(\omega) = \mathrm{CM}(T)$. This category was introduced in [**9**] (see also [**5**]), called $\mathcal{X}_\omega$ in [**9**], in connection with Cohen-Macaulay approximations: if Λ is a commutative Gorenstein ring, then $\mathrm{CM}(\Lambda)$ is the category of (maximal) Cohen-Macaulay modules. Note that by construction and Corollary 4.2 the subcategory $\mathrm{CM}(\omega)$ contains $\mathcal{X}$ for any torsion class $\mathcal{X}/\omega$ in $\mathcal{C}/\omega$ arising from a contravariantly finite resolving subcategory $\mathcal{X}$ of $\mathcal{C}$. Since by [**9**], $\mathrm{CM}(\omega)$ is closed under extensions and kernels of epimorphisms, it is resolving if it contains the projectives. Putting things together we have the following consequence.

COROLLARY 4.4. *Let* ω *be a functorially finite self-orthogonal subcategory of* $\mathcal{C}$. *If* $\mathrm{CM}(\omega)$ *contains the projectives of* $\mathcal{C}$, *then the following are equivalent.*
(i) $\mathrm{CM}(\omega)/\omega$ *is a torsion class in* $\mathcal{C}/\omega$.
(ii) *the inclusion* $\mathrm{CM}(\omega)/\omega \hookrightarrow \mathcal{C}/\omega$ *admits a right adjoint.*

EXAMPLE. Let $\omega = \mathcal{P} \cap \mathcal{I}$ be the full subcategory of $\mathcal{C}$ consisting of the projective-injective objects. We say that the object C has *infinite dominant dimension*, and then we write $\mathrm{dom.dim}C = \infty$, if in any injective resolution $0 \to C \to I_0 \to I_1 \to \cdots$ of C all the injective objects I_i are projective. Let $\mathrm{Dom}(\mathcal{C})$ be the full subcategory of $\mathcal{C}$ consisting of all objects with $\mathrm{dom.dim}C = \infty$. Observe that $\mathrm{CM}(\omega) = \mathrm{Dom}(\mathcal{C})$. If ω is functorially finite in $\mathcal{C}$ and any projective object has infinite dominant dimension, then by Corollary 4.4 we have that $\mathrm{Dom}(\mathcal{C})/\omega$ is a torsion class in $\mathcal{C}/\omega$ iff $\mathrm{Dom}(\mathcal{C})/\omega$ is coreflective in $\mathcal{C}/\omega$. This is equivalent to $\mathrm{Dom}(\mathcal{C})$ being contravariantly finite in $\mathcal{C}$, if $\mathcal{C}$ is in addition Krull-Schmidt.

Dually we denote by $\mathrm{CoCM}(\omega)$ the maximal subcategory of $\mathcal{C}$ which admits ω as an Ext-projective generator. Obviously

$$\mathrm{CoCM}(\omega) = \big\{ C \in \mathcal{C} \mid \exists \text{ exact sequence } \cdots \to T_2 \xrightarrow{f_1} T_1 \xrightarrow{f_0} T^0 \to C \to 0,$$

$$T_s \in \omega,\ \forall s \geq 0: \ \mathrm{Ext}^t_{\mathcal{C}}(\omega,\mathrm{Coker}(f_n)) = 0,\ \forall n \geq 0,\ \forall t \geq 1 \big\}.$$

If $\omega = \mathrm{add}(T)$ for $T \in \mathcal{C}$, then we use the notation $\mathrm{CoCM}(\omega) = \mathrm{CoCM}(T)$. The duals of the above observations also hold. We note only the following.

NOTE. If $(\mathcal{X},\mathcal{Y})$ is a (co)resolving cotorsion pair in $\mathcal{C}$ with $\mathcal{X} \cap \mathcal{Y} = \omega$, then:

$${}^{\perp}\mathrm{CoCM}(\omega) \subseteq \mathcal{X} \subseteq \mathrm{CM}(\omega) \quad \text{and} \quad \mathrm{CM}(\omega)^{\perp} \subseteq \mathcal{Y} \subseteq \mathrm{CoCM}(\omega).$$

Let now $(\mathcal{X}, \mathcal{Y})$ be a good pair in the abelian Krull-Schmidt category $\mathcal{C}$, and assume that $\omega = \mathcal{X} \cap \mathcal{Y}$ is functorially finite in $\mathcal{C}$ and $\mathcal{X}$ is resolving. Then we have that $(\mathcal{X}/\omega, \mathcal{Y}/\omega)$ is a torsion pair in $\mathcal{C}/\omega$. In some cases it is possible to get a description of $\mathcal{Y}$ (see [**5**]). For a subcategory $\mathcal{Z}$ of $\mathcal{C}$ we denote by $\widehat{\mathcal{Z}}$ the full subcategory of $\mathcal{C}$ whose objects are the C such that there is an exact sequence $0 \to Z_n \to \cdots \to Z_1 \to Z_0 \to C \to 0$ with the Z_i in $\mathcal{Z}$. Dually we denote by $\widetilde{\mathcal{Z}}$ the full subcategory of $\mathcal{C}$ whose objects are the C such that there is an exact sequence $0 \to C \to Z_0 \to Z_1 \to \cdots \to Z_m \to 0$ with the Z_i in $\mathcal{Z}$. For instance if $\mathcal{P}$ is the full subcategory of projectives, then $\widehat{\mathrm{CM}(\mathcal{P})}$ is the full subcategory of objects with finite Gorenstein dimension and $\widehat{\mathcal{P}} := \mathcal{P}^{<\infty}$ is the full subcategory of $\mathcal{C}$ consisting of the objects with finite projective dimension. Dually if $\mathcal{I}$ is the full subcategory of injectives, then $\widetilde{\mathcal{I}} := \mathcal{I}^{<\infty}$ is the full subcategory of $\mathcal{C}$ consisting of the objects with finite injective dimension. Recall from [**6**] that the (projective) *Gorenstein dimension* $\text{G-dim}_{\mathcal{P}}\, C$ of an object C in $\mathcal{C}$ is defined inductively as follows. If C is in $\mathrm{CM}(\mathcal{P})$, then $\text{G-dim}_{\mathcal{P}}\, C = 0$. If $t \geq 1$, then $\text{G-dim}_{\mathcal{P}}\, C \leq t$ if there exists an exact sequence $0 \to G_t \to \cdots \to G_1 \to G_0 \to C \to 0$ where $\text{G-dim}_{\mathcal{P}}\, G_i = 0$, for $0 \leq i \leq t$. Then $\text{G-dim}_{\mathcal{P}}\, C = t$ if $\text{G-dim}_{\mathcal{P}}\, C \leq t$ and $\text{G-dim}_{\mathcal{P}}\, C \nleq t-1$. Finally if $\text{G-dim}_{\mathcal{P}}\, C \neq t$ for any $t \geq 0$, then define $\text{G-dim}_{\mathcal{P}}\, C = \infty$. It follows that $\widehat{\mathrm{CM}(\mathcal{P})} = \{C \in \mathcal{C} \mid \text{G-dim}_{\mathcal{P}}\, C < \infty\}$.

PROPOSITION 4.5. *With the above notations and assumptions we have:* $\widehat{\mathcal{X}} = \mathcal{C}$ *iff* $\mathcal{Y} = \widehat{\omega}$. *Dually we have:* $\widetilde{\mathcal{Y}} = \mathcal{C}$ *iff* $\mathcal{X} = \widetilde{\omega}$.

In the not necessarily Krull-Schmidt case we have the following result, which is an easy consequence of our previous results and the results of [**5**].

PROPOSITION 4.6. *If* $\mathcal{X}$ *is a resolving subcategory with a functorially finite Ext-injective cogenerator* ω, *then the following are equivalent:*

(i) $(\mathcal{X}/\omega, \widehat{\omega}/\omega)$ *is a torsion pair in* $\mathcal{C}/\omega$.

(ii) $\widehat{\mathcal{X}} = \mathcal{C}$.

If this the case, then $\mathcal{X} = \mathrm{CM}(\omega)$.

EXAMPLE. Let Λ be a left coherent ring. It is not difficult to see that the resolving subcategory $\mathrm{Flat}(\Lambda)$ of flat right Λ-modules admits the subcategory $\omega := \mathrm{Flat}(\Lambda) \cap \mathrm{PInj}(\Lambda)$ of flat and pure-injective modules as an Ext-injective cogenerator. Hence $\mathrm{Flat}(\Lambda) \subseteq \mathrm{CM}(\omega)$. If, in addition, $\text{w.gl.dim}\Lambda < \infty$, then by Proposition 4.6 we have a resolving cotorsion pair $(\mathrm{Flat}(\Lambda), \widehat{\omega})$ in $\mathrm{Mod}(\Lambda)$ with $\mathrm{Flat}(\Lambda) = \mathrm{CM}(\omega)$. Note that left coherence of Λ forces $\mathrm{Flat}(\Lambda)$ to be covariantly finite [**106**]. Hence by Corollary 3.11, ω is covariantly finite. However $\widehat{\omega}$, hence ω, is not contravariantly finite in general, see the example at the end of this section.

EXAMPLE. Let $\mathcal{C}$ be an abelian category with enough projective and injective objects, and let ω the full subcategory of projective-injective objects of $\mathcal{C}$. Then $\mathcal{C}$ is Frobenius if and only if any projective object of $\mathcal{C}$ has infinite dominant dimension and there exists a cotorsion pair $(\mathrm{Dom}(\omega), \widehat{\omega})$ in $\mathcal{C}$. In this case $\mathcal{C} = \mathrm{Dom}(\mathcal{C}) = \mathrm{CM}(\omega)$.

The above results are true in the more general situation in which $\mathcal{C}$ does not necessarily have enough projectives and we don't necessarily have $\widehat{\mathcal{X}} = \mathcal{C}$. Indeed the following remark shows that it is possible to get a cotorsion pair in a smaller piece of $\mathcal{C}$ and a torsion pair in a smaller piece of $\mathcal{C}/\omega$.

REMARK 4.7. Let $\mathcal{X}$ be a full subcategory of $\mathcal{C}$ which is closed under extensions and kernels of epimorphisms, and assume that $\mathcal{X}$ admits a cogenerator ω with $\mathrm{Ext}^n(\mathcal{X},\omega) = 0$, $\forall n \geq 1$. It follows by the results of [**5**] that $(\mathcal{X},\widehat{\omega})$ is a resolving cotorsion pair in $\widehat{\mathcal{X}}$ with $\mathcal{X} \cap \widehat{\omega} = \omega$, hence a resolving cotorsion pair in $\mathcal{C}$ if $\widehat{\mathcal{X}} = \mathcal{C}$. If in addition ω is an Ext-projective generator of $\mathcal{X}$, then it follows easily that the stable category $\widehat{\mathcal{X}}/\omega$ is pretriangulated and then the pair $(\mathcal{X}/\omega, \widehat{\omega}/\omega)$ is a torsion pair in $\widehat{\mathcal{X}}/\omega$, hence a torsion pair in $\mathcal{C}/\omega$ if in addition ω is functorially finite in $\mathcal{C}$ and $\widehat{\mathcal{X}} = \mathcal{C}$. In particular for any contravariantly finite self-orthogonal subcategory ω of $\mathcal{C}$, the pair $(\mathrm{CM}(\omega),\widehat{\omega})$ is a cotorsion pair in $\widehat{\mathrm{CM}(\omega)}$ and the pair $(\mathrm{CM}(\omega)/\omega, \widehat{\omega}/\omega)$ is a torsion pair in $\widehat{\mathrm{CM}(\omega)}/\omega$. For instance we can take $\omega = \mathcal{P}$ to be the full subcategory of projective objects of $\mathcal{C}$. Then $(\mathrm{CM}(\mathcal{P}), \mathcal{P}^{<\infty})$ is a cotorsion pair in $\widehat{\mathrm{CM}(\mathcal{P})}$ with $\mathrm{CM}(\mathcal{P}) \cap \mathcal{P}^{<\infty} = \mathcal{P}$ and $(\mathrm{CM}(\mathcal{P})/\mathcal{P}, \mathcal{P}^{<\infty}/\mathcal{P})$ is a torsion pair in $\widehat{\mathrm{CM}(\mathcal{P})}/\mathcal{P}$.

Dual remarks hold for subcategories admitting an Ext-projective generator and CoCohen-Macaulay objects.

The following is an example from homological group theory which illustrates the above remark.

EXAMPLE. Let G be a group and let k be a commutative ring of coefficients. We denote by $B = B(G,k)$ the set of functions $G \to k$ which take only finitely many different values in k. Then G acts on B by multiplication and B is free as a k-module. Let $\mathcal{X}$ be the full subcategory of $\mathrm{Mod}(kG)$ consisting of all kG-modules M such that $B \otimes_k M$ is projective. The modules in $\mathcal{X}$ are known in the literature as cofibrant modules, see [**28**] and the example before Definition VII.4.4 for an explanation of the terminology. It is not difficult to see that $\mathcal{X}$ is resolving and $\widehat{\mathcal{X}}$ consists of all kG-modules M such that $B \otimes_k M$ has finite projective dimension. Let ω be the category of projective kG-modules, so $\widehat{\omega}$ is the full subcategory of all kG-modules of finite projective dimension. It is not difficult to see that ω is an Ext-injective cogenerator of $\mathcal{X}$, hence by the above remark, $(\mathcal{X},\widehat{\omega})$ is cotorsion pair in $\widehat{\mathcal{X}}$. We refer to [**28**] for details and more information. Note that if k has finite global dimension, and the group G is of type FP_∞, that is, the trivial G-module k admits an exact resolution by finitely generated projective modules, and belongs to the Kropholler's class $\mathbf{H}\mathfrak{F}$ of hierarchically decomposable groups [**40**], then $\widehat{\mathcal{X}} = \mathrm{Mod}(kG)$. Hence in this case $(\mathcal{X},\widehat{\omega})$ is a cotorsion pair in $\mathrm{Mod}(kG)$.

The finiteness condition $\widehat{\mathcal{X}} = \mathcal{C}$ or $\widetilde{\mathcal{Y}} = \mathcal{C}$ has some interesting consequences on the level of derived categories, as we now explain.

Let $(\mathcal{X}/\omega, \mathcal{Y}/\omega)$ be a torsion pair in $\mathcal{C}/\omega$ arising from a cotorsion pair $(\mathcal{X},\mathcal{Y})$ with $\omega = \mathcal{X}\cap\mathcal{Y}$ functorially finite in $\mathcal{C}$. Then $\mathcal{X}$, $\mathcal{Y}$ and ω are closed under extensions in $\mathcal{C}$, hence they are exact subcategories of $\mathcal{C}$ in the sense of Quillen; obviously ω carries the split exact structure since it is self-orthogonal. Let $\mathbf{D}^b(\mathcal{X})$, $\mathbf{D}^b(\mathcal{Y})$ and $\mathbf{D}^b(\omega)$ be the corresponding bounded derived categories, see [**71**] for the notion of the derived category of an exact category. Then $\mathbf{D}^b(\omega)$ coincides with the bounded homotopy category $\mathcal{H}^b(\omega)$.

COROLLARY 4.8. *We have a commutative diagram of full exact embeddings:*

$$\begin{array}{ccc} \mathbf{D}^b(\omega) & \longrightarrow & \mathbf{D}^b(\mathcal{Y}) \\ \downarrow & & \downarrow \\ \mathbf{D}^b(\mathcal{X}) & \longrightarrow & \mathbf{D}^b(\mathcal{C}) \end{array}$$

(1) *If* $\widehat{\mathcal{X}} = \mathcal{C}$ *(equivalently* $\mathcal{Y} \subseteq \widehat{\omega}$*), then the above functors induce triangle equivalences* $\mathbf{D}^b(\mathcal{X}) \xrightarrow{\approx} \mathbf{D}^b(\mathcal{C})$ *and* $\mathbf{D}^b(\omega) \xrightarrow{\approx} \mathbf{D}^b(\mathcal{Y})$.

(2) *If* $\widetilde{\mathcal{Y}} = \mathcal{C}$ *(equivalently* $\mathcal{X} \subseteq \widetilde{\omega}$*), then the above functors induce triangle equivalences* $\mathbf{D}^b(\mathcal{Y}) \xrightarrow{\approx} \mathbf{D}^b(\mathcal{C})$ *and* $\mathbf{D}^b(\omega) \xrightarrow{\approx} \mathbf{D}^b(\mathcal{X})$.

PROOF. We only prove (1) since (2) is dual. By the dual of Theorem 12.1 in [**71**] the induced functor $\mathbf{D}^b(\mathcal{X}) \to \mathbf{D}^b(\mathcal{C})$ is fully faithful, if any short exact sequence $0 \to A \to B \to X \to 0$ in $\mathcal{C}$ with $X \in \mathcal{X}$ can be embedded in an exact commutative diagram

$$\begin{array}{ccccccccc} 0 & \longrightarrow & X' & \longrightarrow & X_B & \longrightarrow & X & \longrightarrow & 0 \\ & & \downarrow & & \downarrow{\scriptstyle f_B} & & \| & & \\ 0 & \longrightarrow & A & \longrightarrow & B & \longrightarrow & X & \longrightarrow & 0 \end{array}$$

where X' and X_B are in $\mathcal{X}$. This holds by choosing f_B to be a right $\mathcal{X}$-approximation of B; then X' is in $\mathcal{X}$ since $\mathcal{X}$ is closed under kernels of epimorphisms. If $\widehat{\mathcal{X}} = \mathcal{C}$ then the fully faithful functor $\mathbf{D}^b(\mathcal{X}) \to \mathbf{D}^b(\mathcal{C})$ is an equivalence by devissage. Since $\widehat{\mathcal{X}} = \mathcal{C}$ implies that $\mathcal{Y} \subseteq \widehat{\omega}$, again by devissage the full embedding $\mathbf{D}^b(\omega) \hookrightarrow \mathbf{D}^b(\mathcal{Y})$ is an equivalence. □

It is not difficult to see that the above full exact embeddings induce short exact sequences of triangulated categories

$$0 \longrightarrow \mathbf{D}^b(\omega) \longrightarrow \mathbf{D}^b(\mathcal{X}) \longrightarrow \mathcal{T}_r(\mathcal{X}/\omega) \longrightarrow 0$$

and

$$0 \longrightarrow \mathbf{D}^b(\omega) \longrightarrow \mathbf{D}^b(\mathcal{Y}) \longrightarrow \mathcal{T}_l(\mathcal{Y}/\omega) \longrightarrow 0$$

where $\mathcal{T}_r(\mathcal{X}/\omega)$ is the stabilization of the right triangulated torsion class $\mathcal{X}/\omega$ and $\mathcal{T}_l(\mathcal{Y}/\omega)$ is the stabilization of the left triangulated torsion-free class $\mathcal{Y}/\omega$.

We close this section by pointing out some interesting examples of good or cotorsion pairs in the category $\mathrm{Mod}(\Lambda)$ of all right Λ-modules, where Λ is any ring:

EXAMPLE.

(i) $({}^{\perp}\mathcal{Y}, \mathcal{Y})$, where $\mathcal{Y}$ is a covariantly finite coresolving subcategory of $\mathrm{Mod}(\Lambda)$ consisting of pure-injective modules.

(ii) $(\mathcal{X}, \mathcal{X}^{\perp})$, where $\mathcal{X}$ is a contravariantly finite resolving subcategory of $\mathrm{Mod}(\Lambda)$ closed under direct limits.

(iii) $({}^{\perp}T, ({}^{\perp}T)^{\perp})$, where T is a pure-injective module such that any projective module is a submodule of a product of copies of T, e.g. $T = \Lambda$ for a right perfect ring Λ.

(iv) $({}^{\perp}(T^{\perp}), T^{\perp})$, where T is a finitely presented module with right perfect endomorphism ring and such that $\mathrm{Ext}^1_{\Lambda}(T, T) = 0$.

The above assertions follow from Lemma 3.3 and the following facts. Any module admits a special right $\mathcal{X}$-approximation, provided that $\mathcal{X}$ is a resolving contravariantly finite subcategory which is closed under filtered colimits [**106**]. And any module admits a special left $\mathcal{Y}$-approximation, provided that $\mathcal{Y}$ is a coresolving subcategory which consists of pure-injective modules [**78**].

More generally by [**45**], if $\mathcal{A} \subseteq \mathrm{Mod}(\Lambda)$ is any skeletally small subcategory, then $\big({}^{\perp}(\mathcal{A}^{\perp}), \mathcal{A}^{\perp}\big)$ is a cotorsion pair in $\mathrm{Mod}(\Lambda)$. Dually $\big({}^{\perp}\mathcal{B}, ({}^{\perp}\mathcal{B})^{\perp}\big)$ is a cotorsion pair in Mod(Λ, provided that $\mathcal{B}$ consists of pure injective modules. The above cotorsion pairs are (co)resolving, provided that, in addition, $\mathcal{A}$ is closed under syzygies and $\mathcal{B}$ is closed under cosyzygies. By the recent solution of the Flat Cover Conjecture by Bashir, Bican and Enochs [**17**], it follows that for any ring Λ, the pair $\big(\mathrm{Flat}(\Lambda), \mathrm{Flat}(\Lambda)^{\perp}\big)$ is a resolving cotorsion pair in $\mathrm{Mod}(\Lambda)$. As we now explain this example also shows that there exists a cotorsion pair $(\mathcal{X}, \mathcal{Y})$ such that $\mathcal{X} \cap \mathcal{Y}$ is not functorially finite, hence we cannot speak of a torsion pair in $\mathcal{C}/\mathcal{X} \cap \mathcal{Y}$.

EXAMPLE. Consider the flat cotorsion pair $\big(\mathrm{Flat}(\Lambda), \mathrm{Flat}(\Lambda)^{\perp}\big)$ in $\mathrm{Mod}(\Lambda)$ and assume that Λ is a left coherent and right IF-ring (that is any right injective module is flat) which is not right Noetherian. Then it is not difficult to see that the intersection $\omega = \mathrm{Flat}(\Lambda) \cap \mathrm{Flat}(\Lambda)^{\perp}$ is precisely the category of injective right modules, which is known to be contravariantly finite iff Λ is right Noetherian, see [**46**]. Note that ω is covariantly finite since Λ is left coherent.

5. Tilting Modules

An important source of examples of good pairs for which the induced torsion or torsion-free class is easily described emerges from cotilting or tilting modules over an Artin algebra Λ. It is well-known that the category $\mathrm{mod}(\Lambda)$ of finitely generated modules is a Krull-Schmidt category.

Let T be a finitely generated Λ-module with $\mathrm{Ext}^i_{\Lambda}(T,T) = 0, \forall i > 0$. Recall that T is a *cotilting* module if $\mathrm{id}_{\Lambda}T < \infty$ and the injective cogenerator $\mathrm{D}(\Lambda)$ lies in $\widehat{\mathrm{add}(T)}$. Dually T is a *tilting* module if $\mathrm{pd}_{\Lambda}T < \infty$ and the projective generator Λ lies in $\widecheck{\mathrm{add}(T)}$. If ω is a self-orthogonal subcategory then we denote by $\mathcal{X}_{\omega}$ the subcategory

$$\mathcal{X}_{\omega} := \big\{ C \in \mathrm{mod}(\Lambda) \mid \mathrm{Ext}^i_{\Lambda}(C, \omega) = 0, \ \forall i > 0 \big\}.$$

The subcategory $\mathcal{Y}_{\omega}$ is defined dually. If $\omega = \mathrm{add}(T)$, for a self-orthogonal module T, then we use the notations $\mathcal{X}_T$ and $\mathcal{Y}_T$.

NOTE. We warn the reader that the above notation is not standard in the literature. The subcategory above is denoted usually by ${}^{\perp}\omega$. However in this paper we reserved the last notation for the subcategory $\{C \in \mathrm{mod}(\Lambda) \mid \mathrm{Ext}^1_{\Lambda}(C, \omega) = 0\}$.

If T is a cotilting module, then $\mathcal{X}_T \cap \widehat{\mathrm{add}(T)} = \mathrm{add}(T)$, and $(\mathcal{X}_T, \widehat{\mathrm{add}(T)})$ is a good pair with $\mathcal{X}_T$ resolving, hence a cotorsion pair since $\mathrm{mod}(\Lambda)$ is Krull-Schmidt. Therefore $\big(\mathcal{X}_T/\mathrm{add}(T), \widehat{\mathrm{add}(T)}/\mathrm{add}(T)\big)$ is a torsion pair in $\underline{\mathrm{mod}}_T(\Lambda) := \mathrm{mod}(\Lambda)/\mathrm{add}(T)$ which is pretriangulated, since $\mathrm{add}(T)$ is always functorially finite [**13**]. It is well-known that T is cotilting module iff $\mathrm{add}(T)$ is a cogenerator for $\mathcal{X}_T$ and $\widehat{\mathcal{X}_T} = \mathrm{mod}(\Lambda)$ iff $\widehat{\mathrm{CM}(T)} = \mathrm{mod}(\Lambda)$, see [**9**]. We state explicitly the following result which characterizes the cotilting modules in torsion theoretic terms, leaving its dual concerning tilting modules to the reader.

PROPOSITION 5.1. *Let Λ be an Artin algebra and let T be a finitely generated Λ-module with $\mathrm{Ext}^i_\Lambda(T,T)=0, \forall i>0$. Then the following are equivalent.*

(i) *T is a cotilting module.*

(ii) *$(\mathcal{X}_T/\mathrm{add}(T), \widehat{\mathrm{add}(T)}/\mathrm{add}(T))$ is a (hereditary) torsion pair in $\underline{\mathrm{mod}}_T(\Lambda)$.*

PROOF. The implication (i) $\Rightarrow$ (ii) follows by the above considerations. (ii) $\Rightarrow$ (i) By induction we see easily that $\mathrm{Ext}^1_\Lambda(X,Y)=0$, $\forall X\in\mathcal{X}_T, \forall Y\in\widehat{\mathrm{add}(T)}$. Then Proposition 3.4 ensures that $(\mathcal{X}_T,\widehat{\mathrm{add}(T)})$ is a cotorsion pair in $\mathrm{mod}(\Lambda)$. Therefore $\widehat{\mathcal{X}_T}=\widehat{\mathrm{CM}(T)}=\mathrm{mod}(\Lambda)$ by Proposition 4.6. Hence T is a cotilting module. $\square$

Now let T be a tilting and cotilting module. Then we have the torsion pairs:

(i) $(\mathcal{X}/\mathrm{add}(T),\mathcal{Y}/\mathrm{add}(T))$ in $\underline{\mathrm{mod}}_T(\Lambda)$, where $\mathcal{X}=\mathcal{X}_T$ and $\mathcal{Y}=\widehat{\mathrm{add}(T)}$.

(ii) $(\mathcal{Z}/\mathrm{add}(T),\mathcal{W}/\mathrm{add}(T))$ in $\underline{\mathrm{mod}}_T(\Lambda)$, where $\mathcal{Z}=\widecheck{\mathrm{add}(T)}$ and $\mathcal{W}=\mathcal{Y}_T$.

When Λ has finite global dimension, then by [**9**] we have $\mathcal{X}=\mathcal{Z}$ and $\mathcal{Y}=\mathcal{W}$, so that the torsion pairs coincide. When $\mathrm{id}_\Lambda T\le 1$, then $\mathcal{X}$ is the torsion-free class of a torsion pair $(\mathcal{T},\mathcal{X})$ in $\mathrm{mod}(\Lambda)$, and when $\mathrm{pd}_\Lambda T\le 1$, then $\mathcal{Y}$ is the torsion class of a torsion pair $(\mathcal{Y},\mathcal{F})$ in $\mathrm{mod}(\Lambda)$. We have the following close relationship between the torsion pairs, a special case of which was observed by Dlab and Ringel in connection with tilting modules associated to quasi-hereditary algebras, see Theorem 4.3 in [**42**].

PROPOSITION 5.2. *With the above notation, if Λ has finite global dimension and T is a classical tilting-cotilting module, then we have equivalences:*

$$\mathcal{X}/\mathrm{add}(T)\xrightarrow{\approx}\mathcal{F}\quad\textit{and}\quad\mathcal{Y}/\mathrm{add}(T)\xrightarrow{\approx}\mathcal{T}.$$

PROOF. Let $\mathbf{t}:\mathrm{mod}(\Lambda)\to\mathcal{T}$ be the right adjoint of the inclusion $\mathcal{T}\hookrightarrow\mathrm{mod}(\Lambda)$ and let $\mathbf{f}:\mathrm{mod}(\Lambda)\to\mathcal{F}$ be the left adjoint of the inclusion $\mathcal{F}\hookrightarrow\mathrm{mod}(\Lambda)$. Then for any module C there exists a functorial exact sequence $0\to\mathbf{t}(C)\to C\to\mathbf{f}(C)\to 0$. We consider the functor $\mathbf{f}:\mathcal{X}_T\to\mathcal{F}$. Then $\mathbf{f}(T)=0$, hence there is induced a functor $\mathbf{F}:\mathcal{X}_T/\mathrm{add}(T)\to\mathcal{F}$. Let F be a module in $\mathcal{F}$ and consider the right $\mathcal{X}_T$–approximation sequence $0\to Y_F\to X_F\to F\to 0$ of F. Then we know that $Y_F\in\mathcal{Y}$. Hence there exists an extension $0\to T_1\to T_0\to Y_F\to 0$ with $T_i\in\mathrm{add}(T)$. Applying to this sequence the functor $(T,-)$ it follows easily that $\mathrm{Ext}^1_\Lambda(T,Y_F)=0$, that is $Y_F\in\mathcal{T}$. Hence $\mathbf{t}(X_F)=Y_F$ and then by definition $\mathbf{F}(X_F)=F$. Hence $\mathbf{F}$ is surjective on objects. If $\alpha:X_1\to X_2$ is a morphism in $\mathcal{X}_T$ such that $\mathbf{f}(\alpha)=0$, then obviously α factors through $\mathbf{t}(X_2)\in\mathcal{Y}$. This implies that α factors through $\mathrm{add}(T)$, since $(\mathcal{X}_T/\mathrm{add}(T),\mathcal{Y}/\mathrm{add}(T))=0$. Hence $\underline{\alpha}=0$ in $\mathcal{X}_T/\mathrm{add}(T)$. Finally if $\beta:F_1\to F_2$ is a morphism in $\mathcal{F}$, then the pull-back of the extension $0\to Y_{F_2}\to X_{F_2}\to F_2\to 0$ along the composition $X_{F_1}\to F_1\to F_2$, splits since $\mathrm{Ext}^1_\Lambda(X_{F_1},Y_{F_2})=0$. This implies that the composition $X_{F_1}\to F_1\to F_2$ factors through $X_{F_2}\to F_2$ say via a morphism $\alpha:X_{F_1}\to X_{F_2}$. Then by construction $\mathbf{F}(\underline{\alpha})=\beta$. Hence $\mathbf{F}$ is full. We infer that $\mathbf{F}:\mathcal{X}_T/\mathrm{add}(T)\to\mathcal{F}$ is an equivalence. The proof of the second equivalence is dual. $\square$

The above example shows that it can happen that the torsion or torsion-free class is an abelian category, or even a module category. Indeed, keep the above assumptions and suppose in addition that the projective cover of T is injective.

Then as in [**42**] one can show that the regular module Λ is a projective generator of $\mathcal{X}/\mathrm{add}(T)$, hence we have an equivalence $\mathcal{X}/\mathrm{add}(T) \approx \mathrm{mod}(\underline{\mathrm{End}}_T(\Lambda))$. Moreover $\mathrm{D}(\Lambda)$ is an injective cogenerator of $\mathcal{W}/\mathrm{add}(T)$, hence we have an equivalence $\big(\mathcal{W}/\mathrm{add}(T)\big)^{\mathrm{op}} \xrightarrow{\approx} \mathrm{mod}\big(\underline{\mathrm{End}}_T(\mathrm{D}(\Lambda))\big)$. It is also easy to see that the hearts of all the torsion pairs in the above example are trivial.

The triviality of the hearts is a rather typical situation encountered, if the pretriangulated category is not triangulated. However the following example shows that the heart of a torsion pair in a stable non-triangulated category can be a non-zero abelian category.

EXAMPLE. Let $\Gamma = k[x]/(x^2)$, where k is a field and consider the Artin algebra

$$\Lambda = \begin{pmatrix} \Gamma & 0 \\ \Gamma & \Gamma \end{pmatrix}.$$

Then Λ is cotilting module of injective dimension 1. The indecomposable Λ-modules fall into the following three groups, where we represent the Λ-modules via Γ-morphisms, see [**15**].

(α) $\mathcal{X} = \mathrm{CM}(\Lambda) = \mathrm{add}\big(\{(\Gamma \xrightarrow{=} \Gamma), (0 \to \Gamma), (0 \to k), (k \xrightarrow{=} k), (k \hookrightarrow \Gamma)\}\big)$.

(β) $\mathcal{Y} = \mathcal{P}_\Lambda^{<\infty} = \mathrm{add}\big(\{(\Gamma \xrightarrow{=} \Gamma), (0 \to \Gamma), (\Gamma \to 0), (\Gamma \xrightarrow{f} \Gamma)\}\big)$, where f is non-zero and non-invertible. Here $\mathcal{P}_\Lambda^{<\infty}$ denotes the full subcategory of modules with finite projective dimension.

(γ) $\mathcal{U} = \mathrm{add}\big(\{(k \to 0), (\Gamma \xrightarrow{g} k)\}\big)$, where g is non-zero.

Denote by $\mathcal{H}_l = \mathcal{X}/\mathcal{P}_\Lambda \cap \Sigma(\mathcal{Y}/\mathcal{P}_\Lambda)$ the left heart of the induced torsion pair $(\mathcal{X}/\mathcal{P}_\Lambda, \mathcal{Y}/\mathcal{P}_\Lambda)$ in $\underline{\mathrm{mod}}(\Lambda)$. It is not difficult to see that the only indecomposable object in $\mathcal{H}_l$ is given by $(k \hookrightarrow \Gamma)$ with stable endomorphism ring, the field k. It follows that $\mathcal{H}_l$ is a non-zero abelian category. Observe that the right heart is trivial. We don't know if in general the left or right heart of a torsion pair in a stable category is always zero or abelian.

We have seen that the right adjoint $\mathbf{R}$ of the inclusion of a torsion class tends to be left exact. However $\mathbf{R}$ is not right exact in general. Indeed in the above example, since $\mathrm{Hom}_\Lambda[(k \to 0), \Lambda)] = 0$, we have a sequence $(k \to 0) \to 0 \to 0 \to 0$, which induces the right triangle $(k \to 0) \to 0 \to 0 \to 0$ in $\underline{\mathrm{mod}}(\Lambda)$. The right $\mathcal{X}$-approximation of $(k \to 0)$ is $(k \hookrightarrow \Gamma)$, which is not projective. Since there is no exact sequence $(k \hookrightarrow \Gamma) \to P_1 \to P_0 \to 0$, with P_1, P_0 projectives, by the construction of $\mathbf{R}$, we infer that $\mathbf{R}$ is not right exact.

Interesting examples of cotilting modules T occur for quasi-hereditary algebras, where we have that $\mathcal{X}_T$ is the category of good modules [**96**]. Other special cases are $T = \Lambda$, where Λ is a Gorenstein Artin algebra or a commutative complete local Gorenstein ring, and $T = \omega$, where ω is a dualizing bimodule over a Cohen-Macaulay Artin algebra or a commutative complete local Noetherian Cohen-Macaulay ring. In all these cases $\mathcal{X}_T = \mathrm{CM}(T)$ is the category of (maximal) Cohen-Macaulay modules.

We can also speak more generally of ω being a cotilting subcategory of $\mathcal{C}$, meaning that $\mathrm{Ext}^i_\Lambda(\omega, \omega) = 0, \forall i > 0$, that there is some $t \geq 0$ such that $\mathrm{id}_\Lambda T \leq t$, for each $T \in \omega$ and each injective object of $\mathcal{C}$ lies in $\widehat{\omega}$. This gives rise to similar results which generalize easily to the infinite-dimensional case for any ring.

CHAPTER VI

Triangulated Torsion(-Free) Classes in Stable Categories

In this chapter we study the natural question of when the torsion, resp. torsion-free, class of a torsion pair in a stable pretriangulated category, is a triangulated subcategory. This is especially interesting since it implies that the torsion pair is hereditary, resp. cohereditary. We show that if the torsion pair is induced by a cotorsion pair $(\mathcal{X}, \mathcal{Y})$ in an abelian category with enough projectives, resp. injectives, then the torsion class $\mathcal{X}/\mathcal{X} \cap \mathcal{Y}$, resp. the torsion-free class $\mathcal{Y}/\mathcal{X} \cap \mathcal{Y}$, is triangulated if and only if $\mathcal{X} \cap \mathcal{Y}$ coincides with the projectives, resp. injectives. Usually such situations arise from cotorsion triples $(\mathcal{X}, \mathcal{Y}, \mathcal{Z})$ in an abelian category, that is, $(\mathcal{X}, \mathcal{Y})$ and $(\mathcal{Y}, \mathcal{Z})$ are cotorsion pairs. Cotorsion triples are the abelian analogues of TTF-triples in an triangulated category and are studied in the third section. We also give applications to finitely generated modules over an Artin algebra. In this case the notions of good pair (triple) and cotorsion pair (triple) coincide. We would like to stress that in this working setting our results have again strong connections with tilting theory and lead to various torsion-theoretic characterizations of Gorenstein algebras.

1. Triangulated Subcategories

In this section we state and prove some general results on triangulated subcategories of a stable category of an abelian category $\mathcal{C}$, which will be useful in the later sections in connection with (co)torsion pairs and Cohen-Macaulay objects.

We begin with the following observation.

LEMMA 1.1. *Let $\mathcal{X}$ be a full subcategory of $\mathcal{C}$, closed under kernels of epics, and let $\omega \subseteq \mathcal{X}$ be a full subcategory.*

(α) *If ω is an Ext-injective cogenerator of $\mathcal{X}$, then the stable category $\mathcal{X}/\omega$ is right triangulated.*

(β) *If ω is contravariantly finite in $\mathcal{X}$ and any ω-epic in $\mathcal{X}$ is epic, e.g. if ω is an Ext-projective generator of $\mathcal{X}$, then the stable category $\mathcal{X}/\omega$ is left triangulated.*

(γ) *If ω is contravariantly finite in $\mathcal{X}$ and an Ext-injective cogenerator for $\mathcal{X}$, and if any ω-epic in $\mathcal{X}$ is epic, then the stable category $\mathcal{X}/\omega$ is pretriangulated.*

PROOF. (α) Obviously ω is covariantly finite in $\mathcal{X}$. To show that $\mathcal{X}/\omega$ is right triangulated it suffices by [**19**] to show that C is in $\mathcal{X}$ whenever $X_1 \xrightarrow{f} X_2 \xrightarrow{g} C \to 0$ is exact in $\mathcal{C}$ where $X_1, X_2 \in \mathcal{X}$ and f is ω-monic. Since X_1 is embedded in an object in ω, it follows that f is monic. Since ω is a cogenerator for $\mathcal{X}$ and X_2 is in $\mathcal{X}$, there exists an exact sequence $0 \to X_2 \xrightarrow{\kappa} \omega_0 \xrightarrow{\lambda} X_3 \to 0$, where $\omega_0 \in \omega$ and

$X_3 \in \mathcal{X}$. Consider the following exact commutative diagram

$$\begin{array}{ccccccccc} 0 & \longrightarrow & X_1 & \xrightarrow{f} & X_2 & \xrightarrow{g} & C & \longrightarrow & 0 \\ & & \| & & \kappa\downarrow & & \alpha\downarrow & & \\ 0 & \longrightarrow & X_1 & \xrightarrow{h} & \omega_0 & \xrightarrow{\varepsilon} & D & \longrightarrow & 0 \end{array}$$

Then κ induces an isomorphism $\mathrm{Coker}(\kappa) \cong \mathrm{Coker}(\alpha) \cong X_3$. We claim that h is ω-monic. Indeed let $\phi : X_1 \to T$ be a morphism with $T \in \omega$. Since f is ω-monic, there exists $\tau : X_2 \to T$ such that $f \circ \tau = \phi$. Since $\mathrm{Coker}(\kappa) \in \mathcal{X}$ and ω consists of Ext-injectives in $\mathcal{X}$, it follows that κ is ω-monic. Hence there exists $\sigma : \omega_0 \to T$ such that $\kappa \circ \sigma = \tau$. Then $h \circ \sigma = f \circ \kappa \circ \sigma = f \circ \tau = \phi$. Hence ϕ factors through h and then h is ω-monic, i.e. a left ω-approximation of X_1. Since ω is an Ext-injective cogenerator of $\mathcal{X}$, there exists an exact sequence $0 \to X_1 \xrightarrow{\zeta} \omega_0' \xrightarrow{\eta} X \to 0$ with $\omega_0' \in \omega$ and $X \in \mathcal{X}$. Then obviously ζ is a left ω-approximation of X_1. Since h, ζ are both left ω-approximations of X_1, standard arguments show that in the stable category $\mathcal{C}/\omega$ we have an isomorphism $\underline{D} \cong \underline{X}$ and then $D \in \mathcal{X}$, since $X \in \mathcal{X}$. Since $\mathrm{Coker}(\kappa)$ and D are in $\mathcal{X}$ and $\mathcal{X}$ is closed under kernels of epics, we have $C \in \mathcal{X}$.

(β) Since any ω-epic in $\mathcal{X}$ is epic and $\mathcal{X}$ is closed under kernels of epics, we have that any ω-epic in $\mathcal{X}$ has a kernel in $\mathcal{X}$. Then by [**19**], the stable category $\mathcal{X}/\omega$ is left triangulated. Part (γ) follows easily from (α) and (β). □

If $(\mathcal{X}, \mathcal{Y})$ is a cotorsion pair in $\mathcal{C}$ then we know that $\omega = \mathcal{X} \cap \mathcal{Y}$ is an Ext-injective cogenerator of $\mathcal{X}$ and, by Lemma 1.1, the stable category $\mathcal{X}/\omega$ is right triangulated. Therefore it is natural to ask when, for a subcategory $\mathcal{X}$ endowed with an Ext-injective cogenerator ω, the stable category $\mathcal{X}/\omega$ is triangulated. In this connection we have the following result which gives a useful criterion for deciding when such a stable category is triangulated.

PROPOSITION 1.2. *Let $\mathcal{X}$ be a full subcategory of $\mathcal{C}$ closed under extensions and kernels of epics, and let ω be an Ext-injective cogenerator for $\mathcal{X}$. Then the stable category $\mathcal{X}/\omega$ is triangulated iff ω is an Ext-projective generator for $\mathcal{X}$.*

PROOF. If ω is an Ext-projective generator for $\mathcal{X}$, then obviously ω is contravariantly finite in $\mathcal{X}$ and any ω-epic in $\mathcal{X}$ is epic. Then by the above lemma, we have that $\mathcal{X}/\omega$ is pretriangulated. Let $0 \to X \to T \to X' \to 0$ be exact with $T \in \omega$ and $X, X' \in \mathcal{X}$. Since ω is Ext-injective in $\mathcal{X}$, we have that $\Sigma(\underline{X}) = \underline{X'}$ in $\mathcal{X}/\omega$. Since ω is Ext-projective in $\mathcal{X}$, we have that $\Omega\Sigma(\underline{X}) = \Omega(\underline{X'}) = \underline{X}$. It follows that the canonical morphism $\mathrm{Id}_{\mathcal{X}/\omega} \to \Omega\Sigma$ is invertible. Dually the canonical morphism $\Sigma\Omega \to \mathrm{Id}_{\mathcal{X}/\omega}$ is invertible. Hence $\mathcal{X}/\omega$ is triangulated. Conversely if $\mathcal{X}/\omega$ is a triangulated subcategory of $\mathcal{C}/\omega$, then $\Sigma|_{\mathcal{X}/\omega}$ is surjective on objects, so for any object $X \in \mathcal{X}$, there exists a right exact sequence $X' \xrightarrow{\beta} T \xrightarrow{\alpha} X \to 0$ such that β is ω-monic and $T \in \omega$. Since ω is a cogenerator of $\mathcal{X}$, β is monic, so ω is a generator of $\mathcal{X}$. It remains to show that the objects of ω are Ext-projective in $\mathcal{X}$. Let $0 \to X \xrightarrow{f} A \xrightarrow{g} T \to 0$ be an extension with $X \in \mathcal{X}$ and $T \in \omega$. Then A is in $\mathcal{X}$ since the latter is closed under extensions. Since T is Ext-injective in $\mathcal{X}$, the above extension induces a triangle $\underline{X} \xrightarrow{\underline{f}} \underline{A} \to 0 \to \Sigma(\underline{X})$ in $\mathcal{X}/\omega$. Since the latter is triangulated we have that $\underline{f} : \underline{X} \to \underline{A}$ is invertible. Hence there exists a morphism $h : A \to X$ such that the endomorphism $1_X - f \circ h$ of X factors through an object T' in ω, i.e. $1_X - f \circ h = \kappa \circ \lambda$, where $\kappa : X \to T'$ and $\lambda : T' \to X$. Since $T' \in \omega$ and the objects of ω are Ext-injective in $\mathcal{X}$, the morphism κ factors through f. Hence there exists a morphism $\rho : A \to T'$ such that $\kappa = f \circ \rho$. Then $1_X = f \circ h + \kappa \circ \lambda = f \circ h + f \circ \rho \circ \lambda = f \circ (h + \rho \circ \lambda)$. We infer that f is split monic and then T is Ext-projective. Hence the objects of ω are Ext-projective in $\mathcal{X}$. □

For instance if $\mathcal{X}$ is a resolving subcategory of an abelian category $\mathcal{C}$ with enough projectives, and the full subcategory ω of projective objects of $\mathcal{C}$ form an Ext-injective cogenerator of $\mathcal{X}$, then the stable category $\mathcal{X}/\omega$ is triangulated.

EXAMPLE. The full subcategory $\mathcal{X}$ of $\mathrm{Mod}(kG)$ of cofibrant modules in the example after Remark V.4.7, satisfies the assumptions of the above Proposition, where ω is the full subcategory of projective kG-modules. Hence $\mathcal{X}/\omega$ is triangulated.

We leave to the reader the dualization of the above result.

2. Triangulated Torsion(-Free) Classes

Throughout this section we fix an abelian category $\mathcal{C}$ with enough projective objects. As usual we denote by $\mathcal{P}$ the full subcategory of $\mathcal{C}$ consisting of the projective objects. The full subcategory of $\mathcal{C}$ consisting of all objects with finite projective dimension is denoted by $\mathcal{P}^{<\infty}$. When ω is a functorially finite subcategory of $\mathcal{C}$, we investigate when a torsion class in the stable pretriangulated category $\mathcal{C}/\omega$ is triangulated. We show that this happens if and only if ω coincides with the projectives. In this case the torsion class is related to Cohen-Macaulay objects. This connection will be useful in the last section when discussing Gorenstein algebras, which provide a natural source for the existence of nicely behaved (co)torsion pairs.

An important class of triangulated subcategories of the stable category $\mathcal{C}/\mathcal{P}$ emerges from Cohen-Macaulay objects. Let $\mathrm{CM}(\mathcal{P})$ be the full subcategory of Cohen-Macaulay objects in $\mathcal{C}$ with respect to $\mathcal{P}$. Since $\mathcal{P}$ is an Ext-projective generator and an Ext-injective cogenerator of $\mathrm{CM}(\mathcal{P})$, by Proposition 1.2 the stable category $\mathrm{CM}(\mathcal{P})/\mathcal{P}$ modulo projectives is a triangulated subcategory of $\mathcal{C}/\mathcal{P}$.

The next result shows that the example of Cohen-Macaulay objects is universal in the sense that any triangulated torsion class induced by a cotorsion pair consists of Cohen-Macaulay objects. First note that if $(\mathcal{X}, \mathcal{Y})$ is a cotorsion pair in $\mathcal{C}$ then the subcategories $\mathcal{X}$ and $\mathcal{Y}$ are closed under extensions in $\mathcal{C}$. Hence $\mathcal{X}$ and $\mathcal{Y}$ are exact subcategories of $\mathcal{C}$. The admissible exact sequences in $\mathcal{X}$, resp. $\mathcal{Y}$, are the sequences in $\mathcal{X}$, resp. $\mathcal{Y}$, which are exact in $\mathcal{C}$. Observe that if $\mathcal{X}$ is resolving, equivalently $\mathcal{Y}$ is coresolving, then any epimorphism in $\mathcal{X}$ is admissible and any monomorphism in $\mathcal{Y}$ is admissible.

THEOREM 2.1. *Let $(\mathcal{X}, \mathcal{Y})$ be a resolving cotorsion pair in $\mathcal{C}$ and assume that $\omega := \mathcal{X} \cap \mathcal{Y}$ is functorially finite. Then the following are equivalent:*

(i) *The torsion class $\mathcal{X}/\omega$ is triangulated.*
(ii) *The exact subcategory $\mathcal{X}$ of $\mathcal{C}$ is Frobenius.*
(iii) $\omega = \mathcal{P}$.
(iv) *$\mathcal{Y}$ is resolving.*

If $\omega = \mathcal{P}$, then $\mathcal{P}$ is functorially finite, $\mathcal{X} \subseteq \mathrm{CM}(\mathcal{P})$ and $\mathcal{P}^{<\infty} \subseteq \mathcal{Y}$, the coreflection functor $\mathbf{R} : \mathcal{C}/\omega \to \mathcal{X}/\omega$ is left exact and the torsion pair $(\mathcal{X}/\omega, \mathcal{Y}/\omega)$ is hereditary.

Conversely if $\mathcal{P}$ is functorially finite in $\mathcal{C}$ and $(\mathcal{X}/\mathcal{P}, \mathcal{Y}/\mathcal{P})$ is a hereditary torsion pair in $\mathcal{C}/\mathcal{P}$ with $\mathcal{X} \subseteq \mathrm{CM}(\mathcal{P})$, then $(\mathcal{X}, \mathcal{Y})$ is a resolving cotorsion pair in $\mathcal{C}$ with $\mathcal{X} \cap \mathcal{Y} = \mathcal{P}$.

PROOF. (i) $\Rightarrow$ ii) Assume that the torsion class $\mathcal{X}/\omega$ is a triangulated subcategory of $\mathcal{C}/\omega$. By Proposition V.1.3, ω is an Ext-injective cogenerator of $\mathcal{X}$. Hence for any object X in $\mathcal{X}$, there exists an admissible exact sequence $0 \to X \to T \to X' \to 0$ in $\mathcal{X}$ with $T \in \omega$. Since the objects of ω are Ext-injective in $\mathcal{X}$, we infer that the exact category $\mathcal{X}$ has enough injectives and the full subcategory of injective objects of $\mathcal{X}$ coincides with ω. By Proposition 1.2, ω is an Ext-projective generator of $\mathcal{X}$. Hence for any object X in $\mathcal{X}$, there exists an admissible exact sequence $0 \to X' \to T \to X \to 0$ in $\mathcal{X}$ with T in ω. Since the objects of ω are Ext-projective in $\mathcal{X}$, we infer that the exact category $\mathcal{X}$ has enough projectives and the full subcategory of projective objects of $\mathcal{X}$ coincides with ω. We conclude that $\mathcal{X}$ is Frobenius.

(ii) $\Rightarrow$ (iii) Since $\mathcal{X}$ contains the projectives, it follows directly that $\mathcal{P} \subseteq \omega$. Let T be in ω, and consider the exact sequence $0 \to \Omega(T) \to P \to T \to 0$ in $\mathcal{C}$ with P projective. Since $\mathcal{X}$ is resolving, we have that $\Omega(T)$ lies in $\mathcal{X}$. Since any epimorphism in $\mathcal{X}$ is admissible, the above sequence is admissible in $\mathcal{X}$. Since T is projective in $\mathcal{X}$, the sequence splits, so T is projective. We conclude that $\omega = \mathcal{P}$.

(iii) $\Rightarrow$ (iv) Since $\omega = \mathcal{P}$, the subcategory $\mathcal{Y}$ contains the projectives. Let $0 \to C \to Y_1 \to Y_2 \to 0$ be an exact sequence with $Y_1, Y_2 \in \mathcal{Y}$. Since $\omega = \mathcal{P}$ is a generator of $\mathcal{Y}$ and $\mathcal{C}/\mathcal{P}$ is left triangulated, we have a left triangle $\Omega(\underline{Y}_2) \to \underline{C} \to \underline{Y}_1 \to \underline{Y}_2$ in $\mathcal{C}/\mathcal{P}$ with $\Omega(\underline{Y}_2) \in \mathcal{Y}/\mathcal{P}$. Applying $\mathcal{C}/\mathcal{P}(\underline{X}, -)$, with $X \in \mathcal{X}$, to this triangle, we have an exact sequence: $\mathcal{C}/\mathcal{P}(\underline{X}, \Omega(\underline{Y}_2)) \to \mathcal{C}/\mathcal{P}(\underline{X}, \underline{C}) \to \mathcal{C}/\mathcal{P}(\underline{X}, \underline{Y}_1) \to \mathcal{C}/\mathcal{P}(\underline{X}, \underline{Y}_2)$, which implies that $\mathcal{C}/\mathcal{P}(\underline{X}, \underline{C}) = 0$. Hence $\underline{C} \in (\mathcal{X}/\mathcal{P})^{\perp} = \mathcal{Y}/\mathcal{P}$, that is, C lies in $\mathcal{Y}$. Therefore $\mathcal{Y}$ is closed under kernels of epimorphisms. Since $\mathcal{Y}$ is closed under extensions and contains the projectives, we infer that $\mathcal{Y}$ is resolving.

(iv) $\Rightarrow$ (i) Since $\mathcal{Y}$ contains the projectives, we have $\mathcal{P} \subseteq \omega$. Let T be in ω and let $0 \to \Omega(T) \to P \to T \to 0$ be exact with P projective. Since $\mathcal{X}$ and $\mathcal{Y}$ are resolving, we have that $\Omega(T)$ lies in ω. This implies that the sequence splits. Hence T is projective. It follows that $\mathcal{P} = \omega$, and then obviously ω is an Ext-projective generator of $\mathcal{X}$. By Proposition 1.2 we conclude that $\mathcal{X}/\omega$ is triangulated.

Assume now that condition (i) holds. Since $\mathcal{X}$ is resolving and $\mathcal{P}$ is an Ext-injective cogenerator of $\mathcal{X}$, it follows from Section V.4 that $\mathcal{X} \subseteq \mathrm{CM}(\mathcal{P})$. Since $\mathcal{Y}$ is coresolving and contains $\mathcal{P}$, it follows easily that $\mathcal{P}^{<\infty} \subseteq \mathcal{Y}$. It remains to show that the torsion pair $(\mathcal{X}/\mathcal{P}, \mathcal{Y}/\mathcal{P})$ in $\mathcal{C}/\mathcal{P}$ is hereditary. Since $\mathcal{X}$ is resolving and $\omega = \mathcal{P}$ it is easy to check that the inclusion $\mathbf{i} : \mathcal{X}/\mathcal{P} \hookrightarrow \mathcal{C}/\mathcal{P}$ is left exact. By Proposition 3.6 of [**9**], any exact sequence $0 \to A \to B \to C \to 0$ in $\mathcal{C}$ induces an exact sequence $0 \to X_A \to X_B \to X_C \to 0$ in $\mathcal{X}$ where X_A, X_B, X_C are right $\mathcal{X}$-approximations of A, B, C. The last short exact sequence induces a triangle $\Omega\mathbf{R}(\underline{C}) \to \mathbf{R}(\underline{A}) \to \mathbf{R}(\underline{B}) \to \mathbf{R}(\underline{C})$ in $\mathcal{X}/\mathcal{P}$. If B is projective then the above triangle induces an isomorphism $\Omega\mathbf{R}(\underline{C}) \xrightarrow{\cong} \mathbf{R}\Omega(\underline{C}) \cong \mathbf{R}(\underline{A})$. These facts imply that the coreflection functor $\mathbf{R} : \mathcal{C}/\mathcal{P} \to \mathcal{X}/\mathcal{P}$ is left exact. It follows that the idempotent functor $\mathbf{iR} : \mathcal{C}/\mathcal{P} \to \mathcal{C}/\mathcal{P}$ is left exact, so that the torsion pair $(\mathcal{X}/\mathcal{P}, \mathcal{Y}/\mathcal{P})$ is hereditary.

The last assertion follows from Theorem V.3.4. □

REMARK 2.2. (1) If the abelian category $\mathcal{C}$ admits a resolving cotorsion pair $(\mathcal{X}, \mathcal{Y})$ with $\mathcal{X} \cap \mathcal{Y}$ functorially finite and such that the torsion class $\mathcal{X}/\omega$ is triangulated, then, by Theorem 2.1, the category $\mathcal{P}$ of projective objects of $\mathcal{C}$ is functorially finite. This puts some restrictions on $\mathcal{C}$. For instance if the category $\operatorname{Mod}(\Lambda)$ of right Λ-modules over a ring Λ admits such a cotorsion pair, then Λ is left coherent and right perfect. This follows from the fact that Λ is left coherent and right perfect iff the category of projective right Λ-modules is functorially finite.

(2) The above theorem implies that if $\mathcal{C}$ has finite global dimension, then $\mathcal{Y} = \mathcal{C}$ and $\mathcal{X} = \mathcal{P}$. Hence for abelian categories of finite global dimension, the only triangulated torsion class induced by a cotorsion pair is the zero subcategory.

The next result describes some important consequences of the existence of a resolving cotorsion pair $(\mathcal{X}, \mathcal{Y})$ in $\mathcal{C}$ with $\mathcal{X} \cap \mathcal{Y}$ consisting of the projectives, which will be useful in Chapter X in connection with Cohen-Macaulay objects.

PROPOSITION 2.3. *Let $(\mathcal{X}, \mathcal{Y})$ be a resolving cotorsion pair in $\mathcal{C}$ with $\mathcal{X} \cap \mathcal{Y} = \mathcal{P}$. Then we have the following.*

(i) *$C \in \widehat{\mathcal{X}}$ if and only if $Y_C \in \mathcal{P}^{<\infty}$, where Y_C is the kernel of a special right $\mathcal{X}$-approximation of C.*
(ii) *$\mathcal{P}^{<\infty} = \mathcal{Y} \cap \widehat{\mathcal{X}}$ and $\mathcal{I}^{<\infty} \subseteq \mathcal{Y}$.*
(iii) *$\widehat{\mathcal{X}} = \mathcal{C}$ iff $\mathcal{Y} = \mathcal{P}^{<\infty}$.*
(iv) *If $\widehat{\mathcal{X}} = \mathcal{C}$, then $\mathcal{X} = \operatorname{CM}(\mathcal{P})$ and $\mathcal{I}^{<\infty} \subseteq \mathcal{P}^{<\infty} = \mathcal{Y}$.*

PROOF. (i) Since $\mathcal{X}$ contains the projectives it follows that $\mathcal{P}^{<\infty} = \widehat{\mathcal{P}} \subseteq \widehat{\mathcal{X}}$. Assume now that C has a finite exact resolution $0 \to X_{n+1} \to X_n \to \cdots \to X_1 \to X_0 \xrightarrow{\alpha} C \to 0$ by objects from $\mathcal{X}$. If $n = 0$, then consider the pull-back diagram

$$\begin{array}{ccccccccc} 0 & \longrightarrow & Y_C & \longrightarrow & A & \longrightarrow & X_0 & \longrightarrow & 0 \\ & & \| & & \downarrow & & \downarrow{\scriptstyle \alpha} & & \\ 0 & \longrightarrow & Y_C & \xrightarrow{g_C} & X_C & \xrightarrow{f_C} & C & \longrightarrow & 0 \end{array}$$

of the special right $\mathcal{X}$-approximation sequence of C along α. Since X_0 is in $\mathcal{X}$, α factors through f_C and this implies that $A \cong Y_C \oplus X_0$. But from the above diagram it follows that A is in $\mathcal{X}$ since the kernel of α is $X_1 \in \mathcal{X}$ and $\mathcal{X}$ is closed under extensions. We infer that $Y_C \in \mathcal{X} \cap \mathcal{Y} = \mathcal{P}$. By induction on the length $n+1$ of the resolution of C by objects from $\mathcal{X}$, we deduce easily that $\operatorname{pd} Y_C \leq n$. Hence Y_C has finite projective dimension.

(ii), (iii) If C is in $\mathcal{P}^{<\infty}$, then, by Theorem 2.1, we have that C is in $\mathcal{Y}$. Since $\mathcal{X}$ contains the projectives, we have $\widehat{\mathcal{P}} = \mathcal{P}^{<\infty} \subseteq \widehat{\mathcal{X}}$. Hence C lies in $\mathcal{Y} \cap \widehat{\mathcal{X}}$. Conversely if C lies in $\mathcal{Y} \cap \widehat{\mathcal{X}}$, then by (i) we have that $Y_C \in \mathcal{P}^{<\infty}$. However since C is in $\mathcal{Y}$, the right $\mathcal{X}$-approximation of C is projective, hence $Y_C = \Omega(C)$. We infer that C has finite projective dimension. It remains to show that $\mathcal{I}^{<\infty} \subseteq \mathcal{Y}$. This follows trivially from the fact that $\mathcal{Y}$ contains the injectives and, according to Theorem 2.1, $\mathcal{Y}$ is resolving. Part (iii) follows directly from (i) and (ii).

(iv) By (iii) we have $\mathcal{Y} = \mathcal{P}^{<\infty}$. Let C be in $\operatorname{CM}(\mathcal{P})$ and let $0 \to Y_C \to X_C \to C \to 0$ be a special right $\mathcal{X}$-approximation of C. Then by (i) we have that Y_C lies in $\mathcal{P}^{<\infty}$. Since trivially $\operatorname{Ext}^1(X,Y)=0$ for $X \in \operatorname{CM}(\mathcal{P})$ and $Y \in \mathcal{P}^{<\infty}$, the above sequence splits. Hence $C \in \mathcal{X}$ and then $\operatorname{CM}(\mathcal{P}) \subseteq \mathcal{X}$. Finally by Theorem 2.1 we have $\mathcal{X} = \operatorname{CM}(\mathcal{P})$, and by (ii) we have $\mathcal{I}^{<\infty} \subseteq \mathcal{P}^{<\infty}$. □

The following consequence of Theorem 2.1 describes when both the torsion and the torsion-free class are triangulated.

COROLLARY 2.4. *Let $\mathcal{C}$ be an abelian category with enough projective and injective objects. Let $(\mathcal{X},\mathcal{Y})$ be a resolving cotorsion pair in $\mathcal{C}$ with $\omega = \mathcal{X} \cap \mathcal{Y}$ functorially finite. Then the following are equivalent:*

(i) *The torsion class $\mathcal{X}/\omega$ and the torsion-free class $\mathcal{Y}/\omega$ are triangulated.*
(ii) *$\mathcal{X}$ is coresolving and $\mathcal{Y}$ is resolving.*
(iii) *$\mathcal{C}$ is Frobenius and $\omega = \mathcal{P} = \mathcal{I}$.*

PROOF. If (i) holds, then by Theorem 2.1 and its dual we have that $\mathcal{X}$ is coresolving and $\mathcal{Y}$ is resolving, and $\mathcal{P} = \omega = \mathcal{I}$. Hence (i) implies (ii) and (iii), and by Theorem 2.1 (ii) implies (i). Finally if (iii) holds, then $\mathcal{C}/\mathcal{P}$ is triangulated. Since $\mathcal{X}$ is resolving, $\mathcal{X}/\mathcal{P}$ is closed under Ω, and this implies that $\mathcal{X}/\mathcal{P}$ is a triangulated subcategory of $\mathcal{C}/\mathcal{P}$. Since $\mathcal{Y}/\mathcal{P} = (\mathcal{X}/\mathcal{P})^{\perp}$, the same is true for $\mathcal{Y}/\mathcal{P}$. □

Actually in the Frobenius case we have a bijection between resolving cotorsion pairs and hereditary torsion pairs.

PROPOSITION 2.5. *Let $\mathcal{C}$ be a Frobenius category and let $\mathcal{X}$ and $\mathcal{Y}$ be full subcategories of $\mathcal{C}$. Then the following are equivalent:*

(i) *$(\mathcal{X},\mathcal{Y})$ is a (co-)resolving cotorsion pair in $\mathcal{C}$ with $\mathcal{X} \cap \mathcal{Y} = \mathcal{P}$.*
(ii) *$(\mathcal{X}/\mathcal{P},\mathcal{Y}/\mathcal{P})$ is a (co-)hereditary torsion pair in $\mathcal{C}/\mathcal{P}$.*
(iii) *$\mathcal{C}/\mathcal{P}(\mathcal{X},\mathcal{Y}) = 0$, $\mathcal{X}/\mathcal{P}$ and $\mathcal{Y}/\mathcal{P}$ are triangulated subcategories of $\mathcal{C}/\mathcal{P}$ and the latter is generated by $\mathcal{X}/\mathcal{P}$ and $\mathcal{Y}/\mathcal{P}$.*

The map $(\mathcal{X},\mathcal{Y}) \mapsto (\mathcal{X}/\mathcal{P},\mathcal{Y}/\mathcal{P})$ gives a bijection between resolving cotorsion pairs in $\mathcal{C}$ with $\mathcal{X} \cap \mathcal{Y} = \mathcal{P}$ and hereditary torsion pairs in $\mathcal{C}/\mathcal{P}$.

PROOF. The implication (i) ⇒ (ii) follows from Corollary 2.4 and the equivalence (ii) ⇒ (iii) follows from Proposition I.1.9. (ii) ⇒ (i) By Corollary V.2.9 it suffices to show that $\operatorname{Ext}^1(\mathcal{X},\mathcal{Y}) = 0$. Since $\mathcal{C}$ is Frobenius, for any $X \in \mathcal{X}$ and $Y \in \mathcal{Y}$ we have $\operatorname{Ext}^1(X,Y) \cong \mathcal{C}/\mathcal{P}(\Omega(X),Y) \cong \mathcal{C}/\mathcal{P}(X,\Sigma(Y))$. Since $\mathcal{X}$ is closed under Ω we infer that $\operatorname{Ext}^1(\mathcal{X},\mathcal{Y}) = 0$. □

If $(\mathcal{X},\mathcal{Y})$ is a cotorsion pair with $\mathcal{X}$ resolving such that the torsion class $\mathcal{X}/\omega$ is triangulated, then by Theorem 2.1 we know that $\omega = \mathcal{P}$ is functorially finite and the coreflection functor $\mathbf{R} : \mathcal{C}/\mathcal{P} \to \mathcal{X}/\mathcal{P}$ is a left exact functor of pretriangulated categories. It is natural to ask if this functor is the universal left exact functor out of $\mathcal{C}/\mathcal{P}$ to triangulated categories. In this connection we have the following result, which gives a nice description of the coreflection functor. Its dual is left to the reader.

PROPOSITION 2.6. *Let $(\mathcal{X},\mathcal{Y})$ be a resolving cotorsion pair in $\mathcal{C}$ with $\omega = \mathcal{X} \cap \mathcal{Y}$ functorially finite. If the torsion class $\mathcal{X}/\omega$ in $\mathcal{C}/\mathcal{P}$ is triangulated (hence $\omega = \mathcal{P}$), then the following are equivalent.*

(i) *The coreflection functor* $\mathbf{R} : \mathcal{C}/\mathcal{P} \to \mathcal{X}/\mathcal{P}$ *is the universal left exact functor out of* $\mathcal{C}/\mathcal{P}$ *to triangulated categories, i.e. the torsion class* $\mathcal{X}/\mathcal{P}$ *is the left stabilization of* $\mathcal{C}/\mathcal{P}$.

(ii) $\widehat{\mathcal{X}} = \mathcal{C}$.

(iii) $\mathcal{Y} = \mathcal{P}^{<\infty}$.

If (i) *holds, then* $\mathcal{X} = \mathrm{CM}(\mathcal{P})$ *and the coreflection functor* $\mathbf{R} : \mathcal{C}/\mathcal{P} \to \mathrm{CM}(\mathcal{P})/\mathcal{P}$ *is given by* $\mathbf{R}(\underline{C}) = \Omega^{-d}\Omega^{d}(\underline{C})$, *where* $d \geq 0$ *is such that* $\Omega^d(C) \in \mathrm{CM}(\mathcal{P})$.

PROOF. (i) $\Rightarrow$ (ii) By Proposition II.5.2 we have that for any module C there exists $n \geq 0$ such that $\Omega^n(\underline{C}) \in \mathcal{X}/\mathcal{P}$. This implies that $\widehat{\mathcal{X}} = \mathcal{C}$. (ii) $\Rightarrow$ (iii) Follows from Proposition 2.3. (iii) $\Rightarrow$ (i) Follows from Proposition II.5.2. If (i) holds, then the description of $\mathbf{R}$ follows from [**20**]. □

3. Cotorsion Triples

In analogy with the concept of a TTF-triple in an abelian or triangulated category we introduce and investigate the concept of a good or cotorsion triple in this section. We give necessary and sufficient conditions for the existence of a good or cotorsion triple, and we show that any such triple induces two torsion pairs in two different stable pretriangulated categories. In addition the torsion class of the first and the torsion-free class of the second are triangulated and they are triangle equivalent.

Following the way we defined torsion triples in abelian and (pre)triangulated categories with respect to the vanishing of Hom, it is suggestive to use the concept of a cotorsion triple, which can be thought of as a reasonable definition of a torsion triple in $\mathcal{C}$ with respect to the vanishing of Ext^1.

DEFINITION 3.1. A triple $(\mathcal{X}, \mathcal{Y}, \mathcal{Z})$ of subcategories of $\mathcal{C}$ is called a **good**, resp. **cotorsion**, **triple**, if $(\mathcal{X}, \mathcal{Y})$ and $(\mathcal{Y}, \mathcal{Z})$ are good, resp. cotorsion, pairs.

Note that if $\mathcal{C}$ is Krull-Schmidt, for example the category $\mathrm{mod}(\Lambda)$ of finitely generated modules over an Artin algebra Λ, then the notions of good triple and cotorsion triple coincide, since the notions of good pair and cotorsion pair coincide.

The following is an obvious example of a cotorsion triple.

EXAMPLE. Let $\mathcal{C}$ be an abelian category with enough projective and injective objects. Then $(\mathcal{P}, \mathcal{C}, \mathcal{I})$ is a cotorsion triple in $\mathcal{C}$.

However the primary motivating example of a cotorsion triple is the following.

EXAMPLE. Let Λ be an Artin algebra. If Λ is Gorenstein, then by Propositions V.5.1, V.4.6 and their duals it follows that $\mathcal{P}^{<\infty}_{\Lambda} = \mathcal{I}^{<\infty}_{\Lambda}$ is a functorially finite resolving and coresolving subcategory of $\mathrm{mod}(\Lambda)$ and we have good (cotorsion) pairs $(\mathrm{CM}(\Lambda), \mathcal{P}^{<\infty}_{\Lambda})$ and $(\mathcal{I}^{<\infty}_{\Lambda}, \mathrm{CoCM}(\mathrm{D}(\Lambda)))$. Hence

$$\Big(\mathrm{CM}(\Lambda),\ \mathcal{P}^{<\infty}_{\Lambda} = \mathcal{I}^{<\infty}_{\Lambda},\ \mathrm{CoCM}(\mathrm{D}(\Lambda))\Big)$$

is a good (cotorsion) triple in $\mathrm{mod}(\Lambda)$. Moreover $\mathrm{CM}(\Lambda) \cap \mathcal{P}^{<\infty}_{\Lambda} = \mathcal{P}_{\Lambda}$ and $\mathcal{I}^{<\infty}_{\Lambda} \cap \mathrm{CoCM}(\mathrm{D}(\Lambda)) = \mathcal{I}_{\Lambda}$, and we have a hereditary torsion pair $(\mathrm{CM}(\Lambda)/\mathcal{P}_{\Lambda}, \mathcal{P}^{<\infty}_{\Lambda}/\mathcal{P}_{\Lambda})$ in $\underline{\mathrm{mod}}(\Lambda)$ where the torsion class $\mathrm{CM}(\Lambda)/\mathcal{P}_{\Lambda}$ is triangulated, and a cohereditary torsion pair $(\mathcal{I}^{<\infty}_{\Lambda}/\mathcal{I}_{\Lambda}, \mathrm{CoCM}(\mathrm{D}(\Lambda))/\mathcal{I}_{\Lambda})$ in $\overline{\mathrm{mod}}(\Lambda)$ where the torsion-free class

$\mathrm{CoCM}(\mathrm{D}(\Lambda))/\mathcal{I}_\Lambda$ is triangulated. Finally by [**11**] it follows that the Nakayama functor $\mathrm{N}^+ = - \otimes_\Lambda \mathrm{D}(\Lambda) : \mathrm{mod}(\Lambda) \to \mathrm{mod}(\Lambda)$ induces a triangle equivalence $\mathrm{CM}(\Lambda)/\mathcal{P}_\Lambda \xrightarrow{\approx} \mathrm{CoCM}(\mathrm{D}(\Lambda))/\mathcal{I}_\Lambda$ with quasi-inverse induced by the Nakayama functor $\mathrm{N}^- = \mathrm{Hom}_\Lambda(\mathrm{D}(\Lambda), -) : \mathrm{mod}(\Lambda) \to \mathrm{mod}(\Lambda)$.

The following main result of this section generalizes the above observations and gives a method for constructing triangulated torsion(-free) classes in a stable category.

THEOREM 3.2. *Let $\mathcal{C}$ be an abelian category with enough projective and injective objects and assume that $\mathcal{P}$ is covariantly finite and $\mathcal{I}$ is contravariantly finite. For a full subcategory $\mathcal{Y}$ of $\mathcal{C}$, consider the following statements.*

(i) *$\mathcal{Y}$ is a functorially finite resolving and coresolving subcategory of $\mathcal{C}$.*
(ii) *$\mathcal{Y}$ is a resolving and coresolving subcategory of $\mathcal{C}$ and any object of $\mathcal{C}$ admits a special right $\mathcal{Y}$-approximation and a special left $\mathcal{Y}$-approximation.*
(iii) *There exists a cotorsion triple $(\mathcal{X}, \mathcal{Y}, \mathcal{Z})$ in $\mathcal{C}$ with $\mathcal{Y}$ (co)resolving.*
(iv) *There is a hereditary torsion pair $(\mathcal{X}/\mathcal{P}, \mathcal{Y}/\mathcal{P})$ in $\mathcal{C}/\mathcal{P}$ with $\mathcal{X}/\mathcal{P}$ triangulated and a cohereditary torsion pair $(\mathcal{Y}/\mathcal{I}, \mathcal{Z}/\mathcal{I})$ in $\mathcal{C}/\mathcal{I}$ with $\mathcal{Z}/\mathcal{I}$ triangulated.*

Then (ii) $\Leftrightarrow$ (iii) $\Leftrightarrow$ (iv) $\Rightarrow$ (i); *if $\mathcal{C}$ is Krull-Schmidt then all statements are equivalent. If* (iii) *holds, then there is a triangle equivalence*

$$\Phi \ : \ \mathcal{X}/\mathcal{P} \ \xrightarrow{\approx} \ \mathcal{Z}/\mathcal{I}.$$

PROOF. Obviously (ii) implies (i), and, by Lemma V.3.3, (ii) implies (iii). If $\mathcal{C}$ is Krull-Schmidt, then that (i) implies (ii) follows from Theorem V.2.2.

(iii) $\Rightarrow$ (ii), (iv) Consider the cotorsion triple $(\mathcal{X}, \mathcal{Y}, \mathcal{Z})$ in $\mathcal{C}$. Then obviously $\mathcal{Y}$ is functorially finite. We assume first that $\mathcal{Y}$ is coresolving, hence, by Proposition V.2.3, $\mathcal{X}$ is resolving. We show that $\mathcal{P} = \mathcal{X} \cap \mathcal{Y}$. Obviously $\mathcal{P} \subseteq \mathcal{X} \cap \mathcal{Y}$. If $T \in \mathcal{X} \cap \mathcal{Y}$ and $(1) : 0 \to K \to P \to T \to 0$ is exact with P projective, the from the long exact sequence $0 \to \mathcal{C}(T, \mathcal{Z}) \to \mathcal{C}(P, \mathcal{Z}) \to \mathcal{C}(K, \mathcal{Z}) \to \mathrm{Ext}^1(T, \mathcal{Z}) \to \cdots$ and the fact that $\mathcal{X}$ is resolving, we infer that $\mathrm{Ext}^1(K, \mathcal{Z}) = 0$, hence K lies in $\mathcal{Y}$. Since also $X \in \mathcal{X}$ we have that $K \in \mathcal{X} \cap \mathcal{Y}$. Then the sequence (1) splits since $\mathcal{X} \cap \mathcal{Y}$ is self-orthogonal. Hence K is projective and we conclude that $\mathcal{P} = \mathcal{X} \cap \mathcal{Y}$. Then by Proposition 2.3 we have that $\mathcal{Y}$ is resolving, and by Theorem 2.1 we have a hereditary torsion pair $(\mathcal{X}/\mathcal{P}, \mathcal{Y}/\mathcal{P})$ in $\mathcal{C}/\mathcal{P}$ where the torsion class $\mathcal{X}/\mathcal{P}$ is is triangulated. Dual arguments as above show that $\mathcal{Y} \cap \mathcal{Z} = \mathcal{I}$ and we have a cohereditary torsion pair $(\mathcal{Y}/\mathcal{I}, \mathcal{Z}/\mathcal{I})$ in $\mathcal{C}/\mathcal{I}$ where the torsion-free class $\mathcal{Z}/\mathcal{I}$ is triangulated.

(iv) $\Rightarrow$ (iii) By Corollary V.2.9 it suffices to show that $\mathrm{Ext}^1(\mathcal{X}, \mathcal{Y}) = 0 = \mathrm{Ext}^1(\mathcal{Y}, \mathcal{Z})$. Since $\mathcal{X}/\mathcal{P}$ and $\mathcal{Z}/\mathcal{I}$ are triangulated subcategories of $\mathcal{C}/\mathcal{P}$ and $\mathcal{C}/\mathcal{I}$ respectively, we have isomorphisms: $\mathrm{Ext}^1(X, C) \xrightarrow{\cong} \mathcal{C}/\mathcal{P}(\Omega(X), C)$ and $\mathrm{Ext}^1(C, Z) \xrightarrow{\cong} \mathcal{C}/\mathcal{I}(C, \Sigma(Z))$, for any $C \in \mathcal{C}$, for any $X \in \mathcal{X}$ and for any $Z \in \mathcal{Z}$. Since $(\mathcal{X}/\mathcal{P}, \mathcal{Y}/\mathcal{P}) = 0 = (\mathcal{Y}/\mathcal{I}, \mathcal{Z}/\mathcal{I})$ and since, by Proposition V.2.5, $\mathcal{X}$ is resolving and $\mathcal{Z}$ is coresolving, these isomorphisms show that $\mathrm{Ext}^1(\mathcal{X}, \mathcal{Y}) = 0 = \mathrm{Ext}^1(\mathcal{Y}, \mathcal{Z})$. We conclude that $(\mathcal{X}, \mathcal{Y}, \mathcal{Z})$ is a cotorsion triple in $\mathcal{C}$ with $\mathcal{Y}$ (co)resolving.

Now if (iii) holds then the last assertion follows from [**11**]. For completeness we include the argument for the existence of an equivalence $\Phi : \mathcal{X}/\mathcal{P} \to \mathcal{Z}/\mathcal{I}$ and we show in addition that Φ is exact. If $X \in \mathcal{X}$, let $0 \to Z_X \xrightarrow{g_X} Y_X \xrightarrow{f_X} X \to 0$ be a special right $\mathcal{Y}$-approximation sequence of X. We set $\Phi(\underline{X}) = \underline{Z}_X$ which lies in $\mathcal{Z}/\mathcal{I}$ since Z_X lies in $\mathcal{Y}^{\perp} = \mathcal{Z}$. Using that $(\mathcal{Y}/\mathcal{I}, \mathcal{Z}/\mathcal{I}) = 0$, it is easy to see that Φ is a well defined additive functor. Dually for any $Z \in \mathcal{Z}$, let $0 \to Z \xrightarrow{g^Z} Y^Z \xrightarrow{f^Z} X^Z \to 0$ be a special left $\mathcal{Y}$-approximation sequence of Z. As above we have that $X^Z \in {}^{\perp}\mathcal{Y} = \mathcal{X}$. Setting $\Psi(\underline{Z}) = \underline{X}^Z$ and using that $(\mathcal{X}/\mathcal{P}, \mathcal{Y}/\mathcal{P}) = 0$, it is easy to see that we obtain a well defined additive functor $\Psi : \mathcal{Z}/\mathcal{I} \to \mathcal{X}/\mathcal{P}$. Since $\mathrm{Ext}^1(\mathcal{X}, \mathcal{Y}) = 0$, it follows that $0 \to Z_X \xrightarrow{g_X} Y_X \xrightarrow{f_X} X \to 0$ is a special left $\mathcal{Y}$-approximation sequence of Z_X. Hence $\underline{X} \cong \underline{X}^{Z_X}$ in $\mathcal{X}/\mathcal{P}$, i.e. $\underline{X} \cong \Psi\Phi(\underline{X})$. Dually $\underline{Z} \cong \Phi\Psi(\underline{Z})$, $\forall Z \in \mathcal{Z}$. Hence Φ, Ψ are mutually inverse equivalences.

Finally we show that Φ is exact. Let X be in $\mathcal{X}$ and let $0 \to \Omega(X) \to P \to X \to 0$ be an exact sequence with P projective. Let $0 \to Z_{\Omega(X)} \to Y_{\Omega(X)} \to \Omega(X) \to 0$ be a special right $\mathcal{Y}$-approximation sequence of $\Omega(X)$. Then in $\mathcal{Z}/\mathcal{I}$ we have $\Phi\Omega(\underline{X}) = \underline{Z}_{\Omega(X)}$. Let $0 \to Z_{\Omega(X)} \to I \to \Sigma(Z_{\Omega(X)}) \to 0$ be an exact sequence with I injective. The above exact sequences can be embedded in the following exact commutative diagram of short exact sequences:

$$\begin{array}{ccccccccc}
0 & \longrightarrow & Z_{\Omega(X)} & \longrightarrow & Y_{\Omega_X} & \longrightarrow & \Omega(X) & \longrightarrow & 0 \\
 & & \downarrow & & \downarrow & & \downarrow & & \\
0 & \longrightarrow & I & \longrightarrow & I \oplus P & \longrightarrow & P & \longrightarrow & 0 \\
 & & \downarrow & & \downarrow & & \downarrow & & \\
0 & \longrightarrow & \Sigma(Z_{\Omega(X)}) & \longrightarrow & \widetilde{Y} & \longrightarrow & X & \longrightarrow & 0
\end{array}$$

Since $\mathcal{Y}$ is coresolving and contains the projectives, we have that $\widetilde{Y}$ is in $\mathcal{Y}$. Since $\Sigma(Z_{\Omega(X)})$ is in $\mathcal{Z}$ and $\mathrm{Ext}^1(\mathcal{Y}, \mathcal{Z}) = 0$, it follows that the lower exact sequence is a special right $\mathcal{Y}$-approximation sequence of X. It follows that in $\mathcal{Z}/\mathcal{I}$ we have $\Sigma(\underline{Z}_{\Omega(X)}) \cong \underline{Z}_X$, and then $\Sigma\Phi\Omega(\underline{X}) \cong \Phi(\underline{X})$, or equivalently $\Phi\Omega(\underline{X}) \cong \Sigma^{-1}\Phi(\underline{X})$. It is easy to see that this induces a natural isomorphism $\zeta : \Phi\Omega \xrightarrow{\cong} \Sigma^{-1}\Phi$, so that Φ commutes with the loop functors Ω, Σ^{-1}. It remains to show that Φ sends triangles in $\mathcal{X}/\mathcal{P}$ to triangles in $\mathcal{Z}/\mathcal{I}$. Let $0 \to X_1 \to X_2 \to X_3 \to 0$ be an exact sequence in $\mathcal{C}$ with the X_i in $\mathcal{X}$. Then by [**9**], there exists a short exact sequence $0 \to Z_{X_1} \to Z \to Z_{X_3} \to 0$ where $Z \in \mathcal{Z}$ is such that $\underline{Z} \cong \underline{Z}_{X_2}$ in $\mathcal{Z}/\mathcal{I}$. Hence $\Phi(\underline{X}_2) \cong \underline{Z}$ and then using the isomorphism ζ we have a triangle $\Sigma^{-1}\Phi(\underline{X}_3) \to \Phi(\underline{X}_1) \to \Phi(\underline{X}_2) \to \Phi(\underline{X}_3)$ in $\mathcal{Z}/\mathcal{I}$. Hence Φ is exact. $\square$

NOTE. The assumption that $\mathcal{P}$ is covariantly finite and $\mathcal{I}$ is contravariantly finite in Theorem 3.2 and all the remaining results of this section, is only needed to ensure that the stable categories $\mathcal{C}/\mathcal{P}$ and $\mathcal{C}/\mathcal{I}$ are pretriangulated, so that the pairs $(\mathcal{X}/\mathcal{P}, \mathcal{Y}/\mathcal{P})$ and $(\mathcal{Y}/\mathcal{I}, \mathcal{Z}/\mathcal{I})$ are torsion pairs.

If $(\mathcal{X}, \mathcal{Y}, \mathcal{Z})$ is a cotorsion triple in $\mathcal{C}$ with $\mathcal{Y}$ (co)resolving, then, by Proposition 2.3, we have $\mathcal{P}^{<\infty} \subseteq \mathcal{Y} \supseteq \mathcal{I}^{<\infty}$. The following consequence of Proposition 2.6 and its dual, and Theorem 3.2, describes when we have equalities.

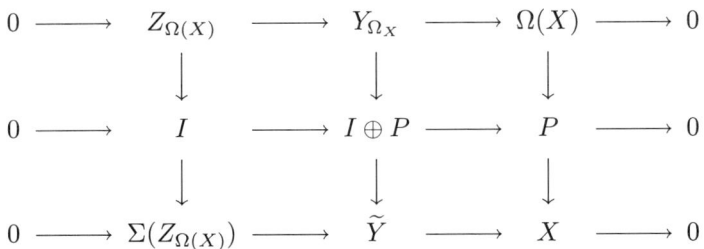

COROLLARY 3.3. *Let $\mathcal{C}$ be an abelian category with enough projective and injective objects and assume that $\mathcal{P}$ is covariantly finite and $\mathcal{I}$ is contravariantly finite. If $(\mathcal{X}, \mathcal{Y}, \mathcal{Z})$ is a cotorsion triple in $\mathcal{C}$ with $\mathcal{Y}$ (co)resolving, then the following are equivalent.*

(i) *The torsion class $\mathcal{X}/\mathcal{P}$ is the left stabilizaton of $\mathcal{C}/\mathcal{P}$ and the torsion-free class $\mathcal{Z}/\mathcal{I}$ is the right stabilization of $\mathcal{C}/\mathcal{I}$.*

(ii) $\widehat{\mathcal{X}} = \mathcal{C} = \widetilde{\mathcal{Z}}$.

(iii) $\mathcal{P}^{<\infty} = \mathcal{Y} = \mathcal{I}^{<\infty}$.

In (i) *holds, then* $\mathcal{X} = \mathrm{CM}(\mathcal{P})$ *and* $\mathcal{Z} = \mathrm{CoCM}(\mathcal{I})$.

We have seen that if $\mathcal{C}$ has enough projective and injective objects, then $(\mathcal{P}, \mathcal{C}, \mathcal{I})$ is a cotorsion triple in $\mathcal{C}$. If $\mathcal{C}$ is Frobenius, then $(\mathcal{C}, \mathcal{P}, \mathcal{C})$ also is a cotorsion triple in $\mathcal{C}$. Motivated by this example we say that a quadruple $(\mathcal{X}, \mathcal{Y}, \mathcal{Z}, \mathcal{W})$ of subcategories is a *cotorsion quadruple* in $\mathcal{C}$, if $(\mathcal{X}, \mathcal{Y}, \mathcal{Z})$ and $(\mathcal{Y}, \mathcal{Z}, \mathcal{W})$ are cotorsion triples. Hence in the Frobenius case we have cotorsion quadruples $(\mathcal{P}, \mathcal{C}, \mathcal{P}, \mathcal{C})$ and $(\mathcal{C}, \mathcal{P}, \mathcal{C}, \mathcal{P})$ in $\mathcal{C}$. Conversely the existence of a cotorsion quadruple in $\mathcal{C}$ implies that $\mathcal{C}$ is Frobenius:

REMARK 3.4. Let $(\mathcal{X}, \mathcal{Y}, \mathcal{Z}, \mathcal{W})$ be a cotorsion quadruple in $\mathcal{C}$ such that one of the involved subcategories is resolving or coresolving. Then by Theorem 3.2 we have $\mathcal{Y} \cap \mathcal{Z} = \mathcal{P} = \mathcal{I}$. Hence $\mathcal{C}$ is Frobenius and we have TTF-triples $(\mathcal{X}/\mathcal{P}, \mathcal{Y}/\mathcal{P}, \mathcal{Z}/\mathcal{P})$ and $(\mathcal{Y}/\mathcal{P}, \mathcal{Z}/\mathcal{P}, \mathcal{W}/\mathcal{P})$ in the triangulated stable category $\mathcal{C}/\mathcal{P}$. In particular there exist triangle equivalences: $\mathcal{X}/\mathcal{P} \xrightarrow{\approx} \mathcal{Z}/\mathcal{P}$ and $\mathcal{Y}/\mathcal{P} \xrightarrow{\approx} \mathcal{W}/\mathcal{P}$.

If $\mathcal{Y}$ is a resolving and coresolving subcategory of $\mathcal{C}$, then by Theorem 3.2, $\mathcal{Y}/\mathcal{P}$ is a left triangulated subcategory of $\mathcal{C}/\mathcal{P}$ and $\mathcal{Y}/\mathcal{I}$ is a right triangulated subcategory of $\mathcal{C}/\mathcal{I}$. Hence we can form the left triangulated quotient $\mathcal{C}/\mathcal{P}/\mathcal{Y}/\mathcal{P}$ and the right triangulated quotient $\mathcal{C}/\mathcal{I}/\mathcal{Y}/\mathcal{I}$. We recall that $\mathcal{C}/\mathcal{P}/\mathcal{Y}/\mathcal{P}$ is the localization of $\mathcal{C}/\mathcal{P}$ at the class $\mathfrak{L}_{\mathcal{Y}}$ of morphisms $\underline{f} : \underline{A} \to \underline{B}$ for which there exists a left triangle $\Omega(\underline{B}) \to \underline{Y} \to \underline{A} \xrightarrow{\underline{f}} \underline{B}$ with Y in $\mathcal{Y}$, that is $\mathcal{C}/\mathcal{P}/\mathcal{Y}/\mathcal{P}$ is obtained by formally inverting the class of morphisms $\mathfrak{L}_{\mathcal{Y}}$ [**50**]. Dually $\mathcal{C}/\mathcal{I}/\mathcal{Y}/\mathcal{I}$ is a localization of $\mathcal{C}/\mathcal{I}$ at the class $\mathfrak{R}_{\mathcal{Y}}$ of morphisms $\underline{f} : \underline{A} \to \underline{B}$ for which there exists a right triangle $\underline{A} \xrightarrow{\underline{f}} \underline{B} \to \underline{Y} \to \Sigma(\underline{A})$ with Y in $\mathcal{Y}$. The following result describes the above quotients and can be regarded as an analogue of Corollary I.2.9.

THEOREM 3.5. *Let $\mathcal{C}$ be an abelian category with enough projective and injective objects and assume that $\mathcal{P}$ is covariantly finite and $\mathcal{I}$ is contravariantly finite. Let $(\mathcal{X}, \mathcal{Y}, \mathcal{Z})$ be a cotorsion triple in $\mathcal{C}$ with $\mathcal{Y}$ (co)resolving, and consider the hereditary torsion pair $(\mathcal{X}/\mathcal{P}, \mathcal{Y}/\mathcal{P})$ in $\mathcal{C}/\mathcal{P}$ and the cohereditary torsion pair $(\mathcal{Y}/\mathcal{I}, \mathcal{Z}/\mathcal{I})$ in $\mathcal{C}/\mathcal{I}$. Then there are triangle equivalences:*

$$\mathcal{C}/\mathcal{P}/\mathcal{Y}/\mathcal{P} \xrightarrow{\approx} \mathcal{X}/\mathcal{P} \quad \text{and} \quad \mathcal{C}/\mathcal{I}/\mathcal{Y}/\mathcal{I} \xrightarrow{\approx} \mathcal{Z}/\mathcal{I}.$$

In particular the localized categories $\mathcal{C}/\mathcal{P}/\mathcal{Y}/\mathcal{P}$ and $\mathcal{C}/\mathcal{I}/\mathcal{Y}/\mathcal{I}$ are triangulated and there is a triangle equivalence $\mathcal{C}/\mathcal{P}/\mathcal{Y}/\mathcal{P} \xrightarrow{\approx} \mathcal{C}/\mathcal{I}/\mathcal{Y}/\mathcal{I}$.

PROOF. By Theorem 3.2 we have that $\mathcal{X}/\mathcal{P}$ is triangulated, in particular $\Sigma(\mathcal{X}/\mathcal{P}_{\Lambda}) = \mathcal{X}/\mathcal{P}$ where Σ is the left adjoint of Ω in $\mathcal{C}/\mathcal{P}$, and $\mathcal{Z}/\mathcal{I}$ is triangulated, in particular $\Omega(\mathcal{Z}/\mathcal{I}) = \mathcal{Z}/\mathcal{I}$ where Ω is the right adjoint of Σ in $\mathcal{C}/\mathcal{I}$. Then the result is a consequence of Proposition II.4.1 and Theorem 3.2. □

We have the following application to stable Grothendieck groups, which follows from Corollaries II.5.7 and II.5.8.

COROLLARY 3.6. *Let $(\mathcal{X}, \mathcal{Y}, \mathcal{Z})$ be a cotorsion triple in $\mathcal{C}$ with $\mathcal{Y}$ (co)resolving. Then we have isomorphisms:*

$$\mathrm{K}_0(\mathcal{C}/\mathcal{P}) \cong \mathrm{K}_0(\mathcal{X}/\mathcal{P}) \oplus \mathrm{K}_0(\mathcal{Y}/\mathcal{P}) \quad \textit{and} \quad \mathrm{K}_0(\mathcal{C}/\mathcal{I}) \cong \mathrm{K}_0(\mathcal{Y}/\mathcal{I}) \oplus \mathrm{K}_0(\mathcal{Z}/\mathcal{I}).$$

If $\mathcal{Y} \subseteq \mathcal{P}^{<\infty} \cap \mathcal{I}^{<\infty}$, they reduce to isomorphisms: $\mathrm{K}_0(\mathcal{Y}/\mathcal{P}) = 0 = \mathrm{K}_0(\mathcal{Y}/\mathcal{I})$ and

$$\mathrm{K}_0(\mathcal{C}/\mathcal{P}) \xrightarrow{\cong} \mathrm{K}_0(\mathcal{X}/\mathcal{P}) \xrightarrow{\cong} \mathrm{K}_0(\mathcal{Z}/\mathcal{I}) \xleftarrow{\cong} \mathrm{K}_0(\mathcal{C}/\mathcal{I}).$$

4. Applications to Gorenstein Artin Algebras

Let Λ be an Artin algebra. In this section we study torsion pairs in the stable module category $\underline{\mathrm{mod}}(\Lambda)$ or $\overline{\mathrm{mod}}(\Lambda)$, such that the torsion or torsion-free class is triangulated, concentrating on specific features of Artin algebras related to Auslander-Reiten theory and tilting. For instance there are interesting connections with tilting theory where the cotilting module is projective or the tilting module is injective. The existence of such (co)tilting modules characterizes the important class of Gorenstein algebras. We recall that an Artin algebra Λ is Gorenstein, if Λ has finite injective dimension both as a left and right Λ-module, equivalently Λ is a cotilting module. We stress that Gorenstein algebras form an important class of Artin algebras since they provide a common generalization of self-injective algebras and algebras of finite global dimension. As another example which is important in representation theory we mention that any gentle algebra is Gorenstein [**52**].

We begin with the following result which characterizes the cotilting modules which are projective in torsion theoretic terms and gives a connection with Gorenstein algebras. Its dual is left to the reader.

PROPOSITION 4.1. *For a Λ-module T the following are equivalent.*

(i) *T is a cotilting module and the torsion class $\mathrm{CM}(T)/\mathrm{add}(T)$ is a triangulated subcategory of $\underline{\mathrm{mod}}_T(\Lambda)$.*
(ii) *T is cotilting projective module.*
(iii) *T is a projective generator and Λ is a Gorenstein algebra.*

PROOF. (i) $\Rightarrow$ (ii) By Theorem 2.1 we have that $\mathrm{add}(T) = \mathcal{P}_\Lambda$, so T is projective. (ii) $\Rightarrow$ (iii) Since T is cotilting, T and Λ have the same number of non-isomorphic indecomposable summands. Hence T is a projective generator and then obviously Λ is a cotilting module, and consequently Λ is Gorenstein. (iii) $\Rightarrow$ (i) The hypothesis implies that Λ is a cotilting module and $\mathrm{add}(T) = \mathcal{P}_\Lambda$. Then (i) follows by Theorem 2.1. □

The following result, which is a consequence of Theorem 3.2, Corollary 3.3 and Proposition 2.6, shows that the category of finitely generated modules over a Gorenstein algebra admits a nicely behaved cotorsion triple. In addition we get that Gorenstein algebras are characterized by universal properties of the induced torsion-(free) classes of the stable module category modulo projectives or injectives.

THEOREM 4.2. *Let Λ be an Artin algebra. Then the following are equivalent.*

(i) *Λ is Gorenstein.*

(ii) *There exists a resolving cotorsion pair* $(\mathcal{X},\mathcal{Y})$ *in* $\mathrm{mod}(\Lambda)$ *with* $\mathcal{X}\cap\mathcal{Y}=\mathcal{P}_\Lambda$ *such that the torsion class* $\mathcal{X}/\mathcal{P}_\Lambda$ *is the left stabilization of* $\underline{\mathrm{mod}}(\Lambda)$.

(iii) *There exists a coresolving cotorsion pair* $(\mathcal{Y},\mathcal{Z})$ *in* $\mathrm{mod}(\Lambda)$ *with* $\mathcal{Y}\cap\mathcal{Z}=\mathcal{I}_\Lambda$ *such that the torsion-free class* $\mathcal{Z}/\mathcal{I}_\Lambda$ *is the right stabilization of* $\overline{\mathrm{mod}}(\Lambda)$.

(iv) $\mathcal{P}_\Lambda^{<\infty}\cap\mathcal{I}_\Lambda^{<\infty}$ *contains a functorially finite resolving and coresolving subcategory* $\mathcal{Y}$ *of* $\mathrm{mod}(\Lambda)$.

(v) *There exists a cotorsion triple* $(\mathcal{X},\mathcal{Y},\mathcal{Z})$ *in* $\mathrm{mod}(\Lambda)$ *with* $\mathcal{Y}$ *resolving such that* $\widehat{\mathcal{X}}=\mathrm{mod}(\Lambda)$.

(vi) *There exists a cotorsion triple* $(\mathcal{X},\mathcal{Y},\mathcal{Z})$ *in* $\mathrm{mod}(\Lambda)$ *with* $\mathcal{Y}$ *coresolving such that* $\widetilde{\mathcal{Z}}=\mathrm{mod}(\Lambda)$.

The following remark identifies the categories $\mathcal{X},\mathcal{Y}$ and $\mathcal{Z}$ appearing in Theorem 4.2, and gives an explicit description of the left and right stabilization functors.

REMARK 4.3. If Λ is Gorenstein, then, with the notation of Theorem 4.2, Corollary 3.3 implies that: $\mathcal{X}=\mathrm{CM}(\Lambda)$, $\mathcal{Z}=\mathrm{CoCM}(\mathrm{D}(\Lambda))$ and $\mathcal{Y}=\mathcal{P}_\Lambda^{<\infty}=\mathcal{I}_\Lambda^{<\infty}$.

Moreover, by Proposition 2.6, the left stabilization functor is the coreflection functor $\mathbf{R}:\underline{\mathrm{mod}}(\Lambda)\to\mathcal{X}/\mathcal{P}_\Lambda$ and is given by $\mathbf{R}(\underline{C})=\Omega^{-d}\Omega^d(\underline{C})$, and the right stabilization functor is the reflection functor $\mathbf{T}:\overline{\mathrm{mod}}(\Lambda)\to\mathcal{Z}/\mathcal{I}_\Lambda$ and is given by $\mathbf{T}(\overline{C})=\Sigma^{-d}\Sigma^d(\overline{C})$, where $d=\mathrm{id}\,{}_\Lambda\Lambda=\mathrm{id}\,\Lambda_\Lambda$.

The following result gives some consequences of the existence of cotorsion triples in connection with Auslander-Reiten theory. For the concept of an Auslander-Reiten triangle we refer to [**57**], [**89**].

COROLLARY 4.4. *Let* $(\mathcal{X},\mathcal{Y},\mathcal{Z})$ *be a good triple in* $\mathrm{mod}(\Lambda)$. *Then the triangulated categories* $\mathcal{X}/\mathcal{P}_\Lambda$ *and* $\mathcal{Z}/\mathcal{I}_\Lambda$ *admit Auslander-Reiten triangles from both sides. In particular* $\mathcal{X}/\mathcal{P}_\Lambda$ *and* $\mathcal{Z}/\mathcal{I}_\Lambda$ *admit a Serre functor (see* [**89**]*).*

PROOF. Since $\mathcal{X}$ is contravariantly finite, $\mathcal{X}$ has right almost split morphisms: if X is indecomposable in $\mathcal{X}$ and $\rho:C\to X$ is a right almost split morphism in $\mathrm{mod}(\Lambda)$, then the composition $\beta:X_C\xrightarrow{f_C}C\xrightarrow{\rho}X$ is a right almost split morphism in $\mathcal{X}$, where f_C is a right $\mathcal{X}$-approximation of C. If X is non-projective, then it is easy to see that $\underline{\beta}:\underline{X_C}\to\underline{X}$ is a right almost split morphism in $\mathcal{X}/\mathcal{P}_\Lambda$. Since $\mathcal{X}/\mathcal{P}_\Lambda$ is Krull-Schmidt, it follows that for any indecomposable object $\underline{X}$ there exists a minimal right almost split morphism $\underline{\alpha}:\underline{X}'\to\underline{X}$ in $\mathcal{X}/\mathcal{P}_\Lambda$. Since $\mathcal{X}/\mathcal{P}_\Lambda$ is triangulated, the triangle $\Omega(\underline{X}'')\to\underline{X}'\xrightarrow{\underline{\alpha}}\underline{X}\to\underline{X}''$ is an Auslander-Reiten triangle, by [**3**], [**57**]. Hence $\mathcal{X}/\mathcal{P}_\Lambda$ has Auslander-Reiten triangles from the right.

By duality we have that $\mathcal{Z}/\mathcal{I}_\Lambda$ has Auslander-Reiten triangles from the left. Since $\mathcal{X}/\mathcal{P}_\Lambda$ and $\mathcal{Z}/\mathcal{I}_\Lambda$ are triangle equivalent, both of them have Auslander-Reiten triangles from left and right, in particular they admit a Serre functor by [**89**]. □

REMARK 4.5. In general the triangle equivalence $\mathcal{X}/\mathcal{P}_\Lambda\xrightarrow{\approx}\mathcal{Z}/\mathcal{I}_\Lambda$ of Theorem 3.2 may not be induced by a functor defined on $\underline{\mathrm{mod}}(\Lambda)$. However there is a triangle equivalence between $\mathcal{X}/\mathcal{P}_\Lambda$ and $\mathcal{Z}/\mathcal{I}_\Lambda$ which enjoys this property, and which has interesting connections with the equivalence $\mathrm{DTr}:\underline{\mathrm{mod}}(\Lambda)\xrightarrow{\approx}\overline{\mathrm{mod}}(\Lambda)$ [**15**].

Let $\mathcal{Y}$ be a functorially finite resolving and coresolving subcategory of $\mathrm{mod}(\Lambda)$. Then DTr restricts to a triangle equivalence $\mathrm{DTr} : \mathcal{X}/\mathcal{P}_\Lambda \xrightarrow{\approx} \mathcal{Z}/\mathcal{I}_\Lambda$ with quasi-inverse induced by TrD, where $\mathcal{X} = {}^\perp\mathcal{Y}$ and $\mathcal{Z} = \mathcal{Y}^\perp$. This was observed in [**11**] in case Λ is a Gorenstein algebra and $\mathcal{Y} = \mathcal{P}^{<\infty}_\Lambda$. Here we present a torsion-theoretic proof in the general case.

Recall that if X is a finitely generated Λ-module, then there is an isomorphism of functors $\mathrm{Ext}^1_\Lambda(-, \mathrm{DTr}(X)) \to \mathrm{D}\underline{\mathrm{Hom}}_\Lambda(\underline{X}, -)$ (see [**15**]). If X is in $\mathcal{X}$, then evaluating the above isomorphism at $\mathcal{Y}$ we have $\mathrm{Ext}^1_\Lambda(\mathcal{Y}, \mathrm{DTr}(X)) \cong \mathrm{D}\underline{\mathrm{Hom}}_\Lambda(X, \mathcal{Y}) = 0$. Hence $\mathrm{DTr}(X)$ lies in $\mathcal{Y}^\perp = \mathcal{Z}$. It follows that DTr restricts to a functor $\mathrm{DTr} : \mathcal{X}/\mathcal{P}_\Lambda \to \mathcal{Z}/\mathcal{I}_\Lambda$ which obviously is an equivalence with quasi-inverse induced by TrD. Moreover the equivalence DTr is exact. Indeed, if $0 \to X_1 \to X_2 \to X_3 \to 0$ is an exact sequence with the X_i in $\mathcal{X}$, then by a result of Auslander-Bridger [**6**] there is an induced exact sequence $0 \to \mathrm{d}(X_3) \to \mathrm{d}(X_2) \to \mathrm{d}(X_1) \xrightarrow{\vartheta} \mathrm{Tr}(X_3) \to \mathrm{Tr}(X_2) \to \mathrm{Tr}(X_3) \to 0$ in $\mathrm{mod}(\Lambda^{\mathrm{op}})$, where $\mathrm{d} = \mathrm{Hom}_\Lambda(-, \Lambda)$ and Tr is the transpose. However since the projectives are Ext-injective in $\mathcal{X}$, the connecting morphism ϑ vanishes. Hence we have an exact sequence $0 \to \mathrm{Tr}(X_3) \to \mathrm{Tr}(X_2) \to \mathrm{Tr}(X_1) \to 0$, which induces an exact sequence $0 \to \mathrm{DTr}(X_1) \to \mathrm{DTr}(X_2) \to \mathrm{DTr}(X_3) \to 0$ in $\mathrm{mod}(\Lambda)$. By the above argument the last sequence lies in $\mathcal{Z}$. This implies easily that $\mathrm{DTr} : \mathcal{X}/\mathcal{P}_\Lambda \xrightarrow{\approx} \mathcal{Z}/\mathcal{I}_\Lambda$ is an exact equivalence of triangulated categories. It is not difficult to see that there is a natural morphism $\varphi : \Phi \to \mathrm{DTr}$ between the equivalences $\Phi, \mathrm{DTr} : \mathcal{X}/\mathcal{P}_\Lambda \to \mathcal{Z}/\mathcal{I}_\Lambda$. In general φ is not invertible. A necessary and sufficient condition for $\varphi_{\underline{X}}$ to be invertible is that the middle part of an almost split sequence starting at any indecomposable non-projective direct summand of $X \in \mathcal{X}$, lies in $\mathcal{Y}$.

REMARK 4.6. Let us denote by τ the Auslander-Reiten translate and by τ^- its inverse, in both the categories $\mathcal{X}/\mathcal{P}_\Lambda$ and $\mathcal{Z}/\mathcal{I}_\Lambda$. If the object $\mathbf{R}(\overline{\mathrm{DTr}(X)})$ is indecomposable, then it is easy to see that $\tau(\underline{X}) \cong \mathbf{R}(\overline{\mathrm{DTr}(X)})$, since then the Auslander-Reiten triangle starting at $\underline{X}$ is the image of the triangle $\Omega(\underline{X}) \to \overline{\mathrm{DTr}(X)} \to \underline{A} \to \underline{X}$ under the functor $\mathbf{R}$, where $0 \to \mathrm{DTr}(X) \to A \to X \to 0$ is an almost split sequence. In particular, if for any indecomposable non-injective Z in $\mathcal{Z}$ its minimal right $\mathcal{X}$-approximation X_Z is indecomposable up to projective summands, then $\tau(\underline{X}) = \mathbf{R}(\overline{\mathrm{DTr}(X)})$ in $\mathcal{X}/\mathcal{P}_\Lambda$ and $\tau(\underline{Z}) = \overline{\mathrm{DTr}\mathbf{R}(Z)}$ in $\mathcal{Z}/\mathcal{I}_\Lambda$. Dually if for any indecomposable non-projective X in $\mathcal{X}$, its minimal left $\mathcal{Z}$-approximation Z^X is indecomposable up to injective summands, then $\tau^-(\overline{X}) = \underline{\mathrm{TrD}\mathbf{S}(X)}$ in $\mathcal{X}/\mathcal{P}_\Lambda$ and $\tau^-(\overline{Z}) = \mathbf{T}(\underline{\mathrm{TrD}(Z)})$ in $\mathcal{Z}/\mathcal{I}_\Lambda$, where $\mathbf{T}$ denotes the reflection $\overline{\mathrm{mod}}(\Lambda) \to \mathcal{Z}/\mathcal{I}_\Lambda$. Note that by the results of [**11**], all the above remarks can be applied to the cotorsion triple $\big(\mathrm{CM}(\Lambda), \mathcal{P}^{<\infty}_\Lambda = \mathcal{I}^{<\infty}_\Lambda, \mathrm{CoCM}(\mathrm{D}(\Lambda))\big)$, where Λ is a Gorenstein algebra.

If $(\mathcal{X}, \mathcal{Y})$ is a cotorsion pair then we know that $\mathcal{X}$ is contravariantly finite and $\mathcal{Y}$ is covariantly finite. The following remark and its corollary shows that if $(\mathcal{X}, \mathcal{Y}, \mathcal{Z})$ is a cotorsion triple, then all the categories $\mathcal{X}, \mathcal{Y}$ and $\mathcal{Z}$ are functorially finite.

REMARK 4.7. Let $(\mathcal{X}, \mathcal{Y}, \mathcal{Z})$ be a cotorsion triple in $\mathrm{mod}(\Lambda)$. We know that $\mathcal{X}/\mathcal{P}_\Lambda = {}^\perp(\mathcal{Y}/\mathcal{P}_\Lambda)$ and $\mathcal{Z}/\mathcal{I}_\Lambda = (\mathcal{Y}/\mathcal{I}_\Lambda)^\perp$ and $(\mathcal{X}/\mathcal{P}_\Lambda)^\perp = \mathcal{Y}/\mathcal{P}_\Lambda$ and ${}^\perp(\mathcal{Z}/\mathcal{I}_\Lambda) = \mathcal{Y}/\mathcal{I}_\Lambda$. The orthogonal subcategories ${}^\perp(\mathcal{X}/\mathcal{P}_\Lambda)$ and $(\mathcal{Z}/\mathcal{I}_\Lambda)^\perp$ also have a computable description. If $C \in \overline{\mathrm{mod}}(\Lambda)$, then $C \in (\mathcal{Z}/\mathcal{I}_\Lambda)^\perp$ iff $\overline{\mathrm{Hom}}_\Lambda(\mathcal{Z}, C) = 0$. Since

$\mathrm{D}\overline{\mathrm{Hom}}_\Lambda(Z,C) \cong \mathrm{Ext}^1(\mathrm{TrD}(C),Z)$, we have $C \in (\mathcal{Z}/\mathcal{I}_\Lambda)^\perp$ iff $\mathrm{TrD}(C) \in \mathcal{Y}/\mathcal{P}_\Lambda$. It follows that the functor $\mathrm{TrD} : (\mathcal{Z}/\mathcal{I}_\Lambda)^\perp \xrightarrow{\approx} \mathcal{Y}/\mathcal{P}_\Lambda$ is an equivalence and similarly the functor $\mathrm{DTr} : {}^\perp(\mathcal{X}/\mathcal{P}_\Lambda) \xrightarrow{\approx} \mathcal{Y}/\mathcal{I}_\Lambda$ is an equivalence.

COROLLARY 4.8. *The subcategory $(\mathcal{Z}/\mathcal{I}_\Lambda)^\perp$ is coreflective in $\overline{\mathrm{mod}}(\Lambda)$ and the subcategory ${}^\perp(\mathcal{X}/\mathcal{P}_\Lambda)$ is reflective in $\underline{\mathrm{mod}}(\Lambda)$. In particular the categories $\mathcal{X}$ and $\mathcal{Z}$ are functorially finite in* $\mathrm{mod}(\Lambda)$.

PROOF. Let $\underline{C}$ be in $\overline{\mathrm{mod}}(\Lambda)$, and let $\underline{Y}_{\mathrm{TrD}(\underline{C})} \to \mathrm{TrD}(\underline{C})$ be the coreflection of $\mathrm{TrD}(\underline{C})$ in $\mathcal{Y}/\mathcal{P}_\Lambda$. Then $\mathrm{DTr}(\underline{Y}_{\mathrm{TrD}(\underline{C})}) \to \underline{C}$ is the coreflection of $\underline{C}$ in $(\mathcal{Z}/\mathcal{I}_\Lambda)^\perp$. Similarly for any $\underline{C}$ in $\underline{\mathrm{mod}}(\Lambda)$, the morphism $\underline{C} \to \mathrm{TrD}(\underline{Y}^{\mathrm{DTr}(\underline{C})})$ is the reflection of $\underline{C}$ in ${}^\perp(\mathcal{X}/\mathcal{P}_\Lambda)$, where $\mathrm{DTr}(\underline{C}) \to \underline{Y}^{\mathrm{DTr}(\underline{C})}$ is the reflection of $\mathrm{DTr}(\underline{C})$ in $\mathcal{Y}/\mathcal{I}_\Lambda$. The last assertion follows from Lemma V.1.1. □

The following result shows that in the category of finitely generated modules over an Artin algebra, the notions of cotorsion pair and cotorsion triple are essentially equivalent.

PROPOSITION 4.9. *If $\mathcal{X},\mathcal{Y}$ and $\mathcal{Z}$ are full subcategories of* $\mathrm{mod}(\Lambda)$*, then the following are equivalent.*

(i) *There exists a resolving cotorsion pair $(\mathcal{X},\mathcal{Y})$ in $\mathrm{mod}(\Lambda)$ with $\mathcal{X}\cap\mathcal{Y} = \mathcal{P}_\Lambda$.*
(ii) *There exists a coresolving cotorsion pair $(\mathcal{Y},\mathcal{Z})$ in $\mathrm{mod}(\Lambda)$ with $\mathcal{Y}\cap\mathcal{Z} = \mathcal{I}_\Lambda$.*
(iii) *There exists a cotorsion triple $(\mathcal{X},\mathcal{Y},\mathcal{Z})$ in $\mathrm{mod}(\Lambda)$ with $\mathcal{Y}$ (co)resolving.*

PROOF. By Theorem 3.2, we have that (iii) implies (i) and (ii). If (i) holds, then $\mathcal{Y}$ is coresolving and covariantly finite. By a result of Krause-Solberg, see [**79**], $\mathcal{Y}$ is contravariantly finite. Since, by Theorem 2.1, $\mathcal{Y}$ is resolving, by a result of Auslander-Reiten, see [**9**], we have a cotorsion pair $(\mathcal{Y},\mathcal{Z})$ in $\mathrm{mod}(\Lambda)$. Hence (iii) holds. The proof that (ii) implies (iii) is similar. □

The above result combined with Theorem 3.2 and the results of [**9**] gives the following nice consequence.

COROLLARY 4.10. *Let Λ be an Artin algebra. Then there are bijections between:*

(i) *Resolving and coresolving contravariantly finite subcategories of* $\mathrm{mod}(\Lambda)$.
(ii) *Resolving and coresolving covariantly finite subcategories of* $\mathrm{mod}(\Lambda)$.
(iii) *Functorially finite resolving and coresolving subcategories of* $\mathrm{mod}(\Lambda)$.
(iv) *Resolving cotorsion pairs $(\mathcal{X},\mathcal{Y})$ in $\mathrm{mod}(\Lambda)$ with $\mathcal{X}\cap\mathcal{Y} = \mathcal{P}_\Lambda$.*
(v) *Coresolving cotorsion pairs $(\mathcal{Y},\mathcal{Z})$ in $\mathrm{mod}(\Lambda)$ with $\mathcal{Y}\cap\mathcal{Z} = \mathcal{I}_\Lambda$.*
(vi) *Cotorsion triples $(\mathcal{X},\mathcal{Y},\mathcal{Z})$ in $\mathrm{mod}(\Lambda)$ with $\mathcal{Y}$ (co)resolving.*

We have seen that for a good triple $(\mathcal{X},\mathcal{Y},\mathcal{Z})$, the categories $\mathcal{X}/\mathcal{P}_\Lambda$ and $\mathcal{Z}/\mathcal{I}_\Lambda$ are always triangle equivalent. The next result shows that if $\mathcal{X} = \mathcal{Z}$, then the algebra Λ is self-injective.

COROLLARY 4.11. *The subcategories $\mathcal{X}$ and $\mathcal{Z}$ coincide iff Λ is self-injective and the TTF-triple $(\mathcal{X}/\mathcal{P}_\Lambda,\mathcal{Y}/\mathcal{P}_\Lambda,\mathcal{Z}/\mathcal{P}_\Lambda)$ in $\underline{\mathrm{mod}}(\Lambda)$ splits. In this case also $\mathcal{Y}/\mathcal{P}_\Lambda$ is triangulated and all the above subcategories are closed under* DTr *and* TrD*, and the stable category admits a decomposition $\underline{\mathrm{mod}}(\Lambda) = \mathcal{X}/\mathcal{P}_\Lambda \times \mathcal{Y}/\mathcal{P}_\Lambda = \mathcal{Y}/\mathcal{I}_\Lambda \times \mathcal{Z}/\mathcal{I}_\Lambda$.*

PROOF. If $\mathcal{X} = \mathcal{Z}$, then $\mathcal{P}_\Lambda = \mathcal{X} \cap \mathcal{Y} = \mathcal{Y} \cap \mathcal{Z} = \mathcal{I}_\Lambda$. Hence Λ is self-injective and then obviously $\mathcal{Y}/\mathcal{P}_\Lambda$ is triangulated. In particular the triple $(\mathcal{X}/\mathcal{P}_\Lambda, \mathcal{Y}/\mathcal{P}_\Lambda, \mathcal{Z}/\mathcal{P}_\Lambda)$ is a TTF-triple in $\underline{\mathrm{mod}}(\Lambda)$. By the discussion after Proposition II.2.3, we have that the TTF-triple $(\mathcal{X}/\mathcal{P}_\Lambda, \mathcal{Y}/\mathcal{P}_\Lambda, \mathcal{Z}/\mathcal{P}_\Lambda)$ in $\underline{\mathrm{mod}}(\Lambda)$ splits. Working as in Remark 4.5 we infer easily that $\mathcal{X}, \mathcal{Y}, \mathcal{Z}$ are closed under DTr, TrD. The converse and the last assertion are easy and are left to the reader. □

EXAMPLE. Let Λ be a self-injective algebra such that the stable category $\underline{\mathrm{mod}}(\Lambda)$ is indecomposable, for instance the group algebra of a finite p-group over a field of characteristic p which divides the order of the group. Then Corollary 4.11 implies that $\mathcal{X} \neq \mathcal{Z}$ for any cotorsion triple $(\mathcal{X}, \mathcal{Y}, \mathcal{Z})$ in $\underline{\mathrm{mod}}(\Lambda)$, unless one of the subcategories $\mathcal{X}$, $\mathcal{Y}$ or $\mathcal{Z}$ coincides with $\mathrm{mod}(\Lambda)$.

If Λ is self-injective, then the following consequence of Theorem 3.2 describes the TTF-triples of $\underline{\mathrm{mod}}(\Lambda)$ in terms of subcategories of $\mathrm{mod}(\Lambda)$.

COROLLARY 4.12. *Let Λ be a self-injective algebra and let $\mathcal{Y}$ be a full subcategory of* $\mathrm{mod}(\Lambda)$. *Then the following are equivalent.*

(i) *$\mathcal{Y}$ is a functorially finite resolving and coresolving subcategory of* $\mathrm{mod}(\Lambda)$.
(ii) *There exists a TTF-triple $(\mathcal{X}/\mathcal{P}_\Lambda, \mathcal{Y}/\mathcal{P}_\Lambda, \mathcal{Z}/\mathcal{P}_\Lambda)$ in* $\underline{\mathrm{mod}}(\Lambda)$.

In the self-injective case the following corollary shows that any cotorsion triple in $\mathrm{mod}(\Lambda)$, equivalently any TTF-triple in $\underline{\mathrm{mod}}(\Lambda)$, produces an infinite number of cotorsion triples in the module category, equivalently an infinite number of TTF-triples in the stable category.

COROLLARY 4.13. *Let Λ be a self-injective algebra and let $(\mathcal{X}, \mathcal{Y}, \mathcal{Z})$ be a good triple in* $\mathrm{mod}(\Lambda)$. *Then $({}^{\perp}\mathcal{X}, \mathcal{X}, \mathcal{Y})$ and $(\mathcal{Y}, \mathcal{Z}, \mathcal{Z}^{\perp})$ are good triples. Iterating this process $\forall n \geq 0$ we get good triples $({}^{(n+2)\perp}\mathcal{Y}, {}^{(n+1)\perp}\mathcal{Y}, {}^{(n)\perp}\mathcal{Y})$ in* $\mathrm{mod}(\Lambda)$, *where ${}^{(n)\perp}\mathcal{Y}$ is the $n-$th iterated left orthogonal of $\mathcal{Y}$ and ${}^{(0)\perp}\mathcal{Y} := \mathcal{Y}$ and good triples $(\mathcal{Y}^{(n)\perp}, \mathcal{Y}^{(n+1)\perp}, \mathcal{Y}^{(n+2)\perp})$ in* $\mathrm{mod}(\Lambda)$, *where $\mathcal{Y}^{(n)\perp}$ is the $n-$th iterated right orthogonal of $\mathcal{Y}$ and $\mathcal{Y}^{(0)\perp} := \mathcal{Y}$.*

The above good triples induce an infinite number of TTF-triples $(\mathcal{T}_n, \mathcal{F}_n, \mathcal{R}_n)$ in $\underline{\mathrm{mod}}(\Lambda)$, *where $\mathcal{F}_n = {}^{(n)\perp}(\mathcal{Y}/\mathcal{P}_\Lambda)$ for $n \leq 0$ and $\mathcal{F}_n = (\mathcal{Y}/\mathcal{P}_\Lambda)^{(n)\perp}$ for $n \geq 0$.*

We close this chapter with an application to the stable Grothendieck group $\mathrm{K}_0(\underline{\mathrm{mod}}(\Lambda))$, which is one of the few known invariants under stable equivalences of algebras.

COROLLARY 4.14. *If Λ is Gorenstein, there are triangle equivalences:*

$$\underline{\mathrm{CM}}(\Lambda) \xrightarrow{\approx} \underline{\mathrm{mod}}(\Lambda)/\mathcal{P}_\Lambda^{<\infty}/\mathcal{P}_\Lambda \xrightarrow{\approx} \overline{\mathrm{mod}}(\Lambda)/\mathcal{I}_\Lambda^{<\infty}/\mathcal{I}_\Lambda \xrightarrow{\approx} \overline{\mathrm{CoCM}}(\mathrm{D}(\Lambda))$$

and isomorphisms

$$\mathrm{K}_0(\underline{\mathrm{mod}}(\Lambda)) \xrightarrow{\cong} \mathrm{K}_0(\underline{\mathrm{CM}}(\Lambda)) \xrightarrow{\cong} \mathrm{K}_0(\overline{\mathrm{CoCM}}(\mathrm{D}(\Lambda))) \xleftarrow{\cong} \mathrm{K}_0(\overline{\mathrm{mod}}(\Lambda)).$$

PROOF. Since Λ is Gorenstein, by Theorem 4.2, we have the hereditary torsion pair $(\underline{\mathrm{CM}}(\Lambda)/\mathcal{P}_\Lambda, \mathcal{P}_\Lambda^{<\infty}/\mathcal{P}_\Lambda)$ in the stable category $\underline{\mathrm{mod}}(\Lambda)$ and the cohereditary torsion pair $(\mathcal{I}_\Lambda^{<\infty}/\mathcal{I}_\Lambda, \underline{\mathrm{CoCM}}(\mathrm{D}(\Lambda)))$ in the stable category $\overline{\mathrm{mod}}(\Lambda)$. Then the assertions follow from Theorem 3.5 and Corollary 3.6. □

CHAPTER VII

Gorenstein Categories and (Co)Torsion Pairs

The results of the previous chapter suggest to study more closely the abelian categories $\mathcal{C}$ admitting a resolving cotorsion pair $(\mathcal{X}, \mathcal{Y})$ with $\mathcal{X} \cap \mathcal{Y} = \mathcal{P}$ such that $\mathcal{X}$-res. dim $\mathcal{C} < \infty$, and/or a coresolving cotorsion pair $(\mathcal{W}, \mathcal{Z})$ with $\mathcal{W} \cap \mathcal{Z} = \mathcal{I}$ such that $\mathcal{Z}$-cores. dim $\mathcal{C} < \infty$. The existence of such cotorsion pairs is strongly connected with cotorsion-theoretic properties of (Co)Cohen-Macaulay objects. In this chapter we introduce and investigate in (co)torsion theoretic terms Gorenstein and Cohen-Macaulay abelian categories, which appear to be the proper generalizations of the category of finitely generated modules over a (commutative Noetherian) Gorenstein and Cohen-Macaulay ring respectively. We give methods for constructing Gorenstein categories out of certain Cohen-Macaulay categories and we study in this context the existence of minimal (Co)Cohen-Macaulay approximations, which behave better than the special (Co)Cohen-Macaulay approximations. We also give (co)torsion theoretic applications to (not necessarily finitely generated) modules over Gorenstein and Cohen-Macaulay rings which admit a Morita self-duality.

1. Dimensions and Cotorsion Pairs

Throughout this section $\mathcal{C}$ denotes an abelian category with enough projective objects. We have seen in Proposition VI.2.3 that if $(\mathcal{X}, \mathcal{Y})$ is a resolving cotorsion pair in $\mathcal{C}$ such that $\mathcal{X} \cap \mathcal{Y} = \mathcal{P}$, then $\mathcal{I}^{<\infty} \subseteq \mathcal{Y} \supseteq \mathcal{P}^{<\infty}$. It is interesting to have conditions ensuring that we have an equality $\mathcal{P}^{<\infty} = \mathcal{I}^{<\infty}$. In this section we study this problem, which is related to the finiteness of several interesting homological dimensions defined on $\mathcal{C}$ and the existence of cotorsion pairs induced by (Co)Cohen-Macaulay objects. We also give applications to complete projective resolutions.

First we need to recall the definition of the resolution dimension. Let $\mathcal{X}$ be a resolving subcategory of $\mathcal{C}$. The *$\mathcal{X}$-resolution dimension* of an object C in $\mathcal{C}$, written $\mathcal{X}$-res. dim C, is defined inductively as follows. If C is in $\mathcal{X}$, then $\mathcal{X}$-res. dim $C = 0$. If $t \geq 1$, then $\mathcal{X}$-res. dim $C \leq t$ if there exists an exact sequence $0 \to X_t \to \cdots \to X_1 \to X_0 \to C \to 0$ where $\mathcal{X}$-res. dim $X_i = 0$, for $0 \leq i \leq t$. Then $\mathcal{X}$-res. dim $C = t$ if $\mathcal{X}$-res. dim $C \leq t$ and $\mathcal{X}$-res. dim $C \nleq t-1$. Finally if $\mathcal{X}$-res. dim $C \neq t$ for any $t \geq 0$, then define $\mathcal{X}$-res. dim $C = \infty$. The *$\mathcal{X}$-resolution dimension* of $\mathcal{C}$ is defined by $\mathcal{X}$-res. dim $\mathcal{C} := \sup\{\mathcal{X}\text{-res. dim } C \mid C \in \mathcal{C}\}$. The $\mathcal{Y}$-coresolution dimension of $\mathcal{C}$, written $\mathcal{Y}$-cores. dim $\mathcal{C}$, for a coresolving subcategory $\mathcal{Y}$ of $\mathcal{C}$, is defined dually.

The following result shows that finiteness of the resolution dimension of $\mathcal{C}$ with respect to a resolving subcategory admitting the projectives as an Ext-injective cogenerator, has several interesting consequences for the category of Cohen-Macaulay objects and the objects of finite projective or injective dimension.

PROPOSITION 1.1. *Let $\mathcal{X}$ be a resolving subcategory of $\mathcal{C}$ having $\mathcal{P}$ as an Ext-injective cogenerator. If $\mathcal{X}$-res.dim$\mathcal{C} = d < \infty$, then $(\mathcal{X}, \mathcal{P}^{<\infty})$ is a resolving cotorsion pair in $\mathcal{C}$ with $\mathrm{CM}(\mathcal{P}) = \mathcal{X}$ and $\mathcal{P}^{<\infty} = \mathcal{I}^{<\infty}$. Moreover if $\mathcal{P}$ is covariantly finite, then $(\mathrm{CM}(\mathcal{P})/\mathcal{P}, \mathcal{P}^{<\infty}/\mathcal{P})$ is a hereditary torsion pair in $\mathcal{C}/\mathcal{P}$ and the torsion class $\mathrm{CM}(\mathcal{P})/\mathcal{P}$ is the left stabilization of $\mathcal{C}/\mathcal{P}$ with stabilization functor given by the coreflection functor $\mathbf{R} : \mathcal{C}/\mathcal{P} \to \mathrm{CM}(\mathcal{P})/\mathcal{P} : \underline{C} \mapsto \mathbf{R}(\underline{C}) = \Omega^{-d}\Omega^{d}(\underline{C})$.*

PROOF. The hypothesis $\mathcal{X}$-res.dim$\mathcal{C} = d < \infty$ implies that $\widehat{\mathcal{X}} = \mathcal{C}$. Hence all assertions, except for the identification $\mathcal{P}^{<\infty} = \mathcal{I}^{<\infty}$, follow from Proposition V.4.6 and Proposition VI.2.6. By Proposition VI.2.3, we have $\mathcal{I}^{<\infty} \subseteq \mathcal{P}^{<\infty}$. We claim that $\mathrm{Ext}^{d+1}(-, Y) = 0$ for any $Y \in \mathcal{P}^{<\infty}$. Indeed this follows easily using induction on the bounded $\mathcal{X}$-resolution length of any object of $\mathcal{C}$, since $\mathrm{Ext}^{n}(X, Y) = 0$, for $n \geq 1$ and $X \in \mathrm{CM}(\mathcal{P})$. Hence $Y \in \mathcal{I}^{<\infty}$. We conclude that $\mathcal{P}^{<\infty} \subseteq \mathcal{I}^{<\infty}$. □

The above result suggests to study more closely the categories $\mathcal{C}$ admitting a resolving cotorsion pair $(\mathcal{X}, \mathcal{Y})$ with $\mathcal{X} \cap \mathcal{Y} = \mathcal{P}$ and such that $\mathcal{X}$-res. dim $\mathcal{C} < \infty$. Note that finiteness of $\mathcal{X}$-res. dim $\mathcal{C}$ implies that $\widehat{\mathcal{X}} = \mathcal{C}$, but the converse is false in general. Hence it is natural to ask under what conditions the converse is true. In this direction we have the following result which gives interesting connections with the finitistic dimensions. First we recall that the *finitistic projective dimension* $\mathrm{FPD}(\mathcal{C})$ of $\mathcal{C}$ is defined by $\mathrm{FPD}(\mathcal{C}) := \sup\{\mathrm{pd}\, C \mid C \in \mathcal{P}^{<\infty}\}$, and the *finitistic injective dimension* $\mathrm{FID}(\mathcal{C})$ of $\mathcal{C}$ is defined by $\mathrm{FID}(\mathcal{C}) := \sup\{\mathrm{id}\, C \mid C \in \mathcal{I}^{<\infty}\}$.

PROPOSITION 1.2. *Let $\mathcal{C}$ be an abelian category with enough projective objects, and let $(\mathcal{X}, \mathcal{Y})$ be a resolving cotorsion pair in $\mathcal{C}$ with $\mathcal{X} \cap \mathcal{Y} = \mathcal{P}$. Then the following statements are equivalent.*

(i) $\mathcal{X}$-res. dim $\mathcal{C} < \infty$.
(ii) $\widehat{\mathcal{X}} = \mathcal{C}$ *and* $\mathrm{FPD}(\mathcal{C}) < \infty$.
(iii) $\sup\{\mathrm{pd}\, Y \mid Y \in \mathcal{Y}\} < \infty$.

If (i) *holds, then $\mathcal{X} = \mathrm{CM}(\mathcal{P})$, $\mathcal{Y} = \mathcal{P}^{<\infty} = \mathcal{I}^{<\infty}$ and: $\mathcal{X}$-res. dim $\mathcal{C} = \mathrm{FPD}(\mathcal{C}) = \mathrm{FID}(\mathcal{C}) < \infty$. Conversely if $\mathrm{CM}(\mathcal{P})$-res. dim $\mathcal{C} < \infty$, then there exists a resolving cotorsion pair $(\mathrm{CM}(\mathcal{P}), \mathcal{P}^{<\infty})$ in $\mathcal{C}$ with $\mathrm{CM}(\mathcal{P}) \cap \mathcal{P}^{<\infty} = \mathcal{P}$, $\mathcal{P}^{<\infty} = \mathcal{I}^{<\infty}$, and $\mathrm{FID}(\mathcal{C}) = \mathrm{FPD}(\mathcal{C}) = \mathrm{CM}(\mathcal{P})$-res. dim $\mathcal{C} =< \infty$.*

PROOF. (i) $\Rightarrow$ (ii) Assume that $\mathcal{X}$-res. dim $\mathcal{C} = d < \infty$. Then obviously $\widehat{\mathcal{X}} = \mathcal{C}$ and by Proposition VI.2.6 we have $\mathcal{Y} = \mathcal{P}^{<\infty}$. Let Y be in $\mathcal{Y}$ and let $0 \to X_d \to \cdots \to X_0 \to Y \to 0$ be an exact sequence with the X_i in $\mathcal{X}$. Since $Y \in \mathcal{Y}$, it follows easily that the X_i are projective, hence $\mathrm{pd}\, Y \leq d$. We infer that $\mathrm{FPD}(\mathcal{C}) \leq d < \infty$.

(ii) $\Rightarrow$ (iii) The proof is trivial since the hypothesis implies that $\mathcal{Y} = \mathcal{P}^{<\infty}$.

(iii) $\Rightarrow$ (i) The hypothesis implies that $\mathcal{Y} \subseteq \mathcal{P}^{<\infty}$, hence $\mathcal{Y} = \mathcal{P}^{<\infty}$ and $\mathrm{FPD}(\mathcal{C}) < \infty$. Let $\mathrm{FPD}(\mathcal{C}) = d$ and consider the special right $\mathcal{X}$-approximation sequence $0 \to Y_C \to X_C \to C \to 0$ of $C \in \mathcal{C}$. By Proposition VI.2.3 we have that Y_C has finite projective dimension, hence $\mathrm{pd}\, Y_C \leq d$. This implies trivially that $\mathcal{X}$-res. dim $C \leq d + 1$. Hence $\mathcal{X}$-res. dim $\mathcal{C} < \infty$.

Assume now that $\mathcal{X}$-res. dim $\mathcal{C} = d < \infty$. Then by the above argument we have $\mathrm{FPD}(\mathcal{C}) \leq d$. By (the proof of) Proposition 1.1 we have $\mathcal{P}^{<\infty} = \mathcal{I}^{<\infty}$ and $\mathrm{id}\, Y \leq d$ for any $Y \in \mathcal{Y}$. Hence $\mathrm{FID}(\mathcal{C}) \leq d < \infty$. An easy induction argument shows that in fact we have $\mathcal{X}$-res. dim $\mathcal{C} = \mathrm{FPD}(\mathcal{C}) = \mathrm{FID}(\mathcal{C})$. □

By Proposition 2.8 it follows that if $\mathcal{X}$-res. dim $\mathcal{C} < \infty$, then $\mathcal{C}$ has finite $\mathcal{P}$-Gorenstein dimension. Recall from Section V.4 that the $\mathcal{P}$-Gorenstein dimension $\text{G-dim}_{\mathcal{P}} C$ of $C \in \mathcal{C}$ is defined by $\text{G-dim}_{\mathcal{P}} C := \mathsf{CM}(\mathcal{P})\text{-res.}\dim C$, and the $\mathcal{P}$-Gorenstein dimension $\text{G-dim}_{\mathcal{P}} \mathcal{C}$ of $\mathcal{C}$ is defined by $\text{G-dim}_{\mathcal{P}} \mathcal{C} := \mathsf{CM}(\mathcal{P})\text{-res.}\dim \mathcal{C}$. We say that $\mathcal{C}$ is **$\mathcal{P}$-Gorenstein** if $\text{G-dim}_{\mathcal{P}} \mathcal{C} < \infty$. Dually we define the *$\mathcal{I}$-Gorenstein dimension* $\text{G-dim}_{\mathcal{I}} C$ of $C \in \mathcal{C}$ by: $\text{G-dim}_{\mathcal{I}} C := \mathsf{CoCM}(\mathcal{I})\text{-cores.}\dim C$. Then the *$\mathcal{I}$-Gorenstein dimension* $\text{G-dim}_{\mathcal{I}} \mathcal{C}$ of $\mathcal{C}$ is defined as follows: $\text{G-dim}_{\mathcal{I}} \mathcal{C} := \mathsf{CoCM}(\mathcal{I})\text{-cores.}\dim \mathcal{C}$. We say that $\mathcal{C}$ is **$\mathcal{I}$-Gorenstein** if $\text{G-dim}_{\mathcal{I}} \mathcal{C} < \infty$.

The following example explains the above terminology.

EXAMPLE. Let Λ be a Noetherian ring and C a finitely generated right Λ-module. If $\mathcal{P}_\Lambda$ is the full subcategory of $\text{mod}(\Lambda)$ consisting of the finitely generated projective right Λ modules and $\mathsf{CM}(\mathcal{P}_\Lambda)$ is the full subcategory of $\text{mod}(\Lambda)$ consisting of the finitely generated Cohen-Macaulay modules, then $\text{G-dim}_{\mathcal{P}} C$ is the Gorenstein dimension of C as defined by Auslander-Bridger [**6**]. By results of Auslander-Reiten, see [**11**], it follows that an Artin algebra Λ is Gorenstein iff $\text{mod}(\Lambda)$ is a $\mathcal{P}_\Lambda$-Gorenstein category. In this case: $\text{G-dim}_{\mathcal{P}_\Lambda} \text{mod}(\Lambda) = \text{id}\,\Lambda_\Lambda = \text{id}\,{}_\Lambda\Lambda$. This common value is also equal to $\sup\{\text{pd}\, I \mid I \in \mathcal{I}_\Lambda\} = \sup\{\text{id}\, P \mid P \in \mathcal{P}_\Lambda\}$.

Closely related to the $\mathcal{P}$-Gorenstein dimension and the $\mathcal{I}$-Gorenstein dimension of an abelian category, are the following dimensions introduced by Gedrich and Gruenberg, see [**51**], in the context of ring theory:

$$\mathsf{spli}(\mathcal{C}) = \sup\{\text{pd}\, I \mid I \in \mathcal{I}\} \qquad \text{and} \qquad \mathsf{silp}(\mathcal{C}) = \sup\{\text{id}\, I \mid P \in \mathcal{P}\}.$$

We have introduced so far several dimensions which are connected to cotorsion pairs where the involved subcategories are related to (Co)Cohen-Macaulay objects and objects of finite projective or injective dimension. So it is natural to ask for the precise relations between these dimensions. In this connection we have the following result, which will be useful in the sequel.

PROPOSITION 1.3. *Let $\mathcal{C}$ be an abelian category with enough projective and injective objects. Then we have the following:*

(i)
$$\text{FPD}(\mathcal{C}) \leq \mathsf{silp}(\mathcal{C}) \leq \text{G-dim}_{\mathcal{P}} \mathcal{C} \leq \text{gl.}\dim \mathcal{C}$$
$$\text{FID}(\mathcal{C}) \leq \mathsf{spli}(\mathcal{C}) \leq \text{G-dim}_{\mathcal{I}} \mathcal{C} \leq \text{gl.}\dim \mathcal{C}$$
and finiteness of any of these dimensions implies that the remaining inequalities on the left are equalities.

(ii) *If $\mathcal{P}^{<\infty} \subseteq \mathcal{I}^{<\infty}$, then:*
$$\text{FPD}(\mathcal{C}) \leq \mathsf{silp}(\mathcal{C}) \leq \text{FID}(\mathcal{C}).$$

(iii) *If $\mathcal{I}^{<\infty} \subseteq \mathcal{P}^{<\infty}$, then:*
$$\text{FID}(\mathcal{C}) \leq \mathsf{spli}(\mathcal{C}) \leq \text{FPD}(\mathcal{C}).$$

(iv) *If $\mathcal{P}^{<\infty} = \mathcal{I}^{<\infty}$, then:*
$$\text{FPD}(\mathcal{C}) = \text{FID}(\mathcal{C}) = \text{G-dim}_{\mathcal{P}} \mathcal{C} = \text{G-dim}_{\mathcal{I}} \mathcal{C} = \mathsf{spli}(\mathcal{C}) = \mathsf{silp}(\mathcal{C}).$$

(v) $\text{G-dim}_{\mathcal{P}} \mathcal{C} = \text{G-dim}_{\mathcal{I}} \mathcal{C}$.

(vi) *If $\mathsf{spli}(\mathcal{C}) < \infty$ and $\mathsf{silp}(\mathcal{C}) < \infty$, then $\mathsf{spli}(\mathcal{C}) = \mathsf{silp}(\mathcal{C})$.*

PROOF. (i) Assume that $\sup\{\operatorname{id} P \mid P \in \mathcal{P}\} = d < \infty$ and let Y be in $\mathcal{C}$ with $\operatorname{pd} Y = n$. Let C be in $\mathcal{C}$ such that $\operatorname{Ext}^n(Y, C) \neq 0$, and let $0 \to \Omega(C) \to P \to C \to 0$ be exact with P projective. Applying to this sequence the functor $\mathcal{C}(Y, -)$, the resulting long exact sequence shows that $\operatorname{Ext}^n(Y, P) \neq 0$. Since $\operatorname{id} P \leq d$, it follows that $n \leq d$. Hence $\operatorname{FPD}(\mathcal{C}) \leq \sup\{\operatorname{id} P \mid P \in \mathcal{P}\}$. Next let $\operatorname{G-dim}_{\mathcal{P}} \mathcal{C} = d < \infty$. Then for any object C in $\mathcal{C}$, we have that $\Omega^d(C)$ is Cohen-Macaulay, in particular $\operatorname{Ext}^n(\Omega^d(C), \mathcal{P}) = 0$, $\forall n \geq 1$. Hence $\operatorname{Ext}^{n+d}(C, \mathcal{P}) = 0$, $\forall n \geq 1$, $\forall C \in \mathcal{C}$. This implies that $\operatorname{id} P \leq d$ for any projective object P. Hence $\sup\{\operatorname{id} P \mid P \in \mathcal{P}\} \leq \operatorname{G-dim}_{\mathcal{P}} \mathcal{C}$. Finally it is obvious that $\operatorname{G-dim}_{\mathcal{P}} \mathcal{C} \leq \operatorname{gl.dim} \mathcal{C}$. Now if $\operatorname{gl.dim} \mathcal{C} < \infty$, then $\operatorname{CM}(\mathcal{P}) = \mathcal{P}$. Hence $\operatorname{gl.dim} \mathcal{C} = \operatorname{G-dim}_{\mathcal{P}} \mathcal{C}$, and then obviously all the dimensions in the first displayed formula are equal. Next if $\operatorname{G-dim}_{\mathcal{P}} \mathcal{C} = d < \infty$, then by (the proof of) Proposition VI.2.3, we have that $\operatorname{FPD}(\mathcal{C}) = \operatorname{G-dim}_{\mathcal{P}} \mathcal{C} = d$, hence also $\sup\{\operatorname{id} P \mid P \in \mathcal{P}\} = d$. Finally if $\sup\{\operatorname{id} P \mid P \in \mathcal{P}\} = d < \infty$, then $\operatorname{FPD}(\mathcal{C}) = d$, since otherwise there exists an object Y with $\operatorname{pd} Y = n > d$ and a projective object P such that $\operatorname{Ext}^n(Y, P) \neq 0$, and this is impossible. The proof of the second displayed formula is similar.

The proof of parts (ii) and (iii) is trivial, and the proof of (iv) follows directly from (ii) and (iii). If one of the Gorenstein dimensions $\operatorname{G-dim}_{\mathcal{P}} \mathcal{C}$ and $\operatorname{G-dim}_{\mathcal{I}} \mathcal{C}$ is finite, then by Proposition 1.2 and its dual, we have $\mathcal{P}^{<\infty} = \mathcal{I}^{<\infty}$. It follows by part (iv) that either both Gorenstein dimensions are infinite or else they are finite and we have $\operatorname{G-dim}_{\mathcal{P}} \mathcal{C} = \operatorname{G-dim}_{\mathcal{I}} \mathcal{C}$, hence part (v) holds. Finally the hypothesis of part (vi) implies that $\mathcal{P}^{<\infty} = \mathcal{I}^{<\infty}$, and the assertion follows by part (iv). □

We have the following characterization of $\mathcal{P}$-Gorenstein categories. The dual characterization concerning $\mathcal{I}$-Gorenstein categories is left to the reader.

THEOREM 1.4. *Let $\mathcal{C}$ be an abelian category with enough projective objects. Then the following are equivalent.*

(i) *$\mathcal{C}$ is $\mathcal{P}$-Gorenstein.*
(ii) *There exists a resolving cotorsion pair $(\mathcal{X}, \mathcal{Y})$ with $\mathcal{X} \cap \mathcal{Y} = \mathcal{P}$ such that $\mathcal{X}\text{-res.}\dim \mathcal{C} < \infty$.*

If $\mathcal{C}$ is $\mathcal{P}$-Gorenstein, then $\mathcal{X} = \operatorname{CM}(\mathcal{P})$, $\mathcal{P}^{<\infty} = \mathcal{I}^{<\infty}$ and:

$$\operatorname{G-dim}_{\mathcal{P}} \mathcal{C} = \operatorname{FPD}(\mathcal{C}) = \operatorname{FID}(\mathcal{C}) = \mathsf{silp}(\mathcal{C}) = \mathsf{spli}(\mathcal{C}) < \infty.$$

If in addition $\mathcal{P}$ is covariantly finite, then $\operatorname{CM}(\mathcal{P})$ is functorially finite.

PROOF. All assertions, except of the last one, follow from Propositions 2.2 and 2.3. However if $\operatorname{G-dim}_{\mathcal{P}} \mathcal{C} = d < \infty$, then obviously $\Omega^d(\mathcal{C}/\mathcal{P}) \subseteq \operatorname{CM}(\mathcal{P})/\mathcal{P}$ and the last assertion follows from Corollary V.3.11. □

We close this section by giving the connection between $\mathcal{P}$-Gorenstein categories and complete projective resolutions. We recall that a *complete projective resolution* of an object C in $\mathcal{C}$ is an acyclic complex $P^\bullet$ with projective components such that: (α) the complex $\mathcal{C}(P^\bullet, Q)$ is acyclic for any projective object Q, and (β) the complex $P^\bullet$ coincides with a projective resolution of C beyond some finite dimension. The notion of a *complete injective resolution* is defined dually.

COROLLARY 1.5. *Let $\mathcal{C}$ be an abelian category with enough projectives. Then the following are equivalent.*

(i) *Any object of* $\mathcal{C}$ *admits a complete projective resolution.*

(ii) $(\mathrm{CM}(\mathcal{P}), \mathcal{P}^{<\infty})$ *is a cotorsion pair in* $\mathcal{C}$*, or equivalently* $\widehat{\mathrm{CM}(\mathcal{P})} = \mathcal{C}$.

In particular $\mathcal{C}$ *is* $\mathcal{P}$*-Gorenstein iff* $\mathrm{FPD}(\mathcal{C}) < \infty$ *and any object of* $\mathcal{C}$ *admits a complete projective resolution.*

PROOF. It is clear from the definition that an object C in $\mathcal{C}$ admits a complete projective resolution iff there exists $n \geq 0$ such that the nth syzygy object $\Omega^n(C)$ of C is Cohen-Macaulay. The equivalence (i) $\Leftrightarrow$ (ii) follow directly from this. The last assertion follows from Proposition 1.2, Theorem 1.4. □

We leave to the reader to formulate the dual version of the above Corollary. Note that if the abelian category $\mathcal{C}$ admits exact infinite coproducts or has a finite number of non-isomorphic soimple objects and any object has finite length, then the existence of complete projective resolutions characterize $\mathcal{P}$-Gorenstein categories. This follows from the fact that $\mathrm{FPD}(\mathcal{C}) < \infty$ in this case, see Proposition 2.4 below.

2. Gorenstein Categories, Cotorsion Pairs and Minimal Approximations

In the previous section we defined an abelian category $\mathcal{C}$ with enough projectives, resp. injectives, to be projectively, resp. injectively, Gorenstein, provided that $\mathcal{C}$ has finite resolution, resp. coresolution, dimension with respect to the full subcategory of Cohen-Macaulay, resp. CoCohen-Macaulay objects. Motivated by the concept of Gorensteinness in commutative algebra and representation theory, we give in this section a more convenient definition of Gorenstein abelian categories and we discuss its cotorsion theoretic consequences. Then, by using our previous results, we show that the three notions of Gorensteinness coincide and in addition we give applications to minimal Cohen-Macaulay approximations of not necessarily finitely generated modules over not necessarily commutative Gorenstein rings.

Recall that by definition a commutative local Noetherian ring or an Artin algebra is called Gorenstein, if $\mathrm{id}\,{}_{\Lambda}\Lambda < \infty$ and $\mathrm{id}\,\Lambda_{\Lambda} < \infty$. Equivalently, Λ is Gorenstein iff $\sup\{\mathrm{pd}\, I \mid I \in \mathcal{I}_{\Lambda}\} < \infty$ and $\sup\{\mathrm{id}\, P \mid P \in \mathcal{P}_{\Lambda}\} < \infty$. This suggests the following notion of Gorensteinness in arbitrary abelian categories.

DEFINITION 2.1. An abelian category with enough projective and injective objects is called **Gorenstein** if

$$\mathsf{spli}(\mathcal{C}) < \infty \quad \text{and} \quad \mathsf{silp}(\mathcal{C}) < \infty.$$

Our next result, which generalizes the situation of commutative Noetherian Gorenstein rings and Gorenstein algebras, gives a variety of useful characterizations of Gorenstein categories in cotorsion theoretic terms. As a consequence we get that $\mathcal{C}$ is $\mathcal{P}$-Gorenstein iff $\mathcal{C}$ is $\mathcal{I}$-Gorenstein iff $\mathcal{C}$ is Gorenstein.

THEOREM 2.2. *Let* $\mathcal{C}$ *be an abelian category with enough projective and injective objects. Then the following statements are equivalent:*

(i) $\mathcal{C}$ *is Gorenstein.*

(ii) $\mathrm{FPD}(\mathcal{C}) < \infty$ *and* $\big(\mathrm{CM}(\mathcal{P}), \mathcal{P}^{<\infty}\big)$ *is a cotorsion pair in* $\mathcal{C}$.

(iii) $\mathrm{FPD}(\mathcal{C}) < \infty$ *and any object of* $\mathcal{C}$ *admits a complete projective resolution.*

(iv) $\mathrm{FID}(\mathcal{C}) < \infty$ *and* $\big(\mathcal{I}^{<\infty}, \mathrm{CoCM}(\mathcal{I})\big)$ *is a cotorsion pair in* $\mathcal{C}$.

(v) $\mathrm{FID}(\mathcal{C}) < \infty$ *and any object of* $\mathcal{C}$ *admits a complete injective resolution.*
(vi) $\mathrm{FPD}(\mathcal{C}) < \infty$ *or* $\mathrm{FID}(\mathcal{C}) < \infty$ *and* $\big(\mathrm{CM}(\mathcal{P}), \mathcal{P}^{<\infty} = \mathcal{I}^{<\infty}, \mathrm{CoCM}(\mathcal{I})\big)$ *is a cotorsion triple in* $\mathcal{C}$.
(vii) $\mathcal{I}^{<\infty} \subseteq \mathcal{P}^{<\infty}$ *and* $\mathsf{silp}(\mathcal{C}) < \infty$.
(viii) $\mathcal{P}^{<\infty} \subseteq \mathcal{I}^{<\infty}$ *and* $\mathsf{spli}(\mathcal{C}) < \infty$.

If $\mathcal{C}$ *is Gorenstein, then we have the following.*
(α) $\mathrm{CM}(\mathcal{P})$ *is covariantly finite iff* $\mathcal{P}$ *is covariantly finite.*
(β) $\mathrm{CoCM}(\mathcal{I})$ *is contravariantly finite iff* $\mathcal{I}$ *is contravariantly finite.*
(γ) *We have the following equalities:*

$$\text{G-dim}_{\mathcal{P}}\,\mathcal{C} = \mathrm{FPD}(\mathcal{C}) = \mathsf{spli}(\mathcal{C}) = \mathsf{silp}(\mathcal{C}) = \mathrm{FID}(\mathcal{C}) = \text{G-dim}_{\mathcal{I}}\,\mathcal{C} < \infty.$$

PROOF. Follows from (the duals of) Propositions 1.2 and 1.3, and Theorem 1.4 and Corollary 1.5. □

As a corollary of Theorem 2.2 we have the following interesting consequence.

COROLLARY 2.3. *If* $(\mathcal{X}, \mathcal{Y}, \mathcal{Z})$ *is a cotorsion triple in* $\mathcal{C}$ *with* $\mathcal{Y}$ *(co)resolving, then* $\mathcal{X}\text{-res.}\dim\mathcal{C} = \mathcal{Z}\text{-cores.}\dim\mathcal{C}$.

PROOF. If $\mathcal{X}\text{-res.}\dim\mathcal{C} < \infty$, then, by Theorem 1.4, $\mathcal{C}$ is $\mathcal{P}$-Gorenstein with $\mathcal{X} = \mathrm{CM}(\mathcal{P})$ and $\mathcal{Y} = \mathcal{P}^{<\infty} = \mathcal{I}^{\infty}$. Then by Theorem 4.2, we have $\mathcal{Z} = \mathrm{CoCM}(\mathcal{I})$ and $\mathcal{Z}\text{-cores.}\dim\mathcal{C} = \mathcal{X}\text{-res.}\dim\mathcal{C} < \infty$. Hence $\mathcal{X}\text{-res.}\dim\mathcal{C} < \infty$ iff $\mathcal{Z}\text{-cores.}\dim\mathcal{C} < \infty$ and then $\mathcal{X}\text{-res.}\dim\mathcal{C} = \mathcal{Z}\text{-cores.}\dim\mathcal{C}$. If $\mathcal{X}\text{-res.}\dim\mathcal{C} = \infty$, then by the above argument we have $\mathcal{Z}\text{-cores.}\dim\mathcal{C} = \infty$. Hence in any case we have $\mathcal{X}\text{-res.}\dim\mathcal{C} = \mathcal{Z}\text{-cores.}\dim\mathcal{C}$. □

The following result shows that the assumptions $\mathrm{FPD}(\mathcal{C}) < \infty$ or $\mathrm{FID}(\mathcal{C}) < \infty$ in Theorem 2.2 can be dropped if $\mathcal{C}$ admits exact infinite products or coproducts, for instance if $\mathcal{C} = \mathrm{Mod}(\Lambda)$ where Λ is a ring, or if $\mathcal{C}$ has a finite number of simple objects and any object has finite length, for instance if $\mathcal{C} = \mathrm{mod}(\Lambda)$ where Λ is an Artin algebra. In these cases $\mathcal{C}$ is Gorenstein iff $\mathcal{P}^{<\infty} = \mathcal{I}^{<\infty}$.

PROPOSITION 2.4. *Let* $\mathcal{C}$ *be an abelian category with enough projective and injective objects. If* $\mathcal{P}^{<\infty} = \mathcal{I}^{<\infty}$, *then* $\mathcal{C}$ *is Gorenstein provided that one of the following conditions holds.*
(i) $\mathcal{C}$ *has exact products or exact coproducts.*
(ii) $\mathcal{C}$ *has a finite number of simple objects and any object has finite length.*

PROOF. Assume first that $\mathcal{C}$ has exact coproducts. We claim that $\mathsf{silp}(\mathcal{C}) = \{\mathrm{id}\,P \mid P \in \mathcal{P}\} < \infty$. Otherwise we can find a strictly increasing sequence $n_1 < n_2 < \cdots$ of positive integers and projective objects P_i such that $\mathrm{id}\,P_i = n_i$. Then $\bigoplus_{i\geq 0} P_i$ is a projective object, hence we have $\mathrm{id}\bigoplus_{i\geq 0} P_i = d$, for some $d < \infty$. This leads to the contradiction that $n_i \leq d$, for any $i \geq 0$. Hence $\mathsf{silp}(\mathcal{C}) < \infty$. Then by Theorem 2.2 we infer that $\mathcal{C}$ is Gorenstein. If $\mathcal{C}$ has exact products, then a similar argument shows that $\mathcal{C}$ is Gorenstein. Finally if $\mathcal{C}$ is Krull-Schmidt category with a finite number of isoclasses of simple objects, then the assertion is easy and is left to the reader. □

Theorem 2.2 suggests the following definition, see also [**20**].

DEFINITION 2.5. A ring Λ is called **right**, resp. **left**, **Gorenstein** if the category $\operatorname{Mod}(\Lambda)$, resp. $\operatorname{Mod}(\Lambda^{\mathrm{op}})$, of right, resp. left, Λ-modules is a Gorenstein category. We say that Λ is Gorenstein, provided that Λ is left and right Gorenstein.

In the literature Gorenstein rings are defined to be the Noetherian rings with finite left and right self-injective dimension. In the above definition we don't use any finiteness conditions; however the following example shows that when we restrict to Noetherian rings, then the above definition agrees with the usual definition.

EXAMPLE. (1) Any ring of finite global dimension and any quasi-Frobenius ring is Gorenstein.

(2) If Λ is an Artin algebra, or more generally a Noetherian ring, with finite self-injective dimension from both sides, then Λ is Gorenstein. Note that, by [**20**], for a Noetherian ring Λ, we have that Λ is left Gorenstein iff Λ is right Gorenstein.

(3) If Λ is the group algebra kG of an $\mathbf{H}\mathfrak{F}$-group G of type FP_∞, where k is a commutative ring of finite global dimension, then, by [**40**], Λ is right Gorenstein.

(4) If Λ is a Noetherian PI Hopf algebra over a field, then, by [**105**], Λ is Gorenstein.

We have the following consequence of Proposition 2.4.

COROLLARY 2.6. *A ring Λ is right Gorenstein if and only if* $\mathbf{P}_\Lambda^{<\infty} = \mathbf{I}_\Lambda^{<\infty}$.

REMARK 2.7. (1) Let Λ be a right Gorenstein ring. Then $\operatorname{CM}(\mathbf{P}_\Lambda)$ is covariantly finite iff Λ is left coherent and right perfect. Similarly $\operatorname{CoCM}(\mathbf{I}_\Lambda)$ is contravariantly finite iff Λ is right Noetherian. In particular if Λ is Artinian, then both $\operatorname{CM}(\mathbf{P}_\Lambda)$ and $\operatorname{CoCM}(\mathbf{I}_\Lambda)$ are functorially finite.

(2) Let Λ be a Noetherian Gorenstein ring. Then $\operatorname{CM}(\Lambda) = \operatorname{CM}(\mathcal{P}_\Lambda)$ is covariantly finite. This follows from the fact that, since Λ is left coherent, the category $\mathcal{P}_\Lambda$ of finitely generated projective right Λ-modules is covariantly finite. Similarly if Λ is Gorenstein with Morita self-duality, then $\operatorname{CoCM}(\mathcal{I}_\Lambda)$ is contravariantly finite.

NOTE. It is an open problem if finiteness of $\mathsf{silp}(\mathcal{C})$ or $\mathsf{spli}(\mathcal{C})$ implies that $\mathcal{C}$ is Gorenstein. We refer to [**26**] for a discussion of the problem, when $\mathcal{C}$ is the module category of an Artin algebra. In this setting the problem is equivalent to the Gorenstein Symmetry Conjecture mentioned in Proposition IV.3.1.

We close this section by investigating the existence of minimal Cohen-Macaulay approximations of modules over a right Gorenstein ring. Recall that any minimal right approximation is special, but in general the converse is false. Our motivation comes from the fact that minimal approximations behave better than the special approximations, for instance they are unique up to isomorphism.

THEOREM 2.8. *Let Λ be a right Gorenstein ring.*

(1) *If Λ is left coherent and right perfect, then:*

- (i) *Any right Λ-module admits a minimal right* $\operatorname{CM}(\mathbf{P}_\Lambda)$*-approximation, and a minimal left* $\mathbf{P}_\Lambda^{<\infty}$*-approximation.*
- (ii) *Any right Λ-module admits a minimal right* $\mathbf{P}_\Lambda^{<\infty}$*-approximation and a minimal left* $\operatorname{CoCM}(\mathbf{I}_\Lambda)$*-approximation.*

(2) *If* Λ *is right Noetherian, then any right* Λ*-module admits a minimal right* $\mathbf{I}_{\Lambda}^{<\infty}$*-approximation and a minimal left* $\mathrm{CoCM}(\mathbf{I}_{\Lambda})$*-approximation.*

PROOF. (1) (i) Consider the subcategory $\mathcal{X}_{\mathbf{P}_{\Lambda}} := \{C \in \mathrm{Mod}(\Lambda) \mid \mathrm{Ext}^n_{\Lambda}(C, \mathbf{P}_{\Lambda}) = 0, \forall n \geq 1\}$. Obviously $\mathrm{CM}(\mathbf{P}_{\Lambda}) \subseteq \mathcal{X}_{\mathbf{P}_{\Lambda}}$. We claim that the above inclusion is an equality. Let $C \in \mathcal{X}_{\mathbf{P}_{\Lambda}}$ and let $0 \to Y_C \to X_C \to C \to 0$ be a special right Cohen-Macaulay approximation of C. Then Y_C has finite projective dimension. It is not difficult to see that $\mathrm{Ext}^n_{\Lambda}(\mathcal{X}_{\mathbf{P}_{\Lambda}}, \mathbf{P}_{\Lambda}^{<\infty}) = 0, \forall n \geq 1$. Hence the above sequence splits, and we infer that C is Cohen-Macaulay. Therefore $\mathrm{CM}(\mathbf{P}_{\Lambda}) = \mathcal{X}_{\mathbf{P}_{\Lambda}}$. Using this fact we shall show that $\mathrm{CM}(\mathbf{P}_{\Lambda})$ is closed under filtered colimits. Let $\{X_i \mid i \in I\}$ be a filtered system of Cohen-Macaulay modules. Then for any right Λ-module Y, we have Roos's spectral sequence

$$E_2^{p,q} = \varprojlim{}^p \mathrm{Ext}^q(X_i, Y) \implies \mathrm{Ext}^n(\varinjlim X_i, Y)$$

Choosing Y to be a projective module, we have that the spectral sequence collapses. Hence we have isomorphisms: $\varprojlim{}^n \mathrm{Hom}_{\Lambda}(X_i, Y) \xrightarrow{\cong} \mathrm{Ext}^n(\varinjlim X_i, Y), \forall n \geq 1$. Since Λ is left coherent and right perfect we have that any projective right Λ-module is pure injective. This implies that $\varprojlim{}^n \mathrm{Hom}_{\Lambda}(X_i, Y) = 0, \forall n \geq 1$. Therefore $\mathrm{Ext}^n(\varinjlim X_i, Y) = 0, \forall n \geq 1$, so $\varinjlim X_i$ is Cohen-Macaulay. Since $\mathrm{CM}(\mathbf{P}_{\Lambda})$ is closed under filtered colimits, by a result of Enochs, see [**47**], we have that any right Λ-module admits a minimal right Cohen-Macaulay approximation and any right Λ-module admits a minimal left $\mathrm{CM}(\mathbf{P}_{\Lambda})^{\perp} = \mathbf{P}_{\Lambda}^{<\infty}$-approximation.

((ii) Since the finitistic projective dimension of Λ is finite, there exists $d \geq 0$ such that $\mathbf{P}_{\Lambda}^{<\infty}$ coincides with the category $\mathbf{P}_{\Lambda}^{\leq d}$ of modules with projective dimension bounded by d. Since Λ is left coherent and right perfect, by a result of Bass, $\mathbf{P}_{\Lambda}^{\leq d}$ is closed under filtered colimits. Hence by the above mentioned result of Enochs, any right Λ-module admits a minimal right $\mathbf{P}_{\Lambda}^{<\infty}$-approximation and a minimal left CoCohen-Macaulay approximation.

(2) Follows as in (ii), using that the finitistic injective dimension of Λ is finite, and the fact that, over a right Noetherian ring, the category of modules with finite injective dimension bounded above, is closed under filtered colimits. □

The following result shows that over a right Gorenstein ring which is not right perfect, there are right modules not admitting minimal right Cohen-Macaulay approximations. So although part (ii) of Theorem 2.8 holds for right Noetherian rings, part (i) does not hold in general.

PROPOSITION 2.9. *Let* Λ *be a right Gorenstein ring. If any right* Λ*-module admits a minimal right Cohen-Macaulay approximation, then* Λ *is right perfect.*

PROOF. Let F be a flat right Λ-module and let $0 \to Y_F \to X_F \xrightarrow{f_F} F \to 0$ be exact, where f_F is a minimal right Cohen-Macaulay approximation of F. Since, by Wakamatsu's Lemma, minimal approximations are special, we have that Y_F has finite projective dimension. By Theorem 2.2, the finitistic projective dimension of Λ is finite. Then by a result of Jensen, see [**66**], any flat module has finite projective dimension. Hence F has finite projective dimension, and this implies that its right Cohen-Macaulay approximation is projective. In particular f_F is a projective cover.

Since the only flat modules admitting a projective cover are the projectives, we infer that F is projective. We conclude that Λ is right perfect. $\square$

The following concrete example shows that, over a non-trivial right Gorenstein ring, there are modules not admitting a minimal right Cohen-Macaulay approximation.

EXAMPLE. Let G be a group with finite virtual cohomological dimension. Then by [**51**] the group ring $\mathbb{Z}G$ satisfies: $\mathsf{silp}(\mathrm{Mod}(\mathbb{Z}G)) = \mathsf{spli}(\mathrm{Mod}(\mathbb{Z}G)) < \infty$, hence $\mathbb{Z}G$ is right Gorenstein. Since obviously $\mathbb{Z}G$ is not right perfect, by Proposition 2.9 we infer that there exists a right $\mathbb{Z}G$-module C which admits a special Cohen-Macaulay approximation, but C does not admit a minimal right Cohen-Macaulay approximation.

3. The Gorenstein Extension of a Cohen-Macaulay Category

Our aim in this section is to give a procedure for constructing Gorenstein categories. Let $\mathcal{C}$ be an abelian category with enough projective and injective objects. A natural generalization of Gorenstein categories is given by the class of abelian categories $\mathcal{C}$ for which there exists an adjoint pair $(F,G) : \mathcal{C} \to \mathcal{C}$ of endofunctors of $\mathcal{C}$, inducing quasi-inverse equivalences $(F,G) : \mathcal{P}^{<\infty} \xrightarrow{\approx} \mathcal{I}^{<\infty}$ between $\mathcal{P}^{<\infty}$ and $\mathcal{I}^{<\infty}$. Our main result shows that the above quasi-inverse equivalences can be normalized in the trivial extension $\mathcal{C} \ltimes F$ of $\mathcal{C}$ by F, see [**48**], in the following sense. If $\mathcal{P}^{<\infty}(\mathcal{C} \ltimes F)$, resp. $\mathcal{I}^{<\infty}(\mathcal{C} \ltimes F)$, is the full subcategory of $\mathcal{C} \ltimes F$ consisting of all objects with finite projective, resp. injective, dimension, then we have an equality: $\mathcal{P}^{<\infty}(\mathcal{C} \ltimes F) = \mathcal{I}^{<\infty}(\mathcal{C} \ltimes F)$. Then, under additional mild assumptions on $\mathcal{C}$, we show that the trivial extension $\mathcal{C} \ltimes F$ is Gorenstein.

We begin by recalling some basic facts about trivial extensions of abelian categories from [**48**]. Let $\mathcal{C}$ be an abelian category and $F : \mathcal{C} \to \mathcal{C}$ a right exact endofuctor. The *trivial extension* $\mathcal{C} \ltimes F$ of $\mathcal{C}$ by F is defined as follows. The objects of $\mathcal{C} \ltimes F$ are pairs (X, f), where $f : F(X) \to X$ is a morphism in $\mathcal{C}$ such that $F(f) \circ f = 0$. A morphism $\alpha : (X, f) \to (Y, g)$ in $\mathcal{C} \ltimes F$ is a morphism $\alpha : X \to Y$ in $\mathcal{C}$ such that: $F(\alpha) \circ g = f \circ \alpha$. Then the category $\mathcal{C} \ltimes F$ is abelian, and we have an adjoint pair of functors (T, U), where $\mathsf{T} : \mathcal{C} \to \mathcal{C} \ltimes F$ and $\mathsf{U} : \mathcal{C} \ltimes F \to \mathcal{C}$. The functor T is defined as follows. If X is an object in $\mathcal{C}$ and if $\alpha : X \to Y$ is a morphism in $\mathcal{C}$, then:

$$\mathsf{T}(X) = (X \oplus F(X), t_X) \text{ where } t_X = \begin{pmatrix} 0 & 1_{F(X)} \\ 0 & 0 \end{pmatrix}, \quad \text{and} \quad \mathsf{T}(\alpha) = \begin{pmatrix} \alpha & 0 \\ 0 & F(\alpha) \end{pmatrix}.$$

The functor U is defined as follows. If (X, f) is an object in $\mathcal{C} \ltimes F$, then $\mathsf{U}(X, f) = X$, and if $\alpha : (X, f) \to (Y, g)$ is a morphism in $\mathcal{C} \ltimes F$, then $\mathsf{U}(\alpha) = \alpha$. It is not difficult to see that if $\mathcal{C}$ has enough projectives, then $\mathcal{C} \ltimes F$ has enough projectives, and any projective object of $\mathcal{C} \ltimes F$ is a direct summand of an object of the form $\mathsf{T}(P)$ where P is projective in $\mathcal{C}$.

Dually the trivial coextension $G \rtimes \mathcal{C}$ of $\mathcal{C}$ by a left exact endofunctor $G : \mathcal{C} \to \mathcal{C}$ is defined as follows. The objects of $G \rtimes \mathcal{C}$ are pairs (X, f), where $f : X \to G(X)$ is a morphism in $\mathcal{C}$ such that $f \circ G(f) = 0$. A morphism $\alpha : (X, f) \to (Y, g)$ in $G \rtimes \mathcal{C}$ is a morphism $\alpha : X \to Y$ in $\mathcal{C}$ such that $f \circ G(\alpha) = \alpha \circ g$. Then the category $G \rtimes \mathcal{C}$ is abelian, and we have an adjoint pair of functors (U, H), where $\mathsf{H} : \mathcal{C} \to G \rtimes \mathcal{C}$

and $\mathsf{U} : G \rtimes \mathcal{C} \to \mathcal{C}$. The functor H is defined as follows. If X is an object in $\mathcal{C}$ and if $\alpha : X \to Y$ is a morphism in $\mathcal{C}$, then:

$$\mathsf{H}(X) = (G(X) \oplus X, s_X) \text{ where } s_X = \begin{pmatrix} 0 & 1_{G(X)} \\ 0 & 0 \end{pmatrix}, \quad \text{and} \quad \mathsf{H}(\alpha) = \begin{pmatrix} G(\alpha) & 0 \\ 0 & \alpha \end{pmatrix}.$$

The functor U is defined as follows. If (X, f) is an object in $G \rtimes \mathcal{C}$, then $\mathsf{U}(X, f) = X$, and if $\alpha : (X, f) \to (Y, g)$ is a morphism in $G \rtimes \mathcal{C}$, then $\mathsf{U}(\alpha) = \alpha$. It is not difficult to see that if $\mathcal{C}$ has enough injectives, then $G \rtimes \mathcal{C}$ has enough injectives, and any injective object of $G \rtimes \mathcal{C}$ is a direct summand of an object of the form $\mathsf{H}(I)$ where I is injective in $\mathcal{C}$.

REMARK 3.1. (1) If (F, G) is an adjoint pair of endofunctors in $\mathcal{C}$, then, by **[48]**, there exists an isomorphism of categories $\mathcal{C} \ltimes F \xrightarrow{\cong} G \rtimes \mathcal{C}$ induced by the adjoint pair (F, G). We refer to the comprehensive treatment **[48]** for more information on trivial (co)extensions.

(2) If $\mathcal{C} = \mathrm{Mod}(\Lambda)$ and $F = - \otimes_\Lambda \omega$ for some Λ-bimodule, then $\mathcal{C} \ltimes F$ is the module category $\mathrm{Mod}(\Lambda \ltimes \omega)$, where $\Lambda \ltimes \omega$ is the trivial extension of Λ by ω. Recall that $\Lambda \ltimes \omega = \Lambda \oplus \omega$ as abelian groups, and the multiplication is defined by $(\lambda_1, m_1)(\lambda_2, m_2) = (\lambda_1\lambda_2, \lambda_1 m_2 + m_1\lambda_2)$.

After these preparations we can state the following main result of this section.

THEOREM 3.2. *Let $\mathcal{C}$ be an abelian category with enough projective and injective objects. Assume that there exists an adjoint pair (F, G) of endofunctors of $\mathcal{C}$ inducing an equivalence $F : \mathcal{P}^{<\infty}(\mathcal{C}) \xrightarrow{\approx} \mathcal{I}^{<\infty}(\mathcal{C})$ with quasi-inverse $G : \mathcal{I}^{<\infty}(\mathcal{C}) \xrightarrow{\approx} \mathcal{P}^{<\infty}(\mathcal{C})$. Then in the trivial extension $\mathcal{C} \ltimes F \xrightarrow{\cong} G \rtimes \mathcal{C}$ we have:*

$$\mathcal{P}^{<\infty}(\mathcal{C} \ltimes F) = \mathcal{I}^{<\infty}(\mathcal{C} \ltimes F).$$

PROOF. Let $\delta : \mathrm{Id}_\mathcal{C} \to GF$ be the unit and $\varepsilon : FG \to \mathrm{Id}_\mathcal{C}$ the counit of the adjoint pair (F, G). To prove the assertion, it suffices to show that any injective object of $\mathcal{C} \ltimes F$ has finite projective dimension and any projective object of $\mathcal{C} \ltimes F$ has finite injective dimension. Since the injective objects of $G \rtimes \mathcal{C}$ are the direct summands of the objects of the form $\mathsf{H}(I)$ where I is injective in $\mathcal{C}$, and since the projective objects of $\mathcal{C} \ltimes F$ are the direct summands of the objects of the form $\mathsf{T}(P)$ where P is projective in $\mathcal{C}$, it suffices to show that for any projective object P of $\mathcal{C}$ and any injective object of $\mathcal{C}$, the object $\mathsf{H}(I)$ has finite projective dimension and the object $\mathsf{T}(P)$ has finite injective dimension. Using the isomorphism of categories $G \rtimes \mathcal{C} \xrightarrow{\cong} \mathcal{C} \ltimes F$, it is not difficult to see that the object $\mathsf{H}(I)$ has the following description in $\mathcal{C} \ltimes F$:

$$\mathsf{H}(I) = \big(G(I) \oplus I, t_I\big), \text{where } t_I = \begin{pmatrix} 0 & \varepsilon_I \\ 0 & 0 \end{pmatrix} : FG(I) \oplus F(I) \to G(I) \oplus I.$$

Since I is injective, by hypothesis, the counit $\varepsilon_I : FG(I) \to I$ is invertible. Then it is easy to see that we have an isomorphism in $\mathcal{C} \ltimes F$:

$$\begin{pmatrix} 1_{G(I)} & 0 \\ 0 & \varepsilon_I \end{pmatrix} : \mathsf{T}(G(I)) \xrightarrow{\cong} \mathsf{H}(I)$$

Since I is injective, $G(I)$ has finite projective dimension. Let $0 \to P_d \to P_{d-1} \to \cdots \to P_1 \to P_0 \to G(I) \to 0$ be a projective resolution of $G(I)$ in $\mathcal{C}$. Since F is an equivalence restricted to $\mathcal{P}^{<\infty}(\mathcal{C})$, it follows that the complex $0 \to F(P_d) \to F(P_{d-1}) \to \cdots \to F(P_1) \to F(P_0) \to FG(I) \to 0$ is exact. Hence $\mathrm{L}_n F(G(I)) = 0$, $\forall n \geq 1$. By [**22**] this implies that we have an isomorphism:

$$\mathrm{Ext}^n_{\mathcal{C}\ltimes F}[\mathsf{T}(G(I)), (Y, g)] \xrightarrow{\cong} \mathrm{Ext}^n_{\mathcal{C}}[G(I), Y], \ \forall (Y, g) \in \mathcal{C} \ltimes F, \ \forall n \geq 1.$$

Since $G(I)$ has finite projective dimension, we infer that the same is true for $\mathsf{T}(G(I))$. This shows that any injective object of $\mathcal{C} \ltimes F$ has finite projective dimension, in other words: $\mathcal{I}^{<\infty}(\mathcal{C} \ltimes F) \subseteq \mathcal{P}^{<\infty}(\mathcal{C} \ltimes F)$. By duality we have that $\mathcal{P}^{<\infty}(\mathcal{C} \ltimes F) \subseteq \mathcal{I}^{<\infty}(\mathcal{C} \ltimes F)$. Hence $\mathcal{P}^{<\infty}(\mathcal{C} \ltimes F) = \mathcal{I}^{<\infty}(\mathcal{C} \ltimes F)$. □

REMARK 3.3. The above proof shows that if there exists an adjoint pair of endofunctors $(F, G) : \mathcal{C} \to \mathcal{C}$ inducing quasi-inverse equivalences between $\mathcal{P}$ and $\mathcal{I}$, then the trivial extension $\mathcal{C} \ltimes F$ is Frobenius.

Recall from [**9**] that an Artin algebra Λ is called *Cohen-Macaulay* if there an adjoint pair of functors $(F, G) : \mathrm{mod}(\Lambda) \to \mathrm{mod}(\Lambda)$ which induces an equivalence between $\mathcal{P}^{<\infty}_{\Lambda}$ and $\mathcal{I}^{<\infty}_{\Lambda}$. Hence it is natural to make the following definition.

DEFINITION 3.4. An abelian category $\mathcal{C}$ with enough projective and injective objects is called **Cohen-Macaulay**, if there exists an adjoint pair (F, G) of endofunctors of $\mathcal{C}$ which induces an equivalence $F : \mathcal{P}^{<\infty}(\mathcal{C}) \xrightarrow{\approx} \mathcal{I}^{<\infty}(\mathcal{C})$ with quasi-inverse $G : \mathcal{I}^{<\infty}(\mathcal{C}) \xrightarrow{\approx} \mathcal{P}^{<\infty}(\mathcal{C})$. In this case the adjoint pair of functors (F, G) is called **dualizing** for $\mathcal{C}$.

REMARK 3.5. If $\mathcal{C}$ is a Gorenstein category, then $\mathcal{C}$ is Cohen-Macaulay with dualizing adjoint pair of functors $(\mathrm{Id}_{\mathcal{C}}, \mathrm{Id}_{\mathcal{C}})$.

If $\mathcal{C}$ is a Cohen-Macaulay category, then the following result, which will be useful later, shows that several important dimensions related to $\mathcal{C}$ are equal.

PROPOSITION 3.6. *Let $\mathcal{C}$ be a Cohen-Macaulay abelian category with dualizing adjoint pair of functors (F, G). Then we have the following:*

$$\mathsf{spli}(\mathcal{C} \ltimes F) = \mathrm{FPD}(\mathcal{C}) = \mathrm{FID}(\mathcal{C}) = \mathsf{silp}(\mathcal{C} \ltimes F) =$$
$$= \sup\{\mathrm{pd}\, G(I) \mid I \in \mathcal{I}\} = \sup\{\mathrm{id}\, F(P) \mid I \in \mathcal{P}\} \leq \min\{\mathsf{silp}(\mathcal{C}), \mathsf{spli}(\mathcal{C})\}.$$

PROOF. We first show that $\mathrm{FPD}(\mathcal{C}) = \sup\{\mathrm{pd}\, G(I) \mid I \in \mathcal{I}\}$. Since for any injective object I, the object $G(I)$ has finite projective dimension, we have $\mathrm{FPD}(\mathcal{C}) \geq \sup\{\mathrm{pd}\, G(I) \mid I \in \mathcal{I}\}$. Let $\sup\{\mathrm{pd}\, G(I) \mid I \in \mathcal{I}\} = d$. If $d = \infty$, then $\mathrm{FPD}(\mathcal{C}) = \sup\{\mathrm{pd}\, G(I) \mid I \in \mathcal{I}\}$. Assume that $d < \infty$, and let X be in $\mathcal{C}$ with $\mathrm{pd}\, X < \infty$. Since $F(X)$ has finite injective dimension, there exists an exact sequence $0 \to F(X) \to I^0 \to \cdots \to I^k \to 0$ where the I^j are injective. Since G is exact in $\mathcal{I}^{<\infty}$, we have an exact sequence $0 \to GF(X) \to G(I^0) \to \cdots \to G(I^k) \to 0$ and $X \cong GF(X)$. Since $\mathrm{pd}\, G(I^j) \leq d$, $\forall j \geq 0$, we infer directly that $\mathrm{pd}\, X \leq d$. This shows that $\mathrm{FPD}(\mathcal{C}) \leq d$. Hence $\mathrm{FPD}(\mathcal{C}) = \sup\{\mathrm{pd}\, G(I) \mid I \in \mathcal{I}\}$. From the proof of Theorem 3.2 we have $\sup\{\mathrm{pd}\, G(I) \mid I \in \mathcal{I}\} = \mathsf{spli}(\mathcal{C} \ltimes F)$. Hence $\mathrm{FPD}(\mathcal{C}) = \sup\{\mathrm{pd}\, G(I) \mid I \in \mathcal{I}\} = \mathsf{spli}(\mathcal{C} \ltimes F)$, and by duality we have: $\mathrm{FID}(\mathcal{C}) = \sup\{\mathrm{id}\, F(P) \mid P \in \mathcal{P}\} = \mathsf{silp}(\mathcal{C} \ltimes F)$. By Theorem 3.2 we have $\mathcal{P}^{<\infty}(\mathcal{C} \ltimes F) = \mathcal{I}^{<\infty}(\mathcal{C} \ltimes F)$. Hence by part (iv) of Proposition 1.3, we have $\mathsf{spli}(\mathcal{C} \ltimes F) = \mathsf{silp}(\mathcal{C} \ltimes F)$. Finally by part (i) of the same proposition it follows that: $\mathrm{FPD}(\mathcal{C}) = \mathrm{FPD}(\mathcal{C}) \leq \min\{\mathsf{spli}(\mathcal{C}), \mathsf{slilp}(\mathcal{C})\}$ and this completes the proof. □

We have the following direct consequence.

COROLLARY 3.7. *Let $\mathcal{C}$ be a Cohen-Macaulay abelian category and let (F, G) be the dualizing adjoint pair of functors for $\mathcal{C}$. Then the following are equivalent:*

(i) $\mathrm{FPD}(\mathcal{C}) < \infty$ *or* $\mathrm{FID}(\mathcal{C}) < \infty$.

(ii) $\mathcal{C} \ltimes F$ *is Gorenstein.*

In particular if $\mathcal{C}$ is Gorenstein, then the categories $\mathcal{C} \ltimes F$ and $\mathcal{C} \ltimes \mathrm{Id}_{\mathcal{C}}$ are Gorenstein.

EXAMPLE. Let Λ be a (right) Gorenstein ring. Since $\Lambda \ltimes \Lambda \xrightarrow{\cong} \Lambda[t]/(t^2)$ and since $\mathrm{Mod}(\Lambda \ltimes \Lambda) = \mathrm{Mod}(\Lambda) \ltimes \mathrm{Id}_{\mathrm{Mod}(\Lambda)}$, we infer that $\Lambda[t]/(t^2)$ is (right) Gorenstein.

If the Cohen-Macaulay category $\mathcal{C}$ admits exact products or coproducts, then it is easy to see that the same is true for the trivial extension $\mathcal{C} \ltimes F$. Therefore we have the following consequence of Propositions 2.4 and 3.6 and Corollary 3.7, which shows in particular that $\mathcal{C}$ has finite finitistic projective and injective dimension.

COROLLARY 3.8. *Let $\mathcal{C}$ be a Cohen-Macaulay abelian category and let (F, G) be the dualizing adjoint pair of functors for $\mathcal{C}$. If $\mathcal{C}$ admits exact (co)products, or if $\mathcal{C}$ has a finite number of non-isomorphic simple objects and any object has finite length, then* $\mathrm{FPD}(\mathcal{C}) = \mathrm{FID}(\mathcal{C}) < \infty$ *and the trivial extension $\mathcal{C} \ltimes F$ is Gorenstein.*

4. Cohen-Macaulay Categories and (Co)Torsion Pairs

We have seen that in a Gorenstein category there exist nicely behaved (co)torsion pairs induced by (Co)Cohen-Macaulay objects. Since a Cohen-Macaulay category can be regarded as a generalization of a Gorenstein category, it is natural to ask if there exist cotorsion pairs induced by suitable subcategories of (relative) (Co)Cohen-Macaulay objects in a Cohen-Macaulay category. We devote this section to this question and its module-theoretic concequences.

Let $\mathcal{C}$ be an abelian category with exact infinite products and coproducts. We recall that an object T in $\mathcal{C}$ is called *product-complete* if $\mathrm{Add}(T) = \mathrm{Prod}(T)$. Since $\mathrm{Add}(T)$ is always contravariantly finite and $\mathrm{Prod}(T)$ is always covariantly finite, it follows that for a product-complete object T the category $\mathrm{Add}(T) = \mathrm{Prod}(T)$ is functorially finite, hence the stable category $\mathcal{C}/\mathrm{Add}(T) = \mathcal{C}/\mathrm{Prod}(T)$ is pretriangulated.

EXAMPLE. If Λ is a ring, then Λ is product-complete iff Λ is left coherent and right perfect. If J is an injective cogenerator of $\mathrm{Mod}(\Lambda)$, then J is product-complete iff Λ is right Noetherian.

The following result shows that there are nicely behaved (co)torsion pairs in a Cohen-Macaulay category $\mathcal{C}$, provided that $\mathcal{C}$ admits a product-complete projective generator or injective cogenerator.

THEOREM 4.1. *Let $\mathcal{C}$ be a Cohen-Macaulay abelian category with exact products and coproducts, and let (F, G) be the dualizing adjoint pair of functors for $\mathcal{C}$.*

(α) *If $\mathcal{C}$ admits a product-complete projective generator and F preserves products, then there exists a product-complete object $\omega \in \mathcal{C}$ such that:*

(i) $(\mathrm{CM}(\omega), \mathcal{I}^{<\infty})$ *is a resolving cotorsion pair in $\mathcal{C}$ such that:* $\widehat{\mathrm{Prod}(\omega)} = \mathcal{I}^{<\infty}$ *and* $\mathrm{Prod}(\omega) = \mathcal{I}^{<\infty} \cap \mathrm{CM}(\omega)$.

(ii) $\big(\mathrm{CM}(\omega)/\operatorname{Prod}(\omega), \mathcal{I}^{<\infty}/\operatorname{Prod}(\omega)\big)$ *is a hereditary torsion pair in the stable pretriangulated category* $\mathcal{C}/\operatorname{Prod}(\omega)$.

(β) *If* $\mathcal{C}$ *admits a product-complete injective cogenerator and* G *preserves coproducts, then there exists a product-complete object* $T \in \mathcal{C}$ *such that:*

(i) $(\mathcal{P}^{<\infty}, \mathrm{CoCM}(T))$ *is a coresolving cotorsion pair in* $\mathcal{C}$ *such that:* $\widetilde{\operatorname{Add}(T)} = \mathcal{P}^{<\infty}$ *and* $\operatorname{Add}(T) = \mathcal{P}^{<\infty} \cap \mathrm{CoCM}(T)$).

(ii) $\big(\mathcal{P}^{<\infty}/\operatorname{Add}(T), \mathrm{CoCM}(T)/\operatorname{Add}(T)\big)$ *is a cohereditary torsion pair in the stable pretriangulated category* $\mathcal{C}/\operatorname{Add}(T)$).

PROOF. (α) Let Q be a product-complete projective generator of $\mathcal{C}$. Setting $\omega := F(Q)$, it follows that $\operatorname{id}\omega = d < \infty$ and therefore $\operatorname{Prod}(\omega) \subseteq \mathcal{I}^{<\infty}$. This implies that $\widehat{\operatorname{Prod}(\omega)} \subseteq \mathcal{I}^{<\infty}$. Now let Y be an object of finite injective dimension. Then $G(Y)$ has finite projective dimension, hence there exists an exact sequence $0 \to P_t \to \cdots \to P_1 \to P_0 \to G(Y) \to 0$ where the P_i are projective. Since Q is a product-complete projective generator, each P_i lies in $\operatorname{Add}(Q) = \operatorname{Prod}(Q)$. Since F preserves products and is exact restricted to $\mathcal{P}^{<\infty}$, we have an exact sequence $0 \to \omega_t \to \cdots \to \omega_1 \to \omega_0 \to FG(Y) \to 0$ in $\mathcal{C}$ where each ω_i lies in $\operatorname{Prod}(\omega)$. Since $FG(Y) \xrightarrow{\cong} Y$ it follows that Y lies in $\widehat{\operatorname{Prod}(\omega)}$. We infer that $\widehat{\operatorname{Prod}(\omega)} = \mathcal{I}^{<\infty}$. Note that since F preserves products and coproducts and $\operatorname{Add}(Q) = \operatorname{Prod}(Q)$, it follows that $\operatorname{Add}(\omega) = \operatorname{Prod}(\omega)$, hence ω is product complete. Consider now the category $\mathcal{X}_\omega := \{C \in \mathcal{C} \mid \operatorname{Ext}^n(C,\omega) = 0, \forall n \geq 1\}$ which contains $\mathrm{CM}(\omega)$. Let $X \in \mathcal{X}_\omega$ and let $0 \to X \xrightarrow{\mu} I \to \Sigma(X) \to 0$ be exact where I is injective. By the above argument there exists a short exact sequence $(E) : 0 \to K \to \omega_0 \to I \to 0$ where the object K lies in $\widehat{\operatorname{Prod}(\omega)}$ and the object ω_0 lies in $\operatorname{Prod}(\omega)$. Forming the pull-back $\mu(E)$ of the extension (E) along the morphism μ and using that $\operatorname{Ext}^n(\mathcal{X}_\omega, \widehat{\operatorname{Prod}(\omega)}) = 0, \forall n \geq 1$, we infer that the extension $\mu(E)$ splits. Hence X factors through ω_0. Since μ is monic, there exists a short exact sequence $0 \to X \xrightarrow{\mu} \omega_0 \to X' \to 0$. Since $\operatorname{Prod}(\omega)$ is covariantly finite, there exists a short exact sequence $0 \to X \xrightarrow{g} \omega^X \to X' \to 0$ where ω^X is a left $\operatorname{Prod}(\omega)$-approximation of X. Applying $\mathcal{C}(-,\omega)$ to this sequence, it follows directly that X' lies in $\mathcal{X}_\omega$. Hence $\operatorname{Prod}(\omega)$ is an injective cogenerator of $\mathcal{X}_\omega$. This implies that $\mathcal{X}_\omega = \mathrm{CM}(\omega)$. Since $\operatorname{id}\omega = d < \infty$, the dth syzygy object $\Omega^d(C)$ of any object $C \in \mathcal{C}$ lies in $\mathrm{CM}(\omega)$. Hence $\widehat{\mathrm{CM}(\omega)} = \mathcal{C}$. Then by Proposition V.4.6 we have a cotorsion pair $(\mathrm{CM}(\omega), \widehat{\operatorname{Prod}(\omega)})$ in $\mathcal{C}$ and as was shown above $\widehat{\operatorname{Prod}(\omega)} = \mathcal{I}^{<\infty}$. Obviously $\operatorname{Prod}(\omega) = \mathrm{CM}(\omega) \cap \mathcal{I}^{<\infty}$. Since $\operatorname{Prod}(\omega)$ is functorially finite, it follows that the stable categorry $\mathcal{C}/\operatorname{Prod}(\omega)$ is pretriangulated. Then by Theorem V.3.7 we have that $\big(\mathrm{CM}(\omega)/\operatorname{Prod}(\omega), \mathcal{I}^{<\infty}/\operatorname{Prod}(\omega)\big)$ is a torsion pair in $\mathcal{C}/\operatorname{Prod}(\omega)$. Finally it is easy to see, using that $\mathrm{CM}(\omega)$ is resolving, that the torsion pair is hereditary. The proof of part (β) is dual and is left to the reader. □

REMARK 4.2. The above result holds with the same proof, if $\mathcal{C}$ is a Krull-Schmidt Cohen-Macaulay category which admits a projective object P such that $\mathcal{P} = \operatorname{add}(P)$ and/or an injective object I such that $\mathcal{I} = \operatorname{add}(I)$.

We have seen that any Gorenstein category is Cohen-Macaulay in an obvious way. It is natural to ask when a Cohen-Macaulay category is Gorenstein. In this connection we have the following result.

PROPOSITION 4.3. *Let $\mathcal{C}$ be an abelian category with exact products or coproducts. If $\mathcal{C}$ admits a product-complete projeective generator or injective cogenerator, then the following are equivalent.*

(i) *$\mathcal{C}$ is Gorenstein.*
(ii) *$\mathcal{C}$ is Cohen-Macaulay and $\mathsf{silp}(\mathcal{C}) < \infty$.*
(iii) *$\mathcal{C}$ is Cohen-Macaulay and $\mathsf{spli}(\mathcal{C}) < \infty$.*

PROOF. Obviously (i) implies (ii) and (iii). Assume now that (ii) holds and let (F, G) be the dualizing adjoint pair of functors for $\mathcal{C}$. Also Q be the product-complete projective generator and let $(\mathrm{CM}(\omega), \widehat{\mathrm{Prod}(\omega)})$ be the cotorsion pair in $\mathcal{C}$ constructed in Theorem 4.1, where $\omega = F(Q)$. Since $\mathsf{silp}(\mathcal{C}) < \infty$, we have $\mathcal{P} \subseteq \widehat{\mathrm{Prod}(\omega)}) = \mathcal{I}^{<\infty}$. Since $\mathrm{CM}(\omega)$ contains $\mathcal{P}$, we infer that $\mathcal{P} \subseteq \mathrm{CM}(\omega) \cap \widehat{\mathrm{Prod}(\omega)} = \mathrm{Prod}(\omega)$. Now let $0 \to K \to P \to \omega \to 0$ be exact with P projective. Since P has finite injective dimension, it follows that K lies in $\mathcal{I}^{<\infty}$. This implies that the above sequence splits and therefore ω is projective. Hence $\omega \in \mathcal{P} = \mathrm{Add}(Q) = \mathrm{Prod}(Q)$. This implies that $\mathrm{Prod}(\omega) \subseteq \mathrm{Prod}(Q) = \mathcal{P}$. It follows that $\mathrm{Prod}(\omega) = \mathcal{P}$, hence $\mathcal{I}^{<\infty} = \widehat{\mathrm{Prod}(\omega)} = \widehat{\mathcal{P}} = \mathcal{P}^{<\infty}$. Then by Theorem 2.2 we conclude that $\mathcal{C}$ is Gorenstein. The proof that (iii) implies (i) is similar and is left to the reader. □

We close this section by discussing the module-theoretic interpretations of the above results. We say that a ring Λ is *right Cohen-Macaulay* if there exists a bimodule ${}_\Lambda\omega_\Lambda$, such that the adjoint pair $(- \otimes_\Lambda \omega, \mathrm{Hom}_\Lambda(\omega, -))$ is dualizing for $\mathrm{Mod}(\Lambda)$, that is, the functor $- \otimes_\Lambda \omega : \mathbf{P}_\Lambda^{<\infty} \to \mathbf{I}_\Lambda^{<\infty}$ is an equivalence with quasi-inverse the functor $\mathrm{Hom}_\Lambda(\omega_\Lambda, -) : \mathbf{I}_\Lambda^{<\infty} \to \mathbf{P}_\Lambda^{<\infty}$. Prominent examples include quasi-Frobenius rings, rings of finite global dimension, and more generally Gorenstein rings. Of course the classical examples are a commutative local Noetherian Cohen-Macaulay ring and, its non-commutative analogue, a Cohen-Macaulay Artin algebra, see [**9**], [**11**]. We have the following consequence of Theorem 4.1 which generalizes results of [**9**], [**11**].

COROLLARY 4.4. *Let Λ be a right Cohen-Macaulay ring and $(F = - \otimes_\Lambda \omega, G = \mathrm{Hom}_\Lambda(\omega_\Lambda, -))$ a dualizing adjoint pair of functors for $\mathrm{Mod}(\Lambda)$. If ${}_\Lambda\omega$ and ω_Λ are finitely presented and Λ is left coherent and right perfect or right Noetherian, then the trivial extension $\Lambda \ltimes \omega$ is right Gorenstein. Moreover if J is an injective cogenerator, then setting $T = \mathrm{Hom}_\Lambda(\omega, J)$, we have cotorsion pairs in $\mathrm{Mod}(\Lambda)$:*

$$\big(\mathrm{CM}(\omega), \mathbf{I}_\Lambda^{<\infty}\big) \quad \text{and} \quad \big(\mathbf{P}_\Lambda^{<\infty}, \mathrm{CoCM}(T)\big)$$

with $\mathbf{I}_\Lambda^{<\infty} = \widehat{\mathrm{Prod}(\omega)}$ and $\mathbf{P}_\Lambda^{<\infty} = \widetilde{\mathrm{Add}(T)}$. Moreover there exist torsion pairs

$$\big(\mathrm{CM}(\omega)/\mathrm{Prod}(\omega), \mathbf{I}_\Lambda^{<\infty}/\mathrm{Prod}(\omega)\big) \quad \text{and} \quad \big(\mathbf{P}_\Lambda^{<\infty}/\mathrm{Add}(T), \mathrm{CoCM}(T)/\mathrm{Add}(T)\big)$$

in the stable pretriangulated categories $\mathrm{Mod}(\Lambda)/\mathrm{Prod}(\omega)$ and $\mathrm{Mod}(\Lambda)/\mathrm{Add}(T)$ respectively.

REMARK 4.5. It is not difficult to see that the analogue of Proposition 2.8 holds for a Cohen-Macaulay ring which satisfies the assumptions of Corollary 4.4. In particular any module admits a minimal right $\mathrm{CM}(\omega)$-approximation and a minimal left $\mathrm{CoCM}(T)$-approximation, and a minimal left $\mathbf{I}_\Lambda^{<\infty}$-approximation and a minimal right $\mathbf{P}_\Lambda^{<\infty}$-approximation.

Let Λ be a ring which admits a Morita self-duality, for instance an Artin algebra or a quasi-Frobenius ring. Then $\mathrm{mod}(\Lambda)$ is abelian with enough projective and injective objects, and it is easy to see that Λ is right Gorenstein or Cohen-Macaulay iff Λ is left Gorenstein or Cohen-Macaulay. Recall from **[9]** that a cotilting, resp. tilting, Λ-module T is called *strong*, if $\widehat{\mathrm{add}(T)} = \mathcal{I}_\Lambda^{<\infty}$, resp. $\widetilde{\mathrm{add}(T)} = \mathcal{P}_\Lambda^{<\infty}$. A bimodule ${}_\Lambda\omega_\Lambda$ is called *dualizing* for Λ, if ω_Λ and ${}_\Lambda\omega$ are (finitely presented) strong cotilting modules and the natural ring map $\Lambda \to \mathrm{End}_\Lambda(\omega)$ is an isomorphism.

We have the following consequence of the above results which extends Proposition 1.3 of **[11]** and Corollary 4.14 of **[22]** from finitely generated over Artin algebras to arbitrary modules over rings with Morita self-duality.

COROLLARY 4.6. *Let Λ be a ring with Morita self-duality. Then Λ is Cohen-Macaulay iff there exists a dualizing Λ-bimodule ${}_\Lambda\omega_\Lambda$. Moreover the trivial extension $\Lambda \ltimes \omega$ is a Gorenstein ring with Morita self-duality.*

PROOF. Let $D : \mathrm{mod}(\Lambda) \to \mathrm{mod}(\Lambda^{op})$ be a Morita duality. If (F, G) is a dualizing adjoint pair for $\mathrm{Mod}(\Lambda)$, then $F = -\otimes\omega_\Lambda$, $G = \mathrm{Hom}_\Lambda(\omega_\Lambda, -)$ and we have seen that $F(\Lambda) = \omega_\Lambda$ is a strong cotilting module and $G(D(\Lambda)) = \mathrm{Hom}_\Lambda(\omega, D(\Lambda))$ is a strong tilting module, hence $DG(D(\Lambda)) = {}_\Lambda\omega$ is a strong cotilting module. Moreover we have $\Lambda = \mathrm{Hom}_\Lambda(F(\Lambda), F(\Lambda)) \cong \mathrm{Hom}(\omega_\Lambda, \omega_\Lambda) = \mathrm{End}_\Lambda(\omega_\Lambda)$, so ${}_\Lambda\omega_\Lambda$ is a dualizing bimodule. Conversely let ${}_\Lambda\omega_\Lambda$ be a dualizing bimodule. Then by Proposition 6.6 of **[11]**, which works in our setting, we have that the functors $F = -\otimes\omega_\Lambda$, $G = \mathrm{Hom}_\Lambda(\omega_\Lambda, -)$ induce quasi-inverse equivalences between $\mathcal{P}_\Lambda^{<\infty}$ and $\mathcal{I}_\Lambda^{<\infty}$. Since Λ admits a Morita self-duality and ω_Λ and ${}_\Lambda\omega$ are finitely presented, it is easy to see that ω_Λ and ${}_\Lambda\omega$ are product-complete. It follows easily from this that the equivalence between $\mathcal{P}_\Lambda^{<\infty}$ and $\mathcal{I}_\Lambda^{<\infty}$ induced by (F, G), extends to an equivalence between $\mathbf{P}_\Lambda^{<\infty}$ and $\mathbf{I}_\Lambda^{<\infty}$. Hence Λ is Cohen-Macaulay. The last assertion follows from Theorem 4.1. □

We close this section with the following connection between Cohen-Macaulay and Gorenstein rings which is a consequence of Proposition 4.3.

COROLLARY 4.7. *If Λ is a ring with Morita self-duality, then the following are equivalent:*

(i) *Λ is Gorenstein.*
(ii) *Λ is Cohen-Macaulay and* $\mathrm{id}\,\Lambda_\Lambda < \infty$.
(iii) *Λ is Cohen-Macaulay and* $\mathrm{id}\,{}_\Lambda\Lambda < \infty$.

CHAPTER VIII

Torsion Pairs and Closed Model Structures

Our aim in this chapter is to present and investigate an interesting connection between torsion and cotorsion pairs on one hand, and closed model structures in the sense of Quillen [**88**] on the other hand. Closed model structures and the associated notion of a closed model category were introduced by Quillen in the late sixties as the proper conceptual framework for doing homotopy theory in more general categories than the category of topological spaces. Since (co)torsion pairs can be regarded as a generalized form of tilting theory, the above connection shows that tilting theory admits a natural homotopical interpretation in the more general context of closed model categories. As a byproduct we obtain new classes of torsion pairs, since closed model structures are omnipresent in algebra and topology. Our main results in this chapter give a classification of cotorsion pairs in an abelian category and a classification of torsion pairs in the stable category of an abelian category, in terms of closed model structures. As a consequence we can obtain a classification of (co)tilting modules in terms of closed model structures. This is investigated more closely in [**27**].

A similar approach to the connection between closed model structures and cotorsion pairs was developed independently by Mark Hovey in a different context, see [**65**]. However our results are quite different.

1. Preliminaries on Closed Model Categories

In this section we collect some basic concepts and results concerning closed model categories.

Let $\mathcal{C}$ be an additive category. Recall that a *closed model structure* on $\mathcal{C}$ in the sense of Quillen is a triple $(\mathfrak{C}, \mathfrak{F}, \mathfrak{W})$ of classes of morphisms, satisfying the following axioms [Mi], $i = 2, ..., 5$. The morphisms in $\mathfrak{F}$ are the *fibrations*, the morphisms in $\mathfrak{C}$ are the *cofibrations*, and the morphisms in $\mathfrak{W}$ are the *weak equivalences*. Similarly we denote by $\mathfrak{T}\mathfrak{F} = \mathfrak{F} \cap \mathfrak{W}$ the class of *trivial fibrations* and by $\mathfrak{T}\mathfrak{C} = \mathfrak{C} \cap \mathfrak{W}$ the class of *trivial cofibrations*.

[M2] [Two out of three axiom] If f, g are composable morphisms in $\mathcal{C}$ and two of f, g and $f \circ g$ are weak equivalences, then so is the third.

[M3] [Retract axiom] If f and g are morphisms in $\mathcal{C}$ and f is a retract of g in the category of morphisms $\mathcal{C}^2$, and if g is a weak equivalence, fibration, or a cofibration, then so is f.

[M4] [Lifting axiom] Consider the following commutative diagram in $\mathcal{C}$:

$$\begin{array}{ccc} A & \xrightarrow{\alpha} & C \\ {\scriptstyle i}\big\downarrow & & \big\downarrow{\scriptstyle p} \\ B & \xrightarrow{\beta} & D \end{array}$$

If either i is a cofibration and p is a trivial fibration, or i is a trivial cofibration and p is a fibration, then there exists a morphism $\lambda : B \to C$ such that $i \circ \lambda = \alpha$ and $\lambda \circ p = \beta$.

[M5] [Factorization axiom] Any morphism α in $\mathcal{C}$ admits a factorization $\alpha = i \circ g$, where i is a cofibration and g is a trivial fibration, and a factorization $\alpha = f \circ p$, where f is a trivial cofibration and p is a fibration.

A *closed model category* is an additive category $\mathcal{C}$ equipped with a closed model structure $(\mathfrak{C}, \mathfrak{F}, \mathfrak{W})$, which satisfies the following axiom:

[M1] $\mathcal{C}$ has kernels and cokernels.

Note that, contrary to the recent treatments of closed model categories [**64**], [**62**], we don't require the existence of all small limits and colimits or functoriality of the factorizations in the axioms [M1] and [M5] respectively, as have been formulated in [**64**], [**62**]. This approach has the advantage that our results are applicable to a wider class of categories which are of interest in representation theory. As an important example we mention the category of finitely presented modules over an Artin algebra.

Consider an additive closed model category $\mathcal{C}$ with closed model structure $(\mathfrak{C}, \mathfrak{F}, \mathfrak{W})$. We denote by Cof the category of *cofibrant* objects, that is the objects C such that $0 \to C$ is a cofibration, and by Fib the category of *fibrant* objects, that is the objects F such that $F \to 0$ is a fibration. Of special importance to us are the following subcategories:

$$\mathsf{TCof} = \{X \in \mathcal{C} \mid 0 \to X \text{ is a trivial cofibration}\} \subseteq \mathsf{Cof}$$

$$\mathsf{TFib} = \{Y \in \mathcal{C} \mid Y \to 0 \text{ is a trivial fibration}\} \subseteq \mathsf{Fib}$$

$$\omega_c = \mathsf{Cof} \cap \mathsf{TFib} \quad \text{and} \quad \omega_f = \mathsf{TCof} \cap \mathsf{Fib}.$$

We call the objects in TCof *trivially cofibrant* and the objects in TFib *trivially fibrant*. Note that $\omega_c, \omega_f \subseteq \mathsf{Cof} \cap \mathsf{Fib}$, that is the objects of the subcategories ω_c, ω_f are fibrant and cofibrant. By the Retract axiom [M2] it follows easily that all the categories $\mathsf{Cof}, \mathsf{Fib}, \mathsf{TCof}, \mathsf{TFib}, \omega_c, \omega_f$ are closed under direct summands.

In what follows we shall use repeatedly the following easy result. For a proof we refer to [**64**], [**62**].

LEMMA 1.1. (i) *If $p : B \to C$ is a trivial fibration, then any morphism $\gamma : X \to C$ from a cofibrant object X to C factors through p.*

(ii) *If $i : A \to B$ is a trivial cofibration, then any morphism $\alpha : A \to Y$ from A to a fibrant object Y factors through i.*

(iii) *Let $0 \to F \to A \xrightarrow{p} B$ be an exact sequence in $\mathcal{C}$. If p is a fibration then $F \in \mathsf{Fib}$, and if p is a trivial fibration, then $F \in \mathsf{TFib}$.*

(iv) *Let $A \xrightarrow{i} B \to C \to 0$ be an exact sequence in $\mathcal{C}$. If i is a cofibration then $C \in \mathsf{Cof}$, and if i is a trivial cofibration, then $C \in \mathsf{TCof}$.*

A very useful property of model categories is that any one of the classes of morphisms which occur in the pairs $\{\mathfrak{C}, \mathfrak{T}\mathfrak{F}\}$, $\{\mathfrak{T}\mathfrak{C}, \mathfrak{F}\}$, characterizes the other via a lifting property of an appropriate commutative square. Moreover the class $\mathfrak{W}$ of weak equivalences is determined by the pair $\{\mathfrak{T}\mathfrak{C}, \mathfrak{T}\mathfrak{F}\}$. To make this precise we need the following definition.

Consider a commutative diagram in $\mathcal{C}$:

$$\begin{array}{ccc} A & \xrightarrow{\alpha} & C \\ {\scriptstyle i}\big\downarrow & & {\scriptstyle p}\big\downarrow \\ B & \xrightarrow{\beta} & D \end{array}$$

DEFINITION 1.2. A **lifting** of the above square is a morphism $\lambda : B \to C$ such that $i \circ \lambda = \alpha$ and $\lambda \circ p = \beta$. If such a lifting exists, then the morphism i is said to have *the left lifting property* with respect to p and p is said to have *the right lifting property* with respect to i.

For the easy proof of the following we refer to [**64**], [**62**].

LEMMA 1.3. (i) *A morphism is a cofibration iff it has the left lifting property with respect to all trivial fibrations.*
(ii) *A morphism is a trivial cofibration iff it has the left lifting property with respect to all fibrations.*
(iii) *A morphism is a fibration iff it has the right lifting property with respect to all trivial cofibrations.*
(iv) *A morphism is a trivial fibration iff it has the right lifting property with respect to all cofibrations.*
(v) *A morphism is a weak equivalence iff it can be factored as a trivial cofibration followed by a trivial fibration.*

For all unexplained concepts and results we use in this chapter, we refer to the standard references [**88**], [**64**], [**62**].

2. Closed Model Structures and Approximation Sequences

An interesting problem for closed model categories is when a closed model structure is determined by the subcategories of (trivial) fibrant or cofibrant objects. This question turns out to be connected with the existence of (co)torsion pairs. In this section we make preparations for this investigation, by first showing that the pairs $(\mathsf{Cof}, \mathsf{TFib})$ and $(\mathsf{TCof}, \mathsf{Fib})$ have properties similar to those of cotorsion pairs. We give a different description of these subcategories in terms of the morphism classes, which suggests how to define closed model structures starting from the subcategories.

We fix throughout an additive closed model category $\mathcal{C}$ with closed model structure $(\mathfrak{C}, \mathfrak{F}, \mathfrak{W})$. In what follows by a *right cofibrant*, resp. *trivially cofibrant*, *approximation* of an object C of $\mathcal{C}$ we mean a right Cof-approximation, resp. TCof-approximation, of C. Similarly by a *left fibrant*, resp. *trivially fibrant*, *approximation* of an object C of $\mathcal{C}$ we mean a left Fib-approximation, resp. TFib-approximation, of C.

We begin with the following basic result which shows that there exists a remarkable similarity between certain exact sequences induced by a closed model category structure and the approximation sequences induced by a cotorsion pair.

PROPOSITION 2.1. *Let $\mathcal{C}$ be a closed model category. Then we have the following.*

(i) *The category* Cof *of cofibrant objects is contravariantly finite in $\mathcal{C}$, and for any object A in $\mathcal{C}$ there exists an exact sequence*

$$0 \longrightarrow Y_A \xrightarrow{g_A} C_A \xrightarrow{f_A} A$$

where f_A is a right cofibrant approximation of A (and a trivial fibration) and the object $Y_A \in \mathsf{TFib}$.

(ii) *The category* Fib *of fibrant objects is covariantly finite in $\mathcal{C}$ and for any object A in $\mathcal{C}$ there exists an exact sequence*

$$A \xrightarrow{g^A} F^A \xrightarrow{f^A} X^A \longrightarrow 0$$

where g^A is a left fibrant approximation of A (and a trivial cofibration) and the object $X^A \in \mathsf{TCof}$.

(iii) *The category* TCof *is contravariantly finite in $\mathcal{C}$, and for any object A in $\mathcal{C}$ there exists an exact sequence*

$$0 \longrightarrow F_A \xrightarrow{\psi_A} X_A \xrightarrow{\phi_A} A$$

where ϕ_A is a right trivially cofibrant approximation of A (and a fibration) and the object $F_A \in \mathsf{Fib}$.

(iv) *The category* TFib *is covariantly finite in $\mathcal{C}$ and for any object A in $\mathcal{C}$ there exists an exact sequence*

$$A \xrightarrow{\psi^A} Y^A \xrightarrow{\phi^A} C^A \longrightarrow 0$$

where ψ^A is a left trivially fibrant approximation of A (and a cofibration) and the object $C^A \in \mathsf{Cof}$.

(v) *The following orthogonality relations hold:*

$$(\mathsf{Cof}/\omega_c, \mathsf{TFib}/\omega_c) = 0 \quad \textit{and} \quad (\mathsf{TCof}/\omega_f, \mathsf{Fib}/\omega_f) = 0$$

i.e. any map $C \to Y$ with $C \in \mathsf{Cof}$ and $Y \in \mathsf{TFib}$, factors through an object in ω_c, and any map $X \to F$ with $F \in \mathsf{Fib}$ and $X \in \mathsf{TCof}$, factors through an object in ω_f.

PROOF. (i), (ii) Let $A \in \mathcal{C}$ and let $0 \to C_A \xrightarrow{f_A} A$ be a factorization of $0 \to A$ into a cofibration followed by a trivial fibration. Then by definition C_A is cofibrant. Consider the exact sequence $0 \to Y_A \xrightarrow{g_A} C_A \xrightarrow{f_A} A$ in $\mathcal{C}$. Since f_A is a trivial fibration, by Lemma 1.1 we have that Y_A is in TFib. Also since f_A is a trivial fibration, the same lemma implies that any morphism $C \to A$ with C cofibrant, factors through f_A. We infer that f_A is a right Cof-approximation of A and by construction f_A is a trivial fibration. Part (ii) is dual.

(iii), (iv) Let A be in $\mathcal{C}$. Then by (ii) there exists a trivial cofibration $f^A : A \to F^A$ with F^A fibrant which is a left fibrant approximation of A. Let $F^A \xrightarrow{c} Y^A \to 0$ be a factorization of $F^A \to 0$ into a cofibration and a trivial fibration. Then Y^A is in TFib and we claim that $\psi^A := f^A \circ c : A \to Y^A$ is a left TFib-approximation of A. Indeed since cofibrations are closed under composition, ψ^A is a cofibration. Let $\alpha : A \to Y$ be a morphism with $Y \in \mathsf{TFib}$. Consider the diagram

$$\begin{array}{ccc} A & \xrightarrow{\alpha} & Y \\ {\scriptstyle \psi^A}\downarrow & & \downarrow \\ Y^A & \longrightarrow & 0 \end{array}$$

Since $Y \to 0$ is a trivial fibration, by the Lifting axiom, α factors through ψ^A. Hence ψ^A is a left TFib-approximation of A and a cofibration, and then $C^A := \operatorname{Coker}(\psi^A)$ is cofibrant as a cokernel of a cofibration. Part (iii) is dual.

(v) Let Y be in TFib and consider the right cofibrant approximation sequence $0 \to F_Y \to C_Y \xrightarrow{f_Y} Y$ of (i). Since any morphism $C \to Y$ with C cofibrant factors through C_Y it suffices to show that C_Y is in ω_c. Since C_Y is already cofibrant it suffices to show that $C_Y \to 0$ is a trivial fibration. Consider the following commutative diagram

$$\begin{array}{ccc} A & \xrightarrow{\alpha} & C_Y \\ {\scriptstyle i}\downarrow & & \downarrow \\ B & \longrightarrow & 0 \end{array}$$

where i is a cofibration. Since Y is in TFib, the morphism $Y \to 0$ is a trivial fibration, hence by the Lifting axiom the morphism $\alpha \circ f_Y : A \to Y$ factors through i, i.e. there exists a morphism $\beta : B \to Y$ such that $i \circ \beta = \alpha \circ f_Y$. Hence we have the commutative square

$$\begin{array}{ccc} A & \xrightarrow{\alpha} & C_Y \\ {\scriptstyle i}\downarrow & & \downarrow{\scriptstyle f_Y} \\ B & \xrightarrow{\beta} & Y \end{array}$$

where i is a cofibration and f_Y is a trivial fibration. Then by the Lifting axiom [M4], α factors through i. Hence $C_Y \to 0$ is a trivial fibration since it has the right lifting property with respect to cofibrations. We infer that $C_F \in \omega_c$. Hence $(\mathsf{Cof}/\omega_c, \mathsf{TFib}/\omega_c) = 0$. The proof that $(\mathsf{TCof}/\omega_f, \mathsf{Fib}/\omega_f) = 0$ is dual. □

In general the subcategories $\mathsf{Cof} \cap \mathsf{TFib}$ and $\mathsf{TCof} \cap \mathsf{Fib}$ are neither contravariantly nor covariantly finite, see for instance the example after Definition 4.8 below. However in certain special cases $\mathsf{Cof} \cap \mathsf{TFib}$ is contravariantly finite and $\mathsf{TCof} \cap \mathsf{Fib}$ is covariantly finite. We thank the referee for the following observation which will be useful later in connection with the classification of cotorsion pairs.

LEMMA 2.2. *Let $(\mathfrak{C}, \mathfrak{F}, \mathfrak{W})$ be a closed model structure in $\mathcal{C}$.*

(1) *If all objects of $\mathcal{C}$ are fibrant, then* $\mathsf{Cof} \cap \mathsf{TFib} = \mathsf{TCof}$. *In particular* $\mathsf{Cof} \cap \mathsf{TFib}$ *is contravariantly finite.*

(2) *If all objects of $\mathcal{C}$ are cofibrant, then* $\mathsf{TCof} \cap \mathsf{Fib} = \mathsf{TFib}$. *In particular* $\mathsf{TCof} \cap \mathsf{Fib}$ *is covariantly finite.*

PROOF. The proof follows directly from Proposition 2.1 and the fact that a map f in $\mathcal{C}$ is a weak equivalence if and only if $\gamma(f)$ is invertible in the homotopy category $\mathsf{Ho}(\mathcal{C})$ of $\mathcal{C}$, where $\gamma : \mathcal{C} \to \mathsf{Ho}(\mathcal{C})$ is the canonical functor, see [**64**]. □

As we shall now see the above results allow us to obtain a homological connection between the full subcategories of (trivially) fibrant or cofibrant objects and the classes of (trivial) fibrations or cofibrations.

For a class $\mathfrak{F}$ of morphisms in a category $\mathcal{C}$, we denote by $\mathcal{P}(\mathfrak{F})$ the full subcategory of $\mathcal{C}$ consisting of all objects P with the property that for any morphism $f : A \to B$ in $\mathfrak{F}$, any morphism $P \to B$ factors through f. Dually for a class $\mathfrak{C}$ of morphisms in $\mathcal{C}$, we denote by $\mathcal{I}(\mathfrak{C})$ the full subcategory of $\mathcal{C}$ consisting of all objects I with the property that for any morphism $g : A \to B$ in $\mathfrak{C}$, any morphism $A \to I$ factors through g. We call the objects in $\mathcal{P}(\mathfrak{F})$ *relative* $\mathfrak{F}$*-projectives* and the objects in $\mathcal{I}(\mathfrak{C})$ *relative* $\mathfrak{C}$*-injectives.*

The following result gives another way of describing the objects from classes of maps, which is more useful for constructing maps on the basis of subcategories.

PROPOSITION 2.3. *For a closed model structure* $(\mathfrak{C}, \mathfrak{F}, \mathfrak{W})$ *in* $\mathcal{C}$ *we have:*

$$\mathsf{Cof} = \mathcal{P}(\mathfrak{T}\mathfrak{F}), \quad \mathsf{Fib} = \mathcal{I}(\mathfrak{T}\mathfrak{C}), \quad \mathsf{TCof} = \mathcal{P}(\mathfrak{F}), \quad \mathsf{TFib} = \mathcal{I}(\mathfrak{C}).$$

PROOF. If C is in Cof, and $p : A \to B$ is a trivial fibration, then by Lemma 1.2((i) any morphism $C \to B$ factors through p. Hence C is $\mathfrak{T}\mathfrak{F}$-projective. Conversely if P is in $\mathcal{P}(\mathfrak{T}\mathfrak{F})$, then by Proposition 2.1, there exists a cofibrant approximation $f_P : C_P \to P$ of P which is a trivial fibration. Hence f_P splits since P is $\mathfrak{T}\mathfrak{F}$-projective. Since Cof is closed under direct summands and C_P is cofibrant, it follows that P is cofibrant. Hence we have proved that $\mathsf{Cof} = \mathcal{P}(\mathfrak{T}\mathfrak{F})$.

Now let X be in TCof and let $p : A \to B$ be a fibration. If $\alpha : X \to B$ is a morphism, then by Lemma 1.3(ii) the square

$$\begin{array}{ccc} 0 & \longrightarrow & A \\ \downarrow & & \downarrow{\scriptstyle p} \\ X & \xrightarrow{\alpha} & B \end{array}$$

admits a lifting since $0 \to X$ is a trivial cofibration. Hence X is in $\mathcal{P}(\mathfrak{F})$. Conversely if P is $\mathfrak{F}$-projective then by Proposition 2.1((iii) there exists a right TCof-approximation $\phi_P : X_P \to P$ and the morphism ϕ_P is a fibration. Hence ϕ_P splits, and this implies that P is in TCof since the latter is closed under direct summands. This proves that $\mathsf{TCof} = \mathcal{P}(\mathfrak{F})$. The remaining equalities are proved in a similar way. □

In particular we see that the classes of morphisms $\mathfrak{F}$ and $\mathfrak{T}\mathfrak{F}$ are projective classes in $\mathcal{C}$ and the classes of morphisms $\mathfrak{C}$ and $\mathfrak{T}\mathfrak{C}$ are injective classes in $\mathcal{C}$, in the sense of relative homological algebra introduced and developed by Eilenberg-Moore [**44**].

The above result suggests the following problem. Under what conditions is a closed model structure $(\mathfrak{C}, \mathfrak{F}, \mathfrak{W})$ in $\mathcal{C}$ determined by its subcategories Cof, Fib, TCof, TFib? Using Proposition 2.3 and the fact that in a closed model category a morphism f is a weak equivalence if and only if f is a composition of a trivial cofibration followed by a trivial cofibration, the above problem splits into two parts.

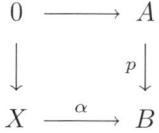

(i) Let $\mathcal{X}, \mathcal{Y}, \mathcal{X}_t, \mathcal{Y}_t$ be full additive subcategories of $\mathcal{C}$. Define $\mathfrak{C}'$ to be the class of $\mathcal{X}_t$-epics, $\mathfrak{F}'$ to be the class of $\mathcal{Y}_t$-monics, and $\mathfrak{W}'$ to be the class of morphisms which can be written as a composition of a $\mathcal{Y}$-monic followed by an $\mathcal{X}$-epic. Under what conditions is the triple $(\mathfrak{C}', \mathfrak{F}', \mathfrak{W}')$ a closed model structure in $\mathcal{C}$?

(ii) If we choose above $\mathcal{X}_t = \mathsf{TCof}$, $\mathcal{X} = \mathsf{Cof}$, $\mathcal{Y}_t = \mathsf{TFib}$, $\mathcal{Y} = \mathsf{Fib}$, for a closed model structure $(\mathfrak{C}, \mathfrak{F}, \mathfrak{W})$ in $\mathcal{C}$, then is it true that $(\mathfrak{C}, \mathfrak{F}, \mathfrak{W})$ coincides with the closed model structure $(\mathfrak{C}', \mathfrak{F}', \mathfrak{W}')$?

In the next sections we will give fairly general answers to the above problem which is connected rather surprisingly with cotorsion pairs.

3. Cotorsion Pairs Arising from Closed Model Structures

Let $\mathcal{C}$ be a closed model category with closed model structure $(\mathfrak{C}, \mathfrak{F}, \mathfrak{W})$, and assume from now on that $\mathcal{C}$ is abelian. In this section we describe when the pairs of subcategories $(\mathsf{Cof}, \mathsf{TFib})$ and/or $(\mathsf{TCof}, \mathsf{Fib})$ are good, or equivalently cotorsion, pairs in $\mathcal{C}$, and also when there are induced torsion pairs in associated stable categories. In view of Proposition 2.1, the missing property is the vanishing of Ext^1. Assuming this property we obtain as a consequence that the closed model structure is determined by the four subcategories.

We begin with the following useful result, valid in any abelian category, which connects the vanishing of the first extension functor with the lifting property of commutative squares defined in Section 1, and provides a tool for checking whether various subcategories in a closed model category are orthogonal with respect to the vanishing of Ext^1.

Let X and Y be in $\mathcal{C}$ and consider the following exact commutative diagram:

$$\begin{array}{ccc}
 & & 0 \\
 & & \downarrow \\
0 & & Y \\
\downarrow & & \downarrow{\scriptstyle c} \\
A & \xrightarrow{\alpha} & C \\
{\scriptstyle i}\downarrow & & \downarrow{\scriptstyle p} \\
B & \xrightarrow{\beta} & D \\
{\scriptstyle d}\downarrow & & \downarrow \\
X & & 0 \\
\downarrow & & \\
0 & &
\end{array}$$

where the sequences $0 \to A \xrightarrow{i} B \xrightarrow{d} X \to 0$ and $0 \to Y \xrightarrow{c} C \xrightarrow{p} D \to 0$ are exact.

EXTENSION-LIFTING LEMMA 3.1. *The following statements are equivalent for fixed objects X and Y in $\mathcal{C}$:*

(i) $\mathrm{Ext}^1(X,Y) = 0$.

(ii) *Any commutative diagram as above admits a lifting: there exists a morphism $\lambda : B \to C$ such that $i \circ \lambda = \alpha$ and $\lambda \circ p = \beta$.*

PROOF. (i) ⇒ (ii) Consider the pull-back diagram

$$\begin{array}{ccccccccc} 0 & \longrightarrow & Y & \xrightarrow{\epsilon} & K & \xrightarrow{\zeta} & B & \longrightarrow & 0 \\ & & \| & & \gamma\downarrow & & \beta\downarrow & & \\ 0 & \longrightarrow & Y & \xrightarrow{c} & C & \xrightarrow{p} & D & \longrightarrow & 0 \end{array}$$

Since $i \circ \beta = \alpha \circ p$, there exists a unique morphism $\phi : A \to K$ such that $\phi \circ \zeta = i$ and $\phi \circ \gamma = \alpha$. Then we have the following exact commutative diagram

$$\begin{array}{ccccccccc} 0 & \longrightarrow & A & \xrightarrow{\phi} & K & \xrightarrow{\xi} & L & \longrightarrow & 0 \\ & & \| & & \zeta\downarrow & & \eta\downarrow & & \\ 0 & \longrightarrow & A & \xrightarrow{i} & B & \xrightarrow{d} & X & \longrightarrow & 0 \end{array}$$

Since the kernel of η is isomorphic to Y, the hypothesis implies that η is split epic, and its kernel $ker(\eta) : Y \to L$ is split monic. By the above diagram $ker(\eta)$ admits a factorization $\epsilon \circ \xi$. It follows that ϵ is split monic or equivalently $\zeta : K \to B$ is split epic. Then by the first diagram above there exists a morphism $\theta : B \to C$ such that $\theta \circ p = \beta$. Then $i \circ \theta \circ p = i \circ \beta = \alpha \circ p \Rightarrow (\alpha - i \circ \theta) \circ p = 0$. Hence there exists a unique morphism $\mu : A \to Y$ such that $\mu \circ c = \alpha - i \circ \theta = 0$. Since $\mathrm{Ext}^1(X,Y) = 0$, the push-out of the extension $0 \to A \xrightarrow{i} B \xrightarrow{d} X \to 0$ along μ, splits. Hence there exists a morphism $\nu : B \to Y$ such that $i \circ \nu = \mu$. Then $i \circ \nu \circ c + i \circ \theta = \alpha$. Setting $\lambda := \theta + \nu \circ c$ we have $\alpha = i \circ \lambda$ and $\lambda \circ p = \theta \circ p + \nu \circ c \circ p = \theta \circ p = \beta$. Hence $\lambda : B \to C$ has the desired property.

(ii) ⇒ (i) Let $0 \to Y \xrightarrow{c} C \xrightarrow{p} X \to 0$ be an extension. Then we have the exact commutative diagram preceding the lemma, if we set $A = C$, $B = A \oplus X$, $\alpha = 1_C$, $\beta = {}^t(p, 1_X)$, $i = (1_A, 0)$, $d = {}^t(0, 1_X)$. Then the hypothesis implies that there exists a morphism ${}^t(\kappa, \lambda) : C \oplus X \to C$ such that ${}^t(\kappa,\lambda) \circ p = 1_X$. From this it follows that $\lambda \circ p = 1_X$. Hence p splits, that is $\mathrm{Ext}^1(X,Y) = 0$. □

By part (v) of Proposition 2.1 the subcategories Cof and TFib are orthogonal with respect to Hom in the stable category $\mathcal{C}/\mathsf{Cof} \cap \mathsf{TFib}$, and the categories TCof and Fib are orthogonal with respect to Hom in the stable category $\mathcal{C}/\mathsf{TCof}\cap\mathsf{Fib}$. The following result gives sufficient conditions for the pairs of subcategories $(\mathsf{Cof}, \mathsf{TFib})$ and $(\mathsf{TCof}, \mathsf{Fib})$ to be orthogonal in $\mathcal{C}$ with respect to Ext^1.

LEMMA 3.2. (i) *If any trivial fibration is epic, then: ${}^{\perp}\mathsf{TFib} \subseteq \mathsf{Cof}$.*

(ii) *If any fibration is epic, then: ${}^{\perp}\mathsf{Fib} \subseteq \mathsf{TCof}$.*

(iii) *If any trivial cofibration is monic, then: $\mathsf{TCof}^{\perp} \subseteq \mathsf{Fib}$.*

(iv) *If any cofibration is monic, then: $\mathsf{Cof}^{\perp} \subseteq \mathsf{TFib}$.*

(v) *If any monic with cokernel in TCof is a trivial cofibration, then:*

$$\mathrm{Ext}^1(\mathsf{TCof}, \mathsf{Fib}) = 0.$$

(vi) *If any epic with kernel in* TFib *is a trivial fibration, then:*

$$\mathrm{Ext}^1(\mathsf{Cof}, \mathsf{TFib}) = 0.$$

PROOF. We prove only (i), (ii), (v), since the proof of (iii), (iv), (vi) is dual.

(i) Let $A \in {}^{\perp}\mathsf{TFib}$ and consider the right cofibrant approximation sequence $0 \to Y_A \to C_A \xrightarrow{f_A} A$ of A. Since f_A is a trivial fibration, f_A is epic. Since $Y_A \in \mathsf{TFib}$ and $A \in {}^{\perp}\mathsf{TFib}$, the sequence splits. Hence A is cofibrant since it is a direct summand of C_A. We infer that ${}^{\perp}\mathsf{TFib} \subseteq \mathsf{Cof}$.

(ii) Let $A \in {}^{\perp}\mathsf{Fib}$ and consider the exact sequence $0 \to F_A \to X_A \xrightarrow{\phi_A} A$ where ϕ_A is a right TCof-approximation of A. Then F_A is fibrant, and since ϕ_A is a fibration, it is epic. Since $A \in {}^{\perp}\mathsf{Fib}$, the sequence splits, and we infer that A is in TCof since it is a direct summand of X_A.

(v) Let $0 \to F \xrightarrow{i} A \to X \to 0$ be an extension with F fibrant and X in TCof. Then by hypothesis i is a trivial cofibration. Since F is fibrant, by Lemma 1.1 the morphism i splits. Hence $\mathrm{Ext}^1(X, F) = 0$. □

The above lemma suggests to search for necessary and sufficient conditions such that any (trivial) fibration is an epimorphism and/or any (trivial) cofibration is a monomorphism. In this connection we have the following result.

LEMMA 3.3. (1) *Assume that* $\mathcal{C}$ *has enough projectives. Then:*

(i) *Any fibration is an epimorphism if and only if any projective object is trivially cofibrant.*

(ii) *Any trivial fibration is an epimorphism if and only if any projective object is cofibrant.*

(2) *Assume that* $\mathcal{C}$ *has enough injectives. Then:*

(i) *Any cofibration is a monomorphism if and only if any injective object is trivially fibrant.*

(ii) *Any trivial cofibration is a monomorphism if and only if any injective object is fibrant.*

PROOF. We prove only part (i) of (1). The proof of the other statements is similar, using Lemma 3.2. First assume that any fibration is an epimorphism. Then by part (ii) of Lemma 3.2 we infer that any projective object is trivially cofibrant. Conversely assume that any projective object is trivially cofibrant. Let $f : B \to C$ be a fibration, and let $\varepsilon : P \twoheadrightarrow C$ be an epimorphism where P is projective. Consider the following commutative diagram

$$\begin{array}{ccc} 0 & \longrightarrow & B \\ \big\downarrow & & \big\downarrow{\scriptstyle f} \\ P & \xrightarrow{\ \varepsilon\ } & C. \end{array}$$

Since any projective object is trivially cofibrant, the map $0 \to P$ is a trivial cofibration. Since any fibration has the right lifting property with respect to all trivial cofibrations, we infer that the above diagram admits a lifting. This implies that f is an epimorphism. □

The next result gives sufficient conditions such that the pair of subcategories $(\mathsf{Cof}, \mathsf{TFib})$, resp. $(\mathsf{TCof}, \mathsf{Fib})$, forms a cotorsion pair in $\mathcal{C}$. Note that if $(\mathsf{Cof}, \mathsf{TFib})$, resp. $(\mathsf{TCof}, \mathsf{Fib})$, is a good pair, then Proposition 2.1 ensures that it is a cotorsion pair. For examples satisfying the conditions below we refer to the next section.

PROPOSITION 3.4. (1) $(\mathsf{Cof}, \mathsf{TFib})$ *is a cotorsion pair in* $\mathcal{C}$, *provided that:*

(i) *Any cofibration is monic.*
(ii) *Any trivial fibration is epic and any epic with kernel in* TFib *is a trivial fibration.*

(2) $(\mathsf{TCof}, \mathsf{Fib})$ *is a cotorsion pair in* $\mathcal{C}$, *provided that:*

(i) *Any fibration is epic.*
(ii) *Any trivial cofibration is monic and any monic with cokernel in* TCof *is a trivial cofibration.*

PROOF. We prove only part (1) since the proof of (2) is dual. By Proposition 2.1 we know that Cof is contravariantly finite and TFib is covariantly finite. By parts (i), (iv), (vi) of Lemma 3.2 we have $\mathsf{Cof}^{\perp} = \mathsf{TFib}$ and ${}^{\perp}\mathsf{TFib} = \mathsf{Cof}$. Hence $(\mathsf{Cof}, \mathsf{TFib})$ is a good pair or equivalently a cotorsion pair in $\mathcal{C}$. □

If $\mathcal{C}$ has enough projective and injective objects, then using Lemma 3.3 and Proposition 3.4 we can prove the following main result of this section, which gives a useful characterization of when $(\mathsf{TCof}, \mathsf{Fib})$ or $(\mathsf{Cof}, \mathsf{TFib})$ is a cotorsion pair.

THEOREM 3.5. *Assume that* $\mathcal{C}$ *has enough projective and injective objects.*
(1) *The following conditions are equivalent.*

(i) $(\mathsf{TCof}, \mathsf{Fib})$ *is a cotorsion pair.*
(ii) *The following statements hold:*
  (a) *Any fibration is epic.*
  (b) *Any trivial cofibration is monic.*
  (c) *Any monic with cokernel in* TCof *is a trivial cofibration.*

(2) *The following conditions are equivalent.*

(i) $(\mathsf{Cof}, \mathsf{TFib})$ *is a cotorsion pair.*
(ii) *The following statements hold:*
  (a) *Any trivial fibration is epic.*
  (b) *Any cofibration is monic.*
  (c) *Any epic with kernel in* TFib *is a trivial fibration.*

PROOF. We prove only part (2) since the proof of (1) is dual. By Proposition 3.4 it suffices to show the (i) implies (ii). Assume that $(\mathsf{Cof}, \mathsf{TFib})$ is a cotorsion pair in $\mathcal{C}$. Then any projective is cofibrant and any injective is trivially fibrant. Hence by Lemma 3.3 any trivial fibration is epic and any cofibration is monic. Hence it suffices to show that any epic with trivially fibrant kernel is a trivial fibration. Let $0 \to Y \to B \xrightarrow{p} A \to 0$ be an extension with Y in TFib. By the Lifting axiom, to show that p is a trivial fibration, it suffices to show that for any commutative square

$$\begin{array}{ccc} K & \xrightarrow{\alpha} & B \\ {\scriptstyle i}\downarrow & & {\scriptstyle p}\downarrow \\ L & \xrightarrow{\beta} & A \end{array}$$

where i is a cofibration, there exists a morphism $\lambda : L \to B$ which factorizes α through i and β through p. Since i is a cofibration, we have an exact sequence $0 \to K \xrightarrow{i} L \to C \to 0$ and C is cofibrant. Since $\mathrm{Ext}^1(C, Y) = 0$, Lemma 3.1 shows that the factorizing morphism λ exists. Hence any epimorphism with kernel in TFib is a trivial fibration. □

Putting together the conditions in (1)(ii) and (2)(ii) of Theorem 3.5, we get the following characterization of the closed model structures $(\mathfrak{C}, \mathfrak{F}, \mathfrak{W})$ in $\mathcal{C}$ such that both pairs $(\mathsf{Cof}, \mathsf{TFib})$ and $(\mathsf{TCof}, \mathsf{Fib})$ are good or equivalently cotorsion in $\mathcal{C}$.

THEOREM 3.6. *Let $\mathcal{C}$ be an abelian closed model category with enough projective and injective objects. Then the following conditions are equivalent.*

- (i) (a) *Any fibration is epic and any epic with kernel in* TFib *is a trivial fibration.*
  - (b) *Any cofibration is monic and any monic with cokernel in* TCof *is a trivial cofibration.*
- (ii) $(\mathsf{Cof}, \mathsf{TFib})$ *and* $(\mathsf{TCof}, \mathsf{Fib})$ *are cotorsion pairs in $\mathcal{C}$.*

Under the assumptions of Theorem 3.6, the results of this chapter, combined with the results of Chapter V, imply the following existence result for torsion pairs in suitable stable pretriangulated categories.

COROLLARY 3.7. *Let $\mathcal{C}$ be an abelian closed model category with enough projective and injective objects. If* $(\mathsf{Cof}, \mathsf{TFib})$ *and* $(\mathsf{TCof}, \mathsf{Fib})$ *are cotorsion pairs in $\mathcal{C}$, then we have the following.*

- (i) *If ω_c is functorially finite in $\mathcal{C}$, then* Cof *is closed under kernels of epimorphisms if and only if* $(\mathsf{Cof}/\omega_c, \mathsf{TFib}/\omega_c)$ *is a torsion pair in the pretriangulated category $\mathcal{C}/\omega_c$.*
- (ii) *If ω_f is functorially finite in $\mathcal{C}$, then* Fib *is closed under cokernels of monomorphisms if and only if* $(\mathsf{TCof}/\omega_f, \mathsf{Fib}/\omega_f)$ *is a torsion pair in the pretriangulated category $\mathcal{C}/\omega_f$.*

Combining Theorem VI.3.2 and the above results, we get the following consequence which gives sufficient conditions for the existence of cotorsion triples arising from closed model structures.

COROLLARY 3.8. *Let $\mathcal{C}$ be an abelian closed model category with enough projective and injective objects. If* $(\mathsf{Cof}, \mathsf{TFib})$ *and* $(\mathsf{TCof}, \mathsf{Fib})$ *are cotorsion pairs in $\mathcal{C}$ and* $\mathsf{TFib} = \mathsf{TCof}$, *then* $(\mathsf{Cof}, \mathsf{TFib} = \mathsf{TCof}, \mathsf{Fib})$ *is a cotorsion triple in $\mathcal{C}$ and the assignments $C \mapsto F_C$ and $F \mapsto C^F$ induce inverse equivalences:*

$$\Phi : \mathsf{Cof}/\omega_c \xrightarrow{\approx} \mathsf{Fib}/\omega_f \quad \textit{and} \quad \Psi : \mathsf{Fib}/\omega_f \xrightarrow{\approx} \mathsf{Cof}/\omega_c.$$

We close this section with the following result which gives a partial answer to the problem posed in Section 2.

PROPOSITION 3.9. *Let $\mathcal{C}$ be an abelian closed model category with closed model structure $(\mathfrak{C}, \mathfrak{F}, \mathfrak{W})$. If $\mathcal{C}$ has enough projective and injective objects and* $(\mathsf{Cof}, \mathsf{TFib})$ *and* $(\mathsf{TCof}, \mathsf{Fib})$ *are cotorsion pairs in $\mathcal{C}$, then the triple of morphisms $(\mathfrak{C}, \mathfrak{F}, \mathfrak{W})$ is completely determined by the subcategories* $\mathsf{Cof}, \mathsf{Fib}, \mathsf{TCof}, \mathsf{TFib}$. *More precisely:*

(i) *$\mathfrak{F}$ is the class of* TCof*-epics with fibrant kernel.*
(ii) *$\mathfrak{C}$ is the class of* TFib*-monics with cofibrant cokernel.*
(iii) *$\mathfrak{T}\mathfrak{F}$ is the class of* Cof*-epics with trivially fibrant kernel.*
(iv) *$\mathfrak{T}\mathfrak{C}$ is the class of* Fib*-monics with trivially cofibrant cokernel.*
(v) *$\mathfrak{W}$ is the class of morphisms which can be written as a composition of a* Fib*-monic with trivially cofibrant cokernel followed by a* Cof*-epic with trivially fibrant kernel.*

PROOF. We prove only assertion (i) since the proof of the other assertions is similar. For a subcategory $\mathcal{X} \subseteq \mathcal{C}$, we denote by $\mathcal{E}(\mathcal{X})$ the class of $\mathcal{X}$-epics. By Proposition 2.3 we have $\mathcal{P}(\mathfrak{F}) = \mathsf{TCof}$. Hence $\mathcal{E}(\mathcal{P}(\mathfrak{F})) = \mathcal{E}(\mathsf{TCof})$. Since obviously $\mathfrak{F} \subseteq \mathcal{E}(\mathcal{P}(\mathfrak{F}))$, we have $\mathfrak{F} \subseteq \mathcal{E}(\mathsf{TCof})$. Since TCof is contravariantly finite and $\mathcal{C}$ has enough projectives, $\mathcal{E}(\mathsf{TCof})$ consists of epics. Since by Lemma 1.2 the kernel of a fibration is fibrant, it suffices to show that any TCof-epic p with fibrant kernel is a fibration. By Lemma 1.3 it suffices to show that p has the right lifting property with respect to trivial cofibrations. This follows from the Extension-Lifting Lemma 3.1 and condition (ii) of Theorem 3.6. □

For a more complete answer to the problem of which closed model structures are determined by objects we refer to Theorems 4.6 and 4.7 below.

REMARK 3.10. Hovey in [**64**] proves independently a version of Theorem 3.6, by using closed model structures compatible with a proper class of short exact sequences in $\mathcal{C}$ in a suitable sense, and he does not use enough projective and injective objects. Moreover he implicitly proves an equivalent version of the first part of the Extension-Lifting Lemma 3.1 in the setting of closed model categories.

4. Closed Model Structures Arising from Cotorsion Pairs

In this section we describe all cotorsion pairs in an abelian closed model category in terms of closed model structures. This description, which is based on the construction of maps in terms of subcategories, will be used in the next section for the classification of all torsion pairs of a stable category. Although all the results of this section are valid (with the necessary modifications) in any abelian category, as in Chapter V, for simplicity we assume throughout this section that $\mathcal{C}$ is an abelian category with enough projective and injective objects.

Projective Closed Model Structures. For the construction of closed model structures in terms of subcategories, the results of the previous sections suggest to start with a pair $(\mathcal{X}, \mathcal{Y})$ of full additive subcategories in $\mathcal{C}$, which are closed under direct summands and isomorphisms. Fixing this setup, our principal aim is to construct a closed model structure in $\mathcal{C}$ out of the triple of subcategories $(\mathcal{X}, \mathcal{Y}, \mathcal{X} \cap \mathcal{Y})$, provided that the subcategories $\mathcal{X}$, $\mathcal{Y}$ and $\mathcal{X} \cap \mathcal{Y}$ satisfy certain finiteness and orthogonality conditions.

We set $\omega := \mathcal{X} \cap \mathcal{Y}$ and consider the following classes of morphisms in $\mathcal{C}$:

(i) $\mathfrak{C}_\omega$ is the class of monics with cokernel in $\mathcal{X}$.
(ii) $\mathfrak{T}\mathfrak{C}_\omega$ is the class of split monics with cokernel in ω.
(iii) $\mathfrak{F}_\omega$ is the class of ω-epics.
(iv) $\mathfrak{T}\mathfrak{F}_\omega$ is the class of epics with kernel in $\mathcal{Y}$.
(v) $\mathfrak{W}_\omega$ is the class of morphisms α which admit a factorization $\alpha = \alpha_1 \circ \alpha_2$ where $\alpha_1 \in \mathfrak{T}\mathfrak{C}_\omega$ and $\alpha_2 \in \mathfrak{T}\mathfrak{F}_\omega$.

We call the morphisms in $\mathfrak{F}_\omega$, resp. $\mathfrak{T}\mathfrak{F}_\omega$, *projective ω-fibrations*, resp. *projective ω-trivial fibrations*. We call the morphisms in $\mathfrak{C}_\omega$, resp. $\mathfrak{T}\mathfrak{C}_\omega$, *projective ω-cofibrations*, resp. *projective ω-trivial cofibrations*. Finally we call the morphisms in $\mathfrak{W}_\omega$, *projective ω-weak equivalences*. Our aim is to investigate when the triple $(\mathfrak{C}_\omega, \mathfrak{F}_\omega, \mathfrak{W}_\omega)$ is a closed model structure in $\mathcal{C}$.

The motivating source for the above definitions is the following example which produces a closed model structure out of a specific cotorsion pair. In a sense this example is the "trivial" case of the more general theory that follows.

EXAMPLE. Let $\mathcal{C}$ be an abelian category with enough projective objects. Then $(\mathcal{P}, \mathcal{C})$ is a cotorsion pair in $\mathcal{C}$ with $\mathcal{P} = \mathcal{P} \cap \mathcal{C}$ contravariantly finite. It follows easily that the class of projective $\mathcal{P}$-(trivial) cofibrations coincides with the class of split monics with projective cokernel. Similarly the class of projective $\mathcal{P}$-(trivial) fibrations coincides with the class of epimorphisms. Finally the class of projective $\mathcal{P}$-weak equivalences coincides with the class of morphisms α which admit a factorization $\alpha = \alpha_1 \circ \alpha_2$, where α_1 is a split monic with projective cokernel and α_2 is an epimorphism. It is easy to see that the triple $(\mathfrak{C}_\mathcal{P}, \mathfrak{C}_\mathcal{P}, \mathfrak{W}_\mathcal{P})$ is a closed model structure in $\mathcal{C}$ with $\mathsf{Cof} = \mathsf{TCof} = \mathcal{P}$ and $\mathsf{TFib} = \mathsf{Fib} = \mathcal{C}$. If $\alpha : A \to B$ is a morphism in $\mathcal{C}$, then we have a factorization $\alpha : A \xrightarrow{(1_A,0)} A \oplus P \xrightarrow{{}^t(\alpha,f)} B$, where $f : P \to B$ is an epimorphism with P projective. It follows that any morphism in $\mathcal{C}$ is a projective $\mathcal{P}$-weak equivalence.

We begin with the following useful result which gives, under some reasonable conditions, a handy description of the projective ω-weak equivalences.

LEMMA 4.1. *Assume that ω is contravariantly finite,* $\operatorname{Ext}^1(\mathcal{X}, \mathcal{Y}) = 0$ *and* $(\mathcal{X}/\omega)^\perp = \mathcal{Y}/\omega$. *If $\mathcal{X}$ contains the projectives and admits ω as a cogenerator, then the following are equivalent for a morphism $\alpha : A \to B$ in $\mathcal{C}$.*
(i) *$\alpha : A \to B$ is in $\mathfrak{W}_\omega$.*
(ii) *$(\underline{X}, \underline{\alpha}) : (\underline{X}, \underline{A}) \to (\underline{X}, \underline{B})$ is invertible in $\mathcal{C}/\omega$, for any $X \in \mathcal{X}$.*

PROOF. We observe first that the contravariant finiteness of ω implies that $\mathcal{C}/\omega$ is left triangulated. Assume that α is in $\mathfrak{W}_\omega$. Then there exists a factorization $A \xrightarrow{(1_A,0)} A \oplus T \xrightarrow{{}^t(\alpha,\kappa)} B$, where T is in ω and ${}^t(\alpha, \kappa)$ is epic with kernel in $\mathcal{Y}$. Since $\operatorname{Ext}^1(\mathcal{X}, \mathcal{Y}) = 0$, the morphism ${}^t(\alpha, \kappa)$ is ω-epic, hence we have a left triangle $\Omega(\underline{B}) \to \underline{Y} \to \underline{A} \xrightarrow{\underline{\alpha}} \underline{B}$ in $\mathcal{C}/\omega$. Applying $(\underline{X}, -)$ to this triangle, we have the following commutative diagram:

$$\begin{array}{ccc} \mathcal{C}(X, A \oplus T) & \xrightarrow{(X,{}^t(\alpha,\kappa))} & \mathcal{C}(X, B) \\ {\scriptstyle \pi_1}\downarrow & & {\scriptstyle \pi_2}\downarrow \\ \mathcal{C}/\omega(\underline{X}, \underline{A}) & \xrightarrow{(\underline{X},\underline{\alpha})} & \mathcal{C}/\omega(\underline{X}, \underline{B}) \end{array} \qquad (\dagger)$$

Since $\operatorname{Ext}^1(\mathcal{X},\mathcal{Y}) = 0$, the morphism $\mathcal{C}(X, {}^t(\alpha,\kappa))$ is epic for any $X \in \mathcal{X}$. By the above diagram this implies that $(\underline{X},\underline{\alpha})$ is epic. However since the cone of $\underline{\alpha}$ is in $\mathcal{Y}/\omega$ it follows that $(\underline{X},\underline{\alpha})$ is monic. Hence $(\underline{X},\underline{\alpha})$ is invertible for any X in $\mathcal{X}$. Conversely assume that this condition holds. First we show that if $\alpha : A \to B$ is an ω-epic, such that $(\underline{X},\underline{\alpha})$ is invertible for any X in $\mathcal{X}$, then α is epic with kernel in $\mathcal{Y}$. Consider the commutative diagram

$$\begin{array}{ccc} \mathcal{C}(X,A) & \xrightarrow{(X,\alpha)} & \mathcal{C}(X,B) \\ {\scriptstyle\pi_1}\downarrow & & {\scriptstyle\pi_2}\downarrow \\ \mathcal{C}/\omega(\underline{X},\underline{A}) & \xrightarrow{(\underline{X},\underline{\alpha})} & \mathcal{C}/\omega(\underline{X},\underline{B}) \end{array} \qquad (\dagger\dagger)$$

Let X be in $\mathcal{X}$ and let $f : X \to B$ be a morphism. Then by the above diagram there exists a unique morphism $\underline{g} : \underline{X} \to \underline{A}$, such that $\underline{g} \circ \underline{\alpha} = \underline{f}$. Hence $f - g \circ \alpha$ factors through a right ω-approximation $\tau_B : T \to B$, say as: $f - g \circ \alpha = \lambda \circ \tau_B$, where $\lambda : T_B \to B$. Since α is an ω-epic, there exists a morphism $\mu : T_B \to A$ such that $\mu \circ \alpha = \tau_B$. Then $f - g \circ \alpha = \lambda \circ \mu \circ \alpha$. Hence $f = (g + \lambda \circ \mu) \circ \alpha$. It follows that any morphism from an object of $\mathcal{X}$ to B factors through α. Since $\mathcal{X}$ contains the projectives, this implies that α is an epimorphism. It remains to show that the kernel Y of α is in $\mathcal{Y}$. Since ω is an (Ext-injective) cogenerator of $\mathcal{X}$ we have that ω is covariantly finite in $\mathcal{X}$ and the cokernel of a left ω-approximation of any object of $\mathcal{X}$ lies in $\mathcal{X}$. In particular the suspension functor $\Sigma : \mathcal{X}/\omega \to \mathcal{X}/\omega$ is defined and we have a natural isomorphism $(\Sigma(\underline{X}),\underline{A}) \xrightarrow{\cong} (\underline{X},\Omega(\underline{A}))$ in $\mathcal{C}/\omega$, for any $X \in \mathcal{X}$ and any object A in $\mathcal{C}$. Now since α is an ω-epic, we have a left triangle $\Omega(\underline{B}) \to \underline{Y} \to \underline{A} \xrightarrow{\underline{\alpha}} \underline{B}$ in $\mathcal{C}/\omega$. Since $(\underline{X},\underline{\alpha})$ is invertible for any $X \in \mathcal{X}$, applying to the above diagram the functor $(\mathcal{X}/\omega, -)$ and using the above adjunction isomorphism, we infer that $\underline{Y} \in (\mathcal{X}/\omega)^{\perp} = \mathcal{Y}/\omega$. Hence the kernel Y of α lies in $\mathcal{Y}$.

Now let $\alpha : A \to B$ be such that $(\underline{X},\underline{\alpha})$ is invertible for any X in $\mathcal{X}$. Consider the ω-epic ${}^t(\alpha,\tau_B) : A \oplus T_B \to B$, where $\tau_B : T_B \to B$ is a right ω-approximation of B. This morphism is obviously isomorphic to α in $\mathcal{C}/\omega$. Hence it enjoys the same property as $\underline{\alpha}$. Then by the above argument we infer that ${}^t(\alpha,\tau_B)$ is epic with kernel in $\mathcal{Y}$. Then α admits a factorization $A \xrightarrow{(1_A,0)} A \oplus T_B \xrightarrow{{}^t(\alpha,\tau_B)} B$, where T_B is in ω and ${}^t(\alpha,\tau_B)$ is in $\mathfrak{T}\mathfrak{F}_\omega$. Hence α is in $\mathfrak{W}_\omega$. □

The following first main result of this section generalizes the previous example and gives necessary and sufficient conditions for the existence of closed model structures in $\mathcal{C}$ arising from cotorsion pairs.

First we recall that the *homotopy category* $\mathsf{Ho}(\mathcal{C})$ of a closed model category structure $(\mathfrak{C},\mathfrak{F},\mathfrak{W})$ is defined as the category of fractions $\mathcal{C}[\mathfrak{W}^{-1}]$, obtained by formally inverting the weak equivalences, see [**88**]. A more convenient description of the homotopy category is that $\mathsf{Ho}(\mathcal{C})$ is equivalent to the factor category $\mathsf{Cof} \cap \mathsf{Fib}/\sim$ of the category of fibrant and cofibrant objects modulo a suitable equivalence relation $\sim$, called the *homotopy relation*, defined on the class of morphisms between fibrant and cofibrant objects.

EXAMPLE. The homotopy category of the closed model structure in $\mathcal{C}$ constructed in the example before Lemma 4.1 is trivial. This follows from the fact that all morphisms are weak equivalences.

THEOREM 4.2. *Let $\mathcal{X}$ and $\mathcal{Y}$ be full subcategories of $\mathcal{C}$ and let $\omega = \mathcal{X} \cap \mathcal{Y}$. Then the following conditions are equivalent:*

(i) *The triple $(\mathfrak{C}_\omega, \mathfrak{F}_\omega, \mathfrak{W}_\omega)$ defines a closed model structure in $\mathcal{C}$ such that: $\mathfrak{C}_\omega$ is the class of cofibrations, $\mathfrak{F}_\omega$ is the class of fibrations, and $\mathfrak{W}_\omega$ is the class of weak equivalences.*

(ii) *$(\mathcal{X}, \mathcal{Y})$ is a cotorsion pair and ω is contravariantly finite in $\mathcal{C}$.*

If (ii) *holds, then $\mathsf{Cof}_\omega = \mathcal{X}$ and $\mathsf{TCof}_\omega = \omega$, and $\mathsf{TFib}_\omega = \mathcal{Y}$ and $\mathsf{Fib}_\omega = \mathcal{C}$. Moreover the associated Quillen homotopy category $\mathsf{Ho}(\mathcal{C})$ is equivalent to the stable right triangulated category $\mathcal{X}/\omega$.*

PROOF. (i) $\Rightarrow$ (ii) If the triple $(\mathfrak{C}_\omega, \mathfrak{F}_\omega, \mathfrak{W}_\omega)$ defines a closed model structure in $\mathcal{C}$, then obviously we have $\mathsf{Cof}_\omega = \mathcal{X}$, $\mathsf{TCof}_\omega = \omega$, $\mathsf{TFib}_\omega = \mathcal{Y}$ and $\mathsf{Fib}_\omega = \mathcal{C}$. Then the assertion follows by Propositions 2.1 and 3.4(1).

(ii) $\Rightarrow$ (i) We check the axioms [Mi], $i = 1, ..., 5$, of a closed model category. First note that [M1] holds since $\mathcal{C}$ as an abelian category has kernels and cokernels.

[M4] Consider a commutative diagram

$$\begin{array}{ccc} A & \xrightarrow{\alpha} & C \\ {\scriptstyle i}\downarrow & & \downarrow{\scriptstyle p} \\ B & \xrightarrow{\beta} & D \end{array}$$

Assume first that i is in $\mathfrak{C}_\omega$ and p is in $\mathfrak{T}\mathfrak{F}_\omega$. Then i is monic with cokernel in $\mathcal{X}$ and p is epic with kernel in $\mathcal{Y}$. Since $\mathrm{Ext}^1(\mathcal{X}, \mathcal{Y}) = 0$, there exists a lifting by the lifting-extension Lemma 3.1. Assume now that i is in $\mathfrak{T}\mathfrak{C}_\omega$ and p is in $\mathfrak{F}_\omega$. Since i is a split monic with cokernel in ω we can assume that i is the inclusion $A \xrightarrow{(1,0)} A \oplus T$ where $T \in \omega$, and then the morphism β is of the form ${}^t(\beta_1, \beta_2) : A \oplus T \to D$. Since p is ω-epic and T is in ω, there exists a morphism $t : T \to C$ such that $t \circ p = \beta_2$. Then the morphism ${}^t(\alpha, t) : A \oplus T \to C$ is a lifting of the above square. Hence axiom [M4] holds.

[M5] Let $\alpha : A \to B$ be a morphism in $\mathcal{C}$. Let $\tau_B : T_B \to B$ be a right ω-approximation of B. Then α can be factorized as follows $A \xrightarrow{(1_A,0)} A \oplus T_B \xrightarrow{{}^t(\alpha,\tau)} B$ where the first morphism is in $\mathfrak{T}\mathfrak{C}_\omega$ and the second is in $\mathfrak{F}_\omega$. Hence α can be written as a composition of a morphism in $\mathfrak{T}\mathfrak{C}_\omega$ followed by a morphism in $\mathfrak{F}_\omega$. To prove that α can be written as a composition of a morphism in $\mathfrak{C}_\omega$ followed by a morphism in $\mathfrak{T}\mathfrak{F}_\omega$, we show first that any monic μ and any epic ϵ has this property. So let $\mu : A \to B$ be monic, and consider the following exact commutative diagram

$$\begin{array}{ccccccccc} 0 & \longrightarrow & A & \xrightarrow{i} & K & \xrightarrow{\xi} & X_C & \longrightarrow & 0 \\ & & \| & & {\scriptstyle p}\downarrow & & \downarrow{\scriptstyle f_C} & & \\ 0 & \longrightarrow & A & \xrightarrow{\mu} & B & \xrightarrow{\beta} & C & \longrightarrow & 0 \end{array}$$

where the second row is induced by pulling-back the first row along the right $\mathcal{X}$-approximation of C. Then by definition i is in $\mathfrak{C}_\omega$. Since $\mathcal{X}$ is contravariantly finite and contains the projectives, the morphism f_C, hence the morphism p, is epic. Since $(\mathcal{X}, \mathcal{Y})$ is a cotorsion pair, the kernel of f_C, hence the kernel of p, lies in $\mathcal{Y}$. It follows that p is in $\mathfrak{T}\mathfrak{F}_\omega$. Hence any monic is a composition of a morphism in $\mathfrak{C}_\omega$

followed by a morphism in $\mathfrak{T}\mathfrak{F}_\omega$. Next let $\beta : B \to C$ be epic. Then consider the following exact commutative diagram:

$$\begin{array}{ccccccccc} 0 & \longrightarrow & K & \longrightarrow & B & \xrightarrow{\epsilon} & C & \longrightarrow & 0 \\ & & {\scriptstyle g^K}\downarrow & & {\scriptstyle i}\downarrow & & \| & & \\ 0 & \longrightarrow & Y^K & \longrightarrow & L & \xrightarrow{p} & C & \longrightarrow & 0 \end{array}$$

where the second row is induced by pushing-out the upper row along the left $\mathcal{Y}$-approximation of K. Then by definition p lies in $\mathfrak{T}\mathfrak{F}_\omega$. Since $\mathcal{Y}$ contains the injectives, g^K is monic. Since $(\mathcal{X}, \mathcal{Y})$ is a cotorsion pair, the cokernel of g^K, hence the cokernel of i, lies in $\mathcal{X}$. Hence i is in $\mathfrak{C}_\omega$ and this shows that any epic is a composition of a morphism in $\mathfrak{C}_\omega$ followed by a morphism in $\mathfrak{T}\mathfrak{F}_\omega$.

Assume now that α is arbitrary and consider the factorization $A \xrightarrow{(1_A,0)} A \oplus B \xrightarrow{{}^t(\alpha,1_B)} B$ of α. Since $(1_A, 0)$ is monic, there exists a factorization $(1_A, 0) : A \xrightarrow{i_1} D \xrightarrow{p_1} A \oplus B$, where $i_1 \in \mathfrak{C}_\omega$ and $p_1 \in \mathfrak{T}\mathfrak{F}_\omega$. Then the morphism $p_1 \circ {}^t(\alpha, 1_B) : D \to B$ is epic, hence it admits a factorization $p_1 \circ {}^t(\alpha, 1_B) : D \xrightarrow{i_2} E \xrightarrow{p_2} B$ where $i_2 \in \mathfrak{C}_\omega$ and $p_2 \in \mathfrak{T}\mathfrak{F}_\omega$. It suffices to show that the monic $i_1 \circ i_2 : A \to E$ has cokernel in $\mathcal{X}$. Since i_1 and i_2 are monics, there exists a short exact sequence $0 \to \mathrm{Coker}(i_1) \to \mathrm{Coker}(i_1 \circ i_2) \to \mathrm{Coker}(i_2) \to 0$. Since $\mathrm{Coker}(i_1), \mathrm{Coker}(i_2) \in \mathcal{X}$ and $\mathcal{X}$ is closed under extensions, it follows that $\mathrm{Coker}(i_1 \circ i_2) \in \mathcal{X}$. Hence $i := i_1 \circ i_2 \in \mathfrak{C}_\omega$, and consequently α admits a factorization $\alpha = i \circ p$ where $i \in \mathfrak{C}_\omega$ and $p := p_2 \in \mathfrak{T}\mathfrak{F}_\omega$.

Before we prove that the remaining axioms [M3] and [M2] hold, we recall from Chapter V, Section 3, that the good pair $(\mathcal{X}, \mathcal{Y})$ satisfies all the properties of Lemma 4.1. Hence a morphism $\alpha : A \to B$ is in $\mathfrak{W}_\omega$ iff $(\underline{X}, \underline{\alpha}) : (\underline{X}, \underline{A}) \to (\underline{X}, \underline{B})$ is invertible for any $X \in \mathcal{X}$.

[M3] Let $\alpha : A \to B$ be a retract of a morphism $\beta : C \to D$ in the category of morphisms. Choosing direct sum decompositions $C \xrightarrow{\cong} A \oplus A'$ and $D \xrightarrow{\cong} B \oplus B'$ it is easy to see that we have a commutative square

$$\begin{array}{ccc} C & \xrightarrow{\cong} & A \oplus A' \\ {\scriptstyle \beta}\downarrow & & \downarrow{\scriptstyle \begin{pmatrix} \alpha & 0 \\ 0 & \alpha' \end{pmatrix}} \\ D & \xrightarrow{\cong} & B \oplus B'. \end{array}$$

If $\beta \in \mathfrak{C}_\omega$, then β is monic with cokernel in $\mathcal{X}$. Then from the diagram above it follows that α is monic and its cokernel lies in $\mathcal{X}$ as a direct summand of $\mathrm{Coker}(\beta)$. If $\beta \in \mathfrak{F}_\omega$, i.e. β is ω-epic, then trivially so is α. Finally if $\beta \in \mathfrak{W}_\omega$, then obviously the same is true for α, using the description of morphisms in $\mathfrak{F}_\omega$.

[M2] We have to show that if two out of the three morphisms $\alpha, \beta, \alpha \circ \beta$ are in $\mathfrak{W}_\omega$, then so is the third. This follows from the description of morphisms of $\mathfrak{W}_\omega$ in Lemma 4.1.

Assume now that (ii) holds and consider the closed model structure of (i). It is not difficult to see that two morphisms $\alpha, \beta : A \to B$ in $\mathcal{C}$ are left homotopic iff $\alpha - \beta$ factors through an object in $\mathcal{Y}$ and they are right homotopic iff $\alpha - \beta$ factors through an object in ω. Since $\mathcal{C} \cap \mathcal{X} = \mathcal{X}$ is the full subcategory of cofibrant-fibrant objects, and a morphism between objects of $\mathcal{X}$ factors through an object in $\mathcal{Y}$ iff it factors through an object in ω, we infer that $\mathsf{Ho}(\mathcal{C})$ is equivalent to the stable category $\mathcal{X}/\omega$ and the canonical functor from $\mathcal{C}$ to the homotopy category $\mathsf{Ho}(\mathcal{C})$ is isomorphic to the functor $\mathcal{C} \to \mathcal{X}/\omega$, $A \mapsto \mathbf{R}(\underline{A})$, where as usual $\mathbf{R} : \mathcal{C}/\omega \to \mathcal{X}/\omega$ is the right adjoint of the inclusion $\mathcal{X}/\omega \hookrightarrow \mathcal{C}/\omega$. □

Let $(\mathcal{X}, \mathcal{Y})$ be a cotorsion pair in $\mathcal{C}$ with $\omega = \mathcal{X} \cap \mathcal{Y}$ contravariantly finite.

DEFINITION 4.3. The closed model structure $(\mathfrak{C}_\omega, \mathfrak{F}_\omega, \mathfrak{W}_\omega)$ of Theorem 4.2 is called the **projective ω-closed model structure** associated to the cotorsion pair $(\mathcal{X}, \mathcal{Y})$.

The following example shows that there are cotorsion pairs $(\mathcal{X}, \mathcal{Y})$ not defining a projective closed model structure.

EXAMPLE. Let Λ be a left coherent right IF-ring which is not right Noetherian, and consider the flat cotorsion pair $\big(\mathrm{Flat}(\Lambda), \mathrm{Flat}(\Lambda)^{\perp}\big)$ in $\mathrm{Mod}(\Lambda)$. Then as in the last example of Section V.4, the subcategory $\omega = \mathrm{Flat}(\Lambda) \cap \mathrm{Flat}(\Lambda)^{\perp}$ is not contravariantly finite. Hence in general there cannot exist a projective ω-closed model structure in $\mathrm{Mod}(\Lambda)$ induced by the flat cotorsion pair.

EXAMPLE. Let G be a group and k be a commutative ring of coefficients, and let B be the set of functions $G \to k$ which take only finitely many different values in k. If $\mathcal{X} = \{X \in \mathrm{Mod}(kG) \mid B \otimes_k X \text{ is projective}\}$ and ω is the full subcategory of projective kG-modules, then by the example after Remark V.4.7, we have a cotorsion pair $(\mathcal{X}, \widehat{\omega})$ in $\widehat{\mathcal{X}} = \{X \in \mathrm{Mod}(kG) \mid \mathrm{pd}_{kG} B \otimes_k X < \infty\}$. Since the subcategory ω of projective kG-modules is contravariantly finite and all closed model theoretic constructions of Theorem 4.2 can be performed in $\widehat{\mathcal{X}}$, Theorem 4.2 produces a closed model structure in $\widehat{\mathcal{X}}$. For this closed model structure the modules in $\mathcal{X}$ are the cofibrant objects and the modules of finite projective dimension are the trivially fibrant objects. The associated Quillen homotopy category is the stable category $\mathcal{X}/\omega$, which is triangulated by Proposition VI.1.2. Note that the existence of this closed model structure was first observed by D. Benson, see Theorem 10.2 in [**28**]. By the example after Remark V.4.7, the above closed model structure extends to a closed model structure in the whole module category $\mathrm{Mod}(kG)$, if k has finite global dimension and the group G is of type FP_∞ and belongs to Kropholler's class of groups $\mathbf{H}\mathfrak{F}$.

As mentioned in Chapter II that the homotopy category of an abelian closed model category $\mathcal{C}$ is pretriangulated. To state our next result we need to recall a definition from [**64**].

DEFINITION 4.4. A closed model structure in $\mathcal{C}$ is called **stable**, if the associated pretriangulated homotopy category is triangulated.

The following characterizes the stable closed model structures arising from cotorsion pairs.

COROLLARY 4.5. *Let $\mathcal{X}$ and $\mathcal{Y}$ be full subcategories of $\mathcal{C}$ and let $\omega = \mathcal{X} \cap \mathcal{Y}$. Then the following conditions are equivalent:*

(i) *The triple $(\mathfrak{C}_\omega, \mathfrak{F}_\omega, \mathfrak{W}_\omega)$ is a stable closed model structure.*

(ii) $(\mathcal{X}, \mathcal{Y})$ *is a cotorsion pair in* $\mathcal{C}$, *and a map is a projective* ω*-fibration* ($=\omega$*-epic) iff it is an epimorphism.*
(iii) $(\mathcal{X}, \mathcal{Y})$ *is a cotorsion pair in* $\mathcal{C}$ *with* $\mathcal{X} \cap \mathcal{Y} = \mathcal{P}$.

PROOF. By Theorem 4.2 we know that the homotopy category of the projective ω-closed model structure is equivalent to $\mathcal{X}/\omega$. As in Theorem VI.2.1 we have that the latter is triangulated iff $\mathcal{P} = \omega$. Hence (i) is equivalent to (iii). Clearly the class of projective ω-fibrations $\mathfrak{F}_\omega$, i.e. of ω-epics, coincides with the class of epimorphisms if and only if $\omega = \mathcal{P}$. Hence (ii) is equivalent to (iii). □

Using Theorem 4.2 we can now prove our second main result of this section, which gives a classification of closed model structures arising from cotorsion pairs in $\mathcal{C}$.

THEOREM 4.6. *The map* $\Phi : (\mathcal{X}, \mathcal{Y}) \longmapsto (\mathfrak{C}_{\mathcal{X}\cap\mathcal{Y}}, \mathfrak{F}_{\mathcal{X}\cap\mathcal{Y}}, \mathfrak{W}_{\mathcal{X}\cap\mathcal{Y}})$ *gives a bijective correspondence between cotorsion pairs* $(\mathcal{X}, \mathcal{Y})$ *in* $\mathcal{C}$ *with* $\mathcal{X} \cap \mathcal{Y}$ *contravariantly finite in* $\mathcal{C}$, *and closed model structures* $(\mathfrak{C}, \mathfrak{F}, \mathfrak{W})$ *in* $\mathcal{C}$ *such that:*

(i) *Any cofibration is monic and any object of* $\mathcal{C}$ *is fibrant.*
(ii) *Trivial fibrations are epics and any epic with kernel in* TFib *is a trivial fibration.*

The inverse bijection is given by $\Psi : (\mathfrak{C}, \mathfrak{F}, \mathfrak{W}) \longmapsto (\mathsf{Cof}, \mathsf{TFib})$.

The above bijection induces a bijective correspondence between cotorsion pairs $(\mathcal{X}, \mathcal{Y})$ *in* $\mathcal{C}$ *with* $\mathcal{X} \cap \mathcal{Y} = \mathcal{P}$, *stable closed model structures satisfying* (i), (ii), *and closed model structures* $(\mathfrak{C}, \mathfrak{F}, \mathfrak{W})$ *satisfying* (i), (ii), *and* (iii), *where:*

(iii) $\mathfrak{F}$ *consists of the epimorphisms.*

PROOF. If $(\mathcal{X}, \mathcal{Y})$ is a cotorsion pair in $\mathcal{C}$ with $\omega = \mathcal{X} \cap \mathcal{Y}$ contravariantly finite in $\mathcal{C}$, then by Theorem 4.2, $\Phi(\mathcal{X}, \mathcal{Y}) = (\mathfrak{C}_\omega, \mathfrak{F}_\omega, \mathfrak{W}_\omega)$ is a closed model structure in $\mathcal{C}$ which satisfies the conditions (i) and (ii). Finally observe that by Theorem 4.2 we have $\Psi\Phi(\mathcal{X}, \mathcal{Y}) = \Psi(\mathfrak{C}_\omega, \mathfrak{F}_\omega, \mathfrak{W}_\omega) = (\mathcal{X}, \mathcal{Y})$.

Conversely let $(\mathfrak{C}, \mathfrak{F}, \mathfrak{W})$ be a closed model structure in $\mathcal{C}$ satisfying the conditions (i) and (ii). Then by Theorem 3.5 we have that $\Psi(\mathfrak{C}, \mathfrak{F}, \mathfrak{W}) = (\mathsf{Cof}, \mathsf{TFib})$ is a cotorsion pair in $\mathcal{C}$. Since any object of $\mathcal{C}$ is fibrant, by Lemma 2.2 it follows that $\mathsf{Cof} \cap \mathsf{TFib} = \mathsf{TCof}$ is contravariantly finite.

It remains to show that $\Phi\Psi(\mathfrak{C}, \mathfrak{F}, \mathfrak{W}) = \Phi(\mathsf{Cof}, \mathsf{TFib}) = (\mathfrak{C}, \mathfrak{F}, \mathfrak{W})$. In other words we have to show that the closed model structures

$$(\mathfrak{C}_{\mathsf{Cof}\cap\mathsf{TFib}},\ \mathfrak{F}_{\mathsf{Cof}\cap\mathsf{TFib}},\ \mathfrak{W}_{\mathsf{Cof}\cap\mathsf{TFib}}) \quad \text{and} \quad (\mathfrak{C},\ \mathfrak{F},\ \mathfrak{W})$$

coincide. We set $\omega := \mathsf{Cof} \cap \mathsf{TFib}$ which is equal to TCof by the above argument. We first prove that $\mathfrak{C} = \mathfrak{C}_\omega$. If i is in $\mathfrak{C}$, then by (i) we have that i is monic. Since the cokernel of a cofibration is cofibrant we have that $\mathrm{Coker}(i)$ lies in Cof. Hence i is a projective ω-cofibration and this shows that $\mathfrak{C} \subseteq \mathfrak{C}_\omega$. Now let $i : A \to B$ be a projective ω-cofibration. Then we have a short exact sequence $0 \to A \xrightarrow{i} B \to C \to 0$ where C is cofibrant. To show that i is a projective ω-cofibration, by Lemma 1.3 it suffices to show i has the left lifting property with respect to all projective ω-trivial fibrations, which by definition are exactly the epimorphisms with kernel in TFib. This follows by the Extension-Lifting Lemma 3.1, since C is cofibrant and $\mathrm{Ext}^1(C, \mathsf{TFib}) = 0$. Next we show that $\mathfrak{F} = \mathfrak{F}_\omega$. Let $p : C \to D$ be a projective

ω-fibration, that is p is an ω-epic. To show that p is a fibration it suffices to show that p has the right lifting property with respect to all trivial cofibrations. Let $i : A \to B$ be a trivial cofibration; then Coker(i) is in $\mathsf{TCof} = \omega$. By Lemma 1.1, any morphism out of A to a fibrant object factors through i. Since all objects are fibrant, it follows that i is split monic. Hence i is a split monic with cokernel in ω, and then without loss of generality we can assume that i is given by $A \xrightarrow{(1_A,0)} A \oplus T$ where $T \in \omega$. Consider a commutative diagram

$$\begin{array}{ccc} A & \xrightarrow{\alpha} & C \\ {\scriptstyle (1_A,0)}\downarrow & & \downarrow{\scriptstyle p} \\ A \oplus T & \xrightarrow{{}^t(\kappa,\lambda)} & D \end{array}$$

Since T is in ω and p is an ω-epic, there exists a morphism $\tau : T \to C$ such that $\tau \circ p = \lambda$. Then the morphism ${}^t(\alpha, \tau) : A \oplus T \to C$ is a lifting of the above square. We infer that $\mathfrak{F}_\omega \subseteq \mathfrak{F}$. Conversely let $p : C \to D$ be a fibration. Let T be in ω and let $\alpha : T \to D$ be a morphism. Since $\omega = \mathsf{TCof}$ we have that $0 \to T$ is a trivial cofibration. Since any fibration has the right lifting property with respect to trivial cofibrations, the following commutative diagram

$$\begin{array}{ccc} 0 & \longrightarrow & C \\ \downarrow & & \downarrow{\scriptstyle p} \\ T & \xrightarrow{\alpha} & D \end{array}$$

admits a lifting, which means that α factors through p. Hence p is an ω-epic, that is $p \in \mathfrak{F}_\omega$, and this shows that $\mathfrak{F} \subseteq \mathfrak{F}_\omega$. We infer that $\mathfrak{F} = \mathfrak{F}_\omega$. It remains to show that $\mathfrak{W} = \mathfrak{W}_\omega$. This follows easily from the fact that in a closed model structure the fibrations together with the cofibrations determine the weak equivalences, see [**62**]. This completes the proof that the functions Φ and Ψ are mutually inverse. □

For later reference we call a closed model structure **projective closed model structure** if it satisfies the conditions (i) and (ii) of Theorem 4.6.

EXAMPLE. Let Λ be a right pure semisimple ring. Then there exists a bijective correspondence between projective closed model structures in Mod(Λ) and cotorsion pairs in Mod(Λ). This follows easily from Theorem 4.6 and the fact that over a right pure semisimle ring, any subcategory of Mod(Λ) which is closed under coproducts and direct summands is contravariantly finite.

Injective Closed Model Structures. What we proved so far in this section can be dualized. For the convenience of the reader we state the dual results, leaving their proof to the reader.

Keeping the set-up of this section we continue to assume that $\mathcal{X}$ and $\mathcal{Y}$ are full additive subcategories of the abelian category $\mathcal{C}$ which are closed under direct summands and isomorphisms, and we let $\omega = \mathcal{X} \cap \mathcal{Y}$.

In a dual manner we consider the following classes of morphisms in $\mathcal{C}$:

(i) $\mathfrak{C}^\omega$ is the class of ω-monics.
(ii) $\mathfrak{TC}^\omega$ is the class of monics with cokernel in $\mathcal{X}$.
(iii) $\mathfrak{F}^\omega$ is the class of epics with kernel in $\mathcal{Y}$.

(iv) $\mathfrak{T}\mathfrak{F}^\omega$ is the class of split epics with kernel in ω.

(v) $\mathfrak{W}^\omega$ is the class of morphisms α which admit a factorization $\alpha = \alpha_1 \circ \alpha_2$ where $\alpha_1 \in \mathfrak{T}\mathfrak{C}^\omega$ and $\alpha_2 \in \mathfrak{T}\mathfrak{F}^\omega$.

We call the morphisms in $\mathfrak{F}^\omega$, resp. $\mathfrak{T}\mathfrak{F}^\omega$, *injective* ω*-fibrations*, resp. *injective* ω*-trivial fibrations*. We call the morphisms in $\mathfrak{C}^\omega$, resp. $\mathfrak{T}\mathfrak{C}^\omega$, *injective* ω*-cofibrations*, resp. *injective* ω*-trivial cofibrations*. Finally we call the morphisms in $\mathfrak{W}^\omega$, *injective* ω*-weak equivalences*.

The following result, which we state without proof, gives the dual version of Theorems 4.2 and 4.6.

THEOREM 4.7. (1) *The following are equivalent:*

(i) *The triple* $(\mathfrak{C}^\omega, \mathfrak{F}^\omega, \mathfrak{W}^\omega)$ *defines a closed model structure on* $\mathcal{C}$ *such that:* $\mathfrak{C}^\omega$ *is the class of cofibrations,* $\mathfrak{F}^\omega$ *is the class of fibrations, and* $\mathfrak{W}^\omega$ *is the class of weak equivalences.*

(ii) $(\mathcal{X}, \mathcal{Y})$ *is a cotorsion pair and* ω *is covariantly finite in* $\mathcal{C}$.

If (ii) *holds, then* $\mathsf{Cof}^\omega = \mathcal{C}$ *and* $\mathsf{TFib}^\omega = \omega$, *and* $\mathsf{Fib}^\omega = \mathcal{Y}$ *and* $\mathsf{TCof}^\omega = \mathcal{X}$. *Moreover the associated Quillen homotopy category* $\mathsf{Ho}(\mathcal{C})$ *is equivalent to the stable left triangulated category* $\mathcal{Y}/\omega$.

(2) *The triple* $(\mathfrak{C}^\omega, \mathfrak{F}^\omega, \mathfrak{W}^\omega)$ *is a stable closed model structure in* $\mathcal{C}$ *if and only if* $(\mathcal{X}, \mathcal{Y})$ *is a cotorsion pair in* $\mathcal{C}$ *and the injective* ω*-cofibrations are the monomorphisms if and only if* $(\mathcal{X}, \mathcal{Y})$ *is a cotorsion pair in* $\mathcal{C}$ *with* $\mathcal{X} \cap \mathcal{Y} = \mathcal{I}$ *(*$\mathcal{I}$*, as always, denotes the full subcategory of injective objects of* $\mathcal{C}$*).*

(3) *The map* $\widehat{\Phi} : (\mathcal{X}, \mathcal{Y}) \longmapsto (\mathfrak{C}^{\mathcal{X}\cap\mathcal{Y}}, \mathfrak{F}^{\mathcal{X}\cap\mathcal{Y}}, \mathfrak{W}^{\mathcal{X}\cap\mathcal{Y}})$ *gives a bijective correspondence between cotorsion pairs* $(\mathcal{X}, \mathcal{Y})$ *in* $\mathcal{C}$ *with* $\mathcal{X} \cap \mathcal{Y}$ *covariantly finite in* $\mathcal{C}$ *and closed model structures* $(\mathfrak{C}, \mathfrak{F}, \mathfrak{W})$ *in* $\mathcal{C}$ *such that:*

(i) *Any fibration is epic and any object of* $\mathcal{C}$ *is cofibrant.*

(ii) *Trivial cofibrations are monics and any monic with cokernel in* TCof *is a trivial cofibration.*

The inverse bijection is given by: $\widehat{\Psi} : (\mathfrak{C}, \mathfrak{F}, \mathfrak{W}) \longmapsto (\mathsf{TCof}, \mathsf{Fib})$.

The above bijection induces a bijection between cotorsion pairs $(\mathcal{X}, \mathcal{Y})$ *in* $\mathcal{C}$ *with* $\mathcal{X} \cap \mathcal{Y} = \mathcal{I}$, *stable closed model structures satisfying* (i), (ii), *and closed model structures* $(\mathfrak{C}, \mathfrak{F}, \mathfrak{W})$ *satisfying* (i), (ii) *and* (iii), *where:*

(iii) $\mathfrak{C}$ *consists of the monomorphisms.*

Let $(\mathcal{X}, \mathcal{Y})$ be a cotorsion pair in $\mathcal{C}$ with $\omega = \mathcal{X} \cap \mathcal{Y}$ covariantly finite.

DEFINITION 4.8. (1) The closed model structure $(\mathfrak{C}^\omega, \mathfrak{F}^\omega, \mathfrak{W}^\omega)$ of part (1) of Theorem 4.7 is called the **injective ω-closed model structure** associated to the pair $(\mathcal{X}, \mathcal{Y})$.

(2) A closed model structure is called **injective closed model structure** if it satisfies the conditions (i) and (ii) of part (3) of Theorem 4.7.

EXAMPLE. Let Λ be a left coherent ring and let $\mathcal{X} = \mathrm{Flat}(\Lambda)$ be the category of flat right Λ-modules. Then $\omega = \mathcal{X} \cap \mathcal{X}^\perp$ which, by Section V.4, is the category of flat and pure-injective modules, is covariantly finite. Since $(\mathcal{X}, \mathcal{X}^\perp)$ is a cotorsion pair, by Theorem 4.7 we get an injective closed model structure in $\mathrm{Mod}(\Lambda)$ where the flat modules are the trivially cofibrant objects, and the modules in $\mathcal{X}^\perp$, widely known as the cotorsion modules, are the fibrant objects. Note that for this closed model structure, ω is the category of cofibrant and trivially fibrant objects, which by the Example after Definition 4.3 is not always contravariantly finite.

EXAMPLE. Let Λ be a right pure semisimple ring. Then there exists a bijective correspondence between injective closed model structures in $\mathrm{Mod}(\Lambda)$ and cotorsion pairs in $\mathrm{Mod}(\Lambda)$. This follows easily from Theorem 4.7 and the fact that over a right pure semisimple ring, any subcategory of $\mathrm{Mod}(\Lambda)$ which is closed under products and direct summands is covariantly finite. The above correspondence combined with the correspondence of the Example after Theorem 4.6, shows that, over a right pure semisimpe ring Λ, there exist bijective correspondences between:

- Cotorsion pairs in $\mathrm{Mod}(\Lambda)$.
- Projective closed model structures in $\mathrm{Mod}(\Lambda)$.
- Injective closed model structures in $\mathrm{Mod}(\Lambda)$.

Combining Theorems 4.6 and 4.7 we have the following consequence.

COROLLARY 4.9. *Assume that $\mathcal{C}$ has enough projective and injective objects. Then $\mathcal{C}$ is Frobenius if and only if $(\mathfrak{C}_\omega, \mathfrak{F}_\omega, \mathfrak{W}_\omega)$ and $(\mathfrak{C}^\omega, \mathfrak{F}^\omega, \mathfrak{W}^\omega)$ are stable closed model structures in $\mathcal{C}$.*

Frobenius and Functorial Closed Model Structures. We have seen that Theorems 4.6 and 4.7 give descriptions of projective or injective closed model structures in $\mathcal{C}$ in terms of suitable cotorsion pairs. Hence it is natural to ask the following question: What is the structure of an abelian category which admits a closed model structure $(\mathfrak{C}, \mathfrak{F}, \mathfrak{W})$ which is both projective and injective? Let us call such a closed model structure $(\mathfrak{C}, \mathfrak{F}, \mathfrak{W})$ **Frobenius**. We have the following result which explains the terminology and in addition shows that a Frobenius closed model structure, if it exists, is uniquely determined and stable.

THEOREM 4.10. *Let $\mathcal{C}$ be an abelian category.*

(i) *$\mathcal{C}$ is Frobenius if and only if it admits a Frobenius closed model structure.*
(ii) *For a closed model structure $(\mathfrak{C}, \mathfrak{F}, \mathfrak{W})$ in $\mathcal{C}$, the following are equivalent:*
  (a) *$(\mathfrak{C}, \mathfrak{F}, \mathfrak{W})$ is Frobenius.*
  (b) *$(\mathfrak{C}, \mathfrak{F}, \mathfrak{W})$ is stable, projective and all objects are cofibrant.*
  (c) *$(\mathfrak{C}, \mathfrak{F}, \mathfrak{W})$ is stable, injective and all objects are fibrant.*
  (d) *$\mathfrak{C}$ is the class of monomorphisms, $\mathfrak{F}$ is the class of epimorphisms, and $\mathfrak{W}$ is the class of stable equivalences, that is, the morphisms which are isomorphisms in the stable category modulo projectives or injectives.*

If $(\mathfrak{C}, \mathfrak{F}, \mathfrak{W})$ is a Frobenius closed model structure in $\mathcal{C}$, then the associated homotopy category is the stable category $\mathcal{C}/\mathcal{P}$.

PROOF. (i) ($\Rightarrow$) Assume that $\mathcal{C}$ is Frobenius. Then we have the cotorsion pairs $(\mathcal{C}, \mathcal{P})$ and $(\mathcal{I}, \mathcal{C})$ in $\mathcal{C}$ with $\mathcal{P} = \mathcal{I}$ functorially finite. It follows easily that the closed model structures arising, via Theorems 4.2 and 4.7, from these cotorsion pairs coincide. Hence the resulting closed model structure in $\mathcal{C}$ is Frobenius.

($\Leftarrow$) Let $(\mathfrak{C}, \mathfrak{F}, \mathfrak{W})$ be a Frobenius closed model structure in $\mathcal{C}$. Then by definition all objects of $\mathcal{C}$ are fibrant and cofibrant, and by Theorems 4.2 and 4.7 we have cotorsion pairs $(\mathcal{C}, \mathsf{TFib})$ and $(\mathsf{TCof}, \mathcal{C})$. Hence $\mathsf{TFib} = \mathcal{I}$ and $\mathsf{TCof} = \mathcal{P}$. By Proposition 2.1, TFib is covariantly finite and TCof is contravariantly finite. Hence $\mathcal{C}$ has enough projective and injective objects. Since all objects are fibrant and cofibrant, by Lemma 2.2 we have $\mathsf{TFib} = \mathsf{TCof}$. Hence $\mathcal{P} = \mathcal{I}$, so that $\mathcal{C}$ is Frobenius.

(ii) (a) $\Rightarrow$ (d) If $(\mathfrak{C}, \mathfrak{F}, \mathfrak{W})$ is a Frobenius closed model structure in $\mathcal{C}$, then, by part (i), it is induced by the cotorsion triple $(\mathcal{C}, \mathcal{P} = \mathcal{I}, \mathcal{C})$; in particular we have $\mathsf{TFib} = \mathcal{I} = \mathcal{P} = \mathsf{TCof}$ and $\mathsf{Cof} = \mathcal{C} = \mathsf{Fib}$. Then the description of the closed model structure $(\mathfrak{C}, \mathfrak{F}, \mathfrak{W})$ follows from Proposition 3.9.

(d) $\Rightarrow$ (a) Since $\mathfrak{C}$ is the class of monomorphisms and $\mathfrak{F}$ is the class of epimorphisms, we infer that all objects of $\mathcal{C}$ are fibrant and cofibrant. By Proposition 2.3 we have $\mathsf{TCof} = \mathcal{P}$ and $\mathsf{TFib} = \mathcal{I}$. Then by Lemma 2.2 we conclude that $\mathcal{P} = \mathcal{I}$, so $\mathcal{C}$ is Frobenius. Then part (i) ensures that the closed model structure $(\mathfrak{C}, \mathfrak{F}, \mathfrak{W})$ is Frobenius.

(a) $\Rightarrow$ (b) Let $(\mathfrak{C}, \mathfrak{F}, \mathfrak{W})$ be a Frobenius closed model structure in $\mathcal{C}$. Then $(\mathfrak{C}, \mathfrak{F}, \mathfrak{W})$ is projective and injective, and, by part (i), all objects are fibrant and cofibrant. By Theorem 4.2 it follows that the associated homotopy category is equivalent to the stable category $\mathcal{C}/\mathcal{P}$. The latter is triangulated since, by part (i), $\mathcal{C}$ is Frobenius. Hence the closed model structure $(\mathfrak{C}, \mathfrak{F}, \mathfrak{W})$ is stable.

(b) $\Rightarrow$ (a) If $(\mathfrak{C}, \mathfrak{F}, \mathfrak{W})$ is a stable projective closed model structure with all objects cofibrant, then by Theorem 4.6 we have that $\mathcal{C}/\mathcal{P}$ is triangulated, and this implies that $\mathcal{C}$ is Frobenius. Hence by part (i) the closed model structure $(\mathfrak{C}, \mathfrak{F}, \mathfrak{W})$ is Frobenius.

The proof of the equivalence (c) $\Leftrightarrow$ (a) is similar and is left to the reader. □

Now let $(\mathcal{X}, \mathcal{Y})$ be a cotorsion pair in $\mathcal{C}$ with $\mathcal{X} \cap \mathcal{Y} := \omega$. If ω is contravariantly finite, then we can define in $\mathcal{C}$ the projective ω-closed structure $(\mathfrak{C}_\omega, \mathfrak{F}_\omega, \mathfrak{W}_\omega)$, and when ω is covariantly finite, we can define in $\mathcal{C}$ the injective ω-closed structure $(\mathfrak{C}^\omega, \mathfrak{F}^\omega, \mathfrak{W}^\omega)$. If ω is functorially finite, so that both closed model structures are defined, then there is an additional closed model structure defined in $\mathcal{C}$ which is induced by ω. Indeed, by [**24**] any functorially finite subcategory ω defines a closed model structure $(\mathfrak{C}(\omega), \mathfrak{F}(\omega), \mathfrak{W}(\omega))$ in $\mathcal{C}$, with cofibrations the ω-monics, fibrations the ω-epics and weak equivalences the morphisms which are isomorphisms in $\mathcal{C}/\omega$. Actually, by [**24**], the map $\omega \mapsto (\mathfrak{C}(\omega), \mathfrak{F}(\omega), \mathfrak{W}(\omega))$ gives a bijection between functorially finite subcategories of $\mathcal{C}$ and closed model structures in $\mathcal{C}$ with all objects fibrant and cofibrant. The inverse is given by $(\mathfrak{C}, \mathfrak{F}, \mathfrak{W}) \mapsto \mathsf{TFib} = \mathsf{TCof}$. This suggests the following definition which will be useful in the next section.

DEFINITION 4.11. (1) If ω is a functorially finite subcategory of $\mathcal{C}$, then the closed model structure $(\mathfrak{C}(\omega), \mathfrak{F}(\omega), \mathfrak{W}(\omega))$ is called the **functorial ω-closed model structure**.

(2) A closed model structure in $\mathcal{C}$ is called **functorial** if all objects of $\mathcal{C}$ are fibrant and cofibrant.

For instance a Frobenius closed model structure is functorial, more precisely it is of the form $(\mathfrak{C}(\mathcal{P}), \mathfrak{F}(\mathcal{P}), \mathfrak{W}(\mathcal{P}))$. Note also that by [**24**], the Quillen homotopy category of a functorial closed model structure is the stable category $\mathcal{C}/\omega_{cf}$, where $\omega_{cf} = \mathsf{TCof} = \mathsf{TFib}$. In particular the functorial ω-closed model structure $(\mathfrak{C}(\omega), \mathfrak{F}(\omega), \mathfrak{W}(\omega))$ arising from a functorially finite subcategory ω, is the stable pretriangulated category $\mathcal{C}/\omega$.

REMARK 4.12. Observe that between the triples $(\mathfrak{C}_\omega, \mathfrak{F}_\omega, \mathfrak{W}_\omega)$, $(\mathfrak{C}^\omega, \mathfrak{F}^\omega, \mathfrak{W}^\omega)$ we have the following relations:

(i) $\mathfrak{F}^\omega = \mathfrak{T}\mathfrak{F}_\omega$ which are epics and $\mathfrak{C}_\omega = \mathfrak{T}\mathfrak{C}^\omega$ which are monics.
(ii) $\mathsf{TCof}_\omega = \mathsf{Cof}_\omega \cap \mathsf{TFib}_\omega$ and $\mathsf{TFib}^\omega = \mathsf{Fib}^\omega \cap \mathsf{TCof}^\omega$.
(iii) $\mathsf{TCof}_\omega = \omega = \mathsf{TFib}^\omega$ and $\mathsf{Fib}_\omega = \mathcal{C} = \mathsf{Cof}^\omega$.

REMARK 4.13. As we pointed out in the Introduction, M. Hovey in [**64**] initiated independently a study of the connections between cotorsion pairs and closed model structures in an abelian category. The closed model structures constructed by Hovey are quite different. First note that Hovey uses two cotorsion pairs to construct a closed model structure, where fibrations are epimorphisms but in general not all objects are fibrant. We use one cotorsion pair to construct a closed model structure, where all objects are fibrant but the fibrations are not in general epimorphisms. The following example illustrates the difference.

Let Λ be an Artin algebra and let T be a finitely presented tilting module. Then by the dual of Proposition V.5.1, we have a cotorsion pair $(\widetilde{\mathrm{add}(T)}, T^\perp)$ in $\mathrm{mod}(\Lambda)$ such that $\widetilde{\mathrm{add}(T)} \cap T^\perp = \mathrm{add}(T)$ is contravariantly finite. Hence our Theorem 4.2 produces a closed model structure where the fibrations are $\mathrm{add}(T)$-epimorphisms. If T is not projective then the fibrations are not epimorphisms, hence the closed model structure we produce does not appear amongst those constructed by Hovey (for any proper class of short exact sequences in the setting of [**64**]).

Conversely if we assume the existence of a closed model structure constructed as in [**64**], then in general we can not recover it by using our Theorem 4.2 since in our case all objects are fibrant.

It is not difficult to see that our closed model structure constructed in Theorem 4.2 coincides with the closed model sructure constructed by Hovey if all objects are fibrant and the class of trivially cofibrant objects coincide with the projective objects of $\mathcal{C}$. In this case the two identical closed model structures are stable.

5. A Classification of (Co)Torsion Pairs

In this section we classify in terms of closed model structures all cotorsion pairs $(\mathcal{X}, \mathcal{Y})$ in $\mathcal{C}$ with the property that $\mathcal{X} \cap \mathcal{Y}$ is functorially finite in $\mathcal{C}$. This classification will be used for the classification of torsion pairs in the stable category of $\mathcal{C}$ modulo a functorially finite subcategory. We investigate briefly also the connection between cotorsion triples and closed model structures. We apply our results to categories of complexes, obtaining in this way the well-known descriptions of the derived category as the homotopy category of suitable closed model structures, and to tilting theory where we obtain a classification of (co)tilting modules over an Artin algebra.

If $(\mathcal{X}, \mathcal{Y})$ is a cotorsion pair in $\mathcal{C}$ with $\omega = \mathcal{X} \cap \mathcal{Y}$ functorially finite, then by Remark 4.11 there are several compatibility conditions between the projective ω-closed model structure $(\mathfrak{C}_\omega, \mathfrak{F}_\omega, \mathfrak{W}_\omega)$ and the injective ω-closed model structure $(\mathfrak{C}^\omega, \mathfrak{F}^\omega, \mathfrak{W}^\omega)$. This suggests the following definition, which can be considered as "glueing" together closed model structures.

DEFINITION 5.1. An (ordered) pair of closed model structures $(\mathfrak{C}_1, \mathfrak{F}_1, \mathfrak{W}_1)$ and $(\mathfrak{C}_2, \mathfrak{F}_2, \mathfrak{W}_2)$ in $\mathcal{C}$ are called **compatible** if:

(i) $\mathfrak{F}_2 = \mathfrak{T}\mathfrak{F}_1$ which are epics and $\mathfrak{C}_1 = \mathfrak{T}\mathfrak{C}_2$ which are monics.

(ii) $\mathsf{TCof}_1 = \mathsf{TFib}_2$.
(iii) $\mathsf{Fib}_1 = \mathcal{C} = \mathsf{Cof}_2$.

Let $(\mathfrak{C}_1, \mathfrak{F}_1, \mathfrak{W}_1)$ and $(\mathfrak{C}_2, \mathfrak{F}_2, \mathfrak{W}_2)$ be a pair of compatible closed model structures in $\mathcal{C}$. It follows easily from the definition and Lemma 1.3 that $\mathfrak{T}\mathfrak{F}_2 \subseteq \mathfrak{F}_1$ and $\mathfrak{T}\mathfrak{C}_1 \subseteq \mathfrak{C}_2$. Then by Proposition 2.3 we have the following relations:

$$\mathsf{TCof}_2 = \mathsf{Cof}_1, \quad \mathsf{TFib}_1 = \mathsf{Fib}_2, \quad \mathsf{TCof}_1 \subseteq \mathsf{Cof}_2, \quad \mathsf{TFib}_2 \subseteq \mathsf{Fib}_1.$$

Moreover by Proposition 2.1, $\mathsf{TCof}_1 = \mathsf{TFib}_2 := \omega$ is a functorially finite subcategory of $\mathcal{C}$. Hence the functorial ω-closed model structure $(\mathfrak{C}(\omega), \mathfrak{F}(\omega), \mathfrak{W}(\omega))$ is defined in $\mathcal{C}$. This motivates us to make the following definition.

DEFINITION 5.2. Let $(\mathfrak{C}_1, \mathfrak{F}_1, \mathfrak{W}_1)$ and $(\mathfrak{C}_2, \mathfrak{F}_2, \mathfrak{W}_2)$ be a pair of compatible closed model structures in $\mathcal{C}$. Their **intersection** $(\mathfrak{C}_1, \mathfrak{F}_1, \mathfrak{W}_1) \cap (\mathfrak{C}_2, \mathfrak{F}_2, \mathfrak{W}_2)$ is defined to be the functorial ω-closed model structure $(\mathfrak{C}(\omega), \mathfrak{F}(\omega), \mathfrak{W}(\omega))$, where $\mathsf{TCof}_1 = \mathsf{TFib}_2 := \omega$.

Using the notion of compatible closed model structures and Theorems 4.6 and 4.7, we arrive at the following consequence which gives a classification of cotorsion pairs $(\mathcal{X}, \mathcal{Y})$ in $\mathcal{C}$ with functorially finite intersection $\mathcal{X} \cap \mathcal{Y}$, in terms of closed model structures.

THEOREM 5.3. *If the abelian category $\mathcal{C}$ has enough projectives and enough injectives, then there exists a bijective correspondence between cotorsion pairs $(\mathcal{X}, \mathcal{Y})$ in $\mathcal{C}$ with $\mathcal{X} \cap \mathcal{Y}$ functorially finite in $\mathcal{C}$ and compatible closed model structures $(\mathfrak{C}_1, \mathfrak{F}_1, \mathfrak{W}_1)$ and $(\mathfrak{C}_2, \mathfrak{F}_2, \mathfrak{W}_2)$ in $\mathcal{C}$. The bijection is given as follows:*

$$(\mathcal{X}, \mathcal{Y}) \longmapsto \Big\{ (\mathfrak{C}_{\mathcal{X}\cap\mathcal{Y}}, \mathfrak{F}_{\mathcal{X}\cap\mathcal{Y}}, \mathfrak{W}_{\mathcal{X}\cap\mathcal{Y}}), \ (\mathfrak{C}^{\mathcal{X}\cap\mathcal{Y}}, \mathfrak{F}^{\mathcal{X}\cap\mathcal{Y}}, \mathfrak{W}^{\mathcal{X}\cap\mathcal{Y}}) \Big\}$$

$$\Big\{ (\mathfrak{C}_1, \mathfrak{F}_1, \mathfrak{W}_1), \ (\mathfrak{C}_2, \mathfrak{F}_2, \mathfrak{W}_2) \Big\} \longmapsto (\mathsf{Cof}_1, \mathsf{Fib}_2).$$

The following consequence of Theorem 5.3 shows that if we have a pair of compatible closed model structures in $\mathcal{C}$, then we obtain a torsion pair in the homotopy category of the closed model structure of their intersection; moreover any torsion pair in the homotopy category is obtained in this way, provided that the torsion and the torsion free subcategories are orthogonal with respect to Ext^1.

THEOREM 5.4. *Assume that the abelian category $\mathcal{C}$ has enough projectives and enough injectives and let ω be a functorially finite subcategory of $\mathcal{C}$. Then the correspondence of Theorem 5.3 induces a bijective correspondence*

$$\Big\{ \textit{torsion pairs } (\mathcal{X}/\omega, \mathcal{Y}/\omega) \textit{ in } \mathcal{C}/\omega \textit{ with } \mathrm{Ext}^1(\mathcal{X}, \mathcal{Y}) = 0 \Big\} \longmapsto$$

$$\Big\{ \textit{compatible closed model structures } (\mathfrak{C}_1, \mathfrak{F}_1, \mathfrak{W}_1) \textit{ and } (\mathfrak{C}_2, \mathfrak{F}_2, \mathfrak{W}_2) \textit{ in } \mathcal{C}$$
$$\textit{such that: } \mathsf{Cof}_1 \textit{ is closed under kernels of epics, or equivalently}$$
$$\mathsf{Fib}_2 \textit{ is closed under cokernels of monics, and: } \mathsf{Cof}_1 \cap \mathsf{Fib}_2 = \omega \Big\}.$$

Note that under the correspondence of Theorem 5.4:

(i) The stable pretriangulated category $\mathcal{C}/\omega$ is the homotopy category of the functorial ω-closed model structure and corresponds to the homotopy category of the intersection $(\mathfrak{C}_1, \mathfrak{F}_1, \mathfrak{W}_1) \cap (\mathfrak{C}_2, \mathfrak{F}_2, \mathfrak{W}_2)$.

(ii) The torsion class $\mathcal{X}/\omega$ is the homotopy category of the projective ω-closed model structure and corresponds to the homotopy category of the closed model structure $(\mathfrak{C}_1, \mathfrak{F}_1, \mathfrak{W}_1)$.
(iii) The torsion-free class $\mathcal{Y}/\omega$ is the homotopy category of the injective ω-closed model structure and corresponds to the homotopy category of the closed model structure $(\mathfrak{C}_2, \mathfrak{F}_2, \mathfrak{W}_2)$.

REMARK 5.5. If $\mathcal{C} = \text{mod}(\Lambda)$ is the category of finitely generated Λ-modules over an Artin algebra Λ, then by a recent result of Krause-Solberg **[79]** it follows that $\mathcal{X} \cap \mathcal{Y}$ is functorially finite in $\text{mod}(\Lambda)$, for any (resolving) cotorsion pair $(\mathcal{X}, \mathcal{Y})$ in $\text{mod}(\Lambda)$. Hence in this case the assumption that $\mathcal{X} \cap \mathcal{Y}$ is functorially finite in Theorems 5.3 and 5.4 can be removed.

It follows by the above result that torsion pairs appear as "glueing" together homotopy categories of compatible closed model structures inside the homotopy category of the closed model structure of their intersection. This gives a conceptual homotopy-theoretic interpretation of the theory developed in Chapters V and VI, and puts it in the proper framework.

The correspondences given by the above theorems combined with the correspondence of Proposition VI.2.5 allow us to deduce the following.

COROLLARY 5.6. *Let $\mathcal{C}$ be a Frobenius abelian category. Then there exist bijective correspondences between:*
(i) *Compatible stable closed model structures in $\mathcal{C}$.*
(ii) *Resolving cotorsion pairs $(\mathcal{X}, \mathcal{Y})$ in $\mathcal{C}$ with $\mathcal{X} \cap \mathcal{Y}$ the projectives $\mathcal{P}$.*
(iii) *Hereditary torsion pairs in $\mathcal{C}/\mathcal{P}$.*

We can define closed model structures in $\mathcal{C}$ starting from a subcategory instead of $\mathcal{X}, \mathcal{Y}$ and ω. Indeed, let $\mathcal{X}$ be a full preresolving subcategory of $\mathcal{C}$. Define $\mathfrak{C}_{\mathcal{X}}$ to be the class of monomorphisms with cokernel in $\mathcal{X}$, $\mathfrak{F}_{\mathcal{X}}$ to be the class of all epics, $\mathfrak{T}\mathfrak{C}_{\mathcal{X}}$ to be the class of all monics with projective cokernel, $\mathfrak{T}\mathfrak{F}_{\mathcal{X}}$ to be the class of epics with kernel in $\mathcal{X}^{\perp}$, and finally $\mathfrak{W}_{\mathcal{X}}$ to be the class of morphisms which can be written as a composition of a morphism in $\mathfrak{T}\mathfrak{C}_{\mathcal{X}}$ followed by a morphism in $\mathfrak{T}\mathfrak{F}_{\mathcal{X}}$.

Dually, for a precoresolving subcategory $\mathcal{Y}$ of $\mathcal{C}$, define $\mathfrak{C}^{\mathcal{Y}}$ to be the class of all monics, $\mathfrak{F}^{\mathcal{Y}}$ to be the class of all epics with kernel in $\mathcal{Y}$, $\mathfrak{T}\mathfrak{C}^{\mathcal{Y}}$ to be the class of all monics with cokernel in ${}^{\perp}\mathcal{Y}$, $\mathfrak{T}\mathfrak{F}^{\mathcal{Y}}$ to be the class of epics with injective kernel, and finally $\mathfrak{W}^{\mathcal{Y}}$ to be the class of morphisms which can be written as a composition of a morphism in $\mathfrak{T}\mathfrak{C}^{\mathcal{Y}}$ followed by a morphism in $\mathfrak{T}\mathfrak{F}^{\mathcal{Y}}$.

THEOREM 5.7. (1) *Assume that $\mathcal{C}$ is Krull-Schmidt or else $\mathcal{C}$ has filtered colimits and $\mathcal{X}$ is closed under filtered colimits. If $\mathcal{X} \cap \mathcal{X}^{\perp} = \mathcal{P}$, then the triple $(\mathfrak{C}_{\mathcal{X}}, \mathfrak{F}_{\mathcal{X}}, \mathfrak{W}_{\mathcal{X}})$ is a (stable) closed model structure in $\mathcal{C}$ if and only if $\mathcal{X}$ is contravariantly finite.*

(2) *Assume that $\mathcal{C}$ is Krull-Schmidt or else $\mathcal{C}$ is Grothendieck and $\mathfrak{T}\mathfrak{C}^{\mathcal{Y}}$ is closed under filtered colimits. If ${}^{\perp}\mathcal{Y} \cap \mathcal{Y} = \mathcal{I}$, then the triple $(\mathfrak{C}^{\mathcal{Y}}, \mathfrak{F}^{\mathcal{Y}}, \mathfrak{W}^{\mathcal{Y}})$ is a (stable) closed model structure in $\mathcal{C}$ if and only if $\mathcal{Y}$ is covariantly finite.*

PROOF. (1) If the preresolving subcategory $\mathcal{X}$ is contravariantly finite, then either of the imposed assumptions on $\mathcal{C}$ and/or $\mathcal{X}$ implies that $(\mathcal{X}, \mathcal{X}^{\perp})$ is a cotorsion pair in $\mathcal{C}$, see the example at the end of Section V.4. Since $\mathcal{X} \cap \mathcal{X}^{\perp} = \mathcal{P}$ is contravariantly finite, by Theorem 4.2 we have a closed model structure $(\mathfrak{C}_{\mathcal{X}}, \mathfrak{F}_{\mathcal{X}}, \mathfrak{W}_{\mathcal{X}})$

and obviously $\mathsf{Cof}_{\mathcal{X}} = \mathcal{X}$, $\mathsf{Fib}_{\mathcal{Y}} = \mathcal{C}$, $\mathsf{TCof}_{\mathcal{X}} = \mathcal{P}$, and $\mathsf{TFib}_{\mathcal{X}} = \mathcal{X}^{\perp}$. The converse follows from Proposition 2.1, since $\mathsf{Cof}_{\mathcal{X}} = \mathcal{X}$. The proof of (2) is dual. □

A Classification of (Co)Tilting Modules. We have seen that for any tilting or cotilting module T over an Artin algebra Λ, there is associated in a natural way a cotorsion pair. Since $\omega = \mathrm{add}(T)$ is functorially finite, Theorems 4.6 and 4.7 can be applied. It follows that any tilting or cotilting module defines natural closed model structures in $\mathrm{mod}(\Lambda)$. We leave to the reader to write down explicitly the associated closed model structures.

We only note that as an application of Theorems 4.6 and 4.7 we can obtain the following classification of cotilting modules in terms of closed model structures. First we recall that a module T is called *basic* if the indecomposable summands in a direct sum decomposition of T are non-isomorphic.

THEOREM 5.8. *If Λ is a basic Artin algebra, then there exist bijective correspondences between:*

(I) *Isomorphism classes of basic cotilting modules T in* $\mathrm{mod}(\Lambda)$.

(II) *Closed model structures in* $\mathrm{mod}(\Lambda)$ *satisfying the following conditions:*
- (a) *Any fibration is epic and any finitely generated Λ-module is cofibrant.*
- (b) *Trivial cofibrations are monics and any monic with trivially cofibrant cokernel is a trivial cofibration.*
- (c) *For any trivial cofibration $K \longrightarrow X$ where X is trivially cofibrant, the module K is trivially cofibrant.*
- (d) *Any fibrant module admits a finite exact resolution by trivially cofibrant modules.*

(III) *Closed model structures in* $\mathrm{mod}(\Lambda)$ *satisfying the following conditions:*
- (a) *Any cofibration is monic and any finitely generated Λ-module is fibrant.*
- (b) *Trivial fibrations are epics and any epic with trivially fibrant kernel is a trivial fibration.*
- (c) *For any trivial fibration $Y \longrightarrow C$ where Y is trivially fibrant, the module C is trivially fibrant.*
- (d) *Any trivially fibrant module admits a finite exact resolution by cofibrant modules.*

The mutually inverse correspondences above are given as follows:

$$T \longmapsto (\mathfrak{C}^{\mathrm{add}(T)}, \mathfrak{F}^{\mathrm{add}(T)}, \mathfrak{W}^{\mathrm{add}(T)}) \qquad \mathbf{(I)} \longrightarrow \mathbf{(II)}$$

$$(\mathfrak{C}, \mathfrak{F}, \mathfrak{W}) \longmapsto T, \ \textit{where}\ \mathsf{TCof} \cap \mathsf{Fib} = \mathrm{add}(T) \qquad \mathbf{(II)} \longrightarrow \mathbf{(I)}$$

$$T \longmapsto (\mathfrak{C}_{\mathrm{add}(T)}, \mathfrak{F}_{\mathrm{add}(T)}, \mathfrak{W}_{\mathrm{add}(T)}) \qquad \mathbf{(I)} \longrightarrow \mathbf{(III)}$$

$$(\mathfrak{C}, \mathfrak{F}, \mathfrak{W}) \longmapsto T, \ \textit{where}\ \mathsf{Cof} \cap \mathsf{TFib} = \mathrm{add}(T) \qquad \mathbf{(III)} \longrightarrow \mathbf{(I)}$$

If T is a basic cotilting module, then we have the following:
- (i) *The closed model structure of part* **(II)** *is stable if and only if T is isomorphic to* $\mathrm{D}(\Lambda)$, *in which case the associated homotopy category is trivial.*
- (ii) *The closed model structure of part* **(III)** *is stable if and only if T is isomorphic to Λ, in which case the associated homotopy category is equivalent to the stable category* $\underline{\mathrm{CM}}(\Lambda)$ *of Cohen-Macaulay modules modulo projectives, and Λ is a Gorenstein algebra.*

PROOF. If T is a basic cotilting Λ-module, then by a basic result of Auslander-Reiten, see [**9**], we have a cotorsion pair $({}^{\perp}T, \widehat{\mathrm{add}(T)})$ in $\mathrm{mod}(\Lambda)$ such that $\widehat{{}^{\perp}T} = \mathrm{mod}(\Lambda)$. Since $\mathrm{add}(T)$ is covariantly finite, by Theorem 4.7 we have the injective $\mathrm{add}(T)$-closed model structure $(\mathfrak{C}^{\mathrm{add}(T)}, \mathfrak{F}^{\mathrm{add}(T)}, \mathfrak{W}^{\mathrm{add}(T)})$ in $\mathrm{mod}(\Lambda)$, where $\mathsf{TCof}^{\mathrm{add}(T)} = {}^{\perp}T$ and $\mathsf{Fib}^{\mathrm{add}(T)} = \widehat{\mathrm{add}(T)}$. In particular the closed model structure satisfies (i) and (ii). Now let $\beta : K \to X$ be a trivial cofibration where X is trivially cofibrant. Then by Lemma 1.1 we have that $\mathrm{Coker}(\beta)$ is trivially cofibrant. Since X is trivially cofibrant and $\mathsf{TCof}^{\mathrm{add}(T)}$ is resolving, we infer that K is trivially cofibrant, i.e. (iii) holds. Finally property (iv) holds since $\widehat{{}^{\perp}T} = \mathrm{mod}(\Lambda)$. Conversely let $(\mathfrak{C}, \mathfrak{F}, \mathfrak{W})$ be a closed model structure in $\mathrm{mod}(\Lambda)$ satisfying (i)-(iv). The first two properties ensure, in view of Theorem 4.7, that $(\mathsf{TCof}, \mathsf{Fib})$ is a cotorsion pair in $\mathrm{mod}(\Lambda)$. Then property (iii) implies that TCof is resolving. Finally let C be a finitely generated Λ-module. By Proposition 2.1 there exists a short exact sequence $0 \to F_C \to X_C \xrightarrow{f} C$ where f is a fibration, F_C is fibrant and X_C is trivially cofibrant. Since fibrations are epics and any fibrant module admits a finite exact resolution by trivially cofibrant modules, it follows that any module admits a finite exact resolution by trivially cofibrant modules. In other words TCof is a contravariantly finite resolving subcategory of $\mathrm{mod}(\Lambda)$ satisfying the property that $\widehat{\mathsf{TCof}} = \mathrm{mod}(\Lambda)$. Then by [**9**] we infer that $\mathsf{TCof} \cap \mathsf{Fib} = \mathrm{add}(T)$ for a cotilting module T. Combining Theorem 4.7 and the results of [**9**] we conclude that the correspondences (**I**) $\longrightarrow$ (**II**) and (**II**) $\longrightarrow$ (**I**) are mutually inverse. Using Theorem 4.6 a similar argument shows that the correspondences (**I**) $\longrightarrow$ (**III**) and (**III**) $\longrightarrow$ (**I**) are mutually inverse.

Finally if T is a basic cotilting module, then by Theorem 4.7 we have that the homotopy category of the closed model structure of part (**II**) is equivalent to the stable category $\widehat{\mathrm{add}(T)}/\mathrm{add}(T)$, which is stable if and only if $\mathrm{add}(T) = \mathcal{I}_{\Lambda}$, that is, if and only if $T = \mathrm{D}(\Lambda)$. In this case the homotopy category is trivial, since $\widehat{\mathcal{I}_{\Lambda}} = \mathcal{I}_{\Lambda}$. Dually, by Theorem 4.6, the homotopy category of the closed model structure of part (**III**) is equivalent to the stable category ${}^{\perp}T/\mathrm{add}(T)$, which is stable if and only if $\mathrm{add}(T) = \mathcal{P}_{\Lambda}$, that is, if and only if $T = \Lambda$. In this case the homotopy category is equivalent to ${}^{\perp}T/\mathrm{add}(T) = \underline{\mathrm{CM}}(\Lambda)$, and the algebra Λ is Gorenstein. □

To state our next result we need a simple lemma which was used implicitly in the proof of Theorem 5.8.

LEMMA 5.9. *Let $\mathcal{C}$ be an abelian category and let $(\mathfrak{C}, \mathfrak{F}, \mathfrak{W})$ be a closed model structure in $\mathcal{C}$.*

(1) *If $(\mathfrak{C}, \mathfrak{F}, \mathfrak{W})$ is projective, then the following are equivalent.*
   (i) *For any trivial fibration $Y \to C$ where Y is trivially fibrant, the object C is trivially fibrant.*
   (ii) Cof *is resolving.*
(2) *If $(\mathfrak{C}, \mathfrak{F}, \mathfrak{W})$ is injective, then the following are equivalent.*
   (i) *For any trivial cofibration $K \to X$ where X is trivially cofibrant, the object K is trivially cofibrant.*
   (ii) Fib *is coresolving.*

PROOF. We prove only part (1) since the proof of part (2) is similar. First let Cof be resolving, and let $p : Y \to C$ be a trivial fibration with Y trivially fibrant. Then by Lemma 1.1 we have that $\operatorname{Ker}(p)$ is trivially fibrant. Since the closed model structure is projective, we have a cotorsion pair $(\mathsf{Cof}, \mathsf{TFib})$ in $\mathcal{C}$ and any trivial fibration is an epimorphism. In particular p is an epimorphism. Since Cof is resolving, the subcategory TFib is coresolving. Since $Y, \operatorname{Ker}(i)$ are trivially fibrant, it follows that the same is true for the object C. Conversely assume that (i) holds, and let $0 \to Y_1 \to Y_2 \xrightarrow{p} C \to 0$ be a short exact sequence with the Y_i trivially fibrant. Since the closed model structure is projective, p is a trivial fibration. Hence by hypothesis, C is trivially fibrant. It follows that TFib is coresolving, or equivalenlty, Cof is resolving. □

Lemma 5.9 suggests the following terminology.

DEFINITION 5.10. Let $\mathcal{C}$ be an abelian category.

(1) A closed model structure is called **resolving** if it is projective and the equivalent conditions of part (1) of Lemma 5.9 hold.

(2) A closed model structure is called **coresolving** if it is injective and the equivalent conditions of part (2) of Lemma 5.9 hold.

Using this terminology we can restate Theorem 5.8 as follows: There are bijective correspondences between:

- Isoclasses of basic cotilting Λ-modules.
- Resolving closed model structures in $\operatorname{mod}(\Lambda)$ with $\mathsf{TFib} \subseteq \widehat{\mathsf{Cof}}$.
- Coresolving closed model structures in $\operatorname{mod}(\Lambda)$ such that $\mathsf{Fib} \subseteq \widehat{\mathsf{TCof}}$.

Under a finitenness condition we can refine the above correspondence. More precisely let Λ be an Artin algebra with finite global dimension. Then by a result of Auslander-Reiten (see [**9**]), the assumption (d) in (**II**) and (**III**) of Theorem 5.8 can be dropped. Hence we have the following consequence of Theorem 5.8 which gives a classification of cotilting Λ-modules in terms of (co)resolving closed model structures.

COROLLARY 5.11. *If* $\operatorname{gl.dim}\Lambda < \infty$, *then the correspondeces of Theorem* 5.8 *induce bijective correspondences between:*

(i) *Isomorphism classes of basic cotilting* Λ*-modules.*
(ii) *Resolving closed model structures in* $\operatorname{mod}(\Lambda)$.
(iii) *Coresolving closed model structures in* $\operatorname{mod}(\Lambda)$.

Note that using Theorems 4.6 and 4.7 we can obtain an analogous classification of tilting modules over an Artin algebra. In particular in the above corollary we can replace cotilting by tilting, provided that Λ has finite global dimension, since in this case the cotilting modules coincide with the tilting modules. We leave the details to the reader. For a classification of (infinitely generated) (co)tilting modules over an arbitrary ring in terms of closed model structures we refer to [**27**]. We note finally that the well-known equivalences and dualities induced by a tilting or cotilting module can be recovered as Quillen equivalences between the above closed model structures.

Cotorsion Triples. We close this section by discussing briefly the connections between closed model structures and cotorsion triples. As a consequence we obtain a closed model theoretic classification of cotorsion triples.

We fix throughout an abelian category $\mathcal{C}$ and assume that $\mathcal{C}$ has enough projective and injective objects. Let $(\mathcal{X}, \mathcal{Y}, \mathcal{Z})$ be a cotorsion triple in $\mathcal{C}$. As in the proof of Theorem VI.3.2 we have that $(\mathcal{X}, \mathcal{Y})$ is a resolving cotorsion pair if and only if $(\mathcal{Y}, \mathcal{Z})$ is a resolving cotorsion pair. Assume in the sequel that one of these conditions holds. Then $\mathcal{X}$ is closed under kernels of epics and $\mathcal{Z}$ is closed under cokernels of monics, and moreover $\mathcal{X} \cap \mathcal{Y} = \mathcal{P}$, hence $\mathcal{X}/\mathcal{P}$ is triangulated, and $\mathcal{Y} \cap \mathcal{Z} = \mathcal{I}$, hence $\mathcal{Z}/\mathcal{I}$ is triangulated. By Theorems 4.6 and 4.7, the cotorsion pairs $(\mathcal{X}, \mathcal{Y})$ and $(\mathcal{Y}, \mathcal{Z})$ induce two stable closed model structures in $\mathcal{C}$:

$$(\mathfrak{C}_{\mathcal{P}}, \mathfrak{F}_{\mathcal{P}}, \mathfrak{W}_{\mathcal{P}}) \quad \text{and} \quad (\mathfrak{C}^{\mathcal{I}}, \mathfrak{F}^{\mathcal{I}}, \mathfrak{W}^{\mathcal{I}})$$

where the first one is resolving and the second is coresolving. By Theorem VI.3.2 these stable closed model structures have (triangle) equivalent homotopy categories, in other words there exists a (triangle) equivalence $\mathcal{X}/\mathcal{P} \xrightarrow{\approx} \mathcal{Z}/\mathcal{I}$. We don't know if this equivalence is induced by a Quillen equivalence in the sense of [**64**]. For a specific situation where this happens we refer to the last chapter.

Combining the above observations with Theorems 4.6 and 4.7, we obtain the following classification of cotorsion triples in $\mathcal{C}$ in terms of closed model structures.

THEOREM 5.12. *Let $\mathcal{C}$ be an abelian category with enough projective and injective objects. Then there exists a bijective correspondence between cotorsion triples $(\mathcal{X}, \mathcal{Y}, \mathcal{Z})$ in $\mathcal{C}$ with $\mathcal{Y}$ (co)resolving, and couples $\{(\mathfrak{C}_1, \mathfrak{F}_1, \mathfrak{W}_1), (\mathfrak{C}_2, \mathfrak{F}_2, \mathfrak{W}_2)\}$ of stable closed model structures in $\mathcal{C}$, such that $(\mathfrak{C}_1, \mathfrak{F}_1, \mathfrak{W}_1)$ is resolving and $(\mathfrak{C}_2, \mathfrak{F}_2, \mathfrak{W}_2)$ is coresolving, and $\mathsf{TFib}_1 = \mathsf{TCof}_2$. The bijective correspondence is given by:*

$$(\mathcal{X}, \mathcal{Y}, \mathcal{Z}) \longmapsto \Big\{(\mathfrak{C}_{\mathcal{X}\cap\mathcal{Y}}, \mathfrak{F}_{\mathcal{X}\cap\mathcal{Y}}, \mathfrak{W}_{\mathcal{X}\cap\mathcal{Y}}),\ (\mathfrak{C}^{\mathcal{Y}\cap\mathcal{Z}}, \mathfrak{F}^{\mathcal{Y}\cap\mathcal{Z}}, \mathfrak{W}^{\mathcal{Y}\cap\mathcal{Z}})\Big\}$$

$$\Big\{(\mathfrak{C}_1, \mathfrak{F}_1, \mathfrak{W}_1),\ (\mathfrak{C}_2, \mathfrak{F}_2, \mathfrak{W}_2)\Big\} \longmapsto \big(\mathsf{Cof}_1, \mathsf{TFib}_1 = \mathsf{TCof}_2, \mathsf{Fib}_2\big).$$

If $\mathcal{P}$ is covariantly finite, then $\mathcal{C}/\mathcal{P}$ is pretriangulated and $(\mathcal{X}/\mathcal{P}, \mathcal{Y}/\mathcal{P})$ is a hereditary torsion pair in $\mathcal{C}/\mathcal{P}$ with $\mathcal{X}/\mathcal{P}$ triangulated. Similarly if $\mathcal{I}$ is contravariantly finite, then $\mathcal{C}/\mathcal{I}$ is pretriangulated and $(\mathcal{Y}/\mathcal{I}, \mathcal{Z}/\mathcal{I})$ is a cohereditary torsion pair in $\mathcal{C}/\mathcal{I}$ with $\mathcal{Z}/\mathcal{I}$ triangulated. By Theorems 4.6 and 4.7, the cotorsion pairs $(\mathcal{X}, \mathcal{Y})$ and $(\mathcal{Y}, \mathcal{Z})$ induce, two additional closed model structures in $\mathcal{C}$:

$$(\mathfrak{C}^{\mathcal{P}}, \mathfrak{F}^{\mathcal{P}}, \mathfrak{W}^{\mathcal{P}}),\ (\mathfrak{C}_{\mathcal{I}}, \mathfrak{F}_{\mathcal{I}}, \mathfrak{W}_{\mathcal{I}})$$

where the first one is coresolving and the second one resolving. Hence if $\mathcal{P}$ is covariantly finite and $\mathcal{I}$ is contravariantly finite, then the cotorsion triple $(\mathcal{X}, \mathcal{Y}, \mathcal{Z})$ induces four closed model structures in $\mathcal{C}$:

$$(\mathfrak{C}_{\mathcal{P}}, \mathfrak{F}_{\mathcal{P}}, \mathfrak{W}_{\mathcal{P}}),\ (\mathfrak{C}^{\mathcal{P}}, \mathfrak{F}^{\mathcal{P}}, \mathfrak{W}^{\mathcal{P}}),\ (\mathfrak{C}_{\mathcal{I}}, \mathfrak{F}_{\mathcal{I}}, \mathfrak{W}_{\mathcal{I}}),\ (\mathfrak{C}^{\mathcal{I}}, \mathfrak{F}^{\mathcal{I}}, \mathfrak{W}^{\mathcal{I}}).$$

Gorenstein Algebras and Categories. A special case of the above situation is the case of the cotorsion triple $\big(\mathrm{CM}(\Lambda), \mathcal{P}^{<\infty}_{\Lambda} = \mathcal{I}^{<\infty}_{\Lambda}, \mathrm{CoCM}(\mathrm{D}(\Lambda))\big)$ in $\mathrm{mod}(\Lambda)$ where Λ is a Gorenstein algebra, or more generally the cotorsion triple $\big(\mathrm{CM}(\mathcal{P}), \mathcal{P}^{<\infty} = \mathcal{I}^{<\infty}, \mathrm{CoCM}(\mathcal{I})\big)$ in a Gorenstein abelian category $\mathcal{C}$. We leave to the reader to write down the four closed model structures. For an arbitrary Artin algebra Λ, working with all (not necessarily finitely generated) modules, we will show in the last chapter that there exists a pair of compatible closed model structures in $\mathrm{Mod}(\Lambda)$, hence four in all, and two of them are stable. For the first stable closed model structure the Cohen-Macaulay modules appear as cofibrant objects and for the second, the CoCohen-Macaulay modules appear as fibrant objects. Actually we prove the existence of such closed model structures in the more general setting of a Nakayama abelian category, which provides a natural generalization of the module category of an Artin algebra.

Cohen-Macaulay Categories and Rings. If $\mathcal{C}$ is a Cohen-Macaulay abelian category with exact (co)products and a product-complete projective generator or injective cogenerator, then the results of this chapter, combined with Theorem VII.4.1, imply that there are four naturally induced closed model structures in $\mathcal{C}$ which are related to relative (Co)Cohen-Macaulay objects and the objects of finite projective or injective dimension. In particular there exist such closed model structures in the module category of a right Cohen-Macaulay ring which is left coherent and right perfect or right Noetherian. We leave to the reader to write down the four closed model structures.

Categories of Complexes. A nice illustration of the above results is the following example which shows that derived categories are homotopy categories of stable closed model structures induced by cotorsion triples in the category of complexes.

Let Λ be a ring and let $\mathsf{C}(\mathrm{Mod}(\Lambda))$ be the category of complexes of right Λ-modules. Let $\mathsf{C}_0(\mathrm{Mod}(\Lambda))$ be the full subcategory of contractible complexes. It is easy to see that the latter is functorially finite, hence by [**24**], the stable category $\mathsf{C}(\mathrm{Mod}(\Lambda))/\mathsf{C}_0(\mathrm{Mod}(\Lambda))$ is identified with the homotopy category $\mathcal{H}(\mathrm{Mod}(\Lambda))$. We denote by $\mathsf{C}_{\mathsf{Ac}}(\mathrm{Mod}(\Lambda))$ the full subcategory of acyclic complexes and by $\mathsf{C}_{\mathsf{P}}(\mathrm{Mod}(\Lambda))$, resp. $\mathsf{C}^{\mathsf{I}}(\mathrm{Mod}(\Lambda))$, the full subcategory of homotopically projective, resp. injective, complexes [**72**]. It is not difficult to see that $\mathsf{C}_{\mathsf{P}}(\mathrm{Mod}(\Lambda)) \cap \mathsf{C}_{\mathsf{Ac}}(\mathrm{Mod}(\Lambda)) = \mathsf{C}_0(\mathrm{Mod}(\Lambda)) = \mathsf{C}_{\mathsf{Ac}}(\mathrm{Mod}(\Lambda)) \cap \mathsf{C}^{\mathsf{I}}(\mathrm{Mod}(\Lambda))$. We denote as in Chapter III by $\mathcal{H}_{\mathsf{P}}(\mathrm{Mod}(\Lambda))$, $\mathcal{H}^{\mathsf{I}}(\mathrm{Mod}(\Lambda))$, $\mathcal{H}_{\mathsf{Ac}}(\mathrm{Mod}(\Lambda))$, the induced homotopy categories. By Corollary IV.1.7 we have a TTF-theory

$$\Big(\mathcal{H}_{\mathsf{P}}(\mathrm{Mod}(\Lambda)), \mathcal{H}_{\mathsf{Ac}}(\mathrm{Mod}(\Lambda))), \mathcal{H}^{\mathsf{I}}(\mathrm{Mod}(\Lambda))\Big)$$

in $\mathcal{H}(\mathrm{Mod}(\Lambda))$, and it is not difficult to see that $\mathrm{Ext}^1(\mathsf{C}_{\mathsf{P}}(\mathrm{Mod}(\Lambda)), \mathsf{C}_{\mathsf{Ac}}(\mathrm{Mod}(\Lambda))) = \mathrm{Ext}^1(\mathsf{C}_{\mathsf{Ac}}(\mathrm{Mod}(\Lambda)), \mathsf{C}^{\mathsf{I}}(\mathrm{Mod}(\Lambda))) = 0$. Hence by Theorem 4.2 we have a closed model structure $(\mathfrak{C}_{\mathsf{P}}, \mathfrak{F}_{\mathsf{P}}, \mathfrak{W}_{\mathsf{P}})$ in $\mathsf{C}(\mathrm{Mod}(\Lambda)$ with corresponding Quillen homotopy category $\mathcal{H}_{\mathsf{P}}(\mathrm{Mod}(\Lambda))$ and a closed model structure $(\mathfrak{C}^{\mathsf{I}}, \mathfrak{F}^{\mathsf{I}}, \mathfrak{W}^{\mathsf{I}})$ in $\mathsf{C}(\mathrm{Mod}(\Lambda))$ with corresponding Quillen homotopy category $\mathcal{H}^{\mathsf{I}}(\mathrm{Mod}(\Lambda))$. By Corollary IV.1.7 both Quillen homotopy categories are triangle equivalent to the unbounded derived category $\mathbf{D}(\mathrm{Mod}(\Lambda))$. This recovers a result of B. Keller [**72**].

It follows that Theorems 4.2 and 4.7 produce the projective and injective closed model structures on the category of complexes investigated in [**64**]. In our terminology the projective model structure of [**64**] is the projective ω-closed model structure $(\mathfrak{C}_{\omega}, \mathfrak{F}_{\omega}, \mathfrak{W}_{\omega})$ associated with the good pair $(\mathsf{C}_{\mathsf{P}}(\mathrm{Mod}(\Lambda)), \mathsf{C}_{\mathsf{Ac}}(\mathrm{Mod}(\Lambda)))$, where $\omega = \mathsf{C}_{\mathsf{P}}(\mathrm{Mod}(\Lambda)) \cap \mathsf{C}_{\mathsf{Ac}}(\mathrm{Mod}(\Lambda))$ is the full subcategory of contractible complexes. For instance for the projective closed model structure we have the following description: the cofibrations are the monomorphisms with cokernel a homotopically projective complex, the fibrations are the epimorphisms, and the weak equivalences are the morphisms which are compositions $\alpha \circ \beta$ where α is split monic with contractible cokernel and β is an epimorphism with acyclic kernel. In particular the weak equivalences are quasi-isomorphisms. The injective closed model structure has a similar description.

Note that the above analysis with the necessary modifications works if we replace the module category with a Grothendieck category (with projectives).

For the history of the development of these and related closed model structures we refer to [**64**].

CHAPTER IX

(Co)Torsion Pairs and Generalized Tate-Vogel Cohomology

Let $\mathcal{C}$ be an abelian category and ω a functorially finite subcategory of $\mathcal{C}$. Our aim in this chapter is to construct and investigate universal (co)homology theories in $\mathcal{C}$ which are induced by a (co)hereditary torsion pair $(\mathcal{X}/\omega, \mathcal{Y}/\omega)$ defined in the pretriangulated stable category $\mathcal{C}/\omega$. These new cohomology theories generalize the Tate-Vogel cohomology in module categories, and are characterized as completions, with respect to the torsion class $\mathcal{X}/\omega$ or the cotorsion-free class $\mathcal{Y}/\omega$ in an appropriate sense, of the relative extension functors induced by the subcategory ω. In this way we generalize results of Mislin. We also study the connections and the interplay between these new (co)homology theories and the relative homology theories in $\mathcal{C}$ induced by subcategories $\mathcal{X}$,$\mathcal{Y}$ and ω, by providing comparison maps between them which fit into a long exact sequence, thus generalizing recent work by Avramov and Martsinkovsky. An especially nice situation occurs when we have torsion pairs induced by a cotorsion triple in $\mathcal{C}$. In this case we show that the complete cohomology theories induced by the two torsion pairs involved are isomorphic.

1. Hereditary Torsion Pairs and Homological Functors

We have seen in Chapter II that any (co)hereditary torsion pair in a pretriangulated category $\mathcal{C}$ gives rise to a torsion pair in the associated triangulated left or right stabilization category of $\mathcal{C}$. In this section, using the left or right stabilization of $\mathcal{C}$, we define universal (co)homology theories in $\mathcal{C}$ with respect to a (co)hereditary torsion pair. In the next section we will apply our results to construct (co)homology theories in an abelian category relative to a (co)hereditary torsion pair for an associated stable category.

We fix throughout a torsion pair $(\mathcal{X}, \mathcal{Y})$ in the pretriangulated category $\mathcal{C}$, and as usual let $\mathbf{R} : \mathcal{C} \to \mathcal{X}$ be the right adjoint of the inclusion $\mathbf{i} : \mathcal{X} \hookrightarrow \mathcal{C}$ and $\mathbf{L} : \mathcal{C} \to \mathcal{Y}$ the left adjoint of the inclusion $\mathbf{j} : \mathcal{Y} \hookrightarrow \mathcal{C}$. We denote by $\mathsf{P} : \mathcal{C} \to \mathcal{T}_l(\mathcal{C})$, resp. $\mathsf{Q} : \mathcal{C} \to \mathcal{T}_r(\mathcal{C})$, the stabilization functor of $\mathcal{C}$, when the latter is a considered as a left, resp. right, triangulated category.

Recall from section II.5 that if the torsion pair $(\mathcal{X}, \mathcal{Y})$ is hereditary, then $\mathcal{X}$ is pretriangulated, and the inclusion functor $\mathbf{i} : \mathcal{X} \hookrightarrow \mathcal{C}$ and the coreflection functor $\mathbf{R} : \mathcal{C} \to \mathcal{X}$ are left exact. In this case there exist unique exact functors $\mathbf{i}^* : \mathcal{T}_l(\mathcal{X}) \to \mathcal{T}_l(\mathcal{C})$ and $\mathbf{R}^* : \mathcal{T}_l(\mathcal{C}) \to \mathcal{T}_l(\mathcal{X})$, such that the following diagrams commute:

$$\begin{array}{ccc} \mathcal{X} & \xrightarrow{\mathbf{i}} & \mathcal{C} \\ {\scriptstyle \mathsf{P}}\downarrow & & \downarrow{\scriptstyle \mathsf{P}} \\ \mathcal{T}_l(\mathcal{X}) & \xrightarrow{\mathbf{i}^*} & \mathcal{T}_l(\mathcal{C}) \end{array} \qquad \begin{array}{ccc} \mathcal{C} & \xrightarrow{\mathbf{R}} & \mathcal{X} \\ {\scriptstyle \mathsf{P}}\downarrow & & \downarrow{\scriptstyle \mathsf{P}} \\ \mathcal{T}_l(\mathcal{C}) & \xrightarrow{\mathbf{R}^*} & \mathcal{T}_l(\mathcal{X}) \end{array}$$

where $\mathsf{P} : \mathcal{X} \to \mathcal{T}_l(\mathcal{X})$ is the stabilization functor of $\mathcal{X}$ when the latter is considered as a left triangulated category. Moreover $\mathbf{R}^*$ is a right adjoint of $\mathbf{i}^*$, and $\big(\mathcal{T}_l(\mathcal{X}), \mathcal{T}_l(\mathcal{Y})\big)$ is a hereditary torsion pair in $\mathcal{T}_l(\mathcal{C})$. Hence we have a short exact sequence of triangulated categories:

$$0 \longrightarrow \mathcal{T}_l(\mathcal{Y}) \xrightarrow{\mathbf{j}^*} \mathcal{T}_l(\mathcal{C}) \xrightarrow{\mathbf{R}^*} \mathcal{T}_l(\mathcal{X}) \longrightarrow 0$$

Dually if the torsion pair $(\mathcal{X}, \mathcal{Y})$ is cohereditary, then $\mathcal{Y}$ is pretriangulated, the inclusion functor $\mathbf{j} : \mathcal{Y} \hookrightarrow \mathcal{C}$ and the reflection functor $\mathbf{L} : \mathcal{C} \to \mathcal{Y}$ are right exact, and there exist unique exact functors $\mathbf{j}^* : \mathcal{T}_r(\mathcal{Y}) \to \mathcal{T}_r(\mathcal{C})$ and $\mathbf{L}^* : \mathcal{T}_r(\mathcal{C}) \to \mathcal{T}_r(\mathcal{Y})$, such that the following diagrams commute:

$$\begin{array}{ccc} \mathcal{Y} & \xrightarrow{\mathbf{j}} & \mathcal{C} \\ {\scriptstyle \mathsf{Q}}\downarrow & & \downarrow{\scriptstyle \mathsf{Q}} \\ \mathcal{T}_r(\mathcal{X}) & \xrightarrow{\mathbf{j}^*} & \mathcal{T}_r(\mathcal{C}) \end{array} \qquad \begin{array}{ccc} \mathcal{C} & \xrightarrow{\mathbf{L}} & \mathcal{Y} \\ {\scriptstyle \mathsf{Q}}\downarrow & & \downarrow{\scriptstyle \mathsf{Q}} \\ \mathcal{T}_r(\mathcal{C}) & \xrightarrow{\mathbf{L}^*} & \mathcal{T}_r(\mathcal{Y}) \end{array}$$

where $\mathsf{Q} : \mathcal{Y} \to \mathcal{T}_r(\mathcal{Y})$ is the stabilization functor of $\mathcal{Y}$ when the latter is considered as a right triangulated category. Moreover $\mathbf{L}^*$ is a left adjoint of $\mathbf{j}^*$, and $\big(\mathcal{T}_r(\mathcal{X}), \mathcal{T}_r(\mathcal{Y})\big)$ is a hereditary torsion pair in $\mathcal{T}_r(\mathcal{C})$. Hence we have a short exact sequence of triangulated categories:

$$0 \longrightarrow \mathcal{T}_r(\mathcal{X}) \xrightarrow{\mathbf{i}^*} \mathcal{T}_r(\mathcal{C}) \xrightarrow{\mathbf{L}^*} \mathcal{T}_r(\mathcal{Y}) \longrightarrow 0$$

The above considerations suggest the following.

DEFINITION 1.1. Let $(\mathcal{X}, \mathcal{Y})$ be a hereditary torsion pair in $\mathcal{C}$. Then the n**−th projective extension bifunctor**

$$\widehat{\mathsf{Ext}}^n_{(\mathcal{X},\mathcal{Y})}(-,-) : \mathcal{C}^{\mathrm{op}} \times \mathcal{C} \longrightarrow \mathcal{A}b, \quad n \in \mathbb{Z}$$

of $\mathcal{C}$ with respect to $(\mathcal{X}, \mathcal{Y})$ is defined as follows:

$$\widehat{\mathsf{Ext}}^n_{(\mathcal{X},\mathcal{Y})}(A,B) := \mathcal{T}_l(\mathcal{X})[\Omega^n \mathbf{R}^*\mathsf{P}(A), \mathbf{R}^*\mathsf{P}(B)] \xrightarrow{\cong} \mathcal{T}_l(\mathcal{X})[\Omega^n \mathsf{P}\mathbf{R}(A), \mathsf{P}\mathbf{R}(B)] \xrightarrow{\cong}$$
$$\varinjlim_{k,k+n\geq 0} \mathcal{X}[\Omega^{k+n}\mathbf{R}(A), \Omega^k \mathbf{R}(B)] \xrightarrow{\cong} \varinjlim_{k,k+n\geq 0} \mathcal{C}[\Omega^{k+n}\mathbf{R}(A), \Omega^k(B)].$$

Let $(\mathcal{X}, \mathcal{Y})$ be a cohereditary torsion pair in $\mathcal{C}$. Then the n**−th injective extension bifunctor**

$$\widetilde{\mathsf{Ext}}^n_{(\mathcal{X},\mathcal{Y})}(-,-) : \mathcal{C}^{\mathrm{op}} \times \mathcal{C} \longrightarrow \mathcal{A}b, \quad n \in \mathbb{Z}$$

of $\mathcal{C}$ with respect to $(\mathcal{X}, \mathcal{Y})$ is defined as follows:

$$\widetilde{\mathsf{Ext}}^n_{(\mathcal{X},\mathcal{Y})}(A,B) := \mathcal{T}_r(\mathcal{Y})[\mathbf{L}^*\mathsf{Q}(A), \Sigma^n \mathbf{L}^*\mathsf{Q}(B)] \xrightarrow{\cong} \mathcal{T}_r(\mathcal{Y})[\mathsf{Q}\mathbf{L}(A), \Sigma^n \mathsf{Q}\mathbf{L}(B)] \xrightarrow{\cong}$$
$$\varinjlim_{k,k+n\geq 0} \mathcal{Y}[\Sigma^k \mathbf{L}(A), \Sigma^{k+n}\mathbf{L}(B)] \xrightarrow{\cong} \varinjlim_{k,k+n\geq 0} \mathcal{C}[\Sigma^k(A), \Sigma^{k+n}\mathbf{L}(B)].$$

EXAMPLE. Let $\mathcal{C}$ be triangulated and let $(\mathcal{X}, \mathcal{Y})$ be a hereditary torsion pair in $\mathcal{C}$. Then $(\mathcal{X}, \mathcal{Y})$ is also cohereditary, and according to Chapter I, $\mathcal{X}$ and $\mathcal{Y}$ are thick subcategories of $\mathcal{C}$; in particular all the involved stabilization functors are triangle equivalences in this case. Then $\widehat{\mathsf{Ext}}^n_{(\mathcal{X},\mathcal{Y})}(A, B) = \mathcal{X}(\Omega^n\mathbf{R}(A), \mathbf{R}(B)) \cong \mathcal{C}(\Omega^n\mathbf{R}(A), B)$ and $\widetilde{\mathsf{Ext}}^n_{(\mathcal{X},\mathcal{Y})}(A, B) = \mathcal{Y}(\mathbf{L}(A), \Sigma^n\mathbf{L}(B)) \cong \mathcal{C}(A, \Sigma^n\mathbf{L}(B))$.

EXAMPLE. The smallest torsion pair $(0, \mathcal{C})$ and the largest torsion pair $(\mathcal{C}, 0)$ in $\mathcal{C}$ are obviously hereditary and cohereditary. In this case we have

$$\widetilde{\mathsf{Ext}}^n_{(\mathcal{C},0)}(-,-) = 0 = \widehat{\mathsf{Ext}}^n_{(0,\mathcal{C})}(-,-) \quad \text{and} \quad \widehat{\mathsf{Ext}}^n_{(\mathcal{C},0)}(-,-) = \widetilde{\mathsf{Ext}}^n_{(0,\mathcal{C})}(-,-).$$

The above example suggests the following definition. The terminology will be justified in the next section.

DEFINITION 1.2. The n-th projective extension bifunctor with respect to the hereditary torsion pair $(\mathcal{C}, 0)$ is called the n-th **Tate-Vogel projective extension bifunctor** of $\mathcal{C}$ and is denoted by $\widehat{\mathsf{Ext}}^n_{\mathcal{C}}(-,-)$. In other words

$$\widehat{\mathsf{Ext}}^n_{\mathcal{C}}(A, B) := \mathcal{T}_l(\mathcal{C})[\Omega^n\mathsf{P}(A), \mathsf{P}(B)].$$

The n-th injective extension bifunctor with respect to the cohereditary torsion pair $(0, \mathcal{C})$ is called the n-th **Tate-Vogel injective extension bifunctor** of $\mathcal{C}$ and is denoted by $\widetilde{\mathsf{Ext}}^n_{\mathcal{C}}(-,-)$. In other words

$$\widetilde{\mathsf{Ext}}^n_{\mathcal{C}}(A, B) := \mathcal{T}_r(\mathcal{C})[\mathsf{Q}(A), \Sigma^n\mathsf{Q}(B)].$$

From now on we assume that the torsion pair $(\mathcal{X}, \mathcal{Y})$ is hereditary. The corresponding results concerning cohereditary torsion pairs follow by duality.

The following result presents some properties of the bifunctor $\widehat{\mathsf{Ext}}^n_{(\mathcal{X},\mathcal{Y})}(-,-)$.

PROPOSITION 1.3. (1) *If A or B is in $\mathcal{Y}$ then* $\widehat{\mathsf{Ext}}^n_{(\mathcal{X},\mathcal{Y})}(A, B) = 0$, $\forall n \in \mathbb{Z}$.

(2) $\widehat{\mathsf{Ext}}^0_{(\mathcal{X},\mathcal{Y})}(A, A) = 0$ *iff* $\mathbf{R}(A) \in \mathcal{P}^{<\infty}(\mathcal{C})$ *iff* $\Omega^n(A) \in \mathcal{Y}$ *for some* $n \geq 0$.

(3) *For any left triangle $\Omega(C) \to A \to B \to C$ in $\mathcal{C}$ and for any $D \in \mathcal{C}$ there exists an infinite long exact sequence:*

$$\cdots \to \widehat{\mathsf{Ext}}^{-1}_{(\mathcal{X},\mathcal{Y})}(A, D) \to \widehat{\mathsf{Ext}}^0_{(\mathcal{X},\mathcal{Y})}(C, D) \to \widehat{\mathsf{Ext}}^0_{(\mathcal{X},\mathcal{Y})}(B, D) \to$$

$$\to \widehat{\mathsf{Ext}}^0_{(\mathcal{X},\mathcal{Y})}(A, D) \to \widehat{\mathsf{Ext}}^1_{(\mathcal{X},\mathcal{Y})}(C, D) \to \widehat{\mathsf{Ext}}^1_{(\mathcal{X},\mathcal{Y})}(B, D) \to \cdots$$

and an infinite long exact sequence:

$$\cdots \to \widehat{\mathsf{Ext}}^{-1}_{(\mathcal{X},\mathcal{Y})}(D, C) \to \widehat{\mathsf{Ext}}^0_{(\mathcal{X},\mathcal{Y})}(D, A) \to \widehat{\mathsf{Ext}}^0_{(\mathcal{X},\mathcal{Y})}(D, B) \to$$

$$\to \widehat{\mathsf{Ext}}^0_{(\mathcal{X},\mathcal{Y})}(D, C) \to \widehat{\mathsf{Ext}}^1_{(\mathcal{X},\mathcal{Y})}(D, A) \to \widehat{\mathsf{Ext}}^1_{(\mathcal{X},\mathcal{Y})}(D, B) \to \cdots$$

PROOF. Parts (1) and (2) follow directly from the definitions. Part (3) follows from the fact that the stabilization $\mathcal{T}_l(\mathcal{X})$ is a triangulated category. □

Observe that the functor $\mathbf{R}^* : \mathcal{T}_l(\mathcal{C}) \to \mathcal{T}_l(\mathcal{X})$ induces a natural morphism $\widehat{\mathsf{Ext}}^n_{\mathcal{C}}(-,-) \to \widehat{\mathsf{Ext}}^n_{(\mathcal{X},\mathcal{Y})}(-,-)$, $\forall n \in \mathbb{Z}$, which is induced by the lifting of the adjunction $\mathbf{iR} \to \mathrm{Id}_{\mathcal{C}}$. Using Proposition 1.3, the glueing triangle $\Omega\mathbf{L}(A) \to \mathbf{R}(A) \to A \to \mathbf{L}(A)$, $\forall A \in \mathcal{C}$, and Corollary II.5.5, we have the following.

COROLLARY 1.4. *Let A, B be objects in $\mathcal{C}$. Then there exists a natural in A and B infinite long exact sequence:*

$$\cdots \to \widehat{\mathsf{Ext}}^n_{\mathcal{C}}(\mathbf{L}(A), B) \to \widehat{\mathsf{Ext}}^n_{\mathcal{C}}(A, B) \to \widehat{\mathsf{Ext}}^n_{(\mathcal{X},\mathcal{Y})}(A, B) \to \cdots$$

In particular the natural map $\widehat{\mathsf{Ext}}^n_{\mathcal{C}}(A,-) \to \widehat{\mathsf{Ext}}^n_{(\mathcal{X},\mathcal{Y})}(A,-)$ is invertible for all $n \in \mathbb{Z}$ iff $\mathbf{L}(A) \in \mathcal{P}^{<\infty}(\mathcal{C})$, that is: $\Omega^n\mathbf{L}(A) = 0$ for some $n \geq 0$.

Finally the natural map $\widehat{\mathsf{Ext}}^n_{\mathcal{C}}(-,-) \to \widehat{\mathsf{Ext}}^n_{(\mathcal{X},\mathcal{Y})}(-,-)$ is invertible for all $n \in \mathbb{Z}$ iff $\mathcal{Y} \subseteq \mathcal{P}^{<\infty}(\mathcal{C})$ iff $\widehat{\mathcal{X}} = \mathcal{C}$.

Part (3) of Proposition 1.3 suggests the following definition, which will be useful in the next section when we will discuss completions. Note that Definition 1.5 below extends the usual definition of Grothendieck of homology theories (or homological functors) defined on an abelian category. Of course if we view an abelian category as pretriangulated with zero loop and suspension functor and with left and right triangles to be the exact sequences, then the two definitions agree.

DEFINITION 1.5. A **left homology theory** on $\mathcal{C}$ is a sequence of additive functors $F_* := \{F_n\}_{n\in\mathbb{Z}}$, $F_n : \mathcal{C} \to \mathcal{A}b$, such that for any left triangle $\Omega(C) \to A \to B \to C$ in $\mathcal{C}$ there exists a long exact sequence

$$\cdots \to F_{-1}(C) \to F_0(A) \to F_0(B) \to F_0(C) \to F_1(A) \to \cdots$$

which is natural with respect to morphisms of left triangles.

A **right homology theory** on $\mathcal{C}$ is a sequence of additive functors $F_* := \{F_n\}_{n\in\mathbb{Z}}$, $F_n : \mathcal{C} \to \mathcal{A}b$, such that for any right triangle $A \to B \to C \to \Sigma(A)$ in $\mathcal{C}$ there exists a long exact sequence

$$\cdots \to F_{-1}(C) \to F_0(A) \to F_0(B) \to F_0(C) \to F_1(A) \to \cdots$$

which is natural with respect to morphisms of right triangles.

We are mainly concerned with left homology theories on the petriangulated category $\mathcal{C}$, called from now on just homology theories. The treatment of right homology theories is dual and is left to the reader. We consider homological functors in an abelian category $\mathcal{C}$ as (left or right) homology theories, viewing $\mathcal{C}$ as a pretriangulated category in the usual way. In the sequel we shall need the following, which extends a well-known concept of completeness of homological functors defined on an abelian category, see [**83**].

DEFINITION 1.6. Let $\mathcal{Y}$ be a full subcategory of $\mathcal{C}$. A homology theory $\{F_n\}_{n\in\mathbb{Z}}$ is called **$\mathcal{Y}$-complete** if $F_n(\mathcal{Y}) = 0$, $\forall n \in \mathbb{Z}$.

The following are examples of homology theories on $\mathcal{C}$.

EXAMPLE. (i) For any $C \in \mathcal{C}$ the functor $\widehat{\mathsf{Ext}}^*_{(\mathcal{X},\mathcal{Y})}(C,-)$ is a $\mathcal{Y}$-complete homology theory on $\mathcal{C}$.

(ii) For any $C \in \mathcal{C}$ we define a sequence of functors $\pi_n^C : \mathcal{C} \to \mathcal{A}b$ as follows. For $n \geq 0$ we set $\pi_n^C = \mathcal{C}(\Omega^n(C), -)$ and $\pi_{-n}^C(C) = \mathcal{C}(C, \Omega^n(-)) \xrightarrow{\cong} \mathcal{C}(\Sigma^n(C), -)$. It is not difficult to see that if Σ is fully faithful, then the sequence of functors π_*^C is a homology theory in $\mathcal{C}$ which is $\mathcal{Y}$-complete iff $C \in \mathcal{X}$.

Let $F_\bullet = \{F_n\}$ be a homology theory on $\mathcal{C}$. Let $\mathcal{A}b^{\mathbb{Z}}$ be the category of graded abelian groups. Its objects are sequences $\{A_n\}_{n\in\mathbb{Z}}$ of abelian groups, and a morphism $\{A_n\}_{n\in\mathbb{Z}} \to \{A_n\}_{n\in\mathbb{Z}}$ is a sequence $\{\alpha_n : A_n \to B_n\}_{n\in\mathbb{Z}}$ indexed by $\mathbb{Z}$. We consider $\mathcal{A}b^{\mathbb{Z}}$ as a graded category equipped with the shift automorphism $\mathsf{S} : \mathcal{A}b^{\mathbb{Z}} \to \mathcal{A}b^{\mathbb{Z}}$ which sends the graded object $\{A_n\}$ to $\{\mathsf{S}(A_n)\}$ which in degree n has the object A_{n-1}. Then a homology theory is a homological functor $F_\bullet : \mathcal{C} \to \mathcal{A}b^{\mathbb{Z}}$ such that $F_\bullet\Omega \cong \mathsf{S}F_\bullet$. From now on we identify homology theories $\mathcal{C} \to \mathcal{A}b$ and homological functors $\mathcal{C} \to \mathcal{A}b^{\mathbb{Z}}$ which commute with Ω and S.

The following gives a connection between homology theories in the left triangulated category $\mathcal{C}$ and homological functors over its stabilization $\mathcal{T}_l(\mathcal{C})$.

LEMMA 1.7. *For any homology theory $F_\bullet : \mathcal{C} \to \mathcal{A}b^{\mathbb{Z}}$, there exists a homological functor $F_\bullet^* : \mathcal{T}_l(\mathcal{C}) \to \mathcal{A}b^{\mathbb{Z}}$, unique up to isomorphism, such that $F_\bullet^*\mathsf{P} \cong F_\bullet$. The functor $F_\bullet \mapsto F_0^*$ establishes an equivalence between the category of homology theories on $\mathcal{C}$ and the category of homological functors $\mathcal{T}_l(\mathcal{C}) \to \mathcal{A}b$.*

PROOF. Defining $F_\bullet^*(C, n) = \mathsf{S}^{-n}F_\bullet(C)$ it is easy to see that $F_\bullet^*$ is a homological functor on $\mathcal{T}_l(\mathcal{C})$. Moreover $F_\bullet^*\mathsf{P}(C) = F_\bullet^*(C, 0) = F_\bullet(C)$, hence $F_\bullet^*\mathsf{P} \cong F_\bullet$. Trivially $F_\bullet^*$ is the unique homological functor on $\mathcal{T}_l(\mathcal{C})$ which extends $F_\bullet$ via the stabilization functor. This shows that the functor $F_\bullet \mapsto F_\bullet^*$ is an equivalence between homology theories on $\mathcal{C}$ and homology theories on $\mathcal{T}_l(\mathcal{C})$. Since the latter is triangulated, the map $F_\bullet^* \mapsto F_0^*$ gives an equivalence between the category of homology theories on $\mathcal{T}_l(\mathcal{C})$ and the category of homological functors $\mathcal{T}_l(\mathcal{C}) \to \mathcal{A}b$. □

2. Torsion Pairs and Generalized Tate-Vogel (Co-)Homology

Let ω be a functorially finite subcategory of an abelian category $\mathcal{C}$. In this section we use (co)hereditary torsion pairs defined in the stable pretriangulated category $\mathcal{C}/\omega$ to define and investigate generalized Tate-Vogel (co)homology functors defined in the abelian category $\mathcal{C}$. Our focus is concentrated on the construction of completions of a (co)homology theory defined on $\mathcal{C}$, relative to the torsion or torsion-free class of a (co)hereditary torsion pair defined on $\mathcal{C}/\omega$. Our main result shows that the relative extension functor $\operatorname{Ext}_\omega^*$ with respect to the subcategory ω admits such completions, which we call generalized Tate-Vogel extension functors. This result generalizes work of Mislin [**83**]. Moreover we show that the completion can be realized as an appropriate (co)homotopy functor in the sense of Eckmann-Hilton [**61**].

Throughout this section we fix an abelian category $\mathcal{C}$. We also fix a full additive functorially finite subcategory $\omega \subseteq \mathcal{C}$. We start with a reasonable pair of full subcategories of $\mathcal{C}$ which induce a hereditary torsion pair in the stable category $\mathcal{C}/\omega$. Throughout this section our working setup is the following.

- [Setup] $(\mathcal{X}, \mathcal{Y})$ is a fixed pair (not necessarily cotorsion) of subcategories of $\mathcal{C}$ such that:

(i) $\mathcal{X}$ is closed under extensions, direct summands and kernels of epimorphisms.
(ii) $\omega := \mathcal{X} \cap \mathcal{Y}$ is functorially finite in $\mathcal{C}$ and any ω-epic is an epimorphism.
(iii) The pair $(\mathcal{X}/\omega, \mathcal{Y}/\omega)$ is a torsion pair in $\mathcal{C}/\omega$.

In particular we don't assume that $\mathrm{Ext}^1_{\mathcal{C}}(\mathcal{X}, \mathcal{Y}) = 0$.

REMARK 2.1. The following are some consequences of the above setup which will be useful in the sequel.

(i) The torsion pair $(\mathcal{X}/\omega, \mathcal{Y}/\omega)$ in $\mathcal{C}/\omega$ is hereditary.
(ii) Any right ω-approximation of an object in $\mathcal{C}$ is an epimorphism.
(iii) $\mathcal{X}$ is contravariantly finite and, since $\omega \subseteq \mathcal{X}$, any right $\mathcal{X}$-approximation of an object in $\mathcal{C}$ is an epimorphism. Hence for any object C in $\mathcal{C}$, there exists a right $\mathcal{X}$-approximation sequence $0 \to Y_C \to X_C \to C \to 0$ which is exact and where the object Y_C lies in $\mathcal{Y}$.

The main examples of the above setup are the following.

EXAMPLE. (i) $(\mathcal{X}, \mathcal{Y})$ is a resolving cotorsion pair in $\mathcal{C}$ such that $\omega = \mathcal{X} \cap \mathcal{Y}$ is functorially finite in $\mathcal{C}$, and any ω-epic is an epimorphism in $\mathcal{C}$.
(ii) $(\mathcal{C}, \mathcal{P})$ where we assume that the full subcategory $\mathcal{P}$ of projective objects of $\mathcal{C}$ is functorially finite. Then the pair $\big(\mathcal{C}, \mathcal{P}\big)$ is not necessarily cotorsion, but $\big(\mathcal{C}/\mathcal{P}, 0\big)$ is a hereditary torsion pair in $\mathcal{C}/\mathcal{P}$.
(iii) $(\mathcal{X}, \mathcal{Y})$ is a cotorsion pair with $\mathcal{X}$ resolving and $\mathcal{X} \cap \mathcal{Y} = \mathcal{P}$. For instance we can take $\mathcal{X} = \mathcal{P}$ and $\mathcal{Y} = \mathcal{C}$. Then $(\mathcal{X}/\mathcal{P}, \mathcal{Y}/\mathcal{P})$ is a hereditary torsion pair in $\mathcal{C}/\mathcal{P}$ with $\mathcal{X}/\mathcal{P}$ triangulated.

As usual we denote by $\mathbf{R} : \mathcal{C}/\omega \to \mathcal{X}/\omega$ the right adjoint of the inclusion $\mathbf{i} : \mathcal{X}/\omega \hookrightarrow \mathcal{C}/\omega$ and by $\mathbf{L} : \mathcal{C}/\omega \to \mathcal{Y}/\omega$ the left adjoint of the inclusion $\mathbf{j} : \mathcal{Y}/\omega \hookrightarrow \mathcal{C}/\omega$. By the assumptions on $\mathcal{X}$ and ω in the above setup, it follows that the stable category $\mathcal{X}/\omega$ is a pretriangulated subcategory of $\mathcal{C}/\omega$ and the functors $\mathbf{i}$ and $\mathbf{R}$ are left exact, i.e. the torsion pair $(\mathcal{X}/\omega, \mathcal{Y}/\omega)$ in $\mathcal{C}/\omega$ is hereditary. We denote by P the stabilization functor of any of the categories $\mathcal{X}/\omega, \mathcal{Y}/\omega, \mathcal{C}/\omega$ with respect to their left triangulation. By Proposition II.5.3, we know that the hereditary torsion pair $(\mathcal{X}/\omega, \mathcal{Y}/\omega)$ in $\mathcal{C}/\omega$ induces a hereditary torsion pair $(\mathcal{T}_l(\mathcal{X}/\omega), \mathcal{T}_l(\mathcal{Y}/\omega))$ in the stabilization $\mathcal{T}_l(\omega)$ of $\mathcal{C}/\omega$ with respect to its left triangulation. In other words we have a short exact sequence of triangulated categories

$$0 \longrightarrow \mathcal{T}_l(\mathcal{Y}/\omega) \xrightarrow{\mathbf{j}^*} \mathcal{T}_l(\mathcal{C}/\omega) \xrightarrow{\mathbf{R}^*} \mathcal{T}_l(\mathcal{X}/\omega) \longrightarrow 0$$

and the functor $\mathbf{R}^*$ admits a fully faithful left adjoint $\mathbf{i}^*$, and the functor $\mathbf{j}^*$ admits a left adjoint.

Now we can use the results from Section 1 to transfer the concept of a projective extension bifunctor with respect to a hereditary torsion pair in the stable pretriangulated category, to the abelian category.

DEFINITION 2.2. The n-th **complete projective extension bifunctor** of $\mathcal{C}$ **with respect to the pair** $(\mathcal{X}, \mathcal{Y})$:

$$\widehat{\mathsf{Ext}}^n_{(\mathcal{X},\mathcal{Y})}(-,-) \, : \, \mathcal{C}^{\mathrm{op}} \times \mathcal{C} \longrightarrow \mathcal{A}b, \quad n \in \mathbb{Z}$$

is defined to be the n-th complete extension bifunctor $\widehat{\mathsf{Ext}}^n_{(\mathcal{X}/\omega,\mathcal{Y}/\omega)}(-,-)$ of $\mathcal{C}/\omega$ with respect to the hereditary torsion pair $(\mathcal{X}/\omega, \mathcal{Y}/\omega)$. In other words

$$\begin{aligned}\widehat{\mathsf{Ext}}^n_{(\mathcal{X},\mathcal{Y})}(A,B) \quad &:= \quad \mathcal{T}_l(\mathcal{X}/\omega)[\Omega^n\mathbf{R}^*\mathsf{P}(\underline{A}), \mathbf{R}^*\mathsf{P}(\underline{B})] \\ &\xrightarrow{\cong} \quad \varinjlim_{k,k+n\geq 0} \mathcal{C}/\omega[\Omega^{k+n}\mathbf{R}(\underline{A}), \Omega^k(\underline{B})].\end{aligned}$$

EXAMPLE-DEFINITION. Let $\widehat{\mathsf{Ext}}^*_{(\mathcal{C},\mathcal{P})}(-,-)$ be the complete projective extension bifunctor with respect to the largest hereditary torsion pair $(\mathcal{C}/\mathcal{P}, 0)$ in $\mathcal{C}/\mathcal{P}$. Since in this case $\mathbf{R} = \mathrm{Id}_{\mathcal{C}/\mathcal{P}}$, we have:

$$\widehat{\mathsf{Ext}}^n_{(\mathcal{C},\mathcal{P})}(A,B) = \mathcal{T}_l(\mathcal{C}/\mathcal{P})[\Omega^n\mathsf{P}(\underline{A}), \mathsf{P}(\underline{B})], \quad n \in \mathbb{Z}$$

In other words the bifunctor $\widehat{\mathsf{Ext}}^*_{(\mathcal{C},\mathcal{P})}(-,-)$ is what is known in the literature as the **projective Tate-Vogel extension bifunctor** and from now on we denote it by $\widehat{\mathsf{Ext}}^*_{TV}(-,-)$.

HISTORICAL REMARK. Tate cohomology, which was introduced by Tate, see [**38**], was designed for finite groups and found applications in number theory. His cohomology subsumed the ordinary cohomology and ordinary homology into a single cohomological functor. Later Farell extended Tate's cohomology to groups of virtually finite cohomological dimension, see [**35**]. The projective version of Tate-Vogel cohomology was introduced in the late eighties by Vogel (unpublished, but see [**55**]), and independently by Mislin (see [**83**]) and Benson-Carlson (see [**30**]), using different methods. This cohomology theory is working for arbitrary groups and in fact for any ring. See also [**20**] for another approach which works for more general categories and is closer to the present construction.

Note that using the short exact sequences of Proposition 1.3 and the fact that $\mathcal{T}_l(\mathcal{P}^{<\infty}/\mathcal{P}) = 0$, it follows that the projective Tate-Vogel extension bifunctors are the complete projective extension bifunctors with respect to any hereditary torsion pair $(\mathcal{X}/\mathcal{P}, \mathcal{Y}/\mathcal{P})$ in $\mathcal{C}/\mathcal{P}$ such that the cotorsion-free class $\mathcal{Y}$ consists of modules of finite projective dimension.

From now on we concentrate on the construction of completions of homological functors. First we need some preparations on functors which are homological with respect to certain classes of extensions. Let $\mathcal{F}_\omega$ be the class of short exact sequences $0 \to A \to B \to C \to 0$ in $\mathcal{C}$ such that the induced sequence $0 \to \mathcal{C}(\omega, A) \to \mathcal{C}(\omega, B) \to \mathcal{C}(\omega, C) \to 0$ is exact. Since ω is contravariantly finite and any ω-epic is an epimorphism, it follows that $\mathcal{F}_\omega$ is an additive subfunctor of the Yoneda extension functor $\mathrm{Ext}^1_{\mathcal{C}}(-,-)$. Let $F_\bullet := \{F_n\}_{n\in\mathbb{Z}}$ be a sequence of functors $\mathcal{C} \to \mathcal{A}b$. We say that $F_\bullet$ is an *ω-homological functor* if $F_n(\omega) = 0$, $\forall n \in \mathbb{Z}$, and for any extension $0 \to A \to B \to C \to 0$ in $\mathcal{F}_\omega$, there exists a long exact sequence

$$\cdots \to F_{-1}(C) \to F_0(A) \to F_0(B) \to F_0(C) \to F_1(A) \to \cdots$$

which is natural with respect to morphisms of extensions in $\mathcal{F}_\omega$. In other words $F_\bullet$ is ω-homological if the induced functor $\underline{F}_\bullet$ is a (left) homology theory in the left triangulated category $\mathcal{C}/\omega$, where each $\underline{F}_n : \mathcal{C}/\omega \to \mathcal{A}b$ is naturally induced from F_n. $F_\bullet$ is called *$\mathcal{Y}$-complete*, if the induced (left) homology theory $\underline{F}_\bullet$ is $\mathcal{Y}/\omega$-complete. This is equivalent to saying that $F_n(\mathcal{Y}) = 0$, $\forall n \in \mathbb{Z}$. The following is the basic example of a $\mathcal{Y}$-complete ω-homological functor.

EXAMPLE. For any object C the sequence of functors $\{\widehat{\mathsf{Ext}}^n_{(\mathcal{X},\mathcal{Y})}(C,-)\}_{n\in\mathbb{Z}}$ is an ω-homological functor, which is $\mathcal{Y}$-complete by construction, that is we have $\widehat{\mathsf{Ext}}^*_{(\mathcal{X},\mathcal{Y})}(C,\mathcal{Y}) = 0$.

To obtain more examples we need to recall some basic facts about relative homology in an abelian category. These facts will be useful in the sequel.

Let $\mathcal{C}$ be an abelian category and ω a full subcategory of $\mathcal{C}$. A complex $C^\bullet$ is called right ω-exact, if the complex $\mathcal{C}(\omega, C^\bullet)$ is exact in $\mathcal{A}b$. A right ω-projective resolution of an object C in $\mathcal{C}$ is a right ω-exact complex $\cdots \to T_1 \to T_0 \xrightarrow{\varepsilon} C \to 0$ where the T_i are in ω. Then ε is a right ω-approximation of C and the kernel $\operatorname{Ker}(\varepsilon)$ is the first relative syzygy of C with respect to ω, and as usual is denoted by $\Omega_\omega(C)$ or $\Omega(C)$. The higher relative syzygies $\Omega^n(C)$ of C are defined inductively. Observe that a right ω-projective resolution is a genuine exact complex, provided that any ω-epic is an epimorphism. We denote by $\operatorname{Rapp}_\omega(\mathcal{C})$ the full subcategory of $\mathcal{C}$ consisting of all objects which admits right ω-resolutions. Note that $\operatorname{Rapp}_\omega(\mathcal{C}) = \mathcal{C}$ iff ω is contravariantly finite in $\mathcal{C}$.

We define, for any object $C \in \mathcal{C}$, the relative Ext functors $\operatorname{Ext}^n_\omega(-,C)$: $\operatorname{Rapp}_\omega(\mathcal{C})^{\mathrm{op}} \to \mathcal{A}b$, for $n \geq 0$, as follows. Let A be an object with right ω-resolution $\cdots \to T_1 \to T_0 \to A \to 0$. Then $\operatorname{Ext}^n_\omega(A,C)$ is the nth cohomology of the complex $0 \to \mathcal{C}(T_0,C) \to \mathcal{C}(T_1,C) \to \cdots$. Observe that that there exists a natural map $\mathcal{C}(A,C) \to \operatorname{Ext}^0_\omega(A,C)$ which is invertible, provided that any ω-epic is an epimorphism. We refer to **[20]** for basic properties of the relative extensions functors. In particular we shall need the following fact. If $0 \to A \to B \to C \to 0$ is a right ω-exact complex, where the involved objects lie in $\operatorname{Rapp}_\omega(\mathcal{C})$, then, for any object $D \in \mathcal{C}$, there exists a long exact sequence $0 \to \operatorname{Ext}^0_\omega(C,D) \to \operatorname{Ext}^0_\omega(B,D) \to \operatorname{Ext}^0_\omega(A,D) \to \operatorname{Ext}^1_\omega(C,D) \to \cdots$. Observe that if ω is contravariantly finite and any ω-epic is an epimorphism, then $\operatorname{Ext}^1_\omega(-,-)$ is the subfunctor $\mathcal{F}_\omega$ of the usual Yoneda Ext functor $\operatorname{Ext}^1_\mathcal{C}(-,-)$ of $\mathcal{C}$ introduced above.

Now we can give a variety of examples of ω-homological functors.

EXAMPLE. (i) Any object C defines an ω-homological functor $\{\operatorname{Ext}^n_\omega(C,-)\}_{n\in\mathbb{Z}}$, where of course we set $\operatorname{Ext}^n_\omega(C,-) = 0, \forall n < 0$.

(ii) Any object C defines a homological functor $\{\underline{\Pi}^n_C(-)\}_{n\in\mathbb{Z}}$, as follows:

$$\underline{\Pi}^n_C(-) =: \begin{cases} \operatorname{Ext}^n_\omega(C,-) & n \geq 1 \\ \mathcal{C}/\omega(C,\Omega^{-n}_\omega(-)) & n \leq 0 \end{cases}$$

In other words $\underline{\Pi}^n_C(-)$ is an analogue of the n-th projective homotopy functor of C introduced by Eckmann-Hilton, see **[61]** where the case $\omega = \mathcal{P}$ or $\omega = \mathcal{I}$ is treated. From now on we call $\underline{\Pi}^n_C(-)$ the n-th *Eckmann-Hilton projective homotopy functor* of C with respect to ω. The *Eckmann-Hilton injective homotopy functor* $\overline{\Pi}^*_C(-)$ of C with respect to ω is defined dually.

Observe that the negative parts of the homological functors in (i), (ii) are ω-complete. However $\underline{\Pi}^*_C(-)$ is ω-complete iff $\operatorname{Ext}^n_\omega(C,\omega) = 0$, $\forall n \geq 1$, and $\operatorname{Ext}^*_\omega(C,-)$ is ω-complete iff $\operatorname{Ext}^n_\omega(C,\omega) = 0$, $\forall n \geq 0$.

We are going to show that for any object C, the $\mathcal{Y}$-complete projective extension functor $\widehat{\mathsf{Ext}}^{*}_{(\mathcal{X},\mathcal{Y})}(C,-)$ is the $\mathcal{Y}$-completion of the extension functor $\mathrm{Ext}^{*}_{\mathcal{X}\cap\mathcal{Y}}(C,-)$, in the sense of the following definition.

DEFINITION 2.3. The **$\mathcal{Y}$-completion** of an ω-homological functor $\{F_n\}_{n\in\mathbb{Z}} : \mathcal{C} \to \mathcal{A}b$ with respect to a full subcategory $\mathcal{Y}$ of $\mathcal{C}$, is a $\mathcal{Y}$-complete homological functor $\{\widehat{F}_n\}_{n\in\mathbb{Z}} : \mathcal{C} \to \mathcal{A}b$ together with a morphism of homological functors $\phi^* : \{F_n\}_{n\in\mathbb{Z}} \to \{\widehat{F}_n\}_{n\in\mathbb{Z}}$, which satisfies the following universal property: any morphism $\alpha^* : \{F_n\}_{n\in\mathbb{Z}} \to \{G_n\}_{n\in\mathbb{Z}}$ to a $\mathcal{Y}$-complete homological functor $\{G_n\}_{n\in\mathbb{Z}}$ factors uniquely through ϕ^*. In this case ϕ^* is called the **$\mathcal{Y}$-completion morphism**.

We first give some preliminary results on $\mathcal{Y}$-complete homological functors defined on $\mathcal{C}$, which will be useful later for the construction of $\mathcal{Y}$-completions. So we fix for a moment a $\mathcal{Y}$-complete homological functor $\{F_n\}_{n\in\mathbb{Z}}$. Since $\omega \subseteq \mathcal{Y}$, it follows that $F_n(\omega) = 0$. Hence each F_n induces a functor $F_n : \mathcal{C}/\omega \to \mathcal{A}b$.

LEMMA 2.4. *There are natural isomorphisms*

(i) $F_{-n} \xrightarrow{\cong} F_0\Omega^n$, $\forall n \geq 0$.

(ii) $F_k \xrightarrow{\cong} F_{k+1}\Omega$ *and* $F_{k-n} \xrightarrow{\cong} F_k\Omega^n$, $\forall k \in \mathbb{Z}$, $\forall n \geq 0$.

(iii) $\forall C \in \mathcal{C}$, $\forall n \geq 0$, $\forall k \in \mathbb{Z}$:

$$\Big(\mathrm{Ext}^n_{\omega}(C,-), F_k\Big) \xrightarrow{\cong} F_k(\Omega^n(C)) \xleftarrow{\cong} \Big(\underline{\Pi}^n_C, F_k\Big).$$

(iv) $\forall C \in \mathcal{C}$, $\forall n < 0$, $\forall k \in \mathbb{Z}$: $(\underline{\Pi}^n_C, F_k) = F_k\Sigma^{-n}(C)$.

(v) $\forall C \in \mathcal{C}$, *the right $\mathcal{X}$-approximation sequence* $0 \to Y_C \to X_C \xrightarrow{f_C} C \to 0$ *induces a commutative diagram of isomorphisms,* $\forall n \geq 0$, $\forall k \in \mathbb{Z}$:

$$\begin{array}{ccc}
(\mathrm{Ext}^n_{\omega}(X_C,-), F_k) & \xrightarrow{(f^n_C, F_k)} & (\mathrm{Ext}^n_{\omega}(C,-), F_k) \\
\cong\downarrow & & \cong\downarrow \\
F_k(\Omega^n(X_C)) & \xrightarrow{F_k(\Omega^n(f_C))} & F_k(\Omega^n(C)) \\
\cong\downarrow & & \cong\downarrow \\
F_{k-n}(X_C) & \xrightarrow{F_{k-n}(f_C)} & F_{k-n}(C)
\end{array}$$

where $f^n_C = \mathrm{Ext}^n_{\omega}(f_C,-)$ *and all the involved morphisms are isomorphisms.*

PROOF. The assertions in (i), (ii) follow easily from the fact that each F_k kills the objects of ω and is half-exact on extensions from $\mathcal{F}_{\omega}$. If $\Omega(C) \rightarrowtail \omega_C \twoheadrightarrow C$ is a right ω-approximation sequence of C, then applying $(-, F_k)$ to the exact sequence of functors $(C,-) \rightarrowtail (\omega_C,-) \to (\Omega(C),-) \twoheadrightarrow \mathrm{Ext}^1_{\omega}(C,-)$ and using Yoneda's Lemma we deduce an isomorphism: $(\mathrm{Ext}^1_{\omega}(C,-), F_k) \xrightarrow{\cong} F_k(\Omega(C))$. Then part (iii) follows by induction. Part (iv) follows from the isomorphisms $[\mathcal{C}/\omega(C, \Omega^n(-)), F_k] \xrightarrow{\cong} [\mathcal{C}/\omega(\Sigma^n(C),-), F_k] \xrightarrow{\cong} F_k(\Sigma^n(C))$. Here Σ is the left adjoint of the loop functor Ω of $\mathcal{C}/\omega$. Part (v) follows from the isomorphisms of the previous parts and the fact that each F_k kills the objects of $\mathcal{Y}$ and is half-exact on extensions from $\mathcal{F}_{\omega}$. □

Since the homological functor $\{\widehat{\mathrm{Ext}}^{m}_{(\mathcal{X},\mathcal{Y})}(A,-)\}_{m\in\mathbb{Z}}$ is $\mathcal{Y}$-complete, we have the following direct consequence.

LEMMA 2.5. *If $A, C \in \mathcal{C}$, then for any $n \geq 0$, $m \in \mathbb{Z}$:*

$$\left(\mathrm{Ext}^n_\omega(C,-), \widehat{\mathrm{Ext}}^{m}_{(\mathcal{X},\mathcal{Y})}(A,-)\right) \xrightarrow{\cong} \varinjlim_{k+m,k+n\geq 0} \mathcal{C}/\omega[\Omega^{k+m}\mathbf{R}(\underline{C}), \Omega^{k+n}(\underline{A})].$$

For any object C we now define a specific morphism $\varphi^*_{C,-} : \mathrm{Ext}^*_\omega(C,-) \to \widehat{\mathrm{Ext}}^{*}_{(\mathcal{X},\mathcal{Y})}(C,-)$ as follows. For $n < 0$ we set $\varphi^n_{C,-} = 0$. For $n \geq 0$, let $f_C : X_C \to C$ be a right $\mathcal{X}$-approximation of C. Then we have an induced morphism $\Omega^n_\omega(f_C) : \Omega^n_\omega(X_C) \to \Omega^n_\omega(C)$, which in turn serves as a representative of a unique morphism

$$\mathsf{P}(\underline{f}_C) \in \varinjlim_{k,k+n\geq 0} \mathcal{C}/\omega[\Omega^{k+n}\mathbf{R}(\underline{C}), \Omega^{k+n}(\underline{C})] \xrightarrow{\cong} \mathcal{T}_l(\mathcal{C}/\omega)[(\Omega^n\mathsf{P}\mathbf{R}(\underline{C}), \Omega^n\mathsf{P}(\underline{C})]$$

and then by Lemma 2.5 this morphism represents a unique morphism

$$\varphi^n_{C,-} \ : \ \mathrm{Ext}^n_\omega(C,-) \longrightarrow \widehat{\mathrm{Ext}}^{n}_{(\mathcal{X},\mathcal{Y})}(C,-).$$

Now we can prove the main result of this section.

THEOREM 2.6. *The morphism $\varphi^*_{C,-} : \mathrm{Ext}^*_{\mathcal{X}\cap\mathcal{Y}}(C,-) \to \widehat{\mathsf{Ext}}^{*}_{(\mathcal{X},\mathcal{Y})}(C,-)$ is the $\mathcal{Y}$-completion of the relative* Ext*-homological functor* $\mathrm{Ext}^*_{\mathcal{X}\cap\mathcal{Y}}(C,-)$.

PROOF. Let $F_* : \mathcal{C} \to \mathcal{A}b$ be a $\mathcal{Y}$-complete homological functor and let $\xi_* : \mathrm{Ext}^*_\omega(C,-) \to F_*$ be a morphism of ω-homological functors. Then we have $\xi_n = 0$ for $n < 0$, and by Lemma 2.5 we have that ξ_n is represented by a unique element $\zeta_n \in F_n(\Omega^n(C))$. We define a morphism $\widehat{\xi}_* : \widehat{\mathsf{Ext}}^{*}_{(\mathcal{X},\mathcal{Y})}(C,-) \to F_*$ as follows. First let $n \geq 0$, let A be an object and let $\alpha \in \widehat{\mathsf{Ext}}^{*}_{(\mathcal{X},\mathcal{Y})}(C,A)$. Choose a representative $\underline{\alpha}_k : \Omega^{n+k}\mathbf{R}(\underline{C}) \to \Omega^k(\underline{A})$ in $\widehat{\mathsf{Ext}}^{*}_{(\mathcal{X},\mathcal{Y})}(C,A)$, where $k \geq 0$. Then $F_{n+k}(\underline{\alpha}_k) : F_{n+k}\Omega^{n+k}\mathbf{R}(\underline{C}) \to F_{n+k}\Omega^k(\underline{A})$. By Lemma 2.4 this morphism is isomorphic to the morphism $F_n(\underline{\alpha}_k) : F_n\Omega^n\mathbf{R}(\underline{C}) \to F_n(\underline{A})$. It is clear that the construction is independent of the choice of the representative $\underline{\alpha}_k$ of α. By Lemma 2.4 the morphism $F_n(\Omega^n(f_C)) : F_n\Omega^n\mathbf{R}(\underline{C}) \to F_n(\Omega^n(\underline{C}))$ is invertible. Consider the composition

$$F_n(\Omega^n(\underline{C})) \xrightarrow{F_n(\Omega^n(f_C))^{-1}} F_n\Omega^n\mathbf{R}(\underline{C}) \xrightarrow{F_n(\underline{\alpha}_k)} F_n(\underline{A})$$

and define $\widehat{\xi}_n(\alpha) := F_n(\underline{\alpha}_n)((F_n\Omega^n(f_C))^{-1}(\zeta_n))$. It is easy to see that the above construction is functorial, hence $\widehat{\xi}_n : \widehat{\mathsf{Ext}}^{n}_{(\mathcal{X},\mathcal{Y})}(C,-) \to F_n$ is a natural morphism for $n \geq 0$. Now we define the morphism $\widehat{\xi}_{-n}$ for $n > 0$. Consider the morphism $\widehat{\xi}_0 : \widehat{\mathsf{Ext}}^{0}_{(\mathcal{X},\mathcal{Y})}(C,-) \to F_0$ and define $\widehat{\xi}_{-n} = \widehat{\xi}_0\Omega^n : \widehat{\mathsf{Ext}}^{0}_{(\mathcal{X},\mathcal{Y})}(C,\Omega^n(-)) \to F_0\Omega^n \cong F_{-n}$, where the last isomorphism follows by Lemma 2.4 and the fact that $\widehat{\mathsf{Ext}}^{0}_{(\mathcal{X},\mathcal{Y})}(C,\Omega^n(-)) = \mathcal{T}_l(\mathcal{C}/\omega)[\mathsf{P}\mathbf{R}(\underline{C}), \mathsf{P}\Omega^n(-)] = \mathcal{T}_l(\mathcal{C}/\omega)[\Omega^{-n}\mathsf{P}\mathbf{R}(\underline{C}), \mathsf{P}(-)] = \widehat{\mathsf{Ext}}^{-n}_{(\mathcal{X},\mathcal{Y})}(C,-)$.

We claim that $\widehat{\xi}_* \circ \varphi^C_* = \xi_*$. For $n < 0$ this is clear since then $\varphi^n_C = \xi_n = 0$. Assume now that $n \geq 0$. Since φ^n_C is represented by $\Omega^n(f_C)$ using the isomorphisms of Lemmas 2.3 and 2.4 it suffices to show that $\widehat{\xi}_n(\Omega^n(f_C)) = \xi_n$. Using the definition of $\widehat{\xi}_n$ we have $\widehat{\xi}_n(\Omega^n(f_C)) = F_n(\Omega^n(f_C))((F_n\Omega^n(f_C))^{-1}(\zeta_n)) = \zeta_n$ which represents the morphism ξ_n. Since the construction is independent of the choices for representatives, we have $\widehat{\xi}_n(\Omega^n(f_C)) = \xi_n$. We infer that $\widehat{\xi}_* \circ \varphi^*_C = \xi_*$. The easy proof that $\widehat{\xi}_*$ is the unique morphism of homological functors with this property is left to the reader. □

REMARK 2.7. Observe that for any $n \in \mathbb{Z}$, $\widehat{\mathsf{Ext}}^n_{(\mathcal{X},\mathcal{Y})}(-,-)$ is a bifunctor, but enjoys the universal property of the above theorem only in the first variable. The discussion that follows suggests that the functor $\widehat{\mathsf{Ext}}^n_{(\mathcal{X},\mathcal{Y})}(-,-)$ can be considered as a *generalized (projective) Tate-Vogel bifunctor.*

From now on we assume that in the setup of the beginning of this section, the abelian category $\mathcal{C}$ has enough projectives and $\omega = \mathcal{X} \cap \mathcal{Y} = \mathcal{P}$. Mislin in [**83**], working in the setting of a module category over a ring Λ, constructed the completion of the usual extension functor $\mathrm{Ext}^*_\Lambda(C,-)$ with respect to the projective modules, using filtered colimits of satellites of half-exact functors. Applying Theorem 2.6 to the largest hereditary torsion pair $(\mathcal{C}/\mathcal{P}, 0)$ in the stable category $\mathcal{C}/\mathcal{P}$ modulo projectives and noting that by definition we have $\widehat{\mathsf{Ext}}^*_{TV}(C,-) = \widehat{\mathsf{Ext}}^*_{(\mathcal{C}/\mathcal{P},0)}(C,-)$, we get the following generalization of Mislin's result as an immediate consequence.

COROLLARY 2.8. *For any object C in $\mathcal{C}$, the Tate-Vogel projective extension functor $\widehat{\mathsf{Ext}}^*_{TV}(C,-)$ is the $\mathcal{P}$-completion of the homological functor $\mathrm{Ext}^*_\mathcal{C}(C,-)$.*

If $X \in \mathcal{X}$ satisfies $\mathrm{Ext}^1_\mathcal{C}(X,\mathcal{Y}) = 0$, then the Eckmann-Hilton projective homotopy functor $\underline{\Pi}^*_X$ is obviously a $\mathcal{Y}$-complete homological functor. It is natural to ask if in this case the canonical morphism $p^*_X : \mathrm{Ext}^*_\mathcal{C}(X,-) \to \underline{\Pi}^*_X(-)$ is the $\mathcal{Y}$-completion of $\mathrm{Ext}^*_\mathcal{C}(X,-)$, where $p^n_X = 0$ for $n < 0$, p^0_X is the canonical map $\mathcal{C}(X,-) \to \mathcal{C}/\mathcal{P}(X,-)$, and p^n_X is the canonical map $\mathrm{Ext}^*_\mathcal{C}(X,-) \to \mathcal{C}/\mathcal{P}(\Omega^n(X),-)$ for $n \geq 1$. This is not true in general. However we have the following.

COROLLARY 2.9. *For any object $X \in \mathcal{X} \cap {}^\perp\mathcal{Y}$, the following are equivalent:*

(i) *$X \in {}^\perp\mathcal{P}$.*

(ii) *The map $p^*_X : \mathrm{Ext}^*_\mathcal{C}(X,-) \to \underline{\Pi}^*_X(-)$ is the $\mathcal{Y}$-completion of $\mathrm{Ext}^*_\mathcal{C}(X,-)$.*

PROOF. (i) ⇒ (ii) By Lemma 5.12 of [**20**] we have that the canonical maps $\mathcal{C}/\mathcal{P}(\underline{X},\underline{B}) \to \mathcal{T}_l(\mathcal{C}/\mathcal{P})[\mathsf{P}(\underline{X}),\mathsf{P}(\underline{B})]$ and $\mathrm{Ext}^n_\mathcal{C}(X,B) \to \mathcal{C}/\mathcal{P}[\Omega^n(\underline{X}),\underline{B}]$ are invertible, $\forall n \geq 1$, $\forall X \in {}^\perp\mathcal{P}$, and any object B. It follows from this that $\widehat{\mathsf{Ext}}^*_{(\mathcal{X},\mathcal{Y})}(X,-) \xrightarrow{\cong} \mathcal{C}/\mathcal{P}[\Omega^*(\underline{X}),-] = \underline{\Pi}^*_X(-)$, and then p^*_X is isomorphic to φ^*_X.

(ii) ⇒ (i) The hypothesis implies that the canonical morphism $p^*_X : \underline{\Pi}^*_X \to \widehat{\mathsf{Ext}}^*_{(\mathcal{X},\mathcal{Y})}(X,-)$ is invertible. In particular the canonical morphism $\mathcal{C}/\mathcal{P}(\underline{X},-) \to \mathcal{T}_l(\mathcal{C}/\mathcal{P})[\mathsf{P}(\underline{X}),\mathsf{P}(-)]$ is invertible. Then $X \in {}^\perp\mathcal{P}$ by Theorem 4.2 of [**20**]. □

The following result shows that, if $(\mathcal{X},\mathcal{Y})$ is a cotorsion pair in $\mathcal{C}$ with $\mathcal{X} \cap \mathcal{Y} = \mathcal{P}$, then the $\mathcal{Y}$-completion of the usual homological functor $\mathrm{Ext}^*_\mathcal{C}(C,-)$ with respect to the hereditary torsion pair $(\mathcal{X}/\mathcal{P}, \mathcal{Y}/\mathcal{P})$ in $\mathcal{C}/\mathcal{P}$ can be realized as the Eckmann-Hilton projective homotopy functor of the special right $\mathcal{X}$-approximation of C.

COROLLARY 2.10. *Let $(\mathcal{X},\mathcal{Y})$ be a resolving cotorsion pair in $\mathcal{C}$ with $\mathcal{X} \cap \mathcal{Y} = \mathcal{P}$. Then the $\mathcal{Y}$-completion of the functor $\mathrm{Ext}^*_\mathcal{C}(C,-)$ is the Eckmann-Hilton projective homotopy functor $\underline{\Pi}^*_{X_C}(-)$ of the special right $\mathcal{X}$-approximation X_C of C.*

PROOF. By Theorem VI.2.1 we have $\mathcal{X} \subseteq {}^{\perp}\mathcal{P}$. Then for any object C, using the above Corollary and the fact that $\mathbf{R}(\underline{C}) = \underline{X}_C$, we have:

$$\widehat{\mathsf{Ext}}^{*}_{(\mathcal{X},\mathcal{Y})}(C,-) \xrightarrow{\cong} \mathcal{T}_l(\mathcal{C}/\mathcal{P})[\Omega^*\mathsf{P}\mathbf{R}(\underline{C}),\mathsf{P}(-)] \xrightarrow{\cong} \underline{\Pi}^*_{X_C}(-).$$

□

REMARK 2.11. All the results of this section can be dualized working, for instance, with pairs $(\mathcal{Y},\mathcal{Z})$ with $\mathcal{Z}$ coresolving and $\mathcal{Y}\cap\mathcal{Z} = \mathcal{I}$, and assuming that $(\mathcal{Y}/\mathcal{I},\mathcal{Z}/\mathcal{I})$ is a cohereditary torsion pair in $\mathcal{C}/\mathcal{I}$. In this case for any object C the usual cohomological functor $\operatorname{Ext}^*_{\mathcal{C}}(-,C)$ admits the $\mathcal{Y}$-completion $\widetilde{\operatorname{Ext}}^{*}_{(\mathcal{Y},\mathcal{Z})}(-,C)$. We leave to the reader to state the dual results.

3. Relative Homology and Generalized Tate-Vogel (Co)Homology

In this section we concentrate on complete projective or injective extension bifunctors with respect to torsion pairs $(\mathcal{X}/\omega,\mathcal{Y}/\omega)$ in $\mathcal{C}/\omega$, arising from cotorsion pairs in $\mathcal{C}$. By the results of Chapter V we know that ω is an Ext-injective cogenerator of $\mathcal{X}$. In this section we are especially interested in the case where ω is also an Ext-projective generator of $\mathcal{X}$. The last condition is very pleasant since, by Theorem VI.2.1, it implies that the torsion class is triangulated. In this case there are defined two natural relative homological theories on $\mathcal{C}$. The first one is represented by the relative extension functor $\operatorname{Ext}^*_{\mathcal{X}}$ induced by the cotorsion subcategory $\mathcal{X}$ and the other one is the Generalized Tate-Vogel (Co)Homology $\widehat{\mathsf{Ext}}^{*}_{(\mathcal{X},\mathcal{Y})}$. We compare these with the relative homological theory represented by the relative extension functor $\operatorname{Ext}^*_{\omega}$ induced by ω, by providing comparison morphisms which fit in an infinite long exact sequence. These results generalize recent investigations by Avramov-Martsinkovsky which deal with relative (co)homological theories induced by modules of finite Gorenstein dimension over a Noetherian ring, see [**16**].

Throughout this section we fix an abelian category $\mathcal{C}$, not necessarily with enough projective or injective objects. We recall that if $\mathcal{A}$ is a full subcategory of $\mathcal{C}$, then an object C of $\mathcal{C}$ is said to be of finite right $\mathcal{A}$-projective dimension, if there is a right $\mathcal{A}$-projective resolution $0 \to X_n \xrightarrow{x_n} X_{n-1} \to \cdots \to X_i \xrightarrow{x_i} X_{i-1} \to \cdots \to X_0 \xrightarrow{x_0} C \to 0$ of C of length n. In this case we write $\mathcal{A}\text{-pd}\,C \leq n$. The least such integer n is called the (relative) $\mathcal{A}$-projective dimension of C and is denoted by $\mathcal{A}\text{-pd}\,C = n$. Otherwise, C is said to have infinite $\mathcal{A}$-projective dimension and we write $\mathcal{A}\text{-pd}\,C = \infty$.

Instead of working directly with cotorsion pairs in $\mathcal{C}$, it is possible to work in the more general setting of cotorsion pairs defined in a smaller piece of $\mathcal{C}$. So we start with a full additive subcategory $\mathcal{X}$ of $\mathcal{C}$ which is closed under extensions, direct summands and kernels of epimorphisms. We assume throughout that $\mathcal{X}$ admits an Ext-injective cogenerator ω. Then by Auslander-Buchweitz theory, see [**5**], any object C in $\widehat{\mathcal{X}}$ admits a special right $\mathcal{X}$-approximation sequence $0 \to Y_C \xrightarrow{g_C} X_C \xrightarrow{f_C} C \to 0$ with $Y_C \in \widehat{\omega}$, and a special left $\widehat{\omega}$-approximation sequence $0 \to C \xrightarrow{g^C} Y^C \xrightarrow{f^C} X^C \to 0$ with X^C in $\mathcal{X}$. In particular the category $\mathcal{X}$ is contravariantly finite in $\widehat{\mathcal{X}}$, hence $\widehat{\mathcal{X}} \subseteq \operatorname{Rapp}_{\mathcal{X}}(\mathcal{C})$, and the category $\widehat{\omega}$ is covariantly finite in $\widehat{\mathcal{X}}$. Moreover we have $\mathcal{X}\cap\widehat{\omega} = \omega$.

The following result which will be useful later, gives sufficient conditions such that the $\mathcal{X}$-resolution dimension and the relative $\mathcal{X}$-projective dimension are equal.

LEMMA 3.1. *Let $\mathcal{X}$ be a full subcategory of $\mathcal{C}$ which is closed under extension, direct summands and kernels of epics. We assume that $\mathcal{X}$ admits an Ext-injective cogenerator ω and any $\mathcal{X}$-epic in $\mathcal{C}$ is an epimorphism. Then $\mathcal{X}\text{-res.}\dim C = \mathcal{X}\text{-pd}\, C$, provided that one of the following conditions holds: (α) C lies in $\widehat{\mathcal{X}}$ or (β) C lies in $\mathcal{C}$ and $\mathcal{X}$ is contravariantly finite in $\mathcal{C}$.*

PROOF. If $\mathcal{X}\text{-pd}\, C = 1$, then there exists a non-split exact sequence $0 \to X_1 \to X_0 \to C \to 0$ in $\mathcal{C}$ with the X_i in $\mathcal{X}$, such that the sequence remains exact after the application of the functor $\mathcal{C}(X, -)$, for any $X \in \mathcal{X}$. This implies that $\mathcal{X}\text{-res.}\dim C = 1$. Conversely if $\mathcal{X}\text{-res.}\dim C = 1$, let $0 \to X_1 \to X_0 \to C \to 0$ be an exact sequence in $\mathcal{C}$ with the X_i in $\mathcal{X}$. Then C admits a special right $\mathcal{X}$-approximation $0 \to Y_C \to X_C \to C \to 0$, where $Y_C \in \widehat{\omega}$ in case (α), and $Y_C \in \mathcal{X}^{\perp}$ in case (β). Therefore we have the following exact commutative diagram

$$\begin{array}{ccccccccc}
0 & \longrightarrow & X_1 & \longrightarrow & X_0 & \longrightarrow & C & \longrightarrow & 0 \\
 & & {\scriptstyle\gamma}\downarrow & & {\scriptstyle\kappa}\downarrow & & \| & & \\
0 & \longrightarrow & Y_C & \xrightarrow{g_C} & X_C & \xrightarrow{f_C} & A & \longrightarrow & 0
\end{array}$$

Since X_1 lies in $\mathcal{X}$, the morphism γ factors through ω. Since ω is an Ext-injective cogenerator of $\mathcal{X}$, there exists a short exact sequence $0 \to X_1 \xrightarrow{\alpha} T \to X_1' \to 0$ with T in ω and X_1' in $\mathcal{X}$, which is a left ω-approximation sequence of X_1. Hence γ admits a factorization $\gamma = \alpha \circ \beta : X_1 \xrightarrow{\alpha} T \xrightarrow{\beta} X_1'$. Pushing out the upper sequence in the diagram along the morphism α, we get a short exact sequence $(\dagger) : 0 \to T \to X' \to C \to 0$. Since $\mathrm{Coker}(\alpha)$ lies in $\mathcal{X}$ and $\mathcal{X}$ is closed under extensions, we infer that X' lies in $\mathcal{X}$. Since $\mathrm{Ext}^1(\mathcal{X}, \omega) = 0$, we infer that the sequence $(\dagger)$ is an $\mathcal{X}$-resolution of C, hence $\mathcal{X}\text{-pd}\, C = 1$. Now by induction we have that $\mathcal{X}\text{-pd}\, C < \infty$ if and only if $\mathcal{X}\text{-res.}\dim C < \infty$, in which case $\mathcal{X}\text{-pd}\, C = \mathcal{X}\text{-res.}\dim C$. □

From now on we assume that ω is contravariantly finite in $\mathcal{C}$ and any ω-epic in $\mathcal{C}$ is an epimorphism. For any object A in $\widehat{\mathcal{X}}$ and any object C in $\mathcal{C}$, we denote by $\mathrm{Ext}^*_{\mathcal{X}}(A, C)$ the relative extension functors with respect to $\mathcal{X}$. Since $\mathcal{X}$ is contravariantly finite in $\widehat{\mathcal{X}}$, they can be computed using exact resolutions of A by objects from $\mathcal{X}$ which have the property that they remain exact after the application of $\mathcal{C}(X, -)$ for any $X \in \mathcal{X}$. Contravariant finiteness of $\mathcal{X}$ in $\widehat{\mathcal{X}}$ ensures that such resolutions exist in $\widehat{\mathcal{X}}$. Since any ω-epic is an epimorphism in $\mathcal{C}$, it follows that $\mathrm{Ext}^0_{\omega}(-,-) = \mathcal{C}(-,-)$ and $\mathrm{Ext}^1_{\omega}(-,-) \hookrightarrow \mathrm{Ext}^1_{\mathcal{C}}(-,-)$. Since any object A of $\widehat{\mathcal{X}}$ is a factor of an object from $\mathcal{X}$, it follows that any $\mathcal{X}$-epic is an epimorphism, and therefore: $\mathrm{Ext}^0_{\mathcal{X}}(X, -) = \mathcal{C}(X, -)$ and $\mathrm{Ext}^1_{\mathcal{X}}(X, -) \hookrightarrow \mathrm{Ext}^1_{\mathcal{C}}(X, -)$, for any $X \in \widehat{\mathcal{X}}$.

Now since ω is an Ext-injective cogenerator of $\mathcal{X}$, it follows that ω is covariantly finite in $\mathcal{X}$ and any ω-monic in $\mathcal{X}$ admits a cokernel in $\mathcal{X}$. Since ω is contravariantly finite in $\mathcal{X}$ and the latter is closed under kernels of epimorphisms, it follows that any ω-epic in $\mathcal{X}$ admits a kernel in $\mathcal{X}$. We conclude that the stable category $\mathcal{X}/\omega$ is pretriangulated. Let $\mathsf{P} : \mathcal{C}/\omega \to \mathcal{T}_l(\mathcal{C}/\omega)$, resp. $\mathsf{P} : \mathcal{X}/\omega \to \mathcal{T}_l(\mathcal{X}/\omega)$, be the stabilization of $\mathcal{C}/\omega$, resp. $\mathcal{X}/\omega$, with respect to its left triangulation. For any object A in $\widehat{\mathcal{X}}$ and any object C in $\mathcal{C}$, we define, for any $n \in \mathbb{Z}$, the following sequence of functors:

$$\widehat{\mathsf{Ext}}^{n}_{\mathcal{X}}(-,-) \ : \ \widehat{\mathcal{X}} \times \mathcal{C}^{\mathrm{op}} \longrightarrow \mathcal{A}b, \quad \widehat{\mathsf{Ext}}^{n}_{\mathcal{X}}(A,C) := \mathcal{T}_l(\mathcal{C}/\omega)[\Omega^n \mathsf{P}(\underline{X}_A), \mathsf{P}(\underline{C})]$$

REMARK 3.2. If $\mathcal{X}$ is part of a resolving cotorsion pair $(\mathcal{X}, \mathcal{Y})$ in $\mathcal{C}$ such that $\omega := \mathcal{X} \cap \mathcal{Y}$ is functorially finite, then the sequence of functors $\widehat{\mathsf{Ext}}^{*}_{\mathcal{X}}(-,-)$ defined above coincides with the complete extension bifunctors with respect to the cotorsion pair $(\mathcal{X}, \mathcal{Y})$. This follows from the fact that $\underline{X}_A = \mathbf{R}(\underline{A})$, where $\mathbf{R}$ is the right adjoint of the inclusion $\mathbf{i} : \mathcal{X}/\omega \hookrightarrow \mathcal{C}/\omega$, and the following natural isomorphisms:

$$\widehat{\mathsf{Ext}}^{n}_{\mathcal{X}}(A,C) = \mathcal{T}_l(\mathcal{C}/\omega)[\Omega^n \mathsf{P}\mathbf{R}(\underline{A}), \mathsf{P}(\underline{C})] \xrightarrow{\cong} \mathcal{T}_l(\mathcal{C}/\omega)[\Omega^n \mathbf{R}^* \mathsf{P}(\underline{A}), \mathsf{P}(\underline{C})] \xrightarrow{\cong}$$

$$\mathcal{T}_l(\mathcal{X}/\omega)[\Omega^n \mathbf{R}^* \mathsf{P}(\underline{A}), \mathbf{R}^* \mathsf{P}(\underline{C})] \xrightarrow{\cong} \widehat{\mathsf{Ext}}^{n}_{(\mathcal{X},\mathcal{Y})}(A,C)$$

where $\mathbf{R}^*$ is the right adjoint of the fully faithful exact functor $\mathbf{i}^* : \mathcal{T}_l(\mathcal{X}/\omega) \hookrightarrow \mathcal{T}_l(\mathcal{C}/\omega)$ induced by the inclusion $\mathbf{i} : \mathcal{X}/\omega \hookrightarrow \mathcal{C}/\omega$.

We construct as in Section 2 a natural morphism

$$\phi^n_{A,-} \ : \ \mathrm{Ext}^n_{\omega}(A,-) \longrightarrow \widehat{\mathsf{Ext}}^{n}_{\mathcal{X}}(A,-), \quad \forall A \in \widehat{\mathcal{X}}, \quad \forall n \geq 0$$

as follows. For any object C in $\mathcal{C}$ and $n \geq 0$, $\phi^{n+1}_{A,C}$ is the composition of the following morphisms:

$$\mathrm{Ext}^{n+1}_{\omega}(A,C) \xrightarrow{\cong} \mathrm{Ext}^1_{\omega}(\Omega^n(A),C) \xrightarrow{\varepsilon_{\Omega^{n+1}(A),C}} \mathcal{C}/\omega[\Omega^{n+1}(\underline{A}), \underline{C}] \xrightarrow{\mathsf{P}_{\Omega^{n+1}(\underline{A}),\underline{C}}}$$

$$\mathcal{T}_l(\mathcal{C}/\omega)[\mathsf{P}(\Omega^{n+1}(\underline{A})), \mathsf{P}(\underline{C})] \xrightarrow{\cong} \mathcal{T}_l(\mathcal{C}/\omega)[\Omega^{n+1}\mathsf{P}(\underline{A}), \mathsf{P}(\underline{C})] \xrightarrow{\psi^{n+1}_{A,C}}$$

$$\mathcal{T}_l(\mathcal{C}/\omega)[\Omega^{n+1}\mathsf{P}(\underline{X}_A), \mathsf{P}(\underline{C})] = \widehat{\mathsf{Ext}}^{n+1}_{\mathcal{X}}(A,C)$$

where, for any $A \in \widehat{\mathcal{X}}$ and any $C \in \mathcal{C}$:

- $\varepsilon_{A,C} : \mathrm{Ext}^1_{\omega}(A,C) \to \mathcal{C}/\omega[\Omega(\underline{A}), \underline{C}]$ is the canonical (epi)morphism induced by applying the functor $\mathcal{C}(-,C)$ to a the right ω-approximation sequence $0 \to \Omega(A) \to \omega_A \to A \to 0$ of A.
- $\psi^{n+1}_{A,C} = \mathcal{T}_l(\mathcal{C}/\omega)[\Omega^{n+1}\mathsf{P}(\underline{f}_A), \mathsf{P}(\underline{C})]$, where $f_A : X_A \to A$ is a special right $\mathcal{X}$-approximation of A.

Finally $\phi^0_{A,-} : \mathrm{Ext}^0_{\omega}(A,-) \to \widehat{\mathsf{Ext}}^{0}_{\mathcal{X}}(A,-)$ is defined as the composition

$$\mathrm{Ext}^0_{\omega}(A,C) = \mathcal{C}(A,C) \to \mathcal{C}/\omega(\underline{A}, \underline{C}) \to \mathcal{T}_l(\mathcal{C}/\omega)[\mathsf{P}(\underline{A}), \mathsf{P}(\underline{C})] \to$$

$$\mathcal{T}_l(\mathcal{C}/\omega)[\mathsf{P}(\underline{X}_A), \mathsf{P}(\underline{C})] = \widehat{\mathsf{Ext}}^{0}_{\mathcal{X}}(A,C).$$

REMARK 3.3. Consider the special right $\mathcal{X}$-approximation sequence $0 \to Y_A \xrightarrow{g_A} X_A \xrightarrow{f_A} A \to 0$ of $A \in \widehat{\mathcal{X}}$. Then the object Y_A lies in $\widehat{\omega}$. This implies that the loop functor Ω of the stable category $\mathcal{C}/\omega$ restricted to the left triangulated subcategory $\mathcal{Y}/\omega$ is locally nilpotent. Hence, by Section II.5, we have that $\mathsf{P}(\underline{Y}) = 0$, for any $Y \in \widehat{\omega}$. Applying the left exact stabilization functor $\mathsf{P} : \mathcal{C}/\omega \to \mathcal{T}_l(\mathcal{C}/\omega)$ to the left triangle $\Omega(\underline{A}) \to \underline{Y}_A \to \underline{X}_A \xrightarrow{\underline{f}_A} \underline{A}$ in $\mathcal{C}/\omega$, it follows that the morphism

$\mathsf{P}(\underline{f}_A) : \mathsf{P}(\underline{X}_A) \to \mathsf{P}(\underline{A})$ is invertible. We infer that, for any $A \in \widehat{\mathcal{X}}$, the map $\psi^n_{A,C}$ is invertible, $\forall n \geq 0$ and $\forall C \in \mathcal{C}$. It follows that we have natural isomorphisms

$$\widehat{\mathsf{Ext}}^n_{\mathcal{X}}(A, C) \xrightarrow{\cong} \mathcal{T}_l(\mathcal{C}/\omega)[\Omega^n \mathsf{P}(\underline{A}), \mathsf{P}(\underline{C})], \quad \forall A \in \widehat{\mathcal{X}}, \ \forall C \in \mathcal{C}.$$

In particular it follows that the ω-homological functor $\widehat{\mathsf{Ext}}^*_{\mathcal{X}}(A, -)$ is $\widehat{\omega}$-complete.

To proceed further we need the following basic result.

LEMMA 3.4. *Let $\mathcal{X}$ be a full subcategory of an abelian category $\mathcal{C}$ which admits an Ext-injective cogenerator ω. If ω is contravariantly finite in $\mathcal{C}$ and any ω-epic in $\mathcal{C}$ is an epimorphism, then the canonical map*

$$\mathsf{P}_{\underline{X},\underline{C}} \ : \ \mathcal{C}/\omega(\underline{X}, \underline{C}) \ \longrightarrow \ \mathcal{T}_l(\mathcal{C}/\omega)[\mathsf{P}(\underline{X}), \mathsf{P}(\underline{C})], \quad \underline{f} \ \longmapsto \ \mathsf{P}(\underline{f})$$

is invertible, for any $X \in \mathcal{X}$ and any $C \in \mathcal{C}$.

PROOF. Since ω is an Ext-injective cogenerator in $\mathcal{X}$ we have $\mathrm{Ext}^1_{\mathcal{C}}(\mathcal{X}, \omega) = 0$. Then the result follows from Lemma 5.12 in [**20**]. □

Consider now the morphism $\varphi^n_{-,-} : \mathrm{Ext}^n_{\omega}(-,-) \to \widehat{\mathsf{Ext}}^n_{\mathcal{X}}(-,-), \forall n \in \mathbb{Z}$ constructed above. The following result collects some basic properties of the bifunctor $\widehat{\mathsf{Ext}}^*_{(\mathcal{X},\mathcal{Y})}(-,-)$ and the connecting morphism $\varphi^*_{-,-}$, and in addition gives a first connection with the relative homological algebra in $\mathcal{C}$ induced by the subcategory $\mathcal{X}$. We recall that $\mathcal{X}$ induces a subfunctor $\mathcal{F}_{\mathcal{X}}$ of $\mathrm{Ext}^1_{\mathcal{C}}(-,-)$ as follows:

$$\mathcal{F}_{\mathcal{X}}(C, A) := \{(E) : 0 \to A \to B \to C \to 0 \ \in \mathrm{Ext}^1_{\mathcal{C}}(C, A) \text{ such that}$$
$$0 \to \mathcal{C}(\mathcal{X}, A) \to \mathcal{C}(\mathcal{X}, B) \to \mathcal{C}(\mathcal{X}, C) \to 0 \text{ is exact}\}.$$

The subfunctor $\mathcal{F}^{\mathcal{X}}$ is defined dually. Observe that if $\omega \subseteq \mathcal{X}$, then $\mathcal{F}_{\mathcal{X}} \subseteq \mathcal{F}_{\omega}$. Part (iv) below was observed by A. Martsinkovsky in a different setting.

PROPOSITION 3.5. *If ω is an Ext-projective generator of $\mathcal{X}$, then:*

(i) *The stable category $\mathcal{X}/\omega$ is a full triangulated subcategory of the stable left triangulated category $\mathcal{C}/\omega$.*
(ii) *The morphism $\varphi^n_{X,-}$ is invertible $\forall n \geq 1$ if and only if X lies in $\mathcal{X}$.*
(iii) *The kernel of $\varphi^0_{-,-}$ is the ideal of $\mathcal{C}$ consisting of all morphisms factorizing through an object of $\widehat{\omega}$.*
(iv) *The kernel of $\phi^1_{-,-}$ is the subfunctor $\mathcal{F}_{\mathcal{X}} \subseteq \mathrm{Ext}^1_{\omega}(-,-)$.*

PROOF. (i) Follows by Proposition VI.1.2.

(ii) Since $\mathcal{X}/\omega$ is triangulated, it is easy to see that the canonical morphism $\mathrm{Ext}^n_{\omega}(X, B) \to \mathcal{C}/\omega[\Omega^n(\underline{X}), \underline{B}]$ is invertible, for any $X \in \mathcal{X}$, for any $B \in \mathcal{C}$ and for any $n \geq 1$. Then the assertion follows from Remark 3.3 and Lemma 3.4.

(iii) Let $\alpha : A \to C$ be a morphism in the kernel of $\varphi^0_{A,C}$, where A is in $\widehat{\mathcal{X}}$ and C is in $\mathcal{C}$. Using Remark 3.3 and Lemma 3.4, it is easy to see that this implies that $\underline{f}_A \circ \underline{\alpha} = 0$. Hence $\underline{\alpha}$ factors through the reflection $\underline{g}^A : \underline{A} \to \underline{Y}^A$ of $\underline{A}$ in $\widehat{\omega}/\omega$. Therefore there exists a morphism $\rho : Y^A \to C$ such that the morphism $\alpha - g^A \circ \rho$ factors through an object T in ω. Hence there exists a factorization $\alpha - g^A \circ \rho = \kappa \circ \lambda : A \xrightarrow{\kappa} T \xrightarrow{\lambda} C$. Since $T \in \widehat{\omega}$, there exists a morphism $\sigma : Y^A \to T$ such that $g^A \circ \sigma = \kappa$. Then $\alpha - g^A \circ \rho = g^A \circ \sigma \circ \lambda$, and this implies that $\alpha = g^A \circ (\rho + \sigma \circ \lambda)$. Hence α factors through an object in $\widehat{\omega}$. The converse is clear since by Remark 3.3, $\mathsf{P}(\underline{Y}) = 0$, for any $Y \in \widehat{\omega}$.

(iv) Let A be in $\widehat{\mathcal{X}}$ and let $(E) : 0 \to C \to B \xrightarrow{\epsilon} A \to 0$ be a short exact sequence in $\mathcal{C}$ which lies $\mathrm{Ext}^1_\omega(A, C)$. We assume that (E), as an element of $\mathrm{Ext}^1_\omega(C, A)$, is represented by a morphism $\rho : \Omega_\omega(A) \to C$, that is there exists a push-out diagram:

$$\begin{array}{ccccccccc}
0 & \longrightarrow & \Omega_\omega(A) & \longrightarrow & \omega_A & \longrightarrow & A & \longrightarrow & 0 \\
 & & {\scriptstyle\rho}\downarrow & & \downarrow & & \| & & \\
0 & \longrightarrow & C & \longrightarrow & B & \xrightarrow{\epsilon} & A & \longrightarrow & 0
\end{array}$$

By Remark 3.3 and Lemma 3.4, it follows that $(E) \in \mathrm{Ker}(\varphi^1_{A,C})$ iff the composition $\underline{f}_{\Omega_\omega(A)} \circ \underline{\rho} = 0$, where $\underline{f}_{\Omega_\omega(A)} : \underline{X}_{\Omega_\omega(A)} \to \Omega_\omega(\underline{A})$ is the coreflection of $\Omega_\omega(\underline{A})$ in $\mathcal{X}/\omega$. Now for any X in $\mathcal{X}$ we have an exact commutative diagram

$$\begin{array}{ccc}
\mathcal{C}(X, A) & \xrightarrow{\vartheta} & \mathrm{Ext}^1_\omega(X, \Omega_\omega(A)) \\
\| & & {\scriptstyle\rho^*_X}\downarrow \\
\mathcal{C}(X, A) & \xrightarrow{\widetilde{\vartheta}} & \mathrm{Ext}^1_\omega(X, C)
\end{array}$$

Since ω is an Ext-injective cogenerator of $\mathcal{X}$, the inclusion $\mathrm{Ext}^1_\omega(X, \Omega_\omega(A)) \hookrightarrow \mathrm{Ext}^1_{\mathcal{C}}(X, \omega_A) = 0$ shows that ϑ is epic. Hence (E) lies in $\mathcal{F}_{\mathcal{X}}$ iff $\widetilde{\vartheta} = 0$ iff $\rho^*_X = 0$. From the proof of part (ii) it follows that we have isomorphisms $\mathrm{Ext}^1_\omega(X, \Omega_\omega(C)) = \mathcal{C}/\omega[\Omega_\omega(\underline{X}), \Omega_\omega(\underline{A})]$ and $\mathrm{Ext}^1_\omega(X, C) = \mathcal{C}/\omega[\Omega_\omega(\underline{X}), \underline{C}]$, and then ρ^*_X is the morphism which sends $\underline{\alpha} : \Omega_\omega(\underline{X}) \to \Omega_\omega(\underline{A})$ to the composition $\underline{\alpha} \circ \underline{\rho} : \Omega_\omega(\underline{X}) \to \underline{C}$. Since $\underline{X}_{\Omega_\omega(A)} \xrightarrow{\cong} \Omega_\omega(\underline{X}_A)$, and since any morphism $\Omega_\omega(\underline{X}) \to \Omega_\omega(\underline{A})$ factors through $\underline{f}_{\Omega_\omega(A)}$, we have: $(E) \in \mathrm{Ker}(\varphi^1_{A,C})$ iff $\underline{f}_{\Omega_\omega(A)} \circ \underline{\rho} = 0$ iff $\underline{\alpha} \circ \underline{\rho} = 0$, for any morphism $\underline{\alpha} : \Omega_\omega(\underline{X}) \to \Omega_\omega(\underline{A})$ with $X \in \mathcal{X}$, iff $\rho^*_X = 0$, $\forall X \in \mathcal{X}$, iff $(E) \in \mathcal{F}_{\mathcal{X}}$. □

Now we can prove the main result of this section, which gives a nice connection between the three involved relative homological theories represented by the bifunctors $\mathrm{Ext}^*_{\mathcal{X}}(-,-), \mathrm{Ext}^*_\omega(-,-)$ and $\widehat{\mathsf{Ext}}^*_{\mathcal{X}}(-,-)$, by providing comparison morphisms which fit in a long exact sequence.

THEOREM 3.6. *Let $\mathcal{X}$ be a full additive subcategory of an abelian category $\mathcal{C}$ which is closed under extensions, direct summands and kernels of epimorphisms. We assume that $\mathcal{X}$ admits an Ext-injective cogenerator ω which is an Ext-projective generator in $\mathcal{X}$. If ω is contravariantly finite in $\mathcal{C}$ and any ω-epic is an epimorphism, then for any object A in $\widehat{\mathcal{X}}$ there exists a long exact sequence of functors $\mathcal{C} \to \mathcal{A}b$:*

$$\begin{array}{ccccccccc}
0 & \longrightarrow & \mathrm{Ext}^1_{\mathcal{X}}(A,-) & \longrightarrow & \mathrm{Ext}^1_\omega(A,-) & \longrightarrow & \widehat{\mathsf{Ext}}^1_{\mathcal{X}}(A,-) & \longrightarrow & \cdots \\
\cdots & \longrightarrow & \mathrm{Ext}^n_{\mathcal{X}}(A,-) & \longrightarrow & \mathrm{Ext}^n_\omega(A,-) & \longrightarrow & \widehat{\mathsf{Ext}}^n_{\mathcal{X}}(A,-) & \longrightarrow & \cdots \\
\cdots & \longrightarrow & \mathrm{Ext}^d_{\mathcal{X}}(A,-) & \longrightarrow & \mathrm{Ext}^d_\omega(A,-) & \longrightarrow & \widehat{\mathsf{Ext}}^d_{\mathcal{X}}(A,-) & \longrightarrow & 0
\end{array}$$

where $d := \mathcal{X}$-res. dim $A = \mathcal{X}$-pd A.

PROOF. Let A be an object in $\widehat{\mathcal{X}}$, and let

$$0 \to Y_A \xrightarrow{g_A} X_A \xrightarrow{f_A} A \to 0 \qquad (\dagger)$$

be a special right $\mathcal{X}$-approximation of A. Then we know that Y_A lies in $\widehat{\omega}$. If A lies in $\widehat{\omega}$, then $X_A \in \mathcal{X} \cap \widehat{\omega} = \omega$. Hence the right $\mathcal{X}$-approximation sequence (†) of A is a right ω-approximation sequence of A. This implies that for any $Y \in \widehat{\omega}$ we have $\Omega_{\mathcal{X}}(Y) \cong \Omega_{\omega}(Y)$, where $\Omega_{\mathcal{X}}(Y)$, resp. $\Omega_{\omega}(Y)$ is the first relative $\mathcal{X}$-syzygy, resp. ω-syzygy, of Y, and an isomorphism $\mathrm{Ext}^1_{\mathcal{X}}(Y,-) \xrightarrow{\cong} \mathrm{Ext}^1_{\omega}(Y,-)$. Since $\widehat{\omega}$ is closed under relative ω-syzygies, this implies that the canonical map $\mathrm{Ext}^n_{\mathcal{X}}(Y,-) \to \mathrm{Ext}^n_{\omega}(Y,-)$ is invertible, for any $Y \in \widehat{\omega}$ and $n \geq 0$. Since $\omega \subseteq \mathcal{X}$, the special right $\mathcal{X}$-approximation sequence (†) induces a long exact sequence of relative ω-extension functors:

$$\cdots \longrightarrow \mathrm{Ext}^n_{\omega}(Y_A,-) \longrightarrow \mathrm{Ext}^{n+1}_{\omega}(A,-) \longrightarrow \mathrm{Ext}^{n+1}_{\omega}(X_A,-) \longrightarrow \cdots \quad (\dagger\dagger)$$

By the above observation we have an isomorphism $\mathrm{Ext}^n_{\omega}(Y_A,-) \xrightarrow{\cong} \mathrm{Ext}^n_{\mathcal{X}}(Y_A,-)$. Since Y_A is the first relative $\mathcal{X}$-syzygy of A, the last extension space is isomorphic to $\mathrm{Ext}^{n+1}_{\mathcal{X}}(A,-)$. Since, by Proposition 3.5, $\mathcal{X}/\omega$ is a triangulated subcategory of $\mathcal{C}/\omega$, we infer that the canonical map $\mathrm{Ext}^{n+1}_{\omega}(X_A,C) \to \mathcal{C}/\omega[\Omega^{n+1}_{\omega}(\underline{X}),\underline{C}]$ is invertible, for any $n \geq 0$ and any $C \in \mathcal{C}$. By Lemma 3.4, the last space is isomorphic to $\mathcal{T}_l(\mathcal{C}/\omega)[\mathsf{P}(\Omega^{n+1}_{\omega}(\underline{X}_A)),\mathsf{P}(\underline{C})] \xrightarrow{\cong} \mathcal{T}_l(\mathcal{C}/\omega)[\Omega^{n+1}\mathsf{P}(\underline{X}_A),\mathsf{P}(\underline{C})]$. Hence

$$\mathrm{Ext}^{n+1}_{\omega}(X_A,C) \xrightarrow{\cong} \mathcal{T}_l(\mathcal{C}/\omega)[\Omega^{n+1}\mathsf{P}(\underline{X}_A),\mathsf{P}(\underline{C})] = \widehat{\mathsf{Ext}}^{n+1}_{\mathcal{X}}(A,C).$$

It is easy to see that the above isomorphisms make the next diagram commutative:

$$\begin{array}{ccccccccc}
\cdots & \longrightarrow & \mathrm{Ext}^n_{\omega}(Y_A,-) & \longrightarrow & \mathrm{Ext}^{n+1}_{\omega}(A,-) & \longrightarrow & \mathrm{Ext}^{n+1}_{\omega}(X_A,-) & \longrightarrow & \cdots \\
 & & \cong\downarrow & & \| & & \cong\downarrow & & \\
\cdots & \longrightarrow & \mathrm{Ext}^{n+1}_{\mathcal{X}}(A,-) & \xrightarrow{\theta^{n+1}_{A,-}} & \mathrm{Ext}^{n+1}_{\omega}(A,-) & \xrightarrow{\phi^{n+1}_{A,-}} & \widehat{\mathsf{Ext}}^{n+1}_{\mathcal{X}}(A,-) & \longrightarrow & \cdots
\end{array}$$

where $\theta^{n+1}_{A,-}$ is the canonical morphism induced by the inclusion $\mathrm{Ext}^1_{\mathcal{X}}(A,-) \hookrightarrow \mathrm{Ext}^1_{\omega}(A,-)$, which in turn is induced by the inclusion $\omega \subseteq \mathcal{X}$, and $\phi^{n+1}_{A,-}$ is the natural morphism defined before. Since we have an identification $\mathrm{Ext}^1_{\mathcal{X}}(A,-) = \mathrm{Coker}\Big(\mathcal{C}(X_A,-) \to \mathcal{C}(Y_A,-)\Big)$, and by Proposition 3.5 we have an exact sequence $0 \to \mathrm{Ext}^1_{\mathcal{X}}(-,-) \to \mathrm{Ext}^1_{\omega}(-,-) \to \widehat{\mathsf{Ext}}^1_{\mathcal{X}}(-,-)$, we have proved the exactness of the desired sequence. Finally we show that the sequence stops at $\widehat{\mathsf{Ext}}^d_{\mathcal{X}}(A,C)$. To prove this, it suffices to show that $\mathrm{Ext}^{d+1}_{\mathcal{X}}(A,C) = 0$. Let $d = \mathcal{X}\text{-res.}\dim A$. Then, by Lemma 3.1, we have $d = \mathcal{X}\text{-pd}\,A$. Hence $\mathrm{Ext}^{d+1}_{\mathcal{X}}(A,C) = 0$. □

We have the following direct consequence.

COROLLARY 3.7. *Let $\mathcal{C}$ be an abelian category with enough projectives and let $\mathcal{X}$ be a full additive subcategory of $\mathcal{C}$ which is closed under extensions, direct summands and kernels of epimorphisms. If $\mathcal{P}$ is an Ext-injective cogenerator of $\mathcal{X}$, then for any object A in $\widehat{\mathcal{X}}$ there exists a long exact sequence of functors $\mathcal{C} \to \mathcal{A}b$:*

$$\begin{array}{ccccccccc}
0 & \longrightarrow & \mathrm{Ext}^1_{\mathcal{X}}(A,-) & \longrightarrow & \mathrm{Ext}^1_{\mathcal{C}}(A,-) & \longrightarrow & \widehat{\mathsf{Ext}}^1_{TV}(A,-) & \longrightarrow & \cdots \\
\cdots & \longrightarrow & \mathrm{Ext}^n_{\mathcal{X}}(A,-) & \longrightarrow & \mathrm{Ext}^n_{\mathcal{C}}(A,-) & \longrightarrow & \widehat{\mathsf{Ext}}^n_{TV}(A,-) & \longrightarrow & \cdots \\
\cdots & \longrightarrow & \mathrm{Ext}^d_{\mathcal{X}}(A,-) & \longrightarrow & \mathrm{Ext}^d_{\mathcal{C}}(A,-) & \longrightarrow & \widehat{\mathsf{Ext}}^d_{TV}(A,-) & \longrightarrow & 0
\end{array}$$

where $d := \mathcal{X}\text{-res.}\dim A = \mathcal{X}\text{-pd}\, A$.

PROOF. The assertion follows from Theorem 3.6, except of the identification $\widehat{\mathsf{Ext}}^*_{TV}(A,-) \xrightarrow{\cong} \widehat{\mathsf{Ext}}^*_{\mathcal{X}}(A,-)$. However this follows from Remark 3.3 and the definition of projective Tate-Vogel Cohomology. □

In a recent paper Avramov-Martsinkovsky [**16**], using complete resolutions, proved that there is a long exact sequence involving Tate-Vogel cohomology, absolute cohomology, and the relative cohomology induced by finitely generated modules of finite Gorenstein dimension over a Noetherian ring. The following result generalizes and gives a simple proof of the result of Avramov-Martsinkovsky, see Theorem 7.1 of [**16**], avoiding complete resolutions.

COROLLARY 3.8. *Let $\mathcal{C}$ be an abelian category with enough projectives. Then for any object A in $\mathcal{C}$ with finite $\mathcal{P}$-Gorenstein dimension, that is A lies in $\widehat{\mathrm{CM}(\mathcal{P})}$, there exists a long exact sequence of functors $\mathcal{C} \to \mathcal{A}b$:*

$$\begin{array}{ccccccccc} 0 & \longrightarrow & \mathrm{Ext}^1_{\mathrm{CM}(\mathcal{P})}(A,-) & \longrightarrow & \mathrm{Ext}^1_{\mathcal{C}}(A,-) & \longrightarrow & \widehat{\mathsf{Ext}}^1_{TV}(A,-) & \longrightarrow & \cdots \\ \cdots & \longrightarrow & \mathrm{Ext}^n_{\mathrm{CM}(\mathcal{P})}(A,-) & \longrightarrow & \mathrm{Ext}^n_{\mathcal{C}}(A,-) & \longrightarrow & \widehat{\mathsf{Ext}}^n_{TV}(A,-) & \longrightarrow & \cdots \\ \cdots & \longrightarrow & \mathrm{Ext}^d_{\mathrm{CM}(\mathcal{P})}(A,-) & \longrightarrow & \mathrm{Ext}^d_{\mathcal{C}}(A,-) & \longrightarrow & \widehat{\mathsf{Ext}}^d_{TV}(A,-) & \longrightarrow & 0 \end{array}$$

where d is the $\mathcal{P}$-Gorenstein (resolution) dimension of A.

Now we apply our results to cotorsion pairs. Let $\mathcal{C}$ be an abelian category with enough projectives, and let $(\mathcal{X},\mathcal{Y})$ be a resolving cotorsion pair in $\mathcal{C}$. Although Theorem 3.9 below holds for cortorsion pairs with $\mathcal{X} \cap \mathcal{Y} = \omega$, for convenience we assume that $\mathcal{X} \cap \mathcal{Y} = \mathcal{P}$. Then the following result shows that there exists an exact sequence of functors as in Theorem 3.6. However the sequence in general does not stop, since in general $\mathcal{Y}$ contains objects of infinite projective dimension.

THEOREM 3.9. *Let $\mathcal{C}$ be an abelian category with enough projective objects. If $(\mathcal{X},\mathcal{Y})$ is a resolving cotorsion pair in $\mathcal{C}$ such that $\mathcal{X} \cap \mathcal{Y} = \mathcal{P}$, then there exists a long exact sequence of bifunctors: $\mathcal{C}^{\mathrm{op}} \times \mathcal{C} \to \mathcal{A}b$:*

$$\begin{array}{ccccccccc} 0 & \longrightarrow & \mathrm{Ext}^1_{\mathcal{X}}(-,-) & \longrightarrow & \mathrm{Ext}^1_{\mathcal{C}}(-,-) & \longrightarrow & \widehat{\mathsf{Ext}}^1_{(\mathcal{X},\mathcal{Y})}(-,-) & \longrightarrow & \cdots \\ \cdots & \longrightarrow & \mathrm{Ext}^n_{\mathcal{X}}(-,-) & \longrightarrow & \mathrm{Ext}^n_{\mathcal{C}}(-,-) & \longrightarrow & \widehat{\mathsf{Ext}}^n_{(\mathcal{X},\mathcal{Y})}(-,-) & \longrightarrow & \cdots \end{array}$$

where $\widehat{\mathsf{Ext}}^n_{(\mathcal{X},\mathcal{Y})}(-,-)$ is the complete projective extension bifunctor of $\mathcal{C}$ with respect to the cotorsion pair $(\mathcal{X},\mathcal{Y})$. Moreover the sequence stops at some stage if and only if the category $\mathcal{C}$ is Gorenstein.

PROOF. By Remark 3.2, the bifunctor $\widehat{\mathsf{Ext}}^*_{(\mathcal{X},\mathcal{Y})}(-,-)$ coincides with the bifunctor $\widehat{\mathsf{Ext}}^*_{\mathcal{X}}(-,-)$. Now the existence of the long exact sequence follows exactly as in Theorem 3.6, using that $\mathcal{X}$ is contravariantly finite in $\mathcal{C}$. Finally if $\mathrm{Ext}^{d+1}_{\mathcal{X}}(-,-) = 0$ for some $d \geq 0$, then $\mathcal{X}\text{-res.}\dim \mathcal{C} < \infty$, hence $\widehat{\mathcal{X}} = \mathcal{C}$. This implies, by Theorem VII.1.4, that $\mathcal{C}$ is Gorenstein. If $\mathrm{Ext}^{d+1}_{\mathcal{C}}(-,-) = 0$ or $\widehat{\mathsf{Ext}}^{d+1}_{(\mathcal{X},\mathcal{Y})}(-,-) = 0$ for some $d \geq 0$, then it is easy to see that $\mathrm{gl.}\dim \mathcal{C} < \infty$, hence $\mathcal{C}$ is Gorenstein. Conversely if $\mathcal{C}$ is Gorenstein with Gorenstein projective dimension $\text{G-dim}_{\mathcal{P}}\, \mathcal{C} = d < \infty$, then, by Theorem VII.1.4 we have $\mathcal{X} = \mathrm{CM}(\mathcal{P})$ and $\mathrm{Ext}^{d+1}_{\mathcal{X}}(-,-) = 0$. □

Let $(\mathcal{X}, \mathcal{Y})$ be a resolving cotorsion pair in $\mathcal{C}$ with $\mathcal{X} \cap \mathcal{Y} = \mathcal{P}$. In addition to the three homological theories represented by the homological bifunctors $\mathrm{Ext}^*_{\mathcal{X}}(-,-)$, $\mathrm{Ext}^*_{\mathcal{C}}(-,-)$ and $\widehat{\mathsf{Ext}}^*_{(\mathcal{X},\mathcal{Y})}(-,-)$, there is a fourth one, namely the Tate-Vogel projective extension bifunctor $\widehat{\mathsf{Ext}}^*_{TV}(-,-)$. We close this section by comparing the bifunctors $\widehat{\mathsf{Ext}}^*_{TV}(-,-)$ and $\widehat{\mathsf{Ext}}^*_{(\mathcal{X},\mathcal{Y})}(-,-)$.

Let C be an object of $\mathcal{C}$ and consider the completion morphism $\zeta^*_{C,-} : \mathrm{Ext}^*_{\mathcal{C}}(C,-) \to \widehat{\mathsf{Ext}}^*_{TV}(C,-)$ of the extension functor $\mathrm{Ext}^*_{\mathcal{C}}(C,-)$ with respect to the subcategory $\mathcal{P}$, and the completion morphism $\phi^*_{C,-} : \mathrm{Ext}^*_{\mathcal{C}}(C,-) \to \widehat{\mathsf{Ext}}^*_{(\mathcal{X},\mathcal{Y})}(C,-)$ of the extension functor $\mathrm{Ext}^*_{\mathcal{C}}(C,-)$ with respect to the subcategory $\mathcal{Y}$. Since $\mathcal{P} \subseteq \mathcal{Y}$, by the universal property of $\widehat{\mathsf{Ext}}^*_{TV}(C,-)$, there exists a unique morphism $\rho^*_{C,-}$ of $\mathcal{P}$-homological functors making the following diagram commutative:

$$\begin{array}{ccc}
\mathrm{Ext}^*_{\mathcal{C}}(C,-) & \xrightarrow{\zeta^*_{C,-}} & \widehat{\mathsf{Ext}}^*_{TV}(C,-) \\
\Big\| & & \Big\downarrow{\scriptstyle \exists!\,\rho^*_{C,-}} \\
\mathrm{Ext}^*_{\mathcal{C}}(C,-) & \xrightarrow{\phi^*_{C,-}} & \widehat{\mathsf{Ext}}^*_{(\mathcal{X},\mathcal{Y})}(C,-)
\end{array}$$

It is not difficult to see that $\rho^*_{C,-}$ coincides with the natural morphism constructed in Corollary 1.4, in particular $\rho^*_{C,-}$ extends to a morphism of bifunctors

$$\rho^*_{-,-} \ : \ \widehat{\mathsf{Ext}}^*_{TV}(-,-) \longrightarrow \widehat{\mathsf{Ext}}^*_{(\mathcal{X},\mathcal{Y})}(-,-)$$

which we call the **comparison morphism**. The following result shows that the comparison morphism measures how far $\mathcal{C}$ is from being Gorenstein.

PROPOSITION 3.10. *The natural morphism $\rho^*_{C,-}$ is invertible iff $C \in \widehat{\mathcal{X}}$. In particular the comparison morphism $\rho^*_{-,-}$ is invertible iff $\mathcal{C}$ is Gorenstein.*

PROOF. Let $\rho^*_{C,-}$ be invertible. Then by Corollary 1.4, Y_C lies in $\mathcal{P}^{<\infty}$, and this implies that C lies in $\widehat{\mathcal{X}}$. Conversely if $C \in \widehat{\mathcal{X}}$, then by Proposition VI.2.3, Y_C has finite projective dimension, and then $\rho^*_{C,-}$ is invertible by Corollary 1.4. □

REMARK 3.11. The results of this section have dual versions concerning categories which admit an Ext-projective generator, for instance CoCohen-Macaulay objects. In particular the dual results can be applied to categories admitting coresolving cotorsion pairs. We leave the formulation of the dual results to the reader.

4. Cotorsion Triples and Complete Cohomology Theories

Throughout this section we fix an abelian category $\mathcal{C}$ and we assume that $\mathcal{C}$ has enough projective and enough injective objects. As we have seen in Chapter VI, torsion pairs $(\mathcal{X}/\mathcal{P}, \mathcal{Y}/\mathcal{P})$ in $\mathcal{C}/\mathcal{P}$ with $\mathcal{X} \cap \mathcal{Y} = \mathcal{P}$, arise usually in practice from cotorsion triples. In this section we study the complete extension bifunctors induced by the cotorsion pairs $(\mathcal{X}, \mathcal{Y})$ and $(\mathcal{Y}, \mathcal{Z})$ involved in a cotorsion triple $(\mathcal{X}, \mathcal{Y}, \mathcal{Z})$ in $\mathcal{C}$. In particular we show that the two (co)homology theories arising from the two cotorsion pairs are isomorphic, and we show that the same is true for the relative homological theories defined by the cotorsion subcategory $\mathcal{X}$ and the cotorsion-free subcategory $\mathcal{Z}$. This generalizes the well-known isomorphism between the projective and injective Tate-Vogel cohomology bifunctors, see [**87**], [**20**].

Throughout we fix a cotorsion triple $(\mathcal{X}, \mathcal{Y}, \mathcal{Z})$ in $\mathcal{C}$ with $\mathcal{Y}$ (co)resolving. Then we know that $\mathcal{X} \cap \mathcal{Y} = \mathcal{P}$ and $\mathcal{Y} \cap \mathcal{Z} = \mathcal{I}$, the torsion class $\mathcal{X}/\mathcal{P}$ in $\mathcal{C}/\mathcal{P}$ is triangulated, and the torsion-free class $\mathcal{Z}/\mathcal{I}$ in $\mathcal{C}/\mathcal{I}$ is triangulated. Hence the results of the previous section can be applied to both cotorsion pairs $(\mathcal{X}, \mathcal{Y})$ and $(\mathcal{Y}, \mathcal{Z})$. We shall show that the complete extension bifunctors $\widehat{\mathsf{Ext}}^n_{(\mathcal{X},\mathcal{Y})}(-,-)$, $\widetilde{\mathsf{Ext}}^n_{(\mathcal{Y},\mathcal{Z})}(-,-) : \mathcal{C}^{\mathrm{op}} \times \mathcal{C} \longrightarrow \mathcal{A}b$ are isomorphic, $\forall n \in \mathbb{Z}$. We recall that by Theorem VI.3.2 there exists a triangle equivalence $\Phi : \mathcal{X}/\mathcal{P} \xrightarrow{\approx} \mathcal{Z}/\mathcal{I}$ with quasi-inverse Ψ.

First we need the following result which shows that the two reasonable possible ways to go from $\mathcal{C}$ to $\mathcal{X}/\mathcal{P}$ and to $\mathcal{Z}/\mathcal{I}$ are isomorphic. As usual we denote by $\mathbf{R} : \mathcal{C}/\mathcal{P} \longrightarrow \mathcal{X}/\mathcal{P}$ the right adjoint of the inclusion $\mathbf{i} : \mathcal{X}/\mathcal{P} \hookrightarrow \mathcal{C}/\mathcal{P}$, and by $\mathbf{T} : \mathcal{C}/\mathcal{I} \longrightarrow \mathcal{Z}/\mathcal{I}$ the left adjoint of the inclusion $\mathbf{k} : \mathcal{Z}/\mathcal{I} \hookrightarrow \mathcal{C}/\mathcal{I}$.

LEMMA 4.1. *There exist commutative diagrams of functors*

$$\begin{array}{ccc} \mathcal{C} & \xrightarrow{\pi} & \mathcal{C}/\mathcal{I} \\ {\scriptstyle\pi}\downarrow & & \downarrow{\scriptstyle\Sigma^{-1}\mathbf{T}} \\ \mathcal{C}/\mathcal{P} & \xrightarrow{\Phi\mathbf{R}} & \mathcal{Z}/\mathcal{I} \end{array} \qquad \begin{array}{ccc} \mathcal{C} & \xrightarrow{\pi} & \mathcal{C}/\mathcal{P} \\ {\scriptstyle\pi}\downarrow & & \downarrow{\scriptstyle\Omega^{-1}\mathbf{R}} \\ \mathcal{C}/\mathcal{I} & \xrightarrow{\Psi\mathbf{T}} & \mathcal{X}/\mathcal{P} \end{array}$$

that is, there exist natural isomorphisms of functors

$$\varphi : \Sigma^{-1}\mathbf{T}\pi \xrightarrow{\cong} \Phi\mathbf{R}\pi \quad \textit{and} \quad \psi : \Psi\mathbf{T}\pi \xrightarrow{\cong} \Omega^{-1}\mathbf{R}\pi.$$

PROOF. Let A be a object in $\mathcal{C}$ and consider the special right $\mathcal{X}$-approximation sequence $0 \longrightarrow Y_A \longrightarrow X_A \xrightarrow{f_A} A \longrightarrow 0$ of A and the special right $\mathcal{Y}$-approximation sequence $0 \longrightarrow Z_{X_A} \longrightarrow Y_{X_A} \xrightarrow{g_{X_A}} X_A \longrightarrow 0$ of X_A. Then by definition the object $\underline{Z}_{X_A}$ in $\mathcal{C}/\mathcal{I}$ is the object $\Phi\mathbf{R}\pi(A)$. The above exact sequences are included in the following exact commutative diagram

$$\begin{array}{ccccccccc} 0 & \longrightarrow & U & \longrightarrow & Y_{X_A} & \longrightarrow & A & \longrightarrow & 0 \\ & & \downarrow & & \downarrow{\scriptstyle g_{X_A}} & & \| & & \\ 0 & \longrightarrow & Y_A & \longrightarrow & X_A & \xrightarrow{f_A} & A & \longrightarrow & 0 \end{array}$$

which induces an exact sequence $0 \longrightarrow Z_{X_A} \longrightarrow U \longrightarrow Y_A \longrightarrow 0$. Since Z_{X_A} is in $\mathcal{Z}$ and Y_A is in $\mathcal{Y}$, the sequence splits, hence $U \cong Z_{X_A} \oplus Y_A$. Since the left square in the above diagram is bicartesian, it induces an exact sequence $0 \longrightarrow Z_{X_A} \oplus Y_A \longrightarrow Y_A \oplus Y_{X_A} \longrightarrow X_A \longrightarrow 0$. Then we have a right triangle $\overline{Z}_{X_A} \oplus \overline{Y}_A \longrightarrow \overline{Y}_A \oplus \overline{Y}_{X_A} \longrightarrow \overline{X}_A \longrightarrow \Sigma(\overline{Z}_{X_A} \oplus \overline{Y}_A)$ in $\mathcal{C}/\mathcal{I}$. Applying to this triangle the right exact reflection functor $\mathbf{T} : \mathcal{C}/\mathcal{I} \longrightarrow \mathcal{Z}/\mathcal{I}$ and using that $\mathbf{T}(\mathcal{Y}/\mathcal{I}) = 0$, we have an isomorphism $\mathbf{T}(\overline{X}_A) \xrightarrow{\cong} \Sigma(\overline{Z}_{X_A}) = \Sigma\Phi\mathbf{R}\pi(A)$. Applying $\mathbf{T}$ to the right triangle $\overline{Y}_A \longrightarrow \overline{X}_A \longrightarrow \overline{A} \longrightarrow \Sigma(\overline{Y}_A)$ in $\mathcal{C}/\mathcal{I}$, we have an isomorphism $\mathbf{T}(\overline{X}_A) \xrightarrow{\cong} \mathbf{T}(\overline{A})$. Combining the above isomorphisms we infer that we have an isomorphism $\mathbf{T}\pi(A) = \mathbf{T}(\overline{A}) \xrightarrow{\cong} \Sigma\Phi\mathbf{R}\pi(A)$, or equivalently an isomorphism $\varphi : \Sigma^{-1}\mathbf{T}\pi(A) \xrightarrow{\cong} \Phi\mathbf{R}\pi(A)$. We leave to the reader the easy proof that φ is functorial as well as the proof of the existence of the isomorphism ψ. □

The following main result of this section shows that the complete projective extension bifunctor induced by the good pair $(\mathcal{X},\mathcal{Y})$ is isomorphic to the complete injective extension bifunctor induced by the good pair $(\mathcal{Y},\mathcal{Z})$.

THEOREM 4.2. *If $(\mathcal{X},\mathcal{Y},\mathcal{Z})$ is a cotorsion triple in $\mathcal{C}$ with $\mathcal{Y}$ (co)resolving, then the complete extension bifunctors*

$$\widehat{\mathsf{Ext}}^n_{(\mathcal{X},\mathcal{Y})}(-,-) \ \ \textit{and} \ \ \widetilde{\mathsf{Ext}}^n_{(\mathcal{Y},\mathcal{Z})}(-,-) \ : \ \mathcal{C}^{\mathrm{op}} \times \mathcal{C} \longrightarrow \mathcal{A}b$$

are isomorphic, for any $n \in \mathbb{Z}$.

PROOF. Using the isomorphisms of Lemma 4.1 and the exactness of Φ, we have for any $n \in \mathbb{Z}$:

$$\begin{aligned}
\widehat{\mathsf{Ext}}^n_{(\mathcal{X},\mathcal{Y})}(A,B) &= \mathcal{X}/\mathcal{P}[\Omega^n\mathbf{R}(\underline{A}),\mathbf{R}(\underline{B})] \\
&\cong \mathcal{Z}/\mathcal{I}[\Phi\Omega^n\mathbf{R}(\underline{A}),\Phi\mathbf{R}(\underline{B})] \\
&\cong \mathcal{Z}/\mathcal{I}[\Sigma^{-n}\Phi\mathbf{R}(\underline{A}),\Sigma^{-1}\mathbf{T}(\underline{B})] \\
&\cong \mathcal{Z}/\mathcal{I}[\Sigma^{-1}\mathbf{T}(\overline{A}),\Sigma^{-1}\Sigma^n\mathbf{T}(\overline{B})] \\
&\cong \mathcal{Z}/\mathcal{I}[\mathbf{T}(\overline{A}),\Sigma^n\mathbf{T}(\overline{B})] \\
&= \widetilde{\mathsf{Ext}}^n_{(\mathcal{Y},\mathcal{Z})}(A,B).
\end{aligned}$$

□

Combining Theorems 4.2 and 2.6, we obtain the following consequence.

COROLLARY 4.3. *Let $(\mathcal{X},\mathcal{Y},\mathcal{Z})$ be a cotorsion triple in $\mathcal{C}$ with $\mathcal{Y}$ (co)resolving. Then the $\mathcal{Y}$-completion of the homological functor $\mathrm{Ext}^*_{\mathcal{C}}(A,-)$ evaluated at B is isomorphic to the $\mathcal{Y}$-completion of the cohomological functor $\mathrm{Ext}^*_{\mathcal{C}}(-,B)$ evaluated at A.*

The following consequence generalizes and gives a simple proof of a result of [**20**] which was obtained independently by Nucinkis [**87**], using satellites.

THEOREM 4.4. *The following statements are equivalent:*

(i) *$\mathcal{C}$ is Gorenstein.*
(ii) *$\mathrm{FPD}(\mathcal{C}) < \infty$ or $\mathrm{FID}(\mathcal{C}) < \infty$ and the projective Tate-Vogel homology bifunctor is isomorphic to the injective Tate-Vogel homology bifunctor:*

$$\widehat{\mathsf{Ext}}^*_{TV}(-,-) \xrightarrow{\cong} \widetilde{\mathsf{Ext}}^*_{TV}(-,-).$$

If $\mathcal{C}$ admits exact products or exact coproducts or if $\mathcal{C}$ has finite number of non-isomorphic simple objects and any object of $\mathcal{C}$ is of finite length, then the assumption $\mathrm{FPD}(\mathcal{C}) < \infty$ or $\mathrm{FID}(\mathcal{C}) < \infty$ in (ii) can be removed.

PROOF. (i) ⇒ (ii) If $\mathcal{C}$ is Gorenstein, then by Theorem VII.2.2 we have a cotorsion triple $(\mathcal{X},\mathcal{Y},\mathcal{Z})$ in $\mathcal{C}$ with $\mathcal{Y} = \mathcal{P}^{<\infty} = \mathcal{I}^{<\infty}$ and $\mathrm{FPD}(\mathcal{C}) = \mathrm{FID}(\mathcal{C}) < \infty$. Hence $\mathcal{T}_l(\mathcal{Y}/\mathcal{P}) = 0 = \mathcal{T}_r(\mathcal{Y}/\mathcal{I})$ and then $\mathcal{T}_l(\mathcal{C}/\mathcal{P}) = \mathcal{X}/\mathcal{P}$ and $\mathcal{T}_r(\mathcal{C}/\mathcal{I}) = \mathcal{Z}/\mathcal{I}$. It follows that we have isomorphisms

$$\widehat{\mathsf{Ext}}^*_{TV}(-,-) \xrightarrow{\cong} \widehat{\mathsf{Ext}}^*_{(\mathcal{X},\mathcal{Y})}(-,-) \xrightarrow{\cong} \widetilde{\mathsf{Ext}}^*_{(\mathcal{Y},\mathcal{Z})}(-,-) \xleftarrow{\cong} \widetilde{\mathsf{Ext}^*}_{TV}(-,-).$$

(ii) ⇒ (i) Let I be an object of finite injective dimension. Then we have $0 = \widetilde{\mathsf{Ext}}^*_{TV}(I,I) \xrightarrow{\cong} \widehat{\mathsf{Ext}}^*_{TV}(I,I)$, hence $\widehat{\mathsf{Ext}}^0_{TV}(I,I) = \mathcal{T}_l(\mathcal{C}/\mathcal{P})[\mathsf{P}(I),\mathsf{P}(I)] = 0$.

However this holds if and only if I has finite projective dimension. Dually if P has finite projective dimension, then P has finite injective dimension. We conclude that $\mathcal{P}^{<\infty} = \mathcal{I}^{<\infty}$. Then by Proposition VII.1.3 and Theorem VII.2.2, we have that $\mathcal{C}$ is Gorenstein. Finally if $\mathcal{C}$ admits exact (co)products or if $\mathcal{C}$ has a finite number of non-isomorphic simple objects and any object of $\mathcal{C}$ has finite length, then the assertion follows from Proposition VII.2.4. □

It is now natural to ask if the relative homological theories induced by the subcategories $\mathcal{X}$ and $\mathcal{Z}$ involved in a cotorsion triple $(\mathcal{X}, \mathcal{Y}, \mathcal{Z})$ are isomorphic. The answer is provided by the following result which shows that the additive subfunctors $\mathcal{F}_{\mathcal{X}}$ and $\mathcal{F}^{\mathcal{Z}}$ of $\mathrm{Ext}^1_{\mathcal{C}}(-,-)$ coincide.

THEOREM 4.5. *Let $(\mathcal{X}, \mathcal{Y}, \mathcal{Z})$ be a cotorsion triple in $\mathcal{C}$ with $\mathcal{Y}$ (co)resolving. Then the subfunctors $\mathcal{F}_{\mathcal{X}}$ and $\mathcal{F}^{\mathcal{Z}}$ of $\mathrm{Ext}^1_{\mathcal{C}}(-,-)$ coincide. In particular there exists an isomorphism of bifunctors*

$$\mathrm{Ext}^*_{\mathcal{X}}(-,-) \xrightarrow{\cong} \mathrm{Ext}^*_{\mathcal{Z}}(-,-).$$

Hence if $0 \to A \to B \to C \to 0$ is a short exact sequence in $\mathcal{C}$, then any morphism $X \to C$ with $X \in \mathcal{X}$ factors through $B \to C$ if and only if any morphism $A \to Z$ with $Z \in \mathcal{Z}$ factors through $A \to B$.

PROOF. Let $\xi^*_{-,-} : \widehat{\mathsf{Ext}}^*_{(\mathcal{X},\mathcal{Y})}(-,-) \to \widetilde{\mathsf{Ext}}^*_{(\mathcal{Y},\mathcal{Z})}(-,-)$ be the isomorphism of complete extension bifunctors contructed in Theorem 4.2. Consider the $\mathcal{Y}$-completion morphism $\varphi^*_{A,-} : \mathrm{Ext}^*_{\mathcal{C}}(A,-) \to \widehat{\mathsf{Ext}}^n_{(\mathcal{X},\mathcal{Y})}(-,-)$ of the usual extension functor $\mathrm{Ext}^*_{\mathcal{C}}(A,-)$, and the $\mathcal{Y}$-completion morphism $\psi^*_{-,B} : \mathrm{Ext}^*_{\mathcal{C}}(-,B) \to \widetilde{\mathsf{Ext}}^*_{(\mathcal{Y},\mathcal{Z})}(-,-)$ of the usual extension functor $\mathrm{Ext}^*_{\mathcal{C}}(-,B)$. It is easy to see from the construction of the completion morphisms and the isomorphism $\xi^*_{-,-}$, that the completion morphisms $\varphi^*_{A,-}$ and $\psi^*_{-,B}$ are compatible with $\xi^*_{-,-}$ in the sense that for all objects A and B in $\mathcal{C}$, the following diagram commutes:

$$\begin{array}{ccc}
\mathrm{Ext}^*_{\mathcal{C}}(A,B) & \xrightarrow{\varphi^*_{A,B}} & \widehat{\mathsf{Ext}}^*_{(\mathcal{X},\mathcal{Y})}(A,B) \\
\| & & \downarrow{\scriptstyle \xi^*_{A,B}}\ \cong \\
\mathrm{Ext}^*_{\mathcal{C}}(A,B) & \xrightarrow{\psi^*_{A,B}} & \widetilde{\mathsf{Ext}}^*_{(\mathcal{Y},\mathcal{Z})}(A,B)
\end{array}$$

Then by Theorem 3.9 and its dual it follows that the long exact sequences of functors

$$\cdots \longrightarrow \mathrm{Ext}^n_{\mathcal{X}}(-,-) \longrightarrow \mathrm{Ext}^n_{\mathcal{C}}(-,-) \longrightarrow \widehat{\mathsf{Ext}}^n_{(\mathcal{X},\mathcal{Y})}(-,-) \longrightarrow \cdots$$

$$\cdots \longrightarrow \mathrm{Ext}^n_{\mathcal{Z}}(-,-) \longrightarrow \mathrm{Ext}^n_{\mathcal{C}}(-,-) \longrightarrow \widetilde{\mathsf{Ext}}^n_{(\mathcal{Y},\mathcal{Z})}(-,-) \longrightarrow \cdots$$

are isomorphic. Hence the relative extension functors $\mathrm{Ext}^*_{\mathcal{X}}(-,-)$ and $\mathrm{Ext}^*_{\mathcal{Z}}(-,-)$ are isomorphic. □

As usual any one of the subfunctors $\mathcal{F}_{\mathcal{X}}$ and $\mathcal{F}^{\mathcal{Z}}$ defines a relative homological theory in $\mathcal{C}$, see [**14**]. The corresponding global dimensions are denoted by $\mathcal{X}$-gl. dim $\mathcal{C}$ and $\mathcal{Z}$-gl. dim $\mathcal{C}$. Let $\mathcal{P}(\mathcal{F}_{\mathcal{X}})$, resp. $\mathcal{P}(\mathcal{F}^{\mathcal{Z}})$, be the full subcategory of $\mathcal{F}_{\mathcal{X}}$-projective, resp. $\mathcal{F}^{\mathcal{Z}}$-projective, objects of $\mathcal{C}$, and let $\mathcal{I}(\mathcal{F}_{\mathcal{X}})$, resp. $\mathcal{I}(\mathcal{F}^{\mathcal{Z}})$, be the full subcategory of $\mathcal{F}_{\mathcal{X}}$-injective, resp. $\mathcal{F}^{\mathcal{Z}}$-injective, objects of $\mathcal{C}$. We have the following direct consequence of Theorem 4.5, which combined with the fact that, by Lemma 3.1, $\mathcal{X}$-gl. dim $= \mathcal{X}$-res. dim $\mathcal{C}$ and $\mathcal{Z}$-gl. dim $= \mathcal{Z}$-cores. dim $\mathcal{C}$, gives a new proof of Corollary VI.4.4.

COROLLARY 4.6. *Let $(\mathcal{X}, \mathcal{Y}, \mathcal{Z})$ be a cotorsion triple in $\mathcal{C}$ with $\mathcal{Y}$ (co)resolving. Then we have the following.*

- (i) *The subfunctors $\mathcal{F}_{\mathcal{X}}$ and $\mathcal{F}^{\mathcal{Z}}$ have enough projective and injective objects, and moreover: $\mathcal{P}(\mathcal{F}_{\mathcal{X}}) = \mathcal{X} = \mathcal{P}(\mathcal{F}^{\mathcal{Z}})$ and $\mathcal{I}(\mathcal{F}^{\mathcal{Z}}) = \mathcal{Z} = \mathcal{I}(\mathcal{F}_{\mathcal{X}})$.*
- (ii) *$\mathcal{X}$-gl. dim $\mathcal{C} = \mathcal{Z}$-gl. dim $\mathcal{C}$, and this common value is finite if and only if $\mathcal{C}$ is Gorenstein.*

REMARK 4.7. By our previous results, for a resolving cotorsion pair $(\mathcal{X}, \mathcal{Y})$ with $\mathcal{X} \cap \mathcal{Y} = \mathcal{P}$, we have the following: $\mathcal{T}_l(\mathcal{Y}/\mathcal{P}) = 0 \Leftrightarrow \mathbf{R}^* : \mathcal{T}_l(\mathcal{C}/\mathcal{P}) \to \mathcal{X}/\mathcal{P}$ is a triangle equivalence $\Leftrightarrow$ $\mathcal{Y}$ consists of objects of finite projective dimension $\Leftrightarrow$ $\mathcal{C}$ is Gorenstein. Dually if we have a coresolving cotorsion pair $(\mathcal{Y}, \mathcal{Z})$ with $\mathcal{Y} \cap \mathcal{Z} = \mathcal{I}$, then the following are equivalent: $\mathcal{T}_r(\mathcal{Y}/\mathcal{I}) = 0 \Leftrightarrow \mathbf{T}^* : \mathcal{T}_r(\mathcal{C}/\mathcal{I}) \to \mathcal{Z}/\mathcal{I}$ is a triangle equivalence $\Leftrightarrow$ $\mathcal{Y}$ consists of objects of finite injective dimension $\Leftrightarrow$ $\mathcal{C}$ is Gorenstein. Hence the existence of such cotorsion pairs $(\mathcal{X}, \mathcal{Y})$ and/or $(\mathcal{Y}, \mathcal{Z})$ and more generally of cotorsion triples $(\mathcal{X}, \mathcal{Y}, \mathcal{Z})$, provides a generalization of the Gorenstein case.

All the above results can be applied to the category of (finitely generated) modules over an Artin algebra. Any Artin algebra Λ admits the cotorsion pair $(\mathcal{P}_\Lambda, \operatorname{mod}(\Lambda))$ with $\mathcal{P}_\Lambda \cap \operatorname{mod}(\Lambda) = \mathcal{P}_\Lambda$ and the cotorsion pair $(\operatorname{mod}(\Lambda), \mathcal{I}_\Lambda)$ with $\operatorname{mod}(\Lambda) \cap \mathcal{I}_\Lambda = \mathcal{I}_\Lambda$. We call these pairs *trivial*. A cotorsion triple $(\mathcal{X}, \mathcal{Y}, \mathcal{Z})$ is trivial if and only if $\mathcal{X} = \mathcal{P}_\Lambda$ or equivalently $\mathcal{Z} = \mathcal{I}_\Lambda$.

In view of the results of this and the previous sections it is natural to pose the following problem:

- Characterize the Artin algebras Λ such that $\operatorname{mod}(\Lambda)$ admits a non trivial resolving cotorsion pair $(\mathcal{X}, \mathcal{Y})$ with $\mathcal{X} \cap \mathcal{Y} = \mathcal{P}_\Lambda$. By Proposition VI.6.9, this is equivalent to asking for a characterization of the Artin algebras which admit a non trivial coresolving cotorsion pair $(\mathcal{Y}, \mathcal{Z})$ in $\operatorname{mod}(\Lambda)$ with $\mathcal{Y} \cap \mathcal{Z} = \mathcal{I}_\Lambda$, or a non-trivial cotorsion triple $(\mathcal{X}, \mathcal{Y}, \mathcal{Z})$ in $\operatorname{mod}(\Lambda)$ with $\mathcal{Y}$ (co)resolving.

Certainly the class of Gorenstein algebras belongs to all the three classes above, but we don't know if there are additional algebras. In the next chapter we shall show that any Artin algebra belongs to any of the above classes, except possibly for the third class, provided we work with the category of all modules.

CHAPTER X

Nakayama Categories and Cohen-Macaulay Cohomology

As we have seen in the previous chapter a hereditary torsion pair in the pretriangulated stable category of an abelian category gives rise to a complete cohomology theory, which play an important role in the investigation of the homological structure of the abelian category. Our aim in this chapter is to show that in any abelian category with exact infinite (co)products and which is endowed with an adjoint pair of suitable Nakayama functors, the stable category of Cohen-Macaulay, resp. CoCohen-Macaulay, objects is the torsion, resp. torsion-free, class of a hereditary, resp. cohereditary, torsion pair in the stable category modulo projectives, resp. injectives. Hence, using the results of the previous chapter, we can construct the complete cohomology theories induced by the Cohen-Macaulay and CoCohen-Macaulay objects. We investigate the basic properties of these new Cohomology theories, which is natural to call Cohen-Macaulay cohomology theories, and we give connections with relative homology and closed model structures.

1. Nakayama Categories and Cohen-Macaulay Objects

Throughout this section we fix an abelian category $\mathcal{C}$ and assume that $\mathcal{C}$ has enough projective and injective objects. As usual we denote by $\mathcal{P}$, resp. $\mathcal{I}$, the full subcategory of projective, resp. injective objects of $\mathcal{C}$. Our aim in this section is to study (Co)Cohen-Macaulay objects in Nakayama abelian categories which, as we shall see in the next section, provide the proper setting for the investigation of Cohen-Macaulay cohomology. In particular working in this context, we shall see that the (stable) categories of Cohen-Macaulay and CoCohen-Macaulay objects are (triangle) equivalent. Our motivation for introducing the notion of a Nakayama category comes from Artin algebras:

An important feature of the category $\mathrm{Mod}(\Lambda)$ of all right modules over an Artin algebra, is the existence of Nakayama functors. Recall that if $\mathrm{D} : \mathrm{mod}(\Lambda) \rightarrow \mathrm{mod}(\Lambda^{\mathrm{op}})$ is the usual duality of Artin algebras, then setting $\mathsf{N}^{+} = - \otimes_{\Lambda} \mathrm{D}(\Lambda)$ and $\mathsf{N}^{-} = \mathrm{Hom}_{\Lambda}(\mathrm{D}(\Lambda), -)$, we obtain an adjoint pair $(\mathsf{N}^{+}, \mathsf{N}^{-})$ of endofunctors of $\mathrm{Mod}(\Lambda)$ which induces an equivalence between the category of projective modules and the category of injective modules. The following notion, which will be useful later as the appropriate context for the study of Cohen-Macaulay cohomology, generalizes this situation from Artin algebras to a broader context.

DEFINITION 1.1. The abelian category $\mathcal{C}$ is said to be a **Nakayama category** if there exists an adjoint pair of functors $(\mathsf{N}^{+}, \mathsf{N}^{-}) : \mathcal{C} \rightarrow \mathcal{C}$ inducing an equivalence

$\mathsf{N}^+ : \mathcal{P} \xrightarrow{\approx} \mathcal{I}$ with quasi-inverse $\mathsf{N}^- : \mathcal{I} \xrightarrow{\approx} \mathcal{P}$. Then the functors N^+ and N^- are called the **Nakayama functors** of $\mathcal{C}$.

The following gives a variety of interesting examples of Nakayama categories.

EXAMPLE. (1) Any Frobenius abelian category is a Nakayama category. In this case we can choose $\mathsf{N}^+ = \mathsf{N}^- = \mathrm{Id}_{\mathcal{C}}$.

(2) As mentioned above, prominent examples of Nakayama categories are the module categories $\mathrm{mod}(\Lambda)$ and $\mathrm{Mod}(\Lambda)$ over an Artin algebra Λ. More generally the categories $\mathrm{mod}(\Lambda)$ and $\mathrm{Mod}(\Lambda)$ are Nakayama categories, when Λ is a ring with a Morita self-duality. More precisely for a ring Λ the following are equivalent:

(i) $\mathrm{Mod}(\Lambda)$ is a Nakayama category and N^- commutes with filtered colimits.
(ii) Λ is a ring with a Morita self-duality.

Indeed if Λ is a ring with Morita self-duality D, then $\mathrm{D}(\Lambda)$ is finitely generated and the functor $- \otimes_\Lambda \mathrm{D}(\Lambda)$ is a product preserving Nakayama functor with right adjoint the Nakayama functor $\mathrm{Hom}_\Lambda(\mathrm{D}(\Lambda), -)$. Conversely if (i) holds, then Λ is right Artinian and the Λ-bimodule $\omega := \mathsf{N}^+(\Lambda)$ is a finitely generated injective cogenerator of $\mathrm{Mod}(\Lambda)$ with $\mathrm{End}(\omega) = \Lambda$. Hence Λ is a ring with Morita self-duality.

(3) Let $\mathcal{C}$ be a dualizing R-variety in the sense of [**7**], where R is a commutative Artin ring. Then $\mathrm{mod}(\mathcal{C})$ and $\mathrm{Mod}(\mathcal{C})$ are Nakayama categories. This include locally bounded categories over a field.

(4) Let $\mathcal{C}$ be a hereditary Ext-finite abelian k-category over a field k. If $\mathcal{C}$ admits a Serre functor in the sense of [**89**], then $\mathcal{C}$ is a Nakayama category .

If $\mathcal{C}$ is a Nakayama category, then we denote always by $(\mathsf{N}^+, \mathsf{N}^-)$ the adjoint pair of Nakayama functors. Moreover we denote always by $\delta : \mathrm{Id}_{\mathcal{C}} \to \mathsf{N}^-\mathsf{N}^+$ the unit, and by $\varepsilon : \mathsf{N}^+\mathsf{N}^- \to \mathrm{Id}_{\mathcal{C}}$ the counit of the adjoint pair $(\mathsf{N}^+, \mathsf{N}^-)$.

The following result, which will be useful later in connection with torsion pairs induced by Cohen-Macaulay objects, shows that the stable categories modulo projectives or injectives of a Nakayama category are pretriangulated, hence we can speak of torsion pairs in them.

LEMMA 1.2. *If $\mathcal{C}$ is a Nakayama category, then $\mathcal{P}$ and $\mathcal{I}$ are functorially finite. In particular the stable categories $\mathcal{C}/\mathcal{P}$ and $\mathcal{C}/\mathcal{I}$ are pretriangulated.*

PROOF. Let C be in $\mathcal{C}$ and let $\mu : \mathsf{N}^+(C) \to I$ be a monomorphism with I injective. We claim that $C \xrightarrow{\delta_C} \mathsf{N}^-\mathsf{N}^+(C) \xrightarrow{\mathsf{N}^-(\mu)} \mathsf{N}^-(I)$ is a left projective approximation of C. First $\mathsf{N}^-(I)$ is projective, since I is injective. Let $f : C \to P$ be a morphism where P is projective. Since $\mathsf{N}^+(P)$ is injective, then there exists a morphism $g : I \to \mathsf{N}^+(P)$ such that $\mu \circ g = \mathsf{N}^+(f)$. Then $\delta_C \circ \mathsf{N}^-(\mu) \circ \mathsf{N}^-(g) = \delta_C \circ \mathsf{N}^-\mathsf{N}^+(f) = f \circ \delta_P$. But δ_P is invertible, since P is projective. Then $\delta_C \circ \mathsf{N}^-(\mu) \circ \mathsf{N}^-(g) \circ \delta_P^{-1} = f$, hence f factors through $\delta_C \circ \mathsf{N}^-(\mu)$. We infer that $\mathcal{P}$ is covariantly finite. By duality we have that $\mathcal{I}$ is contravariantly finite. □

We are going to show that the (stable) categories of Cohen-Macaulay objects and CoCohen-Macaulay objects are (triangle) equivalent, via the Nakayama functors. First we need the following preliminary result. From now on by ${}^{\perp}\mathcal{X}$ we denote the full subcategory $\{C \in \mathcal{C} \mid \mathrm{Ext}^n_{\mathcal{C}}(C, \mathcal{X}) = 0, \ \forall n \geq 1\}$, and similarly for $\mathcal{Y}^{\perp}$.

LEMMA 1.3. *Let $\mathcal{C}$ be a Nakayama category.*

(i) $\forall C \in \mathcal{C}$, $\forall I \in \mathcal{I}$, *there exists a natural isomorphism:*

$$\mathcal{C}(\mathrm{L}_n\mathsf{N}^+(C), I) \xrightarrow{\cong} \mathrm{Ext}^n(C, \mathsf{N}^-(I)).$$

In particular: $C \in {}^{\perp}\mathcal{P}$ *if and only if* $\mathrm{L}_n\mathsf{N}^+(C) = 0$, $\forall n \geq 1$.

(ii) $\forall C \in \mathcal{C}$, $\forall P \in \mathcal{P}$, *there exists a natural isomorphism:*

$$\mathrm{Ext}^n(\mathsf{N}^+(P), C) \xrightarrow{\cong} \mathcal{C}(P, \mathrm{R}^n\mathsf{N}^-(C)).$$

In particular: $C \in \mathcal{I}^{\perp}$ *if and only if* $\mathrm{R}^n\mathsf{N}^-(C) = 0$, $\forall n \geq 1$.

PROOF. We prove only part (i), since (ii) is dual. Let C be an object in $\mathcal{C}$ and and let $0 \to \Omega(C) \to P \to C \to 0$ be an exact sequence in $\mathcal{C}$ with P projective. Then we have the following exact sequence:

$$0 \to \mathrm{L}_1\mathsf{N}^+(C) \to \mathsf{N}^+(\Omega(C)) \to \mathsf{N}^+(P) \to \mathsf{N}^+(C) \to 0$$

Applying the functor $\mathcal{C}(-, I)$ to the above exact sequence with I injective, we have the following exact sequence:

$$0 \to \mathcal{C}(\mathsf{N}^+(C), I) \to \mathcal{C}(\mathsf{N}^+(P), I) \to \mathcal{C}(\mathsf{N}^+(\Omega(C)), I) \to \mathcal{C}(\mathrm{L}_1\mathsf{N}^+(C), I) \to 0$$

Using adjointness the above sequence is isomorphic to the exact sequence:

$$0 \to \mathcal{C}(C, \mathsf{N}^-(I)) \to \mathcal{C}(P, \mathsf{N}^-(I)) \to \mathcal{C}(\Omega(C), \mathsf{N}^-(I)) \to \mathcal{C}(\mathrm{L}_1\mathsf{N}^+(C), I) \to 0$$

from which we infer an isomorphism: $\mathcal{C}(\mathrm{L}_1\mathsf{N}^+(C), I) \xrightarrow{\cong} \mathrm{Ext}^1(C, \mathsf{N}^-(I))$. Replacing C with its higher syzygies, we deduce natural isomorphisms $\mathcal{C}(\mathrm{L}_n\mathsf{N}^+(C), I) \xrightarrow{\cong} \mathrm{Ext}^n(C, \mathsf{N}^-(I))$, $\forall n \geq 1$. Since any projective object is of the form $\mathsf{N}^-(I)$, where I is injective, it follows that $\mathrm{Ext}^n(C, \mathcal{P}) = 0$, $\forall n \geq 1$ iff $\mathsf{N}^+(C) = 0$, $\forall n \geq 1$. □

PROPOSITION 1.4. *Let $\mathcal{C}$ be a Nakayama category. Then the adjoint pair of Nakayama functors $(\mathsf{N}^+, \mathsf{N}^-)$ induces an equivalence $\mathsf{N}^+ : \mathrm{CM}(\mathcal{P}) \xrightarrow{\approx} \mathrm{CoCM}(\mathcal{I})$ with quasi-inverse $\mathsf{N}^- : \mathrm{CoCM}(\mathcal{I}) \xrightarrow{\approx} \mathrm{CM}(\mathcal{P})$, and a triangle equivalence $\mathsf{N}^+ : \mathrm{CM}(\mathcal{P})/\mathcal{P} \xrightarrow{\approx} \mathrm{CoCM}(\mathcal{I})/\mathcal{I}$ with quasi-inverse $\mathsf{N}^- : \mathrm{CoCM}(\mathcal{I})/\mathcal{I} \xrightarrow{\approx} \mathrm{CM}(\mathcal{P})/\mathcal{P}$.*

PROOF. Since $\mathrm{CM}(\mathcal{P}) \subseteq {}^{\perp}\mathcal{P}$ and $\mathrm{CoCM}(\mathcal{I}) \subseteq \mathcal{I}^{\perp}$, the above Lemma ensures that $\mathrm{L}_n\mathsf{N}^+(X) = 0$, $\forall n \geq 1$, $\forall X \in \mathrm{CM}(\mathcal{P})$, and $\mathrm{R}^n\mathsf{N}^-(Z) = 0$, $\forall n \geq 1$, $\forall Z \in \mathrm{CoCM}(\mathcal{I})$. In particular the functor N^+ is exact on exact sequences with right end term a Cohen-Macaulay object, and the functor N^- is exact on exact sequences with left end term a CoCohen-Macaulay object.

Now let X be a Cohen-Macaulay object. Then there exists an exact coresolution $0 \to X \to P^0 \to P^1 \to \cdots$ of X by projective objects. Since the functor N^+ is exact on Cohen-Macaulay objects, the exact sequence $0 \to \mathsf{N}^+(X) \to \mathsf{N}^+(P^0) \to \mathsf{N}^+(P^1) \to \cdots$ is an injective resolution of $\mathsf{N}^+(X)$. Applying the left exact functor N^- to this sequence we have an exact sequence $0 \to \mathsf{N}^-\mathsf{N}^+(X) \to \mathsf{N}^-\mathsf{N}^+(P^0) \to \mathsf{N}^-\mathsf{N}^+(P^1)$. Since the adjunction morphisms $P^i \to \mathsf{N}^-\mathsf{N}^+(P^i)$ are invertible, we infer that the adjunction morphism $X \to \mathsf{N}^-\mathsf{N}^+(X)$ is invertible. Now let $\cdots \to P_{-2} \to P_{-1} \to X \to 0$ be a projective resolution of X. Since the functor N^+ is exact on Cohen-Macaulay objects, we have an exact resolution $\cdots \to \mathsf{N}^+(P_{-2}) \to \mathsf{N}^+(P_{-1}) \to \mathsf{N}^+(X) \to 0$ of $\mathsf{N}^+(X)$ by injectives. Then to show that $\mathsf{N}^+(X)$ is a CoCohen-Macaulay object it suffices to show that if we apply $\mathcal{C}(I, -)$ for any injective object I to the above coresolution, then we still get an exact sequence. Since any injective object of $\mathcal{C}$ is of the form $\mathsf{N}^+(Q)$ where Q is projective, it suffices to show that the sequence $\cdots \to \mathcal{C}(\mathsf{N}^+(Q), \mathsf{N}^+(P_{-2})) \to \mathcal{C}(\mathsf{N}^+(Q), \mathsf{N}^+(P_{-1})) \to \mathcal{C}(\mathsf{N}^+(Q), \mathsf{N}^+(X)) \to 0$ is exact for any projective object Q. The last sequence is isomorphic to the sequence $\cdots \to \mathcal{C}(Q, \mathsf{N}^-\mathsf{N}^+(P_{-2})) \to \mathcal{C}(Q, \mathsf{N}^-\mathsf{N}^+(P_{-1})) \to \mathcal{C}(Q, \mathsf{N}^-\mathsf{N}^+(X)) \to 0$. Since the unit $\mathrm{Id}_{\mathcal{C}} \to \mathsf{N}^-\mathsf{N}^+$ is invertible on Cohen-Macaulay objects, the last sequence is isomorphic to the sequence $\cdots \to \mathcal{C}(Q, P_{-2}) \to \mathcal{C}(Q, P_{-1}) \to \mathcal{C}(Q, X) \to 0$ which is exact, since Q is projective. We infer that N^+ sends Cohen-Macaulay objects to CoCohen-Macaulay objects, and therefore we have a fully faithful functor $\mathsf{N}^+ : \mathrm{CM}(\mathcal{P}) \to \mathrm{CoCM}(\mathcal{I})$. By duality, for any CoCohen-Macaulay object Z, the object $\mathsf{N}^-(Z)$ is Cohen-Macaulay and the adjunction morphism $\mathsf{N}^+\mathsf{N}^-(Z) \to Z$ is invertible. This shows that $\mathsf{N}^+ : \mathrm{CM}(\mathcal{P}) \xrightarrow{\approx} \mathrm{CoCM}(\mathcal{I})$ is an equivalence with quasi-inverse $\mathsf{N}^- : \mathrm{CoCM}(\mathcal{I}) \xrightarrow{\approx} \mathrm{CM}(\mathcal{P})$. Since the functor F is exact on Cohen-Macaulay modules and sends projectives to injectives, it follows that the induced functor $\mathsf{N}^+ : \mathrm{CM}(\mathcal{P})/\mathcal{P} \xrightarrow{\approx} \mathrm{CoCM}(\mathcal{I})/\mathcal{I}$ is a triangle equivalence with quasi-inverse N^-. □

It is useful to point out that as a consequence of the above proposition we have natural isomorphisms

$$\mathsf{N}^+\Omega \xrightarrow{\cong} \Sigma^{-1}\mathsf{N}^+ : \mathrm{CM}(\mathcal{P})/\mathcal{P} \longrightarrow \mathrm{CoCM}(\mathcal{I})/\mathcal{I},$$

$$\mathsf{N}^-\Sigma \xrightarrow{\cong} \Omega^{-1}\mathsf{N}^- : \mathrm{CoCM}(\mathcal{I})/\mathcal{I} \longrightarrow \mathrm{CM}(\mathcal{P})/\mathcal{P}.$$

In general the triangle equivalence of Proposition 1.4 is not induced by an equivalence between the stable categories. As we now explain, there exists an equivalence $\mathcal{C}/\mathcal{P} \xrightarrow{\approx} \mathcal{C}/\mathcal{I}$ which restricts to an equivalence between the stable categories of (Co)Cohen-Macaulay objects. Let A and B be arbitrary objects in $\mathcal{C}$. Let $0 \to \Omega^2(A) \to P_1 \to P_0 \to A \to 0$ be the start of a projective resolution of A, and let $0 \to B \to I_0 \to I_1 \to \Sigma^2(B) \to 0$ be the start of an injective resolution of B. We define the objects $\tau^+(A)$ and $\tau^-(B)$ by the exact sequences:

$$0 \to \tau^+(A) \to \mathsf{N}^+(P_1) \to \mathsf{N}^+(P_0) \to \mathsf{N}^+(A) \to 0$$

$$0 \to \mathsf{N}^-(B) \to \mathsf{N}^-(I^0) \to \mathsf{N}^-(I^1) \to \tau^-(B) \to 0$$

We call the assignments $A \mapsto \tau^+(A)$ and $A \mapsto \tau^-(A)$, the *Auslander-Reiten operators* of the Nakayama category $\mathcal{C}$ with respect to the adjoint pair $(\mathsf{N}^+, \mathsf{N}^-)$. Part (1) of the following observation follows easily from the fact that the Nakayama functors restrict to an equivalence $(\mathsf{N}^+, \mathsf{N}^-)$ between $\mathcal{P}$ and $\mathcal{I}$. Part (2), which shows that the stable equivalences induced by the Nakayama functors N^+, N^- and the operators τ^+ and τ^- are compatible follows from (1) and Proposition 1.4.

LEMMA 1.5. (1) *The assignments $A \mapsto \tau^+(A)$ and $B \mapsto \tau^-(B)$ induce well defined additive functors*

$$\tau^+ : \mathcal{C}/\mathcal{P} \to \mathcal{C}/\mathcal{I} \quad \textit{and} \quad \tau^- : \mathcal{C}/\mathcal{I} \to \mathcal{C}/\mathcal{P}$$

which are mutually inverse equivalences.

(2) *The equivalences τ^+ and τ^- restrict to triangle quasi-inverse equivalences*

$$\tau^+ : \mathrm{CM}(\mathcal{P})/\mathcal{P} \xrightarrow{\approx} \mathrm{CoCM}(\mathcal{I})/\mathcal{I} \quad \textit{and} \quad \tau^- : \mathrm{CoCM}(\mathcal{I})/\mathcal{I} \xrightarrow{\approx} \mathrm{CM}(\mathcal{P})/\mathcal{P}$$

such that the following diagrams are commutative ($\mathbf{i}$ *and* $\mathbf{k}$ *are the inclusions):*

$$\begin{array}{ccc} \mathrm{CM}(\mathcal{P})/\mathcal{P} & \xrightarrow{\mathsf{N}^+} & \mathrm{CoCM}(\mathcal{I})/\mathcal{I} \\ {\scriptstyle \mathbf{i}}\downarrow & & \downarrow{\scriptstyle \Sigma^{-2}} \\ \mathcal{C}/\mathcal{P} & \xrightarrow{\tau^+} & \mathcal{C}/\mathcal{I} \end{array} \qquad \begin{array}{ccc} \mathrm{CoCM}(\mathcal{I})/\mathcal{I} & \xrightarrow{\mathsf{N}^-} & \mathrm{CM}(\mathcal{P})/\mathcal{P} \\ {\scriptstyle \mathbf{k}}\downarrow & & \downarrow{\scriptstyle \Omega^{-2}} \\ \mathcal{C}/\mathcal{I} & \xrightarrow{\tau^-} & \mathcal{C}/\mathcal{P}. \end{array}$$

REMARK 1.6. If $\mathcal{C} = \mathrm{Mod}(\Lambda)$, where Λ is an Artin algebra or more generally a ring with Morita self-duality, then Lemma 1.5 was first observed by H. Krause in [**76**]. Note that the Auslander-Reiten operators τ^+ and τ^- restricted to finitely generated modules coincide with the operators DTr and TrD, where D is the usual duality for Artin algebras and Tr is the Auslander-Bridger transpose, see [**15**]. Note that the operators DTr and TrD play an important role in the representation theory of Artin algebras.

2. (Co)Torsion Pairs Induced by (Co)Cohen-Macaulay Objects

Throughout this section we fix a Nakayama abelian category $\mathcal{C}$ and let $(\mathsf{N}^+, \mathsf{N}^-)$ be the adjoint pair of Nakayama functors for $\mathcal{C}$. Our aim in this section is show that, under some reasonable conditions, the stable category of Cohen-Macaulay objects modulo projectives is the torsion class of a hereditary torsion pair in $\mathcal{C}/\mathcal{P}$ and the stable category of CoCohen-Macaulay objects modulo injectives is the torsion-free class of a cohereditary torsion pair in $\mathcal{C}/\mathcal{I}$. In particular we shall show that the categories of (Co)Cohen-Macaulay objects are functorially finite and are parts of (co)resolving cotorsion pairs in $\mathcal{C}$.

Let $\mathcal{H}(\mathcal{P})$ be the (unbounded) homotopy category of $\mathcal{P}$ and let $\mathcal{H}(\mathcal{I})$ be the (unbounded) homotopy category of $\mathcal{I}$. We denote by $\mathcal{H}_{\mathsf{Ac}}(\mathcal{P})$ the full subcategory of $\mathcal{H}(\mathcal{P})$ consisting of the acyclic complexes of projectives and by $\mathcal{H}_{\mathsf{Ac}}(\mathcal{I})$ the full subcategory of $\mathcal{H}(\mathcal{I})$ consisting of the acyclic complexes of injectives. Note that $\mathcal{H}_{\mathsf{Ac}}(\mathcal{P})$ is the costabilization of $\mathcal{C}/\mathcal{P}$ and $\mathcal{H}_{\mathsf{Ac}}(\mathcal{I})$ is the costabilization of $\mathcal{C}/\mathcal{I}$, see [**20**] for details. Obviously $\mathcal{H}_{\mathsf{Ac}}(\mathcal{P})$ is a thick subcategory of $\mathcal{H}(\mathcal{P})$ and $\mathcal{H}_{\mathsf{Ac}}(\mathcal{I})$ is a thick subcategory of $\mathcal{H}(\mathcal{I})$. We denote by $\mathcal{E}_{\mathcal{P}}(\mathcal{C})$ the full subcategory of $\mathcal{H}_{\mathsf{Ac}}(\mathcal{P})$ consisting of the acyclic complexes of projectives $P^\bullet$ such that the complex $(P^\bullet, P)$ is acyclic for all projective modules P. Dually we denote by $\mathcal{E}_{\mathcal{I}}(\mathcal{C})$ the full subcategory of $\mathcal{H}_{\mathsf{Ac}}(\mathcal{I})$ consisting of the acyclic complexes of injectives $I^\bullet$ such that the complex $(I, I^\bullet)$ is acyclic for all injective modules I. It is easy to see that the functor $P^\bullet \mapsto \mathrm{Ker}(d^0)$ induces an equivalence $\mathcal{E}_{\mathcal{P}}(\mathcal{C}) \to \mathrm{CM}(\mathcal{P})/\mathcal{P}$ and the functor $I^\bullet \mapsto \mathrm{Ker}(d^0)$ induces an equivalence $\mathcal{E}_{\mathcal{I}}(\mathcal{C}) \to \mathrm{CoCM}(\mathcal{I})/\mathcal{I}$. In particular $\mathcal{E}_{\mathcal{P}}(\mathcal{C})$ and $\mathcal{E}_{\mathcal{I}}(\mathcal{C})$ are triangulated.

To prove the main result of this section we need some preliminary observations. We begin with the following easy consequence of Proposition 1.4 and the definitions.

PROPOSITION 2.1. *The adjoint pair* $(\mathsf{N}^+, \mathsf{N}^-)$ *of Nakayama functors induces quasi-inverse triangle equivalences*

$$\mathsf{N}^+ \ : \mathcal{H}(\mathcal{P}) \xrightarrow{\approx} \mathcal{H}(\mathcal{I}), \qquad \mathsf{N}^- \ : \mathcal{H}(\mathcal{I}) \xrightarrow{\approx} \mathcal{H}(\mathcal{P})$$

which induce quasi-inverse triangle equivalences

$$\mathsf{N}^+ \ : \mathcal{E}_{\mathcal{P}}(\mathcal{C}) \xrightarrow{\approx} \mathcal{E}_{\mathcal{I}}(\mathcal{C}), \qquad \mathsf{N}^- \ : \mathcal{E}_{\mathcal{I}}(\mathcal{C}) \xrightarrow{\approx} \mathcal{E}_{\mathcal{P}}(\mathcal{C}).$$

REMARK 2.2. In general the triangle equivalences $\mathsf{N}^+ : \mathcal{H}(\mathcal{P}) \xrightarrow{\approx} \mathcal{H}(\mathcal{I})$ and $\mathsf{N}^- : \mathcal{H}(\mathcal{I}) \xrightarrow{\approx} \mathcal{H}(\mathcal{P})$ do not restrict to equivalences between $\mathcal{H}_{\mathsf{Ac}}(\mathcal{P})$ and $\mathcal{H}_{\mathsf{Ac}}(\mathcal{I})$. For a necessary and sufficient condition for this to happen, we refer to [**20**]. Note that for Gorenstein categories $\mathcal{C}$, we have $\mathcal{H}_{\mathsf{Ac}}(\mathcal{P}) = \mathcal{E}_{\mathcal{P}}(\mathcal{C})$ and $\mathcal{H}_{\mathsf{Ac}}(\mathcal{I}) = \mathcal{E}_{\mathcal{I}}(\mathcal{C})$.

From now on we assume that the Nakayama category $\mathcal{C}$ has exact products and coproducts and a set of compact projective generators. In addition we assume that the Nakayama functor N^- preserves coproducts. These assumptions hold for the module category of a ring with Morita self-duality, for instance an Artin algebra.

To proceed to the main result of this section we need the following preliminary observation.

LEMMA 2.3. *Under the above assumptions, the stable categories $\mathcal{C}/\mathcal{P}$ and $\mathcal{C}/\mathcal{I}$ and their costabilizations $\mathcal{H}_{\mathsf{Ac}}(\mathcal{P})$ and $\mathcal{H}_{\mathsf{Ac}}(\mathcal{I})$ are compactly generated.*

PROOF. Let $\mathcal{Q}$ be a set of compact projective generators for $\mathcal{C}$. Since N^- preserves coproducts, it follows easily that N^+ preserves compact objects. We infer that $\mathcal{P} = \mathrm{Add}(\mathcal{Q})$ and $\mathsf{N}^+(\mathcal{Q})$ is a set of compact injective cogenerators of $\mathcal{C}$. By a well-known result of Freyd [**100**], $\mathcal{C}$ is equivalent to the functor category $\mathrm{Mod}(\mathcal{Q})$. Since the categories $\mathcal{P}$ and $\mathcal{I}$ are functorially finite, we infer, by [**23**], that the Grothendieck category $\mathrm{Mod}(\mathcal{Q})$ is perfect and locally Noetherian and the Grothendieck category $\mathrm{Mod}(\mathcal{Q}^{\mathrm{op}})$ is locally coherent. In particular we have $\mathcal{I} = \mathrm{Add}(\mathsf{N}^+(\mathcal{Q}))$. Let $\mathcal{T}$ be the set of isoclasses of factors of finite direct sums of objects from the compact generating set $\mathcal{Q}$. Then the category $\mathrm{mod}(\mathcal{Q})$ of finitely presented functors $\mathcal{Q}^{\mathrm{op}} \longrightarrow \mathcal{A}b$ is equivalent to $\mathcal{T}$. By Theorems 5.3 and 5.7 of [**24**] we infer that the stable categories $\mathcal{T}/\mathcal{P}$ and $\mathcal{T}/\mathcal{I}$ are compact generating sets in the stable categories $\mathcal{C}/\mathcal{P}$ and $\mathcal{C}/\mathcal{I}$. Using the above analysis and the results of [**67**], it follows that the categories $\mathcal{H}_{\mathsf{Ac}}(\mathcal{P})$ and $\mathcal{H}_{\mathsf{Ac}}(\mathcal{I})$ are compactly generated. □

Now we can prove the main result of this section.

THEOREM 2.4. *Let $\mathcal{C}$ be a Nakayama abelian category with exact products and coproducts and let $(\mathsf{N}^+, \mathsf{N}^-)$ be the adjoint pair of Nakayama functors for $\mathcal{C}$. If $\mathcal{C}$ admits a set of compact projective generators and N^- preserves coproducts, then:*

(i) *There is a hereditary torsion pair $\big(\mathrm{CM}(\mathcal{P})/\mathcal{P}, \mathcal{Y}/\mathcal{P}\big)$ in $\mathcal{C}/\mathcal{P}$ and a cohereditary torsion pair $\big(\mathcal{W}/\mathcal{I}, \mathrm{CoCM}(\mathcal{I})/\mathcal{I}\big)$ of finite type in $\mathcal{C}/\mathcal{I}$.*
(ii) *The inclusion $\mathrm{CM}(\mathcal{P})/\mathcal{P} \hookrightarrow \mathcal{C}/\mathcal{P}$ admits a left adjoint and a right adjoint, and the inclusion $\mathrm{CoCM}(\mathcal{I})/\mathcal{I} \hookrightarrow \mathcal{C}/\mathcal{I}$ admits a right adjoint and a left adjoint.*
(iii) *The torsion class $\mathrm{CM}(\mathcal{P})/\mathcal{P}$ in $\mathcal{C}/\mathcal{P}$ is a compactly generated triangulated category, and the torsion-free class $\mathrm{CoCM}(\mathcal{I})/\mathcal{I}$ in $\mathcal{C}/\mathcal{I}$ is a compactly generated triangulated category.*
(iv) *There is a resolving cotorsion pair $\big(\mathrm{CM}(\mathcal{P}), \mathcal{Y}\big)$ in $\mathcal{C}$. The full subcategory $\mathrm{CM}(\mathcal{P})$ of Cohen-Macaulay objects is functorially finite and the full subcategory $\mathcal{Y}$ is resolving and coresolving. Moreover: $\mathrm{CM}(\mathcal{P}) \cap \mathcal{Y} = \mathcal{P}$ and $\mathcal{P}^{<\infty} = \mathcal{Y} \cap \widehat{\mathrm{CM}(\mathcal{P})}$.*

(v) *There is a coresolving cotorsion pair* $(\mathcal{W}, \mathrm{CoCM}(\mathcal{I}))$ *in* $\mathcal{C}$. *The full subcategory* $\mathrm{CoCM}(\mathcal{I})$ *of (co)Cohen-Macaulay objects is functorially finite and the full subcategory* $\mathcal{W}$ *is resolving and coresolving. Moreover:* $\mathrm{CoCM}(\mathcal{I}) \cap \mathcal{W} = \mathcal{I}$ *and* $\mathcal{I}^{<\infty} = \widetilde{\mathrm{CoCM}}(\mathcal{I}) \cap \mathcal{W}$.

PROOF. (i) If $\mathcal{C}$ is the module category $\mathrm{Mod}(\Lambda)$, where Λ is an Artin algebra, then the fact that the inclusion $\mathrm{CM}(\mathcal{P})/\mathcal{P} \hookrightarrow \mathcal{C}/\mathcal{P}$ admits a right adjoint was first proved by Joergensen in [**67**]. We include a slightly different proof for the general case, and we also prove the dual assertion for CoCohen-Macaulay objects.

Consider the category $\mathcal{H}_{\mathsf{Ac}}(\mathcal{P})$ of acyclic complexes of projective objects of $\mathcal{C}$. Then we have a left exact functor $\mathsf{R} : \mathcal{H}_{\mathsf{Ac}}(\mathcal{P}) \to \mathcal{C}/\mathcal{P}$ of pretriangulated categories defined by sending an acyclic complex $P^\bullet$ of projectives to $\underline{\mathrm{Im}}(d^0)$. By [**67**], the functor R admits a left adjoint $\mathsf{Sp} : \mathcal{C}/\mathcal{P} \to \mathcal{H}_{\mathsf{Ac}}(\mathcal{P})$, called the projective spectrification functor, which is a right exact functor of pretriangulated categories. Consider the full subcategory $\mathcal{E}_{\mathcal{P}}(\mathcal{C}) \hookrightarrow \mathcal{H}_{\mathsf{Ac}}(\mathcal{P})$. The functor $\mathrm{H}^0\mathsf{N}^+ : \mathcal{H}_{\mathsf{Ac}}(\mathcal{P}) \to \mathcal{A}b$ is a homological functor which preserves coproducts, where H^0 is the 0-th cohomology functor, and it is easy to see that $\mathcal{E}_{\mathcal{P}}(\mathcal{C}) = \{P^\bullet \in \mathcal{H}_{\mathsf{Ac}}(\mathcal{P}) \mid \mathrm{H}^0\mathsf{N}^+\Sigma^n(P^\bullet) = 0, \forall n \in \mathbb{Z}\}$. Since the homological functor $\mathrm{H}^0\mathsf{N}^+$ preserves coproducts, then by [**67**] the quotient $\mathcal{H}_{\mathsf{Ac}}(\mathcal{P})/\mathcal{E}_{\mathcal{P}}(\mathcal{C})$ has small hom-sets. Since $\mathcal{E}_{\mathcal{P}}(\mathcal{C})$ is closed under coproducts and $\mathrm{H}^0\mathsf{N}^+$ preserves coproducts, it is easy to see that the quotient functor $q : \mathcal{H}_{\mathsf{Ac}}(\mathcal{P}) \to \mathcal{H}_{\mathsf{Ac}}(\mathcal{P})/\mathcal{E}_{\mathcal{P}}(\mathcal{C})$ preserves coproducts. By a result of Neeman [**86**], a coproduct preserving exact functor from a compactly generated triangulated category to a triangulated category admits a right adjoint. Since $\mathcal{H}_{\mathsf{Ac}}(\mathcal{P})$ is compactly generated and q preserves coproducts, we have that q admits a right adjoint. Then by a well-known result of Verdier [**102**], the inclusion $\mathcal{E}_{\mathcal{P}}(\mathcal{C}) \hookrightarrow \mathcal{H}_{\mathsf{Ac}}(\mathcal{P})$ admits a right adjoint. Identifying the categories $\mathcal{E}_{\mathcal{P}}(\mathcal{C})$ and $\mathrm{CM}(\mathcal{P})/\mathcal{P}$, it follows that the inclusion $\mathrm{CM}(\mathcal{P})/\mathcal{P} \hookrightarrow \mathcal{H}_{\mathsf{Ac}}(\mathcal{P})$ admits a right adjoint $\mathbf{R}^* : \mathcal{H}_{\mathsf{Ac}}(\mathcal{P}) \to \mathrm{CM}(\mathcal{P})/\mathcal{P}$. It is not difficult to see that the functor $\mathsf{Sp} : \mathrm{CM}(\mathcal{P})/\mathcal{P} \to \mathcal{H}_{\mathsf{Ac}}(\mathcal{P})$ is fully faithful and the strict image of $\mathsf{Sp}|_{\mathrm{CM}(\mathcal{P})/\mathcal{P}}$ is identified with the full subcategory $\mathcal{E}_{\mathcal{P}}(\mathcal{C})$. Moreover for any Cohen-Macaulay object X the natural map $\mathcal{C}/\mathcal{P}(\underline{X}, \underline{A}) \to \mathcal{H}_{\mathsf{Ac}}(\mathcal{P})[\mathsf{Sp}(\underline{X}), \mathsf{Sp}(\underline{A})]$ is invertible for any object A in $\mathcal{C}$. In particular $\mathbf{R}^*\mathsf{Sp}|_{\mathrm{CM}(\mathcal{P})/\mathcal{P}} \xrightarrow{\cong} \mathrm{Id}_{\mathrm{CM}(\mathcal{P})/\mathcal{P}}$. It follows that, for any $X \in \mathrm{CM}(\mathcal{P})$ and any $C \in \mathcal{C}$, we have natural isomorphisms:

$$\mathcal{C}/\mathcal{P}(\underline{X}, \underline{A}) \xrightarrow{\mathsf{Sp}} \mathcal{H}_{\mathsf{Ac}}(\mathcal{P})[\mathsf{Sp}(\underline{X}), \mathsf{Sp}(\underline{A})] \xrightarrow{\cong} \mathrm{CM}(\mathcal{P})/\mathcal{P}[\underline{X}, \mathbf{R}^*\mathsf{Sp}(\underline{A})].$$

This shows that the functor $\mathbf{R}^*\mathsf{Sp} : \mathcal{C}/\mathcal{P} \to \mathrm{CM}(\mathcal{P})/\mathcal{P}$ is the right adjoint of the inclusion $\mathrm{CM}(\mathcal{P})/\mathcal{P} \hookrightarrow \mathcal{C}/\mathcal{P}$. Finally by Proposition V.4.1 we infer that the stable category $\mathrm{CM}(\mathcal{P})/\mathcal{P}$ is the torsion class of a hereditary torsion pair $(\mathrm{CM}(\mathcal{P})/\mathcal{P}, \mathcal{Y}/\mathcal{P})$ in $\mathcal{C}/\mathcal{I}$, where $\mathcal{Y}$ is the full subcategory of $\mathcal{C}$ defined by $\mathcal{Y}/\mathcal{P} = \mathrm{CM}(\mathcal{P})/\mathcal{P})^{\perp}$.

We now show that the inclusion $\mathrm{CoCM}(\mathcal{I})/\mathcal{I} \hookrightarrow \mathcal{C}/\mathcal{I}$ admits a left adjoint, using the results of [**67**] and results of Neeman. Consider the full subcategory $\mathcal{E}_{\mathcal{I}}(\mathcal{C}) \hookrightarrow \mathcal{H}_{\mathsf{Ac}}(\mathcal{I})$. The functor $\mathrm{H}^0\mathsf{N}^- : \mathcal{H}_{\mathsf{Ac}}(\mathcal{I}) \to \mathcal{A}b$ is a homological functor which preserves coproducts, where H^0 is the 0-th cohomology functor, and it is easy to see that $\mathcal{E}_{\mathcal{I}}(\mathcal{C}) = \{I^\bullet \in \mathcal{H}_{\mathsf{Ac}}(\mathcal{I}) \mid \mathrm{H}^0\mathsf{N}^-\Sigma^n(I^\bullet) = 0, \forall n \in \mathbb{Z}\}$. Since the homological functor $\mathrm{H}^0\mathsf{N}^-$ preserves coproducts and $\mathcal{H}_{\mathsf{Ac}}(\mathcal{I})$ is compactly generated, then by [**67**] the quotient $\mathcal{H}_{\mathsf{Ac}}(\mathcal{I})/\mathcal{E}_{\mathcal{I}}(\mathcal{C})$ has small hom-sets. Since $\mathcal{E}_{\mathcal{I}}$ is closed under products and $\mathrm{H}^0\mathsf{N}^-$ preserves products, it is easy to see that the quotient functor $q : \mathcal{H}_{\mathsf{Ac}}(\mathcal{I}) \to \mathcal{H}_{\mathsf{Ac}}(\mathcal{I})/\mathcal{E}_{\mathcal{I}}(\mathcal{C})$ preserves products. By a result of Neeman [**86**], a homological product preserving functor from a compactly generated triangulated category to a Grothendieck category is representable. This implies that a product preserving exact functor from a compactly generated triangulated category to a triangulated category admits a left adjoint. Since $\mathcal{H}_{\mathsf{Ac}}(\mathcal{I})$ is compactly generated and q preserves products, we have that q admits a left adjoint. Then by a well-known result of Verdier [**102**], the inclusion $\mathcal{E}_{\mathcal{I}}(\mathcal{C}) \hookrightarrow \mathcal{H}_{\mathsf{Ac}}(\mathcal{I})$ admits a left adjoint. Since $\mathcal{E}_{\mathcal{I}}(\mathcal{C})$ is triangle equivalent to $\mathrm{CoCM}(\mathcal{I})$, working as above we infer that the inclusion $\mathrm{CoCM}(\mathcal{I}) \hookrightarrow \mathcal{C}/\mathcal{I}$ admits a left adjoint. Then by the dual of Proposition V.4.1 we infer that $\mathrm{CoCM}(\mathcal{I})/\mathcal{I}$ is the torsion-free class of a cohereditary torsion pair $\big(\mathcal{W}/\mathcal{I}, \mathrm{CoCM}(\mathcal{I})/\mathcal{I}\big)$ in $\mathcal{C}/\mathcal{I}$. Finally observe that the category $\mathrm{CoCM}(\mathcal{I})$ is closed under all small coproducts in $\mathcal{C}$. This implies that the stable category $\mathrm{CoCM}(\mathcal{I})/\mathcal{I}$ is closed under all small coproducts in $\mathcal{C}/\mathcal{I}$. Hence the torsion pair $\big(\mathcal{W}/\mathcal{I}, \mathrm{CoCM}(\mathcal{I})/\mathcal{I}\big)$ is of finite type.

(ii) Let $\mathbf{R} : \mathcal{C}/\mathcal{P} \to \mathrm{CM}(\mathcal{P})/\mathcal{P}$ be the right adjoint of the inclusion $\mathrm{CM}(\mathcal{P})/\mathcal{P} \hookrightarrow \mathcal{C}/\mathcal{P}$ and let $\mathbf{T} : \mathcal{C}/\mathcal{I} \to \mathrm{CoCM}(\mathcal{I})/\mathcal{I}$ be the left adjoint of the inclusion $\mathrm{CoCM}(\mathcal{I})/\mathcal{I} \hookrightarrow \mathcal{C}/\mathcal{I}$; these functors exist by part (i). For any object C in $\mathcal{C}$, the object $\mathbf{R}\mathsf{N}^-(\overline{C})$ lies in $\mathrm{CM}(\mathcal{P})/\mathcal{P}$. Hence, by Proposition 1.4, the object $\mathsf{N}^+\mathbf{R}\mathsf{N}^-(\overline{C})$ lies in $\mathrm{CM}(\mathcal{I})/\mathcal{I}$. It follows that we can consider the functor $\mathsf{N}^+\mathbf{R}\mathsf{N}^- : \mathcal{C}/\mathcal{I} \to \mathrm{CoCM}(\mathcal{I})/\mathcal{I}$. Then, using Proposition 1.4, we have the following natural isomorphisms, for any object Z in $\mathrm{CoCM}(\mathcal{I})$ and any object A in $\mathcal{C}$:

$$\mathcal{C}/\mathcal{I}[\overline{Z}, \overline{A}] \xrightarrow{\cong} \mathcal{C}/\mathcal{I}[\mathsf{N}^+\mathsf{N}^-(\overline{Z}), \overline{A}] \xrightarrow{\cong} \mathcal{C}/\mathcal{P}[\mathsf{N}^-(\overline{Z}), \mathsf{N}^-(\overline{A})] \xrightarrow{\cong}$$

$$\mathrm{CM}(\mathcal{P})/\mathcal{P}[\mathsf{N}^-(\overline{Z}), \mathbf{R}\mathsf{N}^-(\overline{A})] \xrightarrow{\cong} \mathrm{CoCM}(\mathcal{I})/\mathcal{I}[\overline{Z}, \mathsf{N}^+\mathbf{R}\mathsf{N}^-(\overline{A})]$$

We infer that the functor $\mathsf{N}^+\mathbf{R}\mathsf{N}^- : \mathcal{C}/\mathcal{I} \to \mathrm{CoCM}(\mathcal{I})/\mathcal{I}$ is the right adjoint of the inclusion $\mathrm{CoCM}(\mathcal{I})/\mathcal{I} \hookrightarrow \mathcal{C}/\mathcal{I}$. By duality we have that the functor $\mathsf{N}^-\mathbf{T}\mathsf{N}^+ : \mathcal{C}/\mathcal{P} \to \mathrm{CM}(\mathcal{P})/\mathcal{P}$ is the left adjoint of the inclusion $\mathrm{CM}(\mathcal{P})/\mathcal{P} \hookrightarrow \mathcal{C}/\mathcal{P}$.

(iii) By Lemma 2.3 we know that the stable category $\mathcal{C}/\mathcal{I}$ is compactly generated as a right triangulated category. Let $\mathcal{T}/\mathcal{I}$ be a compact generating set in $\mathcal{C}/\mathcal{I}$ and let $\mathbf{T} : \mathcal{C}/\mathcal{I} \to \mathrm{CoCM}(\mathcal{I})/\mathcal{I}$ be the left adjoint of the inclusion $\mathrm{CoCM}(\mathcal{I})/\mathcal{I} \hookrightarrow \mathcal{C}/\mathcal{I}$. We claim that the set $\mathbf{T}(\mathcal{T}/\mathcal{I})$ is a compact generating set in $\mathrm{CoCM}(\mathcal{I})/\mathcal{I}$. Since $\big(\mathcal{W}/\mathcal{I}, \mathrm{CoCM}(\mathcal{I})/\mathcal{I}\big)$ is of finite type, we have that the inclusion $\mathrm{CoCM}(\mathcal{I})/\mathcal{I} \hookrightarrow \mathcal{C}/\mathcal{I}$ preserves coproducts. Hence, as in Lemma III.1.2, its left adjoint $\mathbf{T}$ preserves compact objects, so $\mathbf{T}(\mathcal{T}/\mathcal{I})$ is a set of compact objects in $\mathrm{CoCM}(\mathcal{I})/\mathcal{I}$. Now for any $T \in \mathcal{T}$ and any CoCohen-Macaulay object Z we have: $\mathrm{CoCM}(\mathcal{I})/\mathcal{I}[\mathbf{T}(\overline{T}), \overline{Z}] = 0 \Rightarrow \mathcal{C}/\mathcal{I}[\overline{T}, \overline{Z}] = 0 \Rightarrow \overline{Z} = 0$. We conclude that $\mathbf{T}(\mathcal{T}/\mathcal{I})$ is a compact generating set and therefore $\mathrm{CoCM}(\mathcal{I})/\mathcal{I}$ is compactly generated. Since $\mathrm{CM}(\mathcal{P})/\mathcal{P}$ is triangle equivalent to $\mathrm{CoCM}(\mathcal{I})/\mathcal{I}$, it follows that $\mathrm{CM}(\mathcal{P})/\mathcal{P}$ is compactly generated as well by the set $\mathsf{N}^-\mathbf{T}(\mathcal{T}/\mathcal{I})$.

(iv), (v) Using (i) and Theorem V.3.7, it follows that we have a resolving cotorsion pair $(\mathrm{CM}(\mathcal{P}), \mathcal{Y})$ in $\mathcal{C}$ with $\mathrm{CM}(\mathcal{P}) \cap \mathcal{Y} = \mathcal{P}$, and by Theorem VI.2.1 we have that $\mathcal{Y}$ is resolving and $\mathcal{P}^{<\infty} = \mathcal{Y} \cap \widehat{\mathrm{CM}(\mathcal{P})}$. In particular $\mathrm{CM}(\mathcal{P})$ is resolving and contravariantly finite, and $\mathcal{Y}$ is coresolving. By Lemma V.1.1 and part (ii) we also have that $\mathrm{CM}(\mathcal{P})$ is covariantly finite. The proof of part (v) is dual. □

Note that in the literature the terminology Cohen-Macaulay modules over a ring refers to Cohen-Macaulay modules which are finitely generated. In our working setting of an abelian category with exact infinite (co)products, the above result suggests to call the objects in $\mathrm{CM}(\mathcal{P})$ big Cohen-Macaulay objects. Also the results of Chapters VI and VII suggest that the objects in $\mathcal{Y}$ provide a generalization of the objects of finite projective dimension. Hence it is reasonable to introduce the following terminology and notation.

DEFINITION 2.5. Let $\mathcal{C}$ be an abelian category with infinite exact (co)products. We call the objects in $\mathrm{CM}(\mathcal{P})$ **big Cohen-Macaulay** objects, and the objects in $\mathrm{CoCM}(\mathcal{I})$ **big CoCohen-Macaulay** objects. We call the objects in $\mathcal{P}^{<\propto} := \mathrm{CM}(\mathcal{P})^{\perp}$ the objects with **virtually finite projective dimension**, and the objects in $\mathcal{I}^{<\propto} := {}^{\perp}\mathrm{CoCM}(\mathcal{I})$ the objects with **virtually finite injective dimension**.

In this notation we have a resolving cotorsion pair $\big(\mathrm{CM}(\mathcal{P}), \mathcal{P}^{<\propto}\big)$ and a coresolving cotorsion pair $\big(\mathcal{I}^{<\propto}, \mathrm{CoCM}(\mathcal{I})\big)$ in $\mathcal{C}$ with $\mathcal{P}^{<\infty} \subseteq \mathcal{P}^{<\propto} \supseteq \mathcal{I}^{<\infty}$ and $\mathcal{P}^{<\infty} \subseteq \mathcal{I}^{<\propto} \supseteq \mathcal{I}^{<\infty}$. Moreover we have: $\mathcal{P}^{<\infty} = \mathcal{P}^{<\propto} \cap \widehat{\mathrm{CM}(\mathcal{P})}$ and $\mathrm{CM}(\mathcal{P}) \cap \mathcal{P}^{<\propto} = \mathcal{P}$, and dually $\mathcal{I}^{<\infty} = \mathcal{I}^{<\propto} \cap \widetilde{\mathrm{CoCM}(\mathcal{I})}$ and $\mathrm{CoCM}(\mathcal{I}) \cap \mathcal{I}^{<\propto} = \mathcal{I}$.

We shall use heavily the above (co)torsion pairs in the next section for the construction of Cohen-Macaulay cohomology theories in a Nakayama category.

3. Cohen-Macaulay Cohomology

Throughout this section we fix a Nakayama abelian category with exact products and coproducts, and we assume that $\mathcal{C}$ admits a set of compact projective generators. Let $(\mathsf{N}^+, \mathsf{N}^-)$ be the Nakayama functors of $\mathcal{C}$, and we assume throughout that the Nakayama functor N^- preserves coproducts. The main result of the previous section combined with the results of the previous chapter shows that we can define new (complete) cohomology theories for $\mathcal{C}$, based on Cohen-Macaulay objects. In particular we can define such cohomology theories for any Artin algebra, and more generally for any ring with Morita self-duality.

The following main result of this section, whose proof follows from our previous results, summarizes the basic properties of these new cohomology theories.

THEOREM 3.1. *For any $n \in \mathbb{Z}$, there are bifunctors*

$$\widehat{\mathsf{Ext}}^n_{CM}(-,-) \ : \ \mathcal{C}^{\mathrm{op}} \times \mathcal{C} \longrightarrow \mathcal{A}b,$$

the **projective Cohen-Macaulay** *bifunctor, and*

$$\widetilde{\mathsf{Ext}}^n_{CM}(-,-) \ : \ \mathcal{C}^{\mathrm{op}} \times \mathcal{C} \longrightarrow \mathcal{A}b,$$

the **injective Cohen-Macaulay** *bifunctor. These are defined by*

$$\widehat{\mathsf{Ext}}^n_{CM}(A,B) := \mathcal{C}/\mathcal{P}[\Omega^n\mathbf{R}(\underline{A}), \underline{B}] \ \xrightarrow{\cong} \ \underline{\mathrm{CM}}(\mathcal{P})[\Omega^n\mathbf{R}(\underline{A}), \mathbf{R}(\underline{B})]$$

$$\widetilde{\mathsf{Ext}}^n_{CM}(A,B) := \mathcal{C}/\mathcal{I}[\underline{A}, \Sigma^n\mathbf{T}(\underline{B})] \ \xrightarrow{\cong} \ \overline{\mathrm{CoCM}}(\mathcal{I})[\mathbf{T}(\underline{A}), \Sigma^n\mathbf{T}(\underline{B})]$$

and they satisfy the following properties.

(i) *The functors $\widehat{\mathsf{Ext}}^*_{CM}(-,-)$ and $\widetilde{\mathsf{Ext}}^*_{CM}(-,-)$ are homological in each variable.*

(ii) $\widehat{\mathsf{Ext}}^*_{CM}(A,B)$ *vanishes if A or B is an object of (virtually) finite projective dimension and* $\widetilde{\mathsf{Ext}}^*_{CM}(A,B)$ *vanishes if A or B is an object of (virtually) finite injective dimension.*

(iii) *For any object C in $\mathcal{C}$, the functor* $\widehat{\mathsf{Ext}}^*_{CM}(C,-)$ *is the completion of* $\mathrm{Ext}^*_{\mathcal{C}}(C,-)$ *with respect to the full subcategory $\mathcal{P}^{\propto}$ of objects with virtually finite projective dimension. Moreover* $\widehat{\mathsf{Ext}}^*_{CM}(C,-)$ *is isomorphic to the Eckmann-Hilton projective homotopy functor* $\underline{\Pi}^*_{X_C}$ *of the special right big Cohen-Macaulay approximation X_C of C.*

(iv) *For any object A in $\mathcal{C}$, the functor* $\widetilde{\mathsf{Ext}}^*_{CM}(-,A)$ *is the completion of* $\mathrm{Ext}^*_{\mathcal{C}}(-,A)$ *with respect to the full subcategory $\mathcal{I}^{\propto}$ of objects with virtually finite injective dimension. Moreover* $\widetilde{\mathsf{Ext}}^*_{CM}(-,A)$ *is isomorphic to the Eckmann-Hilton injective homotopy functor* $\overline{\Pi}^*_{Z^A}$ *of the special left big CoCohen-Macaulay approximation Z^A of A.*

(v) *For any object C, the $\mathcal{P}^{<\propto}$-completion morphism* $\zeta^*_{C,-} : \mathrm{Ext}^*_{\mathcal{C}}(C,-) \to \widehat{\mathsf{Ext}}^*_{CM}(C,-)$*, factors uniquely through the $\mathcal{P}$-completion morphism* $\chi^*_{C,-} : \mathrm{Ext}^*_{\mathcal{C}}(C,-) \to \widehat{\mathsf{Ext}}^*_{TV}(C,-)$*. The resulting* **comparison** *morphism of homological functors* $\rho^*_{C,-} : \widehat{\mathsf{Ext}}^*_{TV}(C,-) \to \widehat{\mathsf{Ext}}^*_{CM}(C,-)$ *is invertible if and only if C has finite projective Gorenstein (resolution) dimension.*

(vi) *For any object A, the $\mathcal{I}^{<\propto}$-completion morphism* $\eta^*_{C,-} : \mathrm{Ext}^*_{\mathcal{C}}(-,A) \to \widetilde{\mathsf{Ext}}^*_{CM}(-,A)$*, factors uniquely through the $\mathcal{I}$-completion morphism* $\psi^*_{-,A} : \mathrm{Ext}^*_{\mathcal{C}}(-,A) \to \widetilde{\mathsf{Ext}}^*_{TV}(-,A)$*. The resulting* **comparison** *morphism of homological functors* $\tau^*_{-,A} : \widetilde{\mathsf{Ext}}^*_{TV}(-,A) \to \widetilde{\mathsf{Ext}}^*_{CM}(-,A)$ *is invertible if and only if A has finite injective Gorenstein (coresolution) dimension.*

(vii) *$\mathcal{C}$ is Gorenstein if and only if the comparison morphism*

$$\rho^*_{-,-} : \widehat{\mathsf{Ext}}^*_{TV}(-,-) \longrightarrow \widehat{\mathsf{Ext}}^*_{CM}(-,-)$$

is invertible if and only if the comparison morphism

$$\tau^*_{-,-} : \widetilde{\mathsf{Ext}}^*_{TV}(-,-) \longrightarrow \widetilde{\mathsf{Ext}}^*_{CM}(-,-)$$

is invertible.

(viii) *There exists an infinite long exact sequence*

$$0 \longrightarrow \mathrm{Ext}^1_{\mathrm{CM}(\mathcal{P})}(-,-) \longrightarrow \mathrm{Ext}^1_{\mathcal{C}}(-,-) \longrightarrow \widehat{\mathsf{Ext}}^1_{CM}(-,-) \longrightarrow \cdots$$

$$\cdots \longrightarrow \mathrm{Ext}^n_{\mathrm{CM}(\mathcal{P})}(-,-) \longrightarrow \mathrm{Ext}^n_{\mathcal{C}}(-,-) \longrightarrow \widehat{\mathsf{Ext}}^n_{CM}(-,-) \longrightarrow \cdots$$

where $\mathrm{Ext}^*_{\mathrm{CM}(\mathcal{P})}(-,-)$ *is the relative extension bifunctor induced by the functorially finite subcategory of big Cohen-Macaulay objects.*

(ix) *There exists an infinite long exact sequence*

$$0 \longrightarrow \mathrm{Ext}^1_{\mathrm{CoCM}(\mathcal{I})}(-,-) \longrightarrow \mathrm{Ext}^1_{\mathcal{C}}(-,-) \longrightarrow \widetilde{\mathsf{Ext}}^1_{CM}(-,-) \longrightarrow \cdots$$

$$\cdots \longrightarrow \mathrm{Ext}^n_{\mathrm{CoCM}(\mathcal{I})}(-,-) \longrightarrow \mathrm{Ext}^n_{\mathcal{C}}(-,-) \longrightarrow \widetilde{\mathsf{Ext}}^n_{CM}(-,-) \longrightarrow \cdots$$

where $\mathrm{Ext}^*_{\mathrm{CoCM}(\mathcal{I})}(-,-)$ *is the relative extension bifunctor induced by the functorially finite subcategory of big CoCohen-Macaulay objects.*

REMARK 3.2. Cohen-Macaulay cohomology provides a new cohomology theory for Nakayama categories, in particular for Artin algebras. However when a Nakayama category is Gorenstein, then, by Theorem 3.1, Cohen-Macaulay cohomology degenerates to the well-known Tate-Vogel Cohomology. In particular Cohen-Macaulay cohomology is trivial for Nakayama categories with finite global dimension.

If the abelian category $\mathcal{C}$ is Gorenstein, then by the results of Chapters VI and VII we know that there exist cotorsion pairs $(\mathrm{CM}(\mathcal{P}), \mathcal{P}^{<\infty})$ and $(\mathcal{I}^{\infty}, \mathrm{CoCM}(\mathcal{I}))$ in $\mathcal{C}$ and moreover we have:

$$\mathcal{P}^{<\propto} := \mathrm{CM}(\mathcal{P})^{\perp} = \mathcal{P}^{<\infty} = \mathcal{I}^{<\infty} = {}^{\perp}\mathrm{CoCM}(\mathcal{I}) := \mathcal{I}^{<\propto}$$

Hence it is natural to study the Nakayama categories for which in the cotorsion pairs $(\mathrm{CM}(\mathcal{P}), \mathcal{P}^{<\propto})$ and $(\mathcal{I}^{\propto}, \mathrm{CoCM}(\mathcal{I}))$ we have an equality: $\mathcal{P}^{<\propto} = \mathcal{I}^{<\propto}$. By the above observations, these categories provide a natural generalization of Gorenstein categories, so the following definition and terminology seems to be reasonable. However we don't know if in this way we obtain a strictly larger class of categories.

DEFINITION 3.3. A Nakayama category $\mathcal{C}$ is called **virtually Gorenstein** if we have an equality: $\mathcal{P}^{<\propto} = \mathcal{I}^{<\propto}$.

The following result gives interesting characterizations of virtually Gorenstein categories.

THEOREM 3.4. *For a Nakayama abelian category $\mathcal{C}$ the following are equivalent.*

(i) *$\mathcal{C}$ is virtually Gorenstein.*
(ii) *The projective and injective Cohen-Macaulay complete extension bifunctors are isomorphic:*

$$\widehat{\mathsf{Ext}}^{*}_{CM}(-,-) \xrightarrow{\cong} \widetilde{\mathsf{Ext}}^{*}_{CM}(-,-).$$

(iii) *There exists a cotorsion triple $(\mathrm{CM}(\mathcal{P}), \mathcal{P}^{<\propto} = \mathcal{I}^{<\propto}, \mathrm{CoCM}(\mathcal{I}))$ in $\mathcal{C}$.*
(iv) *$\mathcal{F}_{\mathrm{CM}(\mathcal{P})} = \mathcal{F}^{\mathrm{CoCM}(\mathcal{I})}$. That is, for any short exact sequence $0 \to A \xrightarrow{g} B \xrightarrow{f} C \to 0$ in $\mathcal{C}$, any morphism $X \to C$ with $X \in \mathrm{CM}(\mathcal{P})$ factors through f if and only if any morphism $A \to Z$ with $Z \in \mathrm{CoCM}(\mathcal{I})$ factors through g.*
(v) *The relative extension bifunctors with respect to the subcategories $\mathrm{CM}(\mathcal{P})$ and $\mathrm{CoCM}(\mathcal{I})$, are isomorphic:*

$$\mathrm{Ext}^{*}_{\mathrm{CM}(\mathcal{P})}(-,-) \xrightarrow{\cong} \mathrm{Ext}^{*}_{\mathrm{CoCM}(\mathcal{I})}(-,-).$$

If this is the case then $\mathrm{CM}(\mathcal{P})\text{-gl.}\dim\mathcal{C} = \mathrm{CoCM}(\mathcal{I})\text{-gl.}\dim\mathcal{C}$ and this common dimension is finite if and only if $\mathcal{C}$ is Gorenstein.

PROOF. Obviously (i) is equivalent to (iii). If $\mathcal{C}$ is virtually Gorenstein, then the isomorphism in (ii) follows from Theorem IX.4.2. Conversely if (ii) holds, then working as in the proof of Theorem IX.4.4 we infer that $\mathcal{P}^{<\propto} = \mathcal{I}^{<\propto}$, hence $\mathcal{C}$ is virtually Gorenstein. Now obviously (iv) is equivalent to (v), and (i) implies (v) by Theorem IX.4.5. Finally if (v) holds then $\mathcal{C}$ is virtually Gorenstein by Corollary IX.4.6. □

It is not difficult to see that the category of big CoCohen-Macaulay objects is closed under coproducts. Since the full subcategory of injective objects of $\mathcal{C}$ is closed under coproducts, it follows that the stable category $\mathrm{CoCM}(\mathcal{I})/\mathcal{I}$ is closed under coproducts in the stable category $\mathcal{C}/\mathcal{I}$. Hence, using the terminology of Definition III.1.1, the torsion pair $\big(\mathcal{I}^{<\infty}/\mathcal{I}, \mathrm{CoCM}(\mathcal{I})/\mathcal{I}\big)$ is of finite type. We don't know if the torsion pair $\big(\mathrm{CM}(\mathcal{P})/\mathcal{P}, \mathcal{P}^{<\infty}/\mathcal{P}\big)$ is always of finite type.[1] The following remark, which shows that this happens for virtually Gorenstein categories, gives some interesting consequences of the finite type property.

REMARK 3.5. Let $\mathcal{C}$ be a virtually Gorenstein Nakayama category. Since we have a cotorsion triple $\big(\mathrm{CM}(\mathcal{P}), \mathcal{P}^{<\infty} = \mathcal{I}^{<\infty}, \mathrm{CoCM}(\mathcal{I})\big)$ in $\mathcal{C}$, it follows that $\mathcal{P}^{<\infty} = \mathcal{I}^{<\infty}$ is closed under coproducts. This implies that the torsion-free class $\mathcal{P}^{<\infty}/\mathcal{P}$ is closed under coproducts in $\mathcal{C}/\mathcal{P}$. Hence the torsion pair $\big(\mathrm{CM}(\mathcal{P})/\mathcal{P}, \mathcal{P}^{<\infty}/\mathcal{P}\big)$ is of finite type. By the Note after Lemma III.1.2, this implies that the inclusion functor $\mathbf{i} : \mathrm{CM}(\mathcal{P})/\mathcal{P} \hookrightarrow \mathcal{C}/\mathcal{P}$ preserves compact objects. Hence $(\mathrm{CM}(\mathcal{P})/\mathcal{P})^b \subseteq (\mathcal{C}/\mathcal{P})^b$.

Now let Λ an Artin algebra which is *virtually Gorenstein*, in the sense that the Nakayama category $\mathrm{Mod}(\Lambda)$ is virtually Gorenstein. Then we always have $\underline{\mathrm{mod}}(\Lambda) \subseteq \underline{\mathrm{Mod}}(\Lambda)^b$. Hence if any compact object in the stable category $\underline{\mathrm{Mod}}(\Lambda)$ is induced by a finitely generated module, i.e. we have $\underline{\mathrm{Mod}}(\Lambda)^b \subseteq \underline{\mathrm{mod}}(\Lambda)$, then $\underline{\mathrm{Mod}}(\Lambda)^b = \underline{\mathrm{mod}}(\Lambda)$. Now let $\mathrm{CM}(\mathbf{P}_\Lambda)$ be the category of big Cohen-Macaulay modules and let $\underline{\mathrm{CM}}(\mathbf{P}_\Lambda)$ be the stable category modulo projectives. It follows from the above analysis that $\underline{\mathrm{CM}}(\mathbf{P}_\Lambda)^b \subseteq \underline{\mathrm{CM}}(\mathbf{P}_\Lambda) \cap \underline{\mathrm{mod}}(\Lambda) = \underline{\mathrm{CM}}(\Lambda)$. Since always have $\underline{\mathrm{CM}}(\Lambda) \subseteq \underline{\mathrm{CM}}(\mathbf{P}_\Lambda)^b$, we conclude that $\underline{\mathrm{CM}}(\mathbf{P}_\Lambda)^b = \underline{\mathrm{CM}}(\Lambda)$. This has some important consequences for the Artin algebra; we refer to [**26**] for more information. In particular it is proved in [**26**] that for a virtually Gorenstein algebra Λ, the condition $\underline{\mathrm{Mod}}(\Lambda)^b \subseteq \underline{\mathrm{mod}}(\Lambda)$ implies the existence of a cotorsion triple $\big(\mathrm{CM}(\mathcal{P}_\Lambda), \mathcal{Y}, \mathrm{CM}(\mathrm{D}(\Lambda))\big)$ in $\mathrm{mod}(\Lambda)$, where $\mathcal{Y} = \mathbf{P}_\Lambda^{<\infty} \cap \mathrm{mod}(\Lambda) = \mathbf{I}_\Lambda^{<\infty} \cap \mathrm{mod}(\Lambda)$. We don't know if any compact object in the stable category $\underline{\mathrm{Mod}}(\Lambda)$ is induced by a finitely generated module. By [**26**] this is true for Gorenstein algebras.

We close the paper with an application of the above results to closed model structures. Let $\mathcal{C}$ be a Nakayama abelian category. Since the categories $\mathcal{P}$ and $\mathcal{I}$ are functorially finite, the results of Chapter VII can be applied for the cotorsion pairs produced by Theorem 2.4. It follows that there are induced four compatible model structures on $\mathcal{C}$. For possible future use and for the convenience of the reader we write down explicitly the closed model structures; we note only that the adjoint pair $(\mathsf{N}^+, \mathsf{N}^-)$ of Nakayama functors of $\mathcal{C}$ gives a Quillen equivalence, in the sense of [**64**], between the contravariant projective CM-closed model structure and the covariant injective CM-closed model structure.

[1] Added in proof: For a detailed analysis of this and related questions, for instance those of REMARK 3.5, in the context of Artin algebras, we refer the interested reader to [**26**].

1. The contravariant projective CM-closed model structure: $(\mathfrak{C}_{\mathcal{P}}, \mathfrak{F}_{\mathcal{P}}, \mathfrak{W}_{\mathcal{P}})$

Here $\mathfrak{C}_{\mathcal{P}}$ is the class of monics with cokernel a big Cohen-Macaulay object. $\mathfrak{T}\mathfrak{C}_{\mathcal{P}}$ is the class of split monics with cokernel a projective object. $\mathfrak{F}_{\mathcal{P}}$ is the class of epimorphisms. $\mathfrak{T}\mathfrak{F}_{\mathcal{P}}$ is the class of epimorphisms with kernel an object of virtually finite projective dimension. Finally $\mathfrak{W}_{\mathcal{P}}$ is the class of morphisms $A \to B$ which induce an isomorphism modulo projectives between the special right big Cohen-Macaulay approximations of A and B. For the closed model structure $(\mathfrak{C}_{\mathcal{P}}, \mathfrak{F}_{\mathcal{P}}, \mathfrak{W}_{\mathcal{P}})$, $\mathrm{CM}(\mathcal{P})$ is the class of cofibrant objects, all objects are fibrant, the trivially cofibrant objects are the projective objects, and the objects of virtually finite projective dimension are the trivially fibrant objects. The associated Quillen homotopy category of $(\mathfrak{C}_{\mathcal{P}}, \mathfrak{F}_{\mathcal{P}}, \mathfrak{W}_{\mathcal{P}})$ is the stable triangulated category $\mathrm{CM}(\mathcal{P})/\mathcal{P}$.

2. The covariant projective CM-closed model structure: $(\mathfrak{C}^{\mathcal{P}}, \mathfrak{F}^{\mathcal{P}}, \mathfrak{W}^{\mathcal{P}})$

Here $\mathfrak{C}^{\mathcal{P}}$ is the class of $\mathcal{P}$-monics. $\mathfrak{T}\mathfrak{C}^{\mathcal{P}}$ is the class of monics with cokernel a big Cohen-Macaulay object. $\mathfrak{F}^{\mathcal{P}}$ is the class of epimorphisms with kernel an object with virtually finite projective dimension. $\mathfrak{T}\mathfrak{F}^{\mathcal{P}}$ is the class of split epimorphisms with kernel a projective object. Finally $\mathfrak{W}^{\mathcal{P}}$ is the class of morphisms $A \to B$ which induce an isomorphism modulo projectives, between the special left $\mathcal{P}^{<\infty}$-approximations of A and B. For the closed model structure $(\mathfrak{C}^{\mathcal{P}}, \mathfrak{F}^{\mathcal{P}}, \mathfrak{W}^{\mathcal{P}})$, all objects are cofibrant, $\mathrm{CM}(\mathcal{P})$ is the class of trivially cofibrant objects, the trivially fibrant objects are the projective objects, and the fibrant objects are the objects of virtually finite projective dimension. The associated Quillen homotopy category of $(\mathfrak{C}^{\mathcal{P}}, \mathfrak{F}^{\mathcal{P}}, \mathfrak{W}^{\mathcal{P}})$ is the stable pretriangulated category $\mathcal{P}^{<\infty}/\mathcal{P}$.

3. The contravariant injective CM-closed model structure: $(\mathfrak{C}_{\mathcal{I}}, \mathfrak{F}_{\mathcal{I}}, \mathfrak{W}_{\mathcal{I}})$

Here $\mathfrak{C}_{\mathcal{I}}$ is the class of monics with cokernel an object of virtually finite injective dimension. $\mathfrak{T}\mathfrak{C}_{\mathcal{I}}$ is the class of split monics with injective cokernel. $\mathfrak{F}_{\mathcal{I}}$ is the class of $\mathcal{I}$-epics. $\mathfrak{T}\mathfrak{F}_{\mathcal{I}}$ is the class of epics with kernel a big CoCohen-Macaulay object. Finally $\mathfrak{W}_{\mathcal{I}}$ is the class of morphisms $A \to B$ which induce an isomorphism modulo injectives, between the special left $\mathcal{I}^{<\infty}$-approximations of A and B. For the closed model structure $(\mathfrak{C}_{\mathcal{I}}, \mathfrak{F}_{\mathcal{I}}, \mathfrak{W}_{\mathcal{I}})$, the cofibrant objects are the objects of virtually finite injective dimension, all objects are fibrant, the trivially cofibrant objects are the injectives, and the trivially fibrant objects are the big CoCohen-Macaulay objects. The associated Quillen homotopy category of $(\mathfrak{C}_{\mathcal{I}}, \mathfrak{F}_{\mathcal{I}}, \mathfrak{W}_{\mathcal{I}})$ is the stable pretriangulated category $\mathcal{I}^{<\infty}/\mathcal{I}$.

4. The covariant injective CM-closed model structure: $(\mathfrak{C}^{\mathcal{I}}, \mathfrak{F}^{\mathcal{I}}, \mathfrak{W}^{\mathcal{I}})$

Here $\mathfrak{C}^{\mathcal{I}}$ is the class of monics. $\mathfrak{T}\mathfrak{C}^{\mathcal{I}}$ is the class of monics with cokernel an object of virtually finite injective dimension. $\mathfrak{F}^{\mathcal{I}}$ is the class of epimorphisms with kernel a big CoCohen-Macaulay object. $\mathfrak{T}\mathfrak{F}^{\mathcal{I}}$ is the class of split epimorphisms with injective kernel. Finally $\mathfrak{W}^{\mathcal{I}}$ is the class of morphisms $A \to B$ which induce an isomorphism modulo injectives, between the special left big CoCohen-Macaulay approximations of A and B. For the closed model structure $(\mathfrak{C}^{\mathcal{I}}, \mathfrak{F}^{\mathcal{I}}, \mathfrak{W}^{\mathcal{I}})$, all objects are cofibrant, $\mathrm{CoCM}(\mathcal{I})$ is the class of fibrant objects, the trivially fibrant objects are the injective objects, and the trivially cofibrant objects are the objects of virtually finite injective dimension. The associated Quillen homotopy category of $(\mathfrak{C}^{\mathcal{I}}, \mathfrak{F}^{\mathcal{I}}, \mathfrak{W}^{\mathcal{I}})$ is the stable triangulated category $\mathrm{CoCM}(\mathcal{I})/\mathcal{I}$.

Bibliography

[1] L. Alonso Tarrío, A. Jeremías López and M.J. Souto Salorio, *Localization in categories of complexes and unbounded resolutions*, Canad. J. Math. **52** (2000), no. 2, 225–247.

[2] L. Alonso Tarrío, A. Jeremías López and M.J. Souto Salorio, *Construction of t-Structures and Equivalences of Derived Categories*, Trans. Amer. Math. Soc. **355** (2003), no. 6, 2523–2543.

[3] I. Assem, A. Beligiannis and N. Marmaridis, *Right Triangulated Categories with Right Semi-equivalence*, in: Algebras and Modules II (Geiranger 1996), CMS Conf. Proc. **24** Amer. Math. Soc. Providence, RI, (1998), 17–37.

[4] M. Auslander, *Coherent Functors*, in: Proceedings of the Conference on Categorical Algebra, La Jolla (1966), 189–231.

[5] M. Auslander and R.O. Buchweitz, *The Homological Theory of Maximal Cohen-Macaulay Approximations*, Mem. Soc. Math. de France **38** (1989), 5–37.

[6] M. Auslander and M. Bridger, *Stable Module Theory*, Memoirs of A.M.S. **94** (1969).

[7] M. Auslander and I. Reiten, *Stable Equivalence of Dualizing R-Varieties*, Advances in Math. **12** (1974), 306–366.

[8] M. Auslander and I. Reiten, *On a Generalized Version of the Nakayama Conjecture*, Proc. Amer. Math. Soc. **52** (1975), 69–74.

[9] M. Auslander and I. Reiten, *Applications of Contravariantly Finite subcategories*, Advances in Math. **86** (1991), 111–152.

[10] M. Auslander and I. Reiten, *Homologically Finite Subcategories*, Proc. ICRA IV, (1992), 1–37.

[11] M. Auslander and I. Reiten, *Cohen-Macaulay and Gorenstein Artin Algebras*, Progress in Mathematics **95** (1991), 221–245.

[12] M. Auslander and I. Reiten, *k-Gorenstein Algebras and Syzygy Modules* J. Pure Appl. Algebra **92** (1994), no. 1, 1–27.

[13] M. Auslander and S.O. Smalø, *Preprojective Modules over Artin Algebras*, J. Algebra **66** (1980), 61–122.

[14] M. Auslander and O. Solberg, *Relative Homology and Representation Theory I: Relative Homology and Homologically Finite Subcategories*, Comm. in Alg., **21**(9), (1993), 2995–3031.

[15] M. Auslander, I. Reiten and S. Smalø, *Representation Theory of Artin Algebras*, Cambridge University Press, (1995).

[16] L. Avramov and A. Martsinkovsky, *Absolute, Relative, and Tate Cohomology of Modules of Finite Gorenstein Dimension*, Proc. London Math. Soc. (3) **85** (2002), no. 2, 393–440.

[17] El Bashir, L. Bican and E. Enochs, *All Modules have Flat Covers*, Bull. London Math. Soc. **33** (2001), no. 4, 385–390.

[18] A. Beilinson, J. Bernstein and P. Deligne, *Faisceaux Pervers*, (French) [Perverse sheaves], Analysis and topology on singular spaces, I (Luminy, 1981), 5–171, Astrisque, **100**, Soc. Math. France, Paris, (1982).

[19] A. Beligiannis and N. Marmaridis, *Left Triangulated Categories Arising from Contravariantly Finite Subcategories*, Comm. in Algebra **22** (1994), 5021–5036.

[20] A. Beligiannis, *The Homological Theory of Contravariantly Finite Subcategories: Auslander-Buchweitz Contexts, Gorenstein Categories and (Co-)Stabilizations*, Comm. Algebra **28**(10), (2000), 4547–4596.

[21] A. BELIGIANNIS, *Relative Homological Algebra and Purity in Triangulated Categories*, J. Algebra **227** (2000), 268–361.

[22] A. BELIGIANNIS, *On the Relative Homology of Cleft Extensions of Rings and Abelian Categories*, J. Pure Appl. Algebra **150** (2000), 237–299.

[23] A. BELIGIANNIS, *On the Freyd Categories of an Additive Category*, Homology Homotopy Appl. **2** (2000), 147-185.

[24] A. BELIGIANNIS, *Homotopy Theory of Modules and Gorenstein Rings*, Math. Scand. **88** (2001), 5–45.

[25] A. BELIGIANNIS, *Auslander-Reiten Triangles, Ziegler Spectra and Gorenstein Rings*, K-Theory **32** (2004), 1–82.

[26] A. BELIGIANNIS, *Cohen-Macaulay Modules, (Co)Torsion Pairs, and Virtually Gorenstein Algebras*, J. Algebra Vol. **288**, No.1, (2005), 137–211.

[27] A. BELIGIANNIS, *Tilting Theory in Abelian Categories and Closed Model Structures*, preprint, University of the Aegean, (2002).

[28] D.J. BENSON, *Complexity and Varieties for Infinite Groups I and II*, J. Algebra **193** (1997), 260–287 and 288–317.

[29] D.J. BENSON, *Cohomology of modules in the principal block of a finite group*, New York J. Math. **1** (1999), 196–205.

[30] D.J. BENSON AND J. CARLSON, *Products in Negative Cohomology*, J. Pure Appl. Algebra **82** (1992), 107–130.

[31] M. BOECKSTEDT AND A. NEEMAN, *Homotopy Limits in Triangulated Categories*, Compositio Math. **86** (1993), 209–234.

[32] A.I. BONDAL, *Representations of Associative Algebras and Coherent Sheaves*, Izv. Akad. Nauk SSSR **53** (1989), 25–44. English transl. in Math. USSR Izv. **34** (1990).

[33] A.I. BONDAL AND M.M. KAPRANOV, *Representable Functors, Serre Functors, and Reconstructions*, (Russian) Izv. Akad. Nauk SSSR Ser. Mat. **53** (1989), no. 6,1183–1205, 1337; translation in Math. USSR-Izv. **35** (1990), (No. 3), 519–541.

[34] F. BORCEUX, *Handbook of Categorical Algebra 1*, Encyclopedia of Mathematics and its Applications **Vol. 50**, Cambridge University Press (1994).

[35] K. BROWN, *Cohomology of Groups*, Graduate Texts in Math. **87**, Springer, Berlin, (1982).

[36] A. BUAN AND O. SOLBERG, *Relative Cotilting Theory and Almost Complete Cotilting Modules*, in: Algebras and Modules II (Geiranger 1996), CMS Conf. Proc. **24** Amer. Math. Soc. Providence, RI, (1998), 77–92.

[37] R.O. BUCHWEITZ, *The Comparison Theorem*, appendix to *Cohen-Macualay Modules on Quadrics* by R.O. Buchweitz, D. Eisenbud and J. Herzog. Singularities, Representation of Algebras, and Vector Bundles, Lecture Notes in Math. **1273** Springer (1987), 96–116.

[38] H. CARTAN AND S. EILENBERG, *Homological Algebra*, Princeton University Press, Princeton, NJ, 1999. xvi+390 pp.

[39] E. CLINE, B. PARSHALL AND L. SCOTT, *Finite-Dimensional Algebras and Highest Weight Categories*, J. Reine Angew. Math. **351** (1988), 88–99.

[40] J. CORNICK AND P.H. KROPHOLLER, *Homological Finiteness Conditions for Modules over Group Algebras*, J. London Math. Soc. **58** (1998), 49–62.

[41] S.E. DICKSON, *A Torsion Theory for Abelian Categories*, Trans. Am. Math. Soc. **121** (1966), 223–235 .

[42] V. DLAB AND C.M. RINGEL, *The Module Theoretic Approach to Quasi-hereditary Algebras*, Proc. ICRA IV, (1992), 200–224.

[43] W.G. DWYER AND J.P.C. GREENLESS, *Complete modules and Torsion Modules*, Amer. J. Math. **124** (2002), no. 1, 199–220.

[44] S. EILENBERG AND J.C. MOORE, *Foundations of Relative Homological Algebra*, Memoirs Amer. Math. Soc. **55**, (1965).

[45] P. EKLOF AND J. TRLIFAJ, *How to Make* Ext *Vanish*, Bull. London Math. Soc. **33** (No.1) (2001), 41–51.

[46] E. ENOCHS, *Injective and Flat Covers and Resolvents*, Israel Journal Math. **39**,(1981), 189–209.

[47] E.E. ENOCHS AND O.M.G. JENDA, *Relative homological algebra*, de Gruyter Expositions in Mathematics, **30**, (2000), xii+339 pp.

[48] R. FOSSUM, P. GRIFFITH AND I. REITEN, *Trivial Extensions of Abelian Categories with Applications to Ring Theory*, Springer L.N.M. **456**, (1975).

[49] P. GABRIEL, *Des Catégories Abeliénnes*, Bull. Soc. Math. France **90**, (1962), 323–448.

[50] P. GABRIEL AND M. ZISMAN, *Calculus of Fractions and Homotopy Theory*, Springer, (1967).

[51] T.V. GEDRICH AND K.W. GRUENBERG, *Complete Cohomological Functors on Groups*, Topol. Appl. **25**, (1987), 203–223.

[52] CH. GEISS AND I. REITEN, *Gentle Algebras are Gorenstein*, Representations of algebras and related topics, 129–133, Fields Inst. Commun., **45**, Amer. Math. Soc., Providence, RI, (2005).

[53] R. GENTLE, *T.T.F. Theories in Abelian Categories*, Comm. Algebra **16**, (1996), 877–908.

[54] R. GENTLE AND G. TODOROV, *Approximations, Adjoint Functors and Torsion Theories*, CMS Conf. Proc. **14** (1993), 205–219.

[55] F. GOICHOT, *Homologie de Tate-Vogel Equivariante*, J. Pure Appl. Algebra **82**, (1992), 39–64.

[56] J. GOLAN, *Torsion Theories*, Longman Scientific Technical, Harlow, (1986).

[57] D. HAPPEL, *Triangulated Categories in the Representation Theory of Finite-Dimensional Algebras*, London Mathematical Society Lecture Note Series, **119**. Cambridge University Press, Cambridge, 1988. x+208 pp.

[58] D. HAPPEL, *Auslander-Reiten Triangles in Derived Categories of Finite-dimensional Algebras*, Proc. Amer. Math. Soc. 112 (1991), (No. 3), 641–648.

[59] D. HAPPEL, *A Characterization of Hereditary Categories with Tilting Object*, Invent. Math. **144** (No. 2) (2001), 381–398.

[60] D. HAPPEL, I. REITEN AND S. SMALØ, *Tilting in Abelian Categories and Quasitilted Algebras*, Memoirs Amer. Math. Soc. **575**, (1996).

[61] P.J. HILTON, *Homotopy Theory And Duality*, Gordon and Breach, (1965).

[62] PH. S. HIRSCHHORN, *Model categories and their localizations*. Mathematical Surveys and Monographs, **99**. American Mathematical Society, Providence, RI, 2003. xvi+457 pp.

[63] M. HOSHINO, Y. KATO AND J.-I. MIYASHI, *On t-Structures and Torsion Theories induced by Compact Objects*, J. Pure Appl. Algebra **167** (2002), no. 1, 15–35.

[64] M. HOVEY, *Model Categories*, Mathematical Surveys and Monographs, vol **63**, American Mathematical Society, Providence, RI, (1998).

[65] M. HOVEY, *Cotorsion Theories, Closed Model Structures, and Representation Theory*, Math. Z. **241** (2002), no. 3, 553–592.

[66] C.U. JENSEN, *On the vanishing of* $\varprojlim^{(i)}$, J. Algebra **15** (1970), 151–166.

[67] P. JØRGENSEN, *Spectra of Modules*, J. Algebra **244** (2001), no. 2, 744–784.

[68] B. KELLER, *Chain Complexes and Stable Categories*, Manuscr. Math. **67** (1990), 379–417.

[69] B. KELLER, *Deriving DG Categories*, Ann. Scient. Ec. Norm. Sup. **27** (1994), 63–102.

[70] B. KELLER, *A Remark on the Generalized Smashing Conjecture*, Manuscripta Math. **84** (1994), 193–198.

[71] B. KELLER, *Derived Categories and their Uses*, Handbook of algebra, Vol. 1, North-Holland, Amsterdam, (1996), 671–701.

[72] B. KELLER, *On the Construction of Derived Equivalences*, in S. KÖNIG AND A. ZIMMERMAN, *Derived Equivalences for Group Rings*, Springer L.N.M. **1685**, Springer, (1998).

[73] B. KELLER, *A-Infinity Algebras in Representation Theory*, Representations of algebras. Vol. I, II, 74–86, Beijing Norm. Univ. Press, Beijing, 2002.

[74] B. KELLER AND D. VOSSIECK, *Sous les Catégories Dérivées*, C. R. Acad. Sci. Paris **305** (1987), 225–228.

[75] B. KELLER AND D. VOSSIECK, *Aisles in Derived Categories*, Bull. Math. Belg. **40** (1988), 239–253.

[76] H. KRAUSE, *Stable equivalence and representation type*, in: I. Reiten, S. Smalø. O. Solberg (eds.) Algebras and modules II, CMS Conference Proceedings, **24** (1998), 387-391.

[77] H. KRAUSE, *Smashing Subcategories and the Telescope Conjecture*, Inventiones Math. **139**, (2000), 99–133.

[78] H. KRAUSE, *The Spectrum of a Module Category*, Habilitatsionsschrift, University of Bielefeld, (1998). Mem. Amer. Math. Soc. **149** (2001), no. 707, x+125 pp.

[79] H. KRAUSE AND O. SOLBERG, *Applications of Cotorsion Pairs*, J. London Math. Soc. (2) **68** (2003), no. 3, 631–650.

[80] N. J. KUHN, *Generic Representation Theory of the Finite Linear Group and the Steenrod Algebra: II*, K-Theory **8** (1994), 395–428.

[81] H.R. MARGOLIS, *Spectra and the Steenrod Algebra*, North Holland Mathematical Library **29**, North Holland, (1983).

[82] H. MILLER, *Finite Localizations*, Boletin de la Sociedad Matematica Mexicana **37** (1992), 383–390.

[83] G. MISLIN, *Tate Cohomology for Arbitrary Groups via Satellites*, Topology Appl. **56** (1994), 293–300.

[84] A. NEEMAN, *The Chromatic Tower of $D(R)$*, Topology **31** (1992), 519–532.

[85] A. NEEMAN, *The Grothendieck Duality Theorem via Bousfield's Techniques and Brown Representability*, Journal A.M.S. **9** (1996), no. 1, 205–236.

[86] A. NEEMAN, *Triangulated Categories*, Annals of Mathematical Studies, Princeton University Press, (2001), 449pp.

[87] B. NUCINKIS, *Complete Cohomology for Arbitrary Rings Using Satellites*, J. Pure Appl. Algebra **131** (1998), no. 3, 297–318.

[88] D. QUILLEN, *Homotopical Algebra*, Springer Lecture Notes in Math. **43**, (1967).

[89] I. REITEN AND M. VAN DEN BERGH, *Hereditary Noetherian Categories Satisfying Serre Duality*, J. Amer. Math. Soc. **15**, No.2, (2002), 295–366.

[90] I. REITEN AND M. VAN DEN BERGH, *Grothendieck Groups and Tilting Objects*, Algebras and Representation Theory, **4** (No. 1), (2001), 1–23.

[91] J. RICKARD, *Morita Theory of Derived Categories*, J. London Math. Soc. **39** (1989), 436–456.

[92] J. RICKARD, *Derived Categories and Stable Equivalence*, J. Pure Appl. Algebra **61** (1989), No. 13, 303–317.

[93] J. RICKARD, *Derived equivalences as Derived Functors*, J. London Math. Soc. (2) **43** (1991), no. 1, 37–48.

[94] J. RICKARD, *Idempotent Modules in the Stable Category*, J. London Math. Soc. **56** (1997), 149–170.

[95] J. RICKARD AND A. SCHOFIELD, *Cocovers and Tilting Modules*, Math. Proc. Camb. Philos. Soc. **106** (1989), 1–5.

[96] C.M. RINGEL, *The Category of Good Modules over a Quasihereditary Algebra has Almost Split Sequences*, Math. Zeit. **208** (1991), 209–225.

[97] L. SALCE, *Cotorsion Theories for Abelian Groups*, (English) Gruppi abeliani e loro relazioni con la teoria dei moduli, Conv. Roma (1977), Symp. math. **23**, (1979), 11–32.

[98] H. SCHUBERT, *Categories*, Springer-Verlag, Berlin, (1973).

[99] D. SIMSON AND A. TYC, *Brown's Theorem for Cohomology Theories on Categories of Chain Complexes*, Ann. Soc. Math. Pol. **XVIII** (1975), 284–296.

[100] B. STENSTRÖM, *Ring of Quotients*, Die Grund. der Math. Wissen. **217**, Springer-Verlag, Berlin, (1975).

[101] J. TRLIFAJ, *Cotorsion Theories induced by Tilting or Cotilting Modules*, Abelian groups, rings and modules (Perth, 2000), 285–300, Contemp. Math., **273**, Amer. Math. Soc., Providence, RI, 2001.

[102] J.L. VERDIER, *Des Catégories Dérives des Catégories Abéliennes*, (French. French summary) [On Derived Categories of Abelian Categories] With a preface by Luc Illusie. Edited and with a note by Georges Maltsiniotis. Astérisque No. **239**, (1996), xii+253 pp. (1997).

[103] T. WAKAMATSU, *Stable Equivalence for Self-injective Algebras and a Generalization of Tilting Modules*, J. Algebra **134** (1990), 298–325.

[104] C. WEIBEL, *An Introduction to Homological Algebra*, Cambridge University Press, (1994).

[105] Q.-S. WU AND J.J. ZHANG, *Noetherian PI Hopf Algebras are Gorenstein*, Trans. Amer. Math. Soc. **355** (2003), no. 3, 1043–1066.

[106] J. XU, *Flat Covers of Modules*, Lecture Notes in Mathematics **1634**, Springer-Verlag, Berlin, 1996. x+161 pp.

Index

$(\mathcal{T}^{\leq 0}, \mathcal{T}^{\geq 0})$, 17
$(\mathcal{X}, \mathcal{Y}, \mathcal{Z})$, 10, 15, 32
$(\mathcal{X}_{\mathcal{P}}, \mathcal{Y}_{\mathcal{P}})$, 45
$(\mathfrak{C}, \mathfrak{F}, \mathfrak{W})$, 132
$(\mathfrak{C}^{\omega}, \mathfrak{F}^{\omega}, \mathfrak{W}^{\omega})$, 151
$(\mathfrak{C}_{\omega}, \mathfrak{F}_{\omega}, \mathfrak{W}_{\omega})$, 146
$G \rtimes \mathcal{C}$, 125
$\mathcal{C}/\mathcal{X}$, 25
$\mathcal{C} \ltimes F$, 125
$\mathcal{C}^b$, 14
$\mathrm{Ext}^n_{\omega}(-, C)$, 170
$\mathrm{FID}(\mathcal{C})$, 118
$\mathrm{FPD}(\mathcal{C})$, 118
$\mathcal{F}^{\mathcal{Z}}$, 184
$\mathcal{F}_{\mathcal{X}}$, 177
$\mathcal{F}_{\omega}$, 169
$\text{G-dim}_{\mathcal{I}}\, \mathcal{C}$, 119
$\text{G-dim}_{\mathcal{P}}\, \mathcal{C}$, 119
$\mathrm{Prod}(T)$, 27
$\mathcal{U} \star \mathcal{V}$, 12
$\mathcal{U}^{\star n}$, 48
$\mathcal{X}(\mathcal{T}), \mathcal{Y}(\mathcal{F})$, 20
$\mathcal{X}$-res. dim $\mathcal{C}$, 117
$\mathcal{X}^{\perp}$, 8, 84
$\mathcal{X}_{\omega}$, 99
$\mathcal{Y}$-completion, 171
$\mathcal{Y}$-cores. dim $\mathcal{C}$, 117
$\mathrm{add}(T)$, 20
$\big(\mathfrak{C}(\omega), \mathfrak{F}(\omega), \mathfrak{W}(\omega)\big)$, 153
$\underrightarrow{\mathrm{holim}}$, 46
$\underleftarrow{\mathrm{holim}}$, 74
$\mathcal{E}_{\mathcal{P}}(\mathcal{C}), \mathcal{E}_{\mathcal{I}}(\mathcal{C})$, 190
$\mathcal{H}(\mathcal{P})$, 50
$\mathcal{H}(\mathbf{P}_{\Lambda})$, 63
$\mathcal{H}(\mathrm{Mod}(\Lambda))$, 62
$\mathcal{H}^b(\mathcal{P}_{\Lambda})$, 19
$\mathcal{H}_{\mathsf{Ac}}(\mathbf{P}_{\Lambda})$, 63
$\mathcal{H}_{\mathsf{Ac}}(\mathrm{Mod}(\Lambda))$, 62
$\mathcal{H}_{\mathsf{P}}(\mathrm{Mod}(\Lambda)), \mathcal{H}^{\mathsf{I}}(\mathrm{Mod}(\Lambda))$, 63
$\mathcal{I}^{<\alpha}$, 194
$\mathcal{I}(\mathfrak{C})$, 137
$\mathcal{I}^{<\infty}$, 96
$\mathcal{I}^{<\infty}(\mathcal{C})$, 38
$\mathcal{I}^{<\infty}_{\Lambda}$, 108
$\mathcal{I}_{\Lambda}$, 26
$\mathcal{P}(\mathfrak{F})$, 137
$\mathcal{P}^{<\infty}$, 96
$\mathcal{P}^{<\infty}(\mathcal{C})$, 38
$\mathcal{P}^{<\alpha}$, 194
$\mathcal{P}_{\Lambda}$, 26
$\mathcal{T}(\mathcal{C})$, 37
$\mathcal{T}_l(\mathcal{C})$, 38
$\mathcal{T}_r(\mathcal{C})$, 38
$\mathfrak{C}^{\omega}$, 150
$\mathfrak{C}_{\omega}$, 144
$\mathfrak{F}^{\omega}$, 150
$\mathfrak{F}_{\omega}$, 144
$\mathfrak{T}\mathfrak{C}^{\omega}$, 150
$\mathfrak{T}\mathfrak{C}_{\omega}$, 144
$\mathfrak{T}\mathfrak{F}^{\omega}$, 151
$\mathfrak{T}\mathfrak{F}_{\omega}$, 144
$\mathfrak{W}^{\omega}$, 151
$\mathfrak{W}_{\omega}$, 144
$\mathrm{Add}(T)$, 27
$\mathrm{CM}(\omega)$, 95
$\mathrm{CoCM}(\omega)$, 95
$\mathrm{Ext}^*_{\mathrm{CM}(\mathcal{P})}(-, -)$, 195
$\mathrm{Ext}^*_{\mathrm{CoCM}(\mathcal{I})}(-, -)$, 195
$\mathrm{K}_0(\mathcal{C})$, 42
$\mathrm{K}_0^{\Delta}(\mathcal{C})$, 41
$\mathrm{K}_0^{\nabla}(\mathcal{C})$, 41
$\mathrm{Loc}^+(\mathcal{P})$, 48
$\mathrm{L}_n\mathsf{N}^+, \mathrm{R}^n\mathsf{N}^-$, 188
$\mathrm{Rapp}_{\omega}(\mathcal{C})$, 170
Cof, 133
$\mathsf{D}_{\mathbb{Q}/\mathbb{Z}}(P)$, 54
Fib, 133
$\mathsf{Ho}(\mathcal{C})$, 145
$\mathsf{N}^+, \mathsf{N}^-$, 187
Sp, 192
TCof, 133
TFib, 133

$\mathsf{spli}(\mathcal{C})$, $\mathsf{silp}(\mathcal{C})$, 119
$\mathsf{thick}(\mathcal{P})$, 66
ω_c, 133
ω_f, 133
$\overline{\Pi}^*_C(-)$, 170
$\overline{\mathrm{Mod}}(\Lambda)$, 27
$\overline{\mathrm{mod}}(\Lambda)$, 27
τ^+, τ^-, 189
$\underline{\Pi}^n_C(-)$, 170
$\underline{\mathrm{Mod}}(\Lambda)$, 27
$\underline{\mathrm{mod}}(\Lambda)$, 26
$\widehat{\mathcal{X}}$, 38
$\widehat{\mathsf{Ext}}^*_{TV}(-,-)$, 169
$\widehat{\mathsf{Ext}}^n_{(\mathcal{X},\mathcal{Y})}(-,-)$, 164
$\widehat{\mathsf{Ext}}^n_{\mathcal{C}}(-,-)$, 165
$\widehat{\mathsf{Ext}}^n_{CM}(-,-)$, 194
$\widetilde{\mathcal{Y}}$, 38
$\widetilde{\mathsf{Ext}}^n_{(\mathcal{X},\mathcal{Y})}(-,-)$, 164
$\widetilde{\mathsf{Ext}}^n_{\mathcal{C}}(-,-)$, 165
$\widetilde{\mathsf{Ext}}^n_{CM}(-,-)$, 194
$s(M)$, 76
$\mathbf{D}(\mathcal{C})$, 19
$\mathbf{D}(\mathrm{Mod}(\Lambda))$, 14
$\mathbf{D}^b(\mathcal{C})$, 19
$\mathbf{I}_\Lambda$, 27
$\mathbf{P}_\Lambda$, 27
$\mathbf{R}, \mathbf{i}, \mathbf{L}, \mathbf{j}, \mathbf{S}, \mathbf{k}, \mathbf{T}$, 15
${}^\perp\mathcal{Y}$, 8, 84
$\mathcal{X}$-epic, 26
$\mathcal{X}$-monic, 26

$\mathcal{A}$-pd , 174

algebra
  Gorenstein
    virtually, 197
approximation
  cofibrant, 134
  fibrant, 134
  left, 25
    special, 87
  minimal left, 28
  minimal right, 28
  right, 25
    special, 87
  trivially cofibrant, 134
  trivially fibrant, 134
Auslander-Reiten operators, 189
axiom
  factorization, 133
  lifting, 133
  retract, 132
  two out of three, 132

category
  closed model, 133
  Cohen-Macaulay, 127
  costabilization, 63
  Frobenius, 68
  Gorenstein, 121
    $\mathcal{I}$-, 119
    $\mathcal{P}$-, 119
  Grothendieck, 14
  homotopy, 145
  Krull-Schmidt, 29
  left triangulated, 22
    Grothendieck group of, 41
  Nakayama, 186
  of n-extensions, 48
  of extensions, 12
  pretriangulated, 24
  right triangulated, 23
    Grothendieck group of, 41
  stabilization, 37
  stable, 25
  triangulated
    compactly generated, 14
closed model
  structure, 132
    coresolving, 159
    Frobenius, 152
    functorial, 153
    functorial ω-, 153
    injective ω-, 151
    projective, 150
    projective ω-, 148
    resolving, 159
    stable, 148
  structures
    compatible, 154
cofibration, 132
  trivial, 132
complete extension bifunctor
  projective, 168
complex
  homotopically injective, 63
  homotopically projective, 63
cotorsion
  (-free) class, 89
  pair, 88
    coresolving, 91
    resolving, 91
  triple, 108

dimension
  $\mathcal{X}$-resolution, 117
  $\mathcal{Y}$-coresolution, 117
  Gorenstein
    $\mathcal{I}$-, 119
    $\mathcal{P}$-, 119
  injective

finitistic, 118
projective
finitistic, 118
dualizing
adjoint pair, 127
bimodule, 131

extension bifunctor
Cohen-Macaulay
injective, 194
projective, 194
injective, 164
Tate-Vogel, 165
projective, 164
Tate-Vogel, 165, 169

fibration, 132
trivial, 132
functor
$\mathcal{Y}$-complete, 170
ω-homological, 169
t-exact, 57
cohomological, 23
exact, 24
homological, 23
left exact, 23
loop, 22
Nakayama, 27, 187
right exact, 23
spectrification, 192
stabilization, 37
suspension, 23

generating set, 14
glueing condition, 32
good
pair, 84
triple, 108
Gorenstein
algebra, 71
category
virtually, 196
dimension, 96
ring, 123

heart, 17
left, 34
right, 34
homology theory
$\mathcal{Y}$-complete, 166
left, 166
right, 166
homotopy
colimit, 46
limit, 74
homotopy functor
injective
Eckmann-Hilton, 170
projective
Eckmann-Hilton, 170

injective
ω-cofibration, 151
ω-fibration, 151
ω-trivial cofibration, 151
ω-trivial fibration, 151
ω-weak equivalence, 151

lifting of a commutative square, 134

minimal morphism
left, 28
right, 28
module
basic, 157
cotilting, 20
strong, 131
product-complete, 27, 128
tilting, 20
partial, 57
strong, 131
Wakamatsu (co)tilting, 71
morphism of pretriangulated categories, 24

object
CoCohen-Macaulay, 95
big, 194
cofibrant, 133
trivially, 133
Cohen-Macaulay, 95
big, 194
compact, 14
dual, 54
fibrant, 133
trivially, 133
tilting, 55
partial, 55
with virtually finite injective dimension, 194
with virtually finite projective dimension, 194

pretriangulation, 23
projective
ω-cofibration, 144
ω-fibration, 144
ω-trivial cofibration, 144
ω-trivial fibration, 144
ω-weak equivalence, 144
property
left lifting, 134
right lifting, 134

ring
Cohen-Macaulay, 130

Gorenstein, 123
IF-, 99
trivial extension, 126

subcategory
cogenerator, 83
Ext-injective, 85
colocalizing, 12
contravariantly finite, 25
coreflective, 9
coresolving, 84
covariantly finite, 25
functorially finite, 25
generator, 83
Ext-projective, 85
homologically finite, 25
left Hom-orthogonal, 8
left Ext-orthogonal, 84
localizing, 12
precoresolving, 84
preresolving, 84
reflective, 9
resolving, 84
right Hom-orthogonal, 8
right Ext-orthogonal, 84
strict, 8
thick, 12
$(\mathcal{X}, \mathcal{Y})$-stable, 19

t-structure, 17
torsion class, 8, 11, 32
torsion pair
in a pretriangulated category, 32
cohereditary, 33
hereditary, 33
split, 33
in a triangulated category, 11
cohereditary, 12
compactly generated, 47
hereditary, 12
non-degenerate, 53, 54
of finite type, 43
in an abelian category, 8
cohereditary, 9
hereditary, 9
of finite type, 43
tilting, 19
torsion-free class, 8, 11, 32
triangles
left, 22
right, 23
triangulation
left, 22
right, 23
trivial (co)extension
category, 125
trivial extension
ring, 126
TTF-class
in a pretriangulated category, 32
in a triangulated category, 15
in an abelian category, 10
TTF-triple
in a pretriangulated category, 32
in a triangulated category, 15
in an abelian category, 10

Wakamatsu's Lemma, 28
weak
cokernel, 23
generator, 50
kernel, 23
weak equivalence, 132
WT-Conjecture, 71

Editorial Information

To be published in the *Memoirs*, a paper must be correct, new, nontrivial, and significant. Further, it must be well written and of interest to a substantial number of mathematicians. Piecemeal results, such as an inconclusive step toward an unproved major theorem or a minor variation on a known result, are in general not acceptable for publication.

Papers appearing in *Memoirs* are generally at least 80 and not more than 200 published pages in length. Papers less than 80 or more than 200 published pages require the approval of the Managing Editor of the Transactions/Memoirs Editorial Board.

As of February 28, 2007, the backlog for this journal was approximately 15 volumes. This estimate is the result of dividing the number of manuscripts for this journal in the Providence office that have not yet gone to the printer on the above date by the average number of monographs per volume over the previous twelve months, reduced by the number of volumes published in four months (the time necessary for preparing a volume for the printer). (There are 6 volumes per year, each usually containing at least 4 numbers.)

A Consent to Publish and Copyright Agreement is required before a paper will be published in the *Memoirs*. After a paper is accepted for publication, the Providence office will send a Consent to Publish and Copyright Agreement to all authors of the paper. By submitting a paper to the *Memoirs*, authors certify that the results have not been submitted to nor are they under consideration for publication by another journal, conference proceedings, or similar publication.

Information for Authors

Memoirs are printed from camera copy fully prepared by the author. This means that the finished book will look exactly like the copy submitted.

Initial submission. The AMS uses Centralized Manuscript Processing for initial submissions. Authors should submit a PDF file using the Initial Manuscript Submission form found at `www.ams.org/cgi-bin/peertrack/submission.pl`, or send one copy of the manuscript to the following address: Centralized Manuscript Processing, MEMOIRS OF THE AMS, 201 Charles Street, Providence, RI 02904-2294 USA. If a paper copy is being forwarded to the AMS, indicate that it is for it Memoirs and include the name of the corresponding author, contact information such as email address or mailing address, and the name of an appropriate Editor to review the paper (see the list of Editors below).

The paper must contain a *descriptive title* and an *abstract* that summarizes the article in language suitable for workers in the general field (algebra, analysis, etc.). The *descriptive title* should be short, but informative; useless or vague phrases such as "some remarks about" or "concerning" should be avoided. The *abstract* should be at least one complete sentence, and at most 300 words. Included with the footnotes to the paper should be the 2000 *Mathematics Subject Classification* representing the primary and secondary subjects of the article. The classifications are accessible from `www.ams.org/msc/`. The list of classifications is also available in print starting with the 1999 annual index of *Mathematical Reviews*. The Mathematics Subject Classification footnote may be followed by a list of *key words and phrases* describing the subject matter of the article and taken from it. Journal abbreviations used in bibliographies are listed in the latest *Mathematical Reviews* annual index. The series abbreviations are also accessible from `www.ams.org/publications/`. To help in preparing and verifying references, the AMS offers MR Lookup, a Reference Tool for Linking, at `www.ams.org/mrlookup/`.

Electronically prepared manuscripts. The AMS encourages electronically prepared manuscripts, with a strong preference for $\mathcal{A}_{\mathcal{M}}\mathcal{S}$-LaTeX. To this end, the Society has prepared $\mathcal{A}_{\mathcal{M}}\mathcal{S}$-LaTeX author packages for each AMS publication. Author packages include instructions for preparing electronic manuscripts, samples, and a style file that generates the particular design specifications of that publication series. Though $\mathcal{A}_\mathcal{M}\mathcal{S}$-LaTeX is the highly preferred format of TeX, author packages are also available in $\mathcal{A}_\mathcal{M}\mathcal{S}$-TeX.

Authors may retrieve an author package from the AMS website starting from `www.ams.org/tex/` or via FTP to `ftp.ams.org` (login as `anonymous`, enter username as password, and type `cd pub/author-info`). The *AMS Author Handbook* and the *Instruction Manual* are available in PDF format following the author packages link from `www.ams.org/tex/`. The author package can also be obtained free of charge by sending email to `tech-support@ams.org` (Internet) or from the Publication Division, American Mathematical Society, 201 Charles St., Providence, RI 02904-2294, USA. When requesting an author package, please specify $\mathcal{A}_\mathcal{M}\mathcal{S}$-LaTeX or $\mathcal{A}_\mathcal{M}\mathcal{S}$-TeX and the publication in which your paper will appear. Please be sure to include your complete mailing address.

After acceptance. The final version of the electronic file should be sent to the Providence office (this includes any TeX source file, any graphics files, and the DVI or PostScript file) immediately after the paper has been accepted for publication.

Before sending the source file, be sure you have proofread your paper carefully. The files you send must be the EXACT files used to generate the proof copy that was accepted for publication. For all publications, authors are required to send a printed copy of their paper, which exactly matches the copy approved for publication, along with any graphics that will appear in the paper.

Accepted electronically prepared files can be submitted via the web at `www.ams.org/submit-book-journal/`, sent via FTP, or sent on CD-Rom or diskette to the Electronic Prepress Department, American Mathematical Society, 201 Charles Street, Providence, RI 02904-2294 USA. TeX source files, DVI files, and PostScript files can be transferred over the Internet by FTP to the Internet node `ftp.ams.org` (130.44.1.100). When sending a manuscript electronically via CD-Rom or diskette, please be sure to include a message identifying the paper as a Memoir.

Electronically prepared manuscripts can also be sent via email to `pub-submit@ams.org` (Internet). In order to send files via email, they must be encoded properly. (DVI files are binary and PostScript files tend to be very large.)

Electronic graphics. Comprehensive instructions on preparing graphics are available at `www.ams.org/jourhtml/`. A few of the major requirements are given here.

Submit files for graphics as EPS (Encapsulated PostScript) files. This includes graphics originated via a graphics application as well as scanned photographs or other computer-generated images. If this is not possible, TIFF files are acceptable as long as they can be opened in Adobe Photoshop or Illustrator. No matter what method was used to produce the graphic, it is necessary to provide a paper copy to the AMS.

Authors using graphics packages for the creation of electronic art should also avoid the use of any lines thinner than 0.5 points in width. Many graphics packages allow the user to specify a "hairline" for a very thin line. Hairlines often look acceptable when proofed on a typical laser printer. However, when produced on a high-resolution laser imagesetter, hairlines become nearly invisible and will be lost entirely in the final printing process.

Screens should be set to values between 15% and 85%. Screens which fall outside of this range are too light or too dark to print correctly. Variations of screens within a graphic should be no less than 10%.

Inquiries. Any inquiries concerning a paper that has been accepted for publication should be sent to `memo-query@ams.org` or directly to the Electronic Prepress Department, American Mathematical Society, 201 Charles St., Providence, RI 02904-2294 USA.

Editors

This journal is designed particularly for long research papers, normally at least 80 pages in length, and groups of cognate papers in pure and applied mathematics. Papers intended for publication in the *Memoirs* should be addressed to one of the following editors. The AMS uses Centralized Manuscript Processing for initial submissions to AMS journals. Authors should follow instructions listed on the Initial Submission page found at www.ams.org/memo/memosubmit.html.

Algebra to ALEXANDER KLESHCHEV, Department of Mathematics, University of Oregon, Eugene, OR 97403-1222; email: ams@noether.uoregon.edu

Algebra and its application to MINA TEICHER, Emmy Noether Research Institute for Mathematics, Bar-Ilan University, Ramat-Gan 52900, Israel; email: teicher@macs.biu.ac.il

Algebraic geometry to DAN ABRAMOVICH, Department of Mathematics, Brown University, Box 1917, Providence, RI 02912; email: amsedit@math.brown.edu

Algebraic number theory to V. KUMAR MURTY, Department of Mathematics, University of Toronto, 100 St. George Street, Toronto, ON M5S 1A1, Canada; email: murty@math.toronto.edu

Algebraic topology to ALEJANDRO ADEM, Department of Mathematics, University of British Columbia, Room 121, 1984 Mathematics Road, Vancouver, British Columbia, Canada V6T 1Z2; email: adem@math.ubc.ca

Combinatorics to JOHN R. STEMBRIDGE, Department of Mathematics, University of Michigan, Ann Arbor, Michigan 48109-1109; email: FRS@umich.edu

Complex analysis and harmonic analysis to ALEXANDER NAGEL, Department of Mathematics, University of Wisconsin, 480 Lincoln Drive, Madison, WI 53706-1313; email: nagel@math.wisc.edu

Differential geometry and global analysis to LISA C. JEFFREY, Department of Mathematics, University of Toronto, 100 St. George St., Toronto, ON Canada M5S 3G3; email: jeffrey@math.toronto.edu

Dynamical systems and ergodic theory to AMIE WILKINSON, Department of Mathematics, Northwestern University, 2033 Sheridan Road, Evanston, IL 60208-2730; email: transactions@math.northwestern.edu

Functional analysis and operator algebras to DIMITRI SHLYAKHTENKO, Department of Mathematics, University of California, Los Angeles, CA 90095; email: shlyakht@math.ucla.edu

Geometric analysis to WILLIAM P. MINICOZZI II, Department of Mathematics, Johns Hopkins University, 3400 N. Charles St., Baltimore, MD 21218; email: trans@math.jhu.edu

Geometric analysis to MLADEN BESTVINA, Department of Mathematics, University of Utah, 155 South 1400 East, JWB 233, Salt Lake City, Utah 84112-0090; email: bestvina@math.utah.edu

Harmonic analysis, representation theory, and Lie theory to ROBERT J. STANTON, Department of Mathematics, The Ohio State University, 231 West 18th Avenue, Columbus, OH 43210-1174; email: stanton@math.ohio-state.edu

Logic to STEFFEN LEMPP, Department of Mathematics, University of Wisconsin, 480 Lincoln Drive, Madison, Wisconsin 53706-1388; email: lempp@math.wisc.edu

Partial differential equations to GUSTAVO PONCE, Department of Mathematics, South Hall, Room 6607, University of California, Santa Barbara, CA 93106; email: ponce@math.ucsb.edu

Partial differential equations and dynamical systems to PETER POLACIK, School of Mathematics, University of Minnesota, Minneapolis, MN 55455; email: polacik@math.umn.edu

Probability and statistics to KRZYSZTOF BURDZY, Department of Mathematics, University of Washington, Box 354350, Seattle, Washington 98195-4350; email: burdzy@math.washington.edu

Real analysis and partial differential equations to DANIEL TATARU, Department of Mathematics, University of California, Berkeley, Berkeley, CA 94720; email: tataru@math.berkeley.edu

All other communications to the editors should be addressed to the Managing Editor, ROBERT GURALNICK, Department of Mathematics, University of Southern California, Los Angeles, CA 90089-1113; email: guralnic@math.usc.edu.

Titles in This Series

883 **Apostolos Beligiannis and Idun Reiten,** Homological and homotopical aspects of torsion theories, 2007

882 **Lars Inge Hedberg and Yuri Netrusov,** An axiomatic approach to function spaces, spectral synthesis, and Luzin approximation, 2007

881 **Tao Mei,** Operator valued Hardy spaces, 2007

880 **Bruce C. Berndt, Geumlan Choi, Youn-Seo Choi, Heekyoung Hahn, Boon Pin Yeap, Ae Ja Yee, Hamza Yesilyurt, and Jinhee Yi,** Ramanujan's forty identities for the Rogers-Ramanujan functions, 2007

879 **O. García-Prada, P. B. Gothen, and V. Muñoz,** Betti numbers of the moduli space of rank 3 parabolic Higgs bundles, 2007

878 **Alessandra Celletti and Luigi Chierchia,** KAM stability and celestial mechanics, 2007

877 **María J. Carro, José A. Raposo, and Javier Soria,** Recent developments in the theory of Lorentz spaces and weighted inequalities, 2007

876 **Gabriel Debs and Jean Saint Raymond,** Borel liftings of Borel sets: Some decidable and undecidable statements, 2007

875 **C. Krattenthaler and T. Rivoal,** Hypergéométrie et fonction zêta de Riemann, 2007

874 **Sonia Natale,** Semisolvability of semisimple Hopf algebras of low dimension, 2007

873 **A. J. Duncan,** Exponential genus problems in one-relator products of groups, 2007

872 **Anthony V. Geramita, Tadahito Harima, Juan C. Migliore, and Yong Su Shin,** The Hilbert function of a level algebra, 2007

871 **Pascal Auscher,** On necessary and sufficient conditions for L^p-estimates of Riesz transforms associated to elliptic operators on $\mathbb{R}^n$ and related estimates, 2007

870 **Takuro Mochizuki,** Asymptotic behaviour of tame harmonic bundles and an application to pure twistor D-modules, Part 2, 2007

869 **Takuro Mochizuki,** Asymptotic behaviour of tame harmonic bundles and an application to pure twistor D-modules, Part 1, 2007

868 **Gelu Popescu,** Entropy and multivariable interpolation, 2006

867 **Vilmos Totik,** Metric properties of harmonic measures, 2006

866 **William Craig,** Semigroups underlying first-order logic, 2006

865 **Nathanial P. Brown,** Invariant means and finite representation theory of $C*$-algebras, 2006

864 **John M. Lee,** Fredholm operators and Einstein metrics on conformally compact manifolds, 2006

863 **M. Lübke and A. Teleman,** The Universal Kobayashi-Hitchin correspondence on Hermitian manifolds, 2006

862 **Alberto Canonaco,** The Beilinson complex and canonical rings of irregular surfaces, 2006

861 **Leon A. Takhtajan and Lee-Peng Teo,** Weil-Petersson metric on the universal Teichmüller space, 2006

860 **Thomas M. Fiore,** Pseudo limits, biadjoints and pseudo algebras: Categorical foundations of conformal field theory, 2006

859 **N. Arcozzi, R. Rochberg, and E. Sawyer,** Carleson measures and interpolating sequences for Besov spaces on complex balls, 2006

858 **Enrico Valdinoci, Berardino Sciunzi, and Vasile Ovidiu Savin,** Flat level set regularity of p-Laplace phase transitions, 2006

857 **Donatella Danielli, Nocola Garofalo, and Duy-Minh Nhieu,** Non-doubling Ahlfors measures, perimeter measures, and the characterization of the trace spaces of Sobolev functions in Carnot-Carathéodory spaces, 2006

856 **Vladimir Bolotnikov and Harry Dym,** On boundary interpolation for matrix valued Schur functions, 2006

855 **Yevgenia Kashina, Yorck Sommerhäuser, and Yongchang Zhu,** On higher Frobenius-Schur indicators, 2006

854 **Noam Greenberg,** The role of true finiteness in the admissible recursively enumerable degrees, 2006

853 **Joachim Krieger,** Stability of spherically symmetric wave maps, 2006

852 **Viorel Barbu, Irena Lasiecka, and Roberto Triggiani,** Tangential boundary stabilization of Navier-Stokes equations, 2006

851 **Jie Wu,** On maps from loop suspensions to loop spaces and the shuffle relations on the Cohen groups, 2006

850 **Siegfried Echterhoff, S. Kaliszewski, John Quigg, and Iain Raeburn,** A categorical approach to imprimitivity theorems for C^*-dynamical systems, 2006

849 **Katsuhiko Kuribayashi, Mamoru Mimura, and Tetsu Nishimoto,** Twisted tensor products related to the cohomology of the classifying spaces of loop groups, 2006

848 **Bob Oliver,** Equivalences of classifying spaces completed at the prime two, 2006

847 **Eric T. Sawyer and Richard L. Wheeden,** Hölder continuity of weak solutions to subelliptic equations with rough coefficients, 2006

846 **Victor Beresnevich, Detta Dickinson, and Sanju Velani,** Measure theoretic laws for lim–sup sets, 2006

845 **Ehud Friedgut, Vojtech Rödl, Andrzej Ruciński, and Prasad V. Tetali,** A Sharp threshold for random graphs with a monochromatic triangle in every edge coloring, 2006

844 **Amadeu Delshams, Rafael de la Llave, and Tere M. Seara,** A geometric mechanism for diffusion in Hamiltonian systems overcoming the large gap problem: Heuristics and rigorous verification on a model, 2006

843 **Denis V. Osin,** Relatively hyperbolic groups: Intrinsic geometry, algebraic properties, and algorithmic problems, 2006

842 **David P. Blecher and Vrej Zarikian,** The calculus of one-sided M-ideals and multipliers in operator spaces, 2006

841 **Enrique Artal Bartolo, Pierrette Cassou-Noguès, Ignacio Luengo, and Alejandro Melle Hernández,** Quasi-ordinary power series and their zeta functions, 2005

840 **Sławomir Kołodziej,** The complex Monge-Ampère equation and pluripotential theory, 2005

839 **Mihai Ciucu,** A random tiling model for two dimensional electrostatics, 2005

838 **V. Jurdjevic,** Integrable Hamiltonian systems on complex Lie groups, 2005

837 **Joseph A. Ball and Victor Vinnikov,** Lax-Phillips scattering and conservative linear systems: A Cuntz-algebra multidimensional setting, 2005

836 **H. G. Dales and A. T.-M. Lau,** The second duals of Beurling algbras, 2005

835 **Kiyoshi Igusa,** Higher complex torsion and the framing principle, 2005

834 **Kenı́chi Ohshika,** Kleinian groups which are limits of geometrically finite groups, 2005

833 **Greg Hjorth and Alexander S. Kechris,** Rigidity theorems for actions of product groups and countable Borel equivalence relations, 2005

832 **Lee Klingler and Lawrence S. Levy,** Representation type of commutative Noetherian rings III: Global wildness and tameness, 2005

831 **K. R. Goodearl and F. Wehrung,** The complete dimension theory of partially ordered systems with equivalence and orthogonality, 2005